NITRIDE WIDE BANDGAP SEMICONDUCTOR MATERIAL AND ELECTRONIC DEVICES

NITRIDE WIDE BANDGAP SEMICONDUCTOR MATERIAL AND ELECTRONIC DEVICES

YUE HAO
JIN-FENG ZHANG
JIN-CHENG ZHANG

CRC Press
Taylor & Francis Group
Boca Raton London New York

CRC Press is an imprint of the
Taylor & Francis Group, an **informa** business

CRC Press
Taylor & Francis Group
6000 Broken Sound Parkway NW, Suite 300
Boca Raton, FL 33487-2742

CRC Press is an imprint of Taylor & Francis Group, an Informa business

Printed on acid-free paper
Version Date: 20160831

International Standard Book Number-13: 978-1-4987-4512-3 (Hardback)

Library of Congress Cataloging-in-Publication Data

Names: Hao, Yue, 1958 March- author. | Zhang, Jinfeng, 1977- author. | Zhang, Jincheng, 1976- author.
Title: Nitride wide bandgap semiconductor material and electronic devices / Yue Hao, Jin Feng Zhang, and Jin Cheng Zhang.
Description: Boca Raton : CRC Press, Taylor & Francis Group, 2017. | Includes bibliographical references and index.
Identifiers: LCCN 2016018391 | ISBN 9781498745123 (alk. paper)
Subjects: LCSH: Nitrides. | Wide gap semiconductors--Design and construction. | Epitaxy. | Microelectronics--Materials.
Classification: LCC TK7871.15.N57 H36 2017 | DDC 621.3815/2--dc23
LC record available at https://lccn.loc.gov/2016018391

Visit the Taylor & Francis Web site at
http://www.taylorandfrancis.com

and the CRC Press Web site at
http://www.crcpress.com

Printed and bound in the United States of America by Sheridan

Contents

Preface

Since 2000, there has been a rapid development in III-nitride semiconductor heterostructures and electronic devices, and they have been continually upgraded in terms of power density, efficiency, frequency, and bandwidth. In 2005, nitride semiconductor microwave power devices began to enter the market, and the commercial products of nitride semiconductor power switch devices appeared in around 2010. Because of continuous progress of the techniques and high performance manifested in various applications, nitride semiconductor electronic devices are attracting more and more attention. Except for the electronic devices, nitride semiconductor has gained remarkable success in the fields of new generation of solid-state lighting devices and short-wavelength laser devices and detectors. Therefore, the point of view that "III-nitride semiconductor is the most important semiconductor after silicon" has been the general consensus in both academia and the industry.

With respect to the conceptions and structures of heterostructures and electronic devices, nitride semiconductors show no essential difference with the second-generation semiconductors such as GaAs and InP. However, in the fields of physical characteristics, preparation process, operational mechanisms, and performances, nitride semiconductors exhibit many remarkable features: two-dimensional electron gas of high density is induced by strong built-in polarization effects, materials are mainly in the form of single-crystal films and grown via heteroepitaxy, and unique current collapse effect and inverse piezoelectric polarization effect appear and influence the device stability and reliability. Consequently, researchers working on nitride semiconductor electronic devices and other relevant fields have long been expecting a monograph that deals with nitride semiconductor heterostructures and electronic devices in detail, as no such book is available in the domestic market.

For this purpose, the authors have made a great effort to provide a systematical introduction of III-nitride semiconductor materials and electronic devices in this book, which is based on the relevant research experiences and achievements of a research team led by the first author (Yue Hao). The team is affiliated to National Key Lab of Wide Band-Gap Semiconductor Technology in Xidian University, Xi'an, China; the team began research on III-nitride semiconductor materials and electronic devices in 1997 and is among those that took the lead in the research field in China.

This book consists of mainly two large parts dealing with nitride materials and nitride electronic devices, respectively. Chapters 2 through 8 covers basic properties of nitride semiconductor materials, method of III-nitride heteroepitaxy, electrical properties of nitride heterojunctions, growth and optimization of AlGaN/GaN, and InAlN/GaN heterojunction and investigation of material defects. Chapters 9 through 13 contains principle and optimization, preparation processes and performances, electrical and thermal degradation of GaN HEMTs, and enhancement-mode GaN HEMTs and GaN MOS-HEMTs. Chapter 14 proposes the directions for further development of nitride semiconductor material and electronic devices. This book can therefore be used as a reference by those who engage in scientific research and teaching of relevant subjects.

A number of teachers and doctoral and graduate students including those in the authors' team proposed brilliant ideas and/or made great efforts in the research of nitride semiconductor and electronic devices, and their contributions are reflected in the references of each chapter. Especially, we owe much gratitude to Dr. Peijun Ma for his efficient management of the research projects in our laboratory. Professor Xiaohua Ma also deserves high recognition for his great contributions in the collaborative nitride device research in our team.

Meanwhile, the laboratory where the authors work has developed a long-term cooperation with many scientific research institutes, colleges, and universities both at home and abroad, so this book also includes some collaborative research results to make it a complete work. The authors sincerely appreciate all the experts, peers, and friends who offered support and help and are grateful to their colleagues and students who strived together to get this book published. In addition, the authors thank a number of scientific research programs of China for their long-term support such as the National Natural Science Foundation (especially the Key Programs of the National Natural Science Foundation of China with Nos. 61334002 and 60736033), the National High-Technology Research and Program (863 program), the National Key Basic Research Development Program (973 program), the National Science and Technology Major Project, and the National Defense Science and Technology and Research Program. We truly hope that in scientific research, product development, and teaching and training of nitride semiconductor materials and electronic devices, this book can serve as an academic reference and an inspiration for readers, thereby promoting the development of nitride semiconductor material and electronic devices.

This book is translated from Chinese into the English version by Professor Ming Zhang, the School of Foreign Languages, Xidian University, and polished by the authors.

Mistakes and deficiencies are inevitable in this book because of the limited ability of the authors and the translator. The readers' valuable proposals, suggestions, and criticisms are welcome.

Yue Hao, Jin-Feng Zhang, and Jin-Cheng Zhang
Xidian University

Introduction

This book is the outcome of more than ten years' research work of the authors. It systematically introduces physical characteristics and implementations of III-nitride wide-band-gap semiconductor material and electronic devices, with emphasis on high electron mobility transistors (HEMTs). This book contains 14 chapters discussing the basic properties of nitride material; methods and mechanisms of nitride hetero-epitaxy; electrical properties of HEMT materials; growth, optimization, and defects analysis of AlGaN/GaN and InAlN/GaN heterojunctions; principles and optimization of GaN HEMT devices; preparation process, performances, and electric and thermal degradation analysis of GaN HEMT devices; enhancement-mode GaN HEMT devices; and integrated circuits; GaN MOS-HEMT devices; and finally several important directions for further technology development in this academic field.

This book is appropriate for graduates and researchers in the field of semiconductor science and technology.

Authors

Yue Hao earned both his BS and MS in science, specializing in semiconductor physics and devices at Xidian University, Xi'an, China, in 1982 and 1986, respectively, and his doctorate in computational mathematics at Xi'an Jiaotong University, Xi'an, China, in 1991. He has been with the Department of Technical Physics, Xidian University, since 1986 and became a professor in 1993. He has long been engaged in scientific research and nurturing of talent in the fields of wide-band-gap semiconductor materials (gallium nitride and silicon carbide) and devices, micro-nanometer semiconductor devices, and highly reliable integrated circuits. He has authored and coauthored over 300 publications, including 5 books, and holds over 80 Chinese invention patents. He has been the executive director of the Chinese Association of Electronics from 2002, the editor-in-chief of *Acta Electronica Sinica* from 2015, and the chief scientist for the national major basic research project of China from 2008. He was also the chairman, cochairman, and the international advisory committee member of some international conferences such as International Conference of Nitride Semiconductor (ICNS 2015), the 10th International Conference on New Diamond and Nano Carbons (NDNC 2016), and the Asia-Pacific Workshop on Widegap Semiconductors (APWS 2007). He won many awards in the field of science and technology including the Ho Leung Ho Lee Science and Technology Progress Award in 2010. At present, he is an academician in the Chinese Academy of Sciences.

Jin-Feng Zhang earned her BS degree in electronic science and technology and her PhD in microelectronics and solid-state electronics from Xidian University in 2000 and 2006, respectively. She has been with the School of Microelectronics, Xidian University, since 2006 and became an associate professor in 2008. She was a one-year academic visitor in the University of Sheffield, Sheffield, England, in 2009. Her research interests mainly involve the physics of nitride semiconductor heterojunctions and electronic devices, and the physical property and defect of nitride polar and non-polar materials. She has authored and coauthored over 40 publications, including a book, and holds 5 patents. She also received the Provincial/Ministerial Scientific and Technological Award three times.

Jin-Cheng Zhang earned his BS, MS, and PhD degrees in microelectronics and solid-state electronics from Xidian University in 1998, 2001, and 2004, respectively. Since 2001, he has been with the School of Technical Physics (previously the Department of Technical Physics), Xidian University, and became a professor in the School of Microelectronics in 2009. In 2014, he was a visiting scholar in the Interuniversity Microelectronics Centre. His research interests include the growth of nitride semiconductor, high-power GaN HEMTs, high-performance GaN-based LEDs, and solar cells. He has authored and coauthored over 200 publications, including 3 books, and holds over 30 patents. He gave over 10 invited presentations at international conferences such as International Conference of Nitride Semiconductor (ICNS), European Materials Research Society (E-MRS), and Collaborative Conference on Crystal Growth (3CG). He was also the committee member and session chair of many international conferences such as ICNS-11, the 7th Asia-Pacific Workshop on Widegap Semiconductors (APWS 2015), the 10th International Conference on New Diamond and Nano Carbons (NDNC 2016), the 16th Conference on Defects-Recognition, and Imaging and Physics in Semiconductors (DRIP XVI). In 2009, he received the State Technological Invention Award of China for development of Pulse-MOCVD system and high-performance GaN materials. He also received the Provincial/Ministerial Scientific and Technological Award for five times and some other awards including New Century Excellent Talent of Ministry of Education of China (2007).

1 Introduction

As one of the most important and influential advanced technologies of the twentieth century, the influence and significance of semiconductor technology has extended into the twenty-first century. As the foundation of information society, the semiconductor technology has been pushing the human society forward while changing the production, daily life, interpersonal communication, and even the way of thinking of mankind. Semiconductor materials have been playing a critical role in the development of the semiconductor technology. The first transistor invented in 1947 was based on the semiconductor of germanium (Ge) with a room temperature band gap of 0.66 eV. The first integrated circuit born in 1958 was actually of a mixed type, and the first true monolithic integrated circuit appeared in 1961 and employed germanium. Silicon (Si) with a room temperature band gap of 1.12 eV replaced germanium in 1965 to be the major material for semiconductor integrated circuits, and nowadays, remains the principal semiconductor material in microelectronics. The vast majority of the semiconductor industries, either the integrated circuit or the photovoltaic cell industry, are still based on silicon. Silicon and germanium are generally referred to as the first-generation semiconductors for their comparatively long history of development. The second-generation semiconductors including both gallium arsenide (GaAs, with a room temperature band gap of 1.42 eV) and indium phosphide (InP, with a room temperature band gap of 1.35 eV) were introduced in the 1970s and primarily adopted in ultrahigh-speed devices, microwave power devices, and integrated circuits. InP integrated circuits were not commercially available until 1997. Major progresses in the third-generation (wide band gap) semiconductors such as gallium nitride (GaN, with a room temperature band gap of 3.45 eV) and silicon carbide (SiC, with a room temperature band gap of 3.25 eV for 4H-SiC) have been made since the end of the twentieth century.

Apart from a wider band gap, GaAs has six times higher electron mobility and twice greater saturated drift velocity than silicon (see Table 1.1), and therefore GaAs devices are more suitable for high-frequency operations. GaAs field-effect transistors also have the advantages of low noise, high efficiency, and good linearity. However, GaAs has poor thermal conductivity and breakdown field than the third-generation semiconductors GaN and SiC and is limited in power characteristics. The maximum output power of the GaAs metal–semiconductor field-effect transistor (MESFET) was 1.4 W/mm@8 GHz in the early 1980s (Macksey and Doerbeck 1981). Various attempts to improve performance by various methods only yielded a limited increase in power

Table 1.1 Physical Parameters for Semiconductors

SEMICONDUCTORS	Si	GaAs	4H-SiC	GaN
Band gap E_g (eV)	1.12	1.42	3.25	3.45
Relative dielectric constant ε	11.4	13.1	9.7	8.9
Electron mobility μ_e [cm^2/(V·s)]	1400	8500	1020	1000(GaN) 2000(AlGaN/GaN)
Breakdown field E_C (MV/cm)	0.3	0.4	3.0	3.3
Thermal conductivity κ [W/(cm·K)]	1.5	0.5	4.9	2.0
Electron saturation velocity v_{sat} (×10^{7}cm/s)	1.0	2.0	2.0	2.0
Baliga figure of merit (low-frequency) $\varepsilon\mu_e E_c^3$	1	16	600	1450

density with a maximum power density of 1.57 W/mm@1.1 GHz (Chen et al. 1991) obtained at a compromised operation frequency.

In the early 1990s, a record high power density of 1.8 W/mm@30 GHz was achieved in the InP metal–insulator–semiconductor field-effect transistor (MISFET) with Si_3N_4 as gate dielectric (Saunier et al. 1990). But InP MISFETs found little practical use due to unstable current–voltage characteristics induced by unavoidable high density interface states. The later developed InP high electron mobility transistor (HEMT) achieved a microwave power density of 1.45 W/mm@30 GHz coupled with excellent current–voltage characteristics (Aina et al. 1992). The superiority in microwave power characteristics of InP to GaAs devices is mainly attributed to the slightly higher breakdown field and electron saturation velocity of InP. However, InP devices arc not good enough in microwave high-power applications (Wu 1997). Instead, they are most advantageous at operation frequencies of 100 GHz or even higher, especially in high-speed and low-power mixed analog and digital integrated circuits (Hao et al. 2000).

The focus of research on compound semiconductor electronic devices has been the wide band gap semiconductor devices since the early 1990s. Generally speaking, wide band gap semiconductors refer to the semiconductors with a band gap greater than 2 eV (Baliga 1989). The Baliga figure of merit listed in Table 1.1 is a common parameter for characterizing the capability of the semiconductors in high-frequency high-power applications, and it can be seen that the wide band gap semiconductors GaN and SiC enjoy superb material properties such as a large band gap, a high breakdown voltage, and a high electron saturation velocity and are well suited for microwave/millimeter wave high-power applications.

Research shows that both the cut-off frequency f_T and the maximum oscillating frequency f_{max} of SiC MESFET microwave power devices are within 20 GHz, so SiC MESFETs are suitable for frequencies below 7 GHz. Among Group III nitride semiconductors including GaN, aluminum nitride (AlN), indium nitride (InN), and their alloys, both GaN and AlN are wide band gap semiconductors though InN has a relatively narrow band gap (~0.7 eV). The prevailing GaN electronic devices are HEMTs based on GaN heterostructures (such as AlGaN/GaN or InAlN/GaN) with

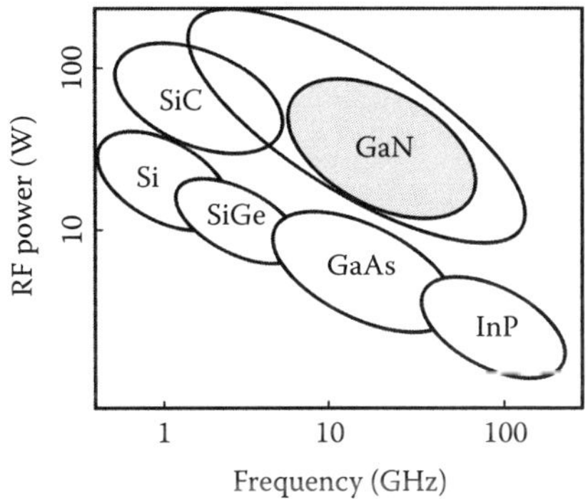

Figure 1.1 Application areas of typical semiconductors. (Reproduced from Nanishi, Y., Present status and challenges of AlGaN/GaN HFETs, *Proceedings of the 7th International Conference on Solid-State and Integrated Circuits Technology*, Beijing, China, 2004.)

the two-dimensional electron gas (2DEG) of high mobility and very high carrier sheet density, making GaN HEMTs better suited for high-frequency high-power applications. GaN HEMTs show excellent heat dissipation characteristics when they are grown on high thermal conductivity substrates such as SiC and diamond and also enjoy both high performance and low cost when grown on mature low-cost large-diameter silicon substrates. Therefore, GaN HEMTs are deemed as the most ideal microwave power devices (Figure 1.1).

With a band gap range completely covering the entire visible spectrum, Group III nitrides also find wide applications in short wavelength photoelectronic devices that are beyond the capability of conventional semiconductors in addition to high-frequency power applications. In fact, it was the blue light-emitting diode (LED) research that initiated the breakthroughs in Group III nitride semiconductor preparation technologies. Taking the most mature GaN technology for example, the crystalline melting point is over 2500°C (Harafuji et al. 2004), but the decomposition point is in the vicinity of 900°C. Therefore, the conventional melting technique for silicon preparation is not applicable for monocrystal GaN preparation. Of the two GaN preparation methods of bulk crystal growth and heteroepitaxial film growth, the latter saw a breakthrough first with high quality GaN epitaxial films effectively obtained by metalorganic chemical vapor deposition (MOCVD) and molecular beam epitaxy (MBE). The rapid growth of the GaN-based blue LED industry in the 1990s also boosted the GaN electronic device research and development.

Previous researches on Group III nitride electronic devices attempted on various device structures such as MESFET and heterojunction field-effect transistor (HFET), among which the AlGaN/GaN HFET quickly established itself as the mainstream structure for GaN electronic devices. Because of the strong spontaneous polarization and piezoelectric polarization of nitride yields in undoped AlGaN/GaN and InAlN/GaN heterostructures and high-density 2DEGs with significantly higher mobility than that of bulk electrons, the GaN HFETs are more often referred to as GaN HEMTs.

The high conductivity of the 2DEG channel coupled with the high breakdown voltage of GaN makes GaN HEMTs a popular topic in microwave power device researches. The first GaN HEMT was born in 1993 (Khan et al. 1993), and 3 years later, the microwave power characteristics was achieved (Wu et al. 1996). Then, the output power density was improved from the initial 1.1 W/mm@2 GHz to 32.2 W/mm@4 GHz and 30.6 W/mm@8 GHz (Wu et al. 2004), and further to 41.4 W/mm@4 GHz in 2006 (Wu et al. 2006). A total output power of 230 W@2 GHz was achieved in a single GaN HEMT with a gate width of 48 mm (Okamoto et al. 2004).

Such a dramatic progress of nitride electronic devices is attributed first to the continuous improvement of AlGaN/GaN material quality and HEMT technology. Then, the general employment of SiC substrate with better heat dissipation performance contributes significantly to increasing output power. The deposition of dielectric films for surface passivation suppresses the trap-induced current collapse and leads to better microwave performance and reliability. Progress in GaN heterostructures has been made along with the development of heterojunction material epitaxy and device technology. Some techniques such as the insertion of AlN interlayer at AlGaN/GaN heterointerface, the addition of GaN cap layer on surface of AlGaN layer, and the introduction of AlGaN or InGaN back barrier under the GaN channel became popular. Meanwhile, the HEMT structure optimizations such as the gate recess and the field plate also gained wide application. All these improvements pushed the continuous progress of GaN HEMT performance.

The operation frequency, gain, and power including efficiency of GaN HEMTs have also seen gradual increase, with a cut-off frequency of 343 GHz (Shinohara et al. 2011) and a W band output power density of 1.7 W/mm@95 GHz (Brown et al. 2011) reported in 2011. Rapid progress was also witnessed in GaN HEMTs on silicon substrate in an attempt to combine the nitride and silicon technologies. The demand for GaN-based high-speed digital circuits has promoted the development of enhancement-mode GaN HEMTs and enhancement/depletion-mode HEMT circuit units. And the requirement for smaller gate leakage currents has stimulated the research on the GaN metal–oxide–semiconductor HEMT (MOS-HEMT). Better high-frequency characteristics have been achieved in InAlN/GaN HEMT and MIS-HEMTs in recent years by using the nearly lattice-matched InAlN/GaN heterostructure that features a higher 2DEG density and a smaller barrier layer thickness.

Commercial GaN HEMT microwave power devices (on SiC and silicon substrates) were available in 2006 but were far from exerting the potential of GaN HEMTs with the product performance way below the laboratory results of the same period due to many unresolved issues in GaN HEMT performance and reliability. GaN HEMT microwave power device research borrowed much from the material and device physics and fabrication technology of the second-generation semiconductors GaAs, InP, and their heterostructures. The market demand and similar technology did accelerate the rapid technical growth and the commercialization of nitride materials and devices, but the lack of in-depth researches on the fundamental problems caused to a certain extent a market demand ahead of technological development and the technology ahead of fundamental research. As compared to the sophisticated silicon and GaAs devices, nitride semiconductors are facing some major basic problems including the high defect density, control of strong polarization, and surface state of heteroepitaxial nitride materials as well as the large leakage current and poor device reliability under high operating voltage. Many problems remain to be investigated in the long-term future, for example, nitride epitaxy on varied substrates; material defect behaviors and characterization; device structure optimization; polarization mechanisms; and polarization applications (polarization engineering) of the heterostructure, gate structures, and the realization of enhancement-mode devices.

In the late 1990s, research on nitride semiconductor electronic materials and devices was initiated in China, and significant progresses were achieved through constant effort with fruitful achievements in AlGaN/GaN HEMTs and MIS-HEMTs, enhancement-mode HEMTs, and lattice-matched InAlN/GaN HEMTs. Most particularly, the support of the National Science and Technology Major Project has led to increasingly higher performance and reliability of GaN HEMT microwave power devices and monolithic microwave integrated circuits, the successful development of C-band internal matching power transistors with a continuous wave power of 60 W, and the engineering practice of S-band GaN power transistors with a pulsed power close to 100 W. Further, more fundamental research and engineering practice are needed to accelerate the development of China's nitride semiconductor technology and industry.

References

Aina, O., M. Burgess, M. Mattingly, A. Meerschaert, J. M. O'Connor, M. Tong, A. Ketterson, and I. Adesida. 1992. A 1.45-W/mm, 30-GHz InP-channel power HEMT. *Electron Device Letters* 13(5):300–302. doi: 10.1109/55.145060.

Baliga, B. J. 1989. Power semiconductor device figure of merit for high-frequency applications. *Electron Device Letters* 10(10):455–457. doi: 10.1109/55.43098.

Brown, D. F., A. Williams, K. Shinohara, A. Kurdoghlian, I. Milosavljevic, P. Hashimoto, R. Grabar et al. 2011. W-band power performance of AlGaN/GaN DHFETs with regrown n+ GaN ohmic contacts by MBE. *IEEE International Electron Devices Meeting*, Washington, DC.

Chen, C.-L., F. W. Smith, B. J. Clifton, L. J. Mahoney, M. J. Manfra, and A. R. Calawa. 1991. High-power-density GaAs MISFETs with a low-temperature-grown epitaxial layer as the insulator. *Electron Device Letters* 12(6):306–308.

Hao, Y., J. Peng, and Y. Yang. 2000. *Silicon Carbide Wide-Bandgap Semiconductor Technology*. Beijing, China: Science Press.

Harafuji, K., T. Tsuchiya, and K. Kawamura. 2004. Molecular dynamics simulation for evaluating melting point of wurtzite-type GaN crystal. *Journal of Applied Physics* 96(5):2501–2512. doi: 10.1063/1.1772878.

Khan, M. A., A. Bhattarai, J. N. Kuznia, and D. T. Olson. 1993. High electron mobility transistor based on a GaN-$Al_xGa_{1-x}N$ heterojunction. *Applied Physics Letters* 63(9):1214–1215.

Macksey, H. M. and F. H. Doerbeck. 1981. GaAs FETs having high output power per unit gate width. *Electron Device Letters* EDL-2(6):147–148.

Nanishi, Y. 2004. Present status and challenges of AlGaN/GaN HFETs. *Proceedings of the 7th International Conference on Solid-State and Integrated Circuits Technology*, Beijing, China.

Okamoto, Y., Y. Ando, K. Hataya, T. Nakayama, H. Miyamoto, T. Inoue, M. Senda et al. 2004. Improved power performance for a recessed-gate AlGaN-GaN heterojunction FET with a field-modulating plate. *IEEE Transactions on Microwave Theory and Techniques* 52(11):2536–2540.

Saunier, P., R. Nguyen, L. J. Messick, and M. A. Khatibzadeh. 1990. An InP MISFET with a power density of 1.8 W/mm at 30 GHz. *IEEE Electron Device Letters* 11(1):48–49. doi: 10.1109/55.46927.

Shinohara, K., D. Regan, A. Corrion, D. Brown, S. Burnham, P. J. Willadsen, I. Alvarado-Rodriguez et al. 2011. Deeply-scaled self-aligned-gate GaN DH-HEMTs with ultrahigh cutoff frequency. *IEEE International Electron Devices Meeting*, Washington, DC.

Wu, Y. F. 1997. AlGaN/GaN microwave power high-mobility-transistors. PhD Dissertation, University of California, Santa Barbara, CA.

Wu, Y. F., B. P. Keller, S. Keller, D. Kapolnek, S. P. Denbaars, and U. K. Mishra. 1996. Measured microwave power performance of AlGaN/GaN MODFET. *IEEE Electron Device Letters* 17(9):455–457. doi: 10.1109/55.536291.

Wu, Y. F., M. Moore, A. Saxler, T. Wisleder, and P. Parikh. 2006. 40-W/mm double field-plated GaN HEMTs. *Device Research Conference*, June 26–28, 2006, Piscataway, NJ.

Wu, Y. F., A. Saxler, M. Moore, R. P. Smith, S. Sheppard, P. M. Chavarkar, T. Wisleder, U. K. Mishra, and P. Parikh. 2004. 30-W/mm GaN HEMTs by field plate optimization. *IEEE Electron Device Letters* 25(3):117–119. doi: 10.1109/led.2003.822667.

2

Properties of Group III Nitride Semiconductor Materials

Group III nitride semiconductor materials contain mainly the binary compounds of AlN, GaN, and InN, and the ternary and quaternary alloys (AlGaN, InGaN, and AlInGaN) composed by them. Since the material structure of most Group III nitride electronic devices is the heterojunction, this chapter focuses on the heterojunction-related material properties such as the crystal structure, band structure, electron transport, and polarization effect. Some methods for nitride property measurements are also introduced.

2.1 Crystal Structure and Band Structure of III Nitride

2.1.1 GaN, AlN, and InN

Nitride semiconductor crystals are usually in two different structures: the hexagonal wurtzite and the cubic zinc-blende, as shown in Figure 2.1 (Holt and Yacobi 2007).

The crystal structure is determined mainly by crystal ionicity. Both covalent bonds and ionic bonds exist between the compound semiconductor crystal atoms. The more ionic the bonds, the stronger ionicity the crystal has and the easier it is for the wurtzite structure to form. All nitrides are strong ionic crystals and hence under room temperature and atmospheric pressure the wurtzite structure is the dominant and thermodynamically stable, whereas the zinc-blende structure is metastable. Wurtzite GaN is a close-packed hexagonal crystal with a $P6_3$ mc space group and only one close-packed plane (0001), each lattice cell having 12 atoms (6 gallium atoms and 6 nitrogen atoms) and in-plane and axial lattice constants being $a = 0.3189$ nm and $c = 0.5185$ nm (Leszczynski et al. 1993), respectively. Although GaN is typically in the wurtzite structure of hexagonal symmetry, zinc-blende GaN of cubic symmetry also exists under certain conditions (Figure 2.1). Zinc-blende GaN belongs to the face-centered cubic (FCC) lattice system of the $F\bar{4}3m$ space group, formed by stacking two face-centered cubes along the body diagonal with a translation of one-fourth the diagonal. Its close-packed atom plane is (111), each lattice cell having eight atoms (four gallium atoms and four nitrogen atoms) and lattice constant being about 0.4520 nm. Generally speaking, wurtzite III nitrides are more stable and representative. Wurtzite III nitrides are employed in the vast majority of nitride semiconductor researches up to date, while little attention has been paid to zinc-blende III nitrides. Accordingly, only wurtzite structure nitrides are involved in this book unless otherwise specified.

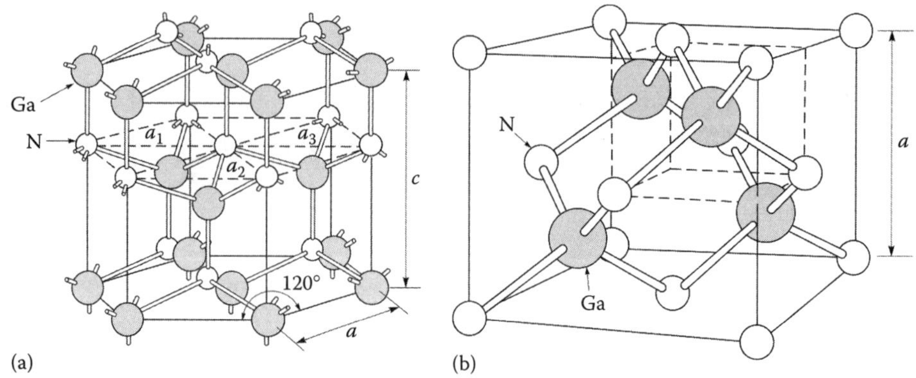

Figure 2.1 Two crystal structures of GaN: (a) wurtzite and (b) zinc-blende. (Reproduced from Holt, D. B. and Yacobi, B. G., *Extended Defects in Semiconductors*, Cambridge University Press, New York, 2007.)

In general, different crystal orientations in the crystal lattice are distinguished by the orientation indices (the relatively prime integers of the same ratio as the projections of a vector along the orientation on the coordinate axes), while different crystal planes by the crystal plane indices (the relatively prime integers of the same ratio as the reciprocals of the intercepts of the plane on the coordinate axes). The hexagonal crystal structure is usually described by a four-axis coordinate system with three axes (X_1, X_2, X_3) on the same basal plane at an angle of 120° to each other in the unit of lattice constant a (thus the name a-axes) and the z-axis perpendicular to the basal plane in the unit of lattice constant c (thus the name c-axis). In such a coordinate system, both the orientation and crystal plane indices are denoted by four numbers as $[uvtw]$ and $(hkil)$, respectively. Since the first three numbers are related by $u+v=-t$ and $h+k=-i$, the third number is often omitted, thus the denotations $[uvw]$ and (hkl) are also used and equivalent to the orientation and crystal plane indices in a three-axis system established by the X_1, X_2, and Z-axes. For example, the $[11\bar{2}0]$ orientation ($\bar{2}$ means the projection is on the negative side of the corresponding X_3-axis) and the $(1\bar{1}00)$ plane can also be denoted as the $[110]$ orientation and the $(1\bar{1}0)$ plane. The orientations or planes with the same lattice point arrangement but in different spatial orientations can be represented by the equivalent orientation indices $\langle uvw \rangle$ and the equivalent crystal plane indices $\{hkl\}$. For instance, the indices of all the six side planes of the hexagonal crystal $(10\bar{1}0)$, $(\bar{1}010)$, $(1\bar{1}00)$, $(\bar{1}100)$, $(01\bar{1}0)$, and $(0\bar{1}10)$ belong in the $\{1\bar{1}00\}$ family. The orientations and planes of the same indices are perpendicular to each other, for example, $[0001]\perp(0001)$.

In the theories of crystal X-ray diffraction, lattice vibration, and crystal electronics, it is simpler to use the reciprocal lattice to describe the lattice structure. The coordinate axes of the reciprocal space, in which the lattices are called reciprocal lattices, are k_x, k_y, and k_z corresponding to the X, Y, and Z-axes of the direct space where the lattices are called direct lattices (Figures 2.2 and 2.3). Each crystal structure has both direct and reciprocal lattices. Suppose the primitive basis vectors of the direct lattice is

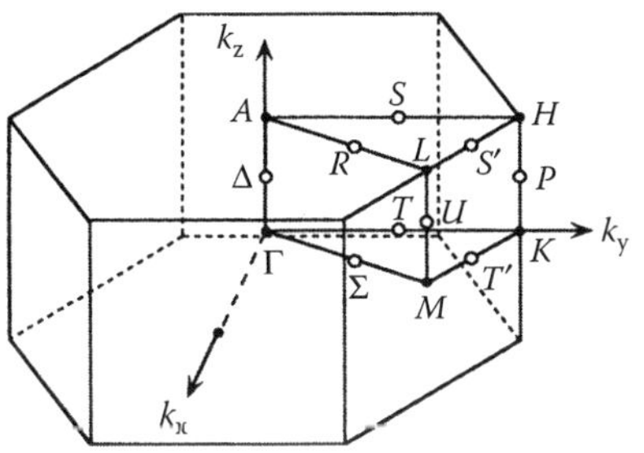

Figure 2.2 Brillouin zone of wurtzite structure. (Reproduced from Suzuki, M. et al., *Phys. Rev. B*, 52, 8132, 1995.)

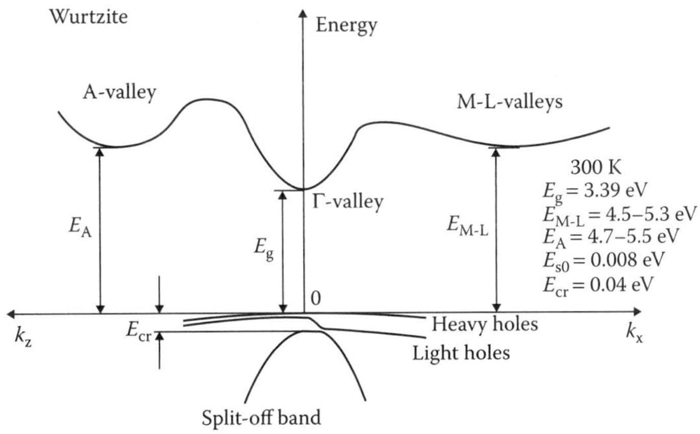

Figure 2.3 Band structure of wurtzite GaN. (Levinshtein, M.E. et al.: *Properties of Advanced Semiconductor Materials GaN, AlN, InN, BN, SiC, SiGe*, 1st edn., 2001. Copyright Wiley-VCH Verlag GmbH & Co. KGaA. Reproduced with permission.)

$\boldsymbol{a}_1$, $\boldsymbol{a}_2$, $\boldsymbol{a}_3$ (i.e., the unit vectors of X, Y, Z), and then the orientations of the reciprocal-lattice coordinate axes can be determined by its unit vectors $\boldsymbol{b}_1$, $\boldsymbol{b}_2$, and $\boldsymbol{b}_3$ (i.e., the primitive basis vectors of the reciprocal lattice).

$$b_1 = 2\pi \frac{\boldsymbol{a}_2 \times \boldsymbol{a}_3}{\boldsymbol{a}_1 \cdot \boldsymbol{a}_2 \times \boldsymbol{a}_3}, \quad b_2 = 2\pi \frac{\boldsymbol{a}_3 \times \boldsymbol{a}_1}{\boldsymbol{a}_1 \cdot \boldsymbol{a}_2 \times \boldsymbol{a}_3}, \quad b_3 = 2\pi \frac{\boldsymbol{a}_1 \times \boldsymbol{a}_2}{\boldsymbol{a}_1 \cdot \boldsymbol{a}_2 \times \boldsymbol{a}_3} \tag{2.1}$$

By its definition, the reciprocal lattice is the crystal lattice in the Fourier space related to the direct space. Each reciprocal-lattice point in the reciprocal space can be given by the vector $\boldsymbol{k}$ (known as the reciprocal-lattice vector)

$$\boldsymbol{k} = h_1 \boldsymbol{b}_1 + h_2 \boldsymbol{b}_2 + h_3 \boldsymbol{b}_3 \tag{2.2}$$

where h_1, h_2, h_3 take integer values. In the reciprocal space with the origin and primitive basis vectors of the reciprocal lattice given, the minimum zone enclosed by the perpendicular bisecting planes of all the reciprocal-lattice vectors with the origin in it is referred to as the first Brillouin zone (generally the Brillouin zone for short) as shown in Figure 2.2.

All other crystal properties are determined by the crystal structure. Shown in Figure 2.2 is the Brillouin zone of wurtzite GaN, whose direct space still adopts the orthogonal *X*, *Y*, and *Z*-axes, as a result of which the k_x, k_y, and k_z axes in the reciprocal space are also orthogonal to each other. The simplified band structure along the k_x and k_z axes of the Brillouin zone is shown in Figure 2.3. The conduction band reaches the bottom while the valence band reaches the peak at the Γ-point, thus a direct band gap; the second lowest valley of the conduction band is the M–L valley, and the third is the A valley; due to crystal symmetry and spin–orbit interaction, the valence band breaks into three bands: the heavy-hole, the light-hole, and the split-off bands. AlN and InN have similar band structures, but it should be noted that the third lowest valley of AlN conduction band is the K valley; the InN band gap was reported in the early investigations to be 1.9–2.05 eV, but with the improvement of InN preparation techniques and material quality, it has been corrected to be 0.64–1.0 eV based on recent experimental observations and theoretical researches. The crystal structure and band structure parameters for wurtzite GaN, AlN, and InN are listed in Table 2.1.

Table 2.1 Parameters of the Crystal Structure and the Band Structure of Wurtzite GaN, AlN, and InN

	GaN	AlN	InN
Lattice constants *a* (nm)	0.3189	0.3112	0.3533
Lattice constants *c* (nm)	0.5185	0.4982	0.5693
Room temperature band gap E_g (300 K) (eV)	3.39	6.026	1.970 (Wu et al. 2003) 0.641 (Bougrov et al. 2001)
Temperature characteristic of band gap $E_g = E_g(0) - AT^2/(T+B)$ (eV)	$E_g(0) = 3.47$ eV $A = 7.7 \times 10^{-4}$ $B = 600$ (Siklitsky)	$E_g(0) = 6.13$ eV $A = 1.799 \times 10^{-3}$ $B = 1462$ (Wu et al. 2003)	$E_g(0) = 1.994$ eV, $A = 2.45 \times 10^{-4}$, $B = 624$ (Wu et al. 2003) or $E_g(0) = 0.69$ eV, $A = 4.1 \times 10^{-4}$, $B = 454$ (Bougrov et al. 2001)
Electron affinity (eV)	4.1	0.6	5.8
Effective densities of states of conduction band N_C (cm^{-3})	$N_C = 4.3 \times 10^{14} \times T^{3/2}$ N_C (300) $= 2.3 \times 10^{18}$	$N_C = 1.2 \times 10^{15} \times T^{3/2}$ N_C (300) $= 6.3 \times 10^{18}$	$N_C = 1.76 \times 10^{14} \times T^{3/2}$ N_C (300) $= 9 \times 10^{17}$
Effective densities of states of valence band N_V (cm^{-3})	$N_V = 8.9 \times 10^{15} \times T^{3/2}$ N_V (300) $= 4.6 \times 10^{19}$	$N_V = 9.4 \times 10^{16} \times T^{3/2}$ N_V (300) $= 4.8 \times 10^{20}$	$N_V = 10^{16} \times T^{3/2}$ N_V (300) $= 5.3 \times 10^{19}$
Electron effective mass	$0.20m_0$	$0.40m_0$	$0.11m_0$
Hole effective mass	$m_{hh} = 1.4m_0$ (heavy hole) $m_{lh} = 0.3m_0$ (Light hole) $m_{sh} = 0.6m_0$ (split-off hole)	k_z:$m_{hz} = 3.53m_0$ k_x:$m_{hx} = 10.42m_0$ (heavy hole) k_z:$m_{lz} = 3.53m_0$ k_x:$m_{lx} = 0.24m_0$ (light hole) k_z:$m_{soz} = 0.25m_0$ k_x:$m_{sox} = 3.81m_0$ (split-off hole)	$m_{hh} = 1.63m_0$ (heavy hole) $m_{lh} = 0.27m_0$ (light hole) $m_{sh} = 0.65m_0$ (split-off hole)

Sources: Levinshtein, M.E. et al.: *Properties of Advanced Semiconductor Materials GaN, AlN, InN, BN,SiC, SiGe*, 1st edn., 2001. Copyright Wiley-VCH Verlag GmbH & Co. KGaA. Reproduced with permission; Guo, Q. and Yoshida, A., *Jap. J. Appl. Phys.*, 33(part 1, 5A), 2453, 1994; Wu, J. et al., *J. Appl. Phys.*, 94(7), 4457, 2003; Bougrov, V. et al.: *Properties of Advanced Semiconductor Materials GaN, AlN, InN, BN, SiC, SiGe*, Rumyantsev, S.L., Levinshtein, M.E., and Shur, M.S. (eds.). 2001. Copyright Wiley-VCH Verlag GmbH & Co. KGaA. Reproduced with permission.

Note: $E_g(0)$ is the band gap at 0 K, and m_0 is the electron rest mass.

2.1.2 Lattice Constants and Band Gaps of Nitride Alloys

The lattice constant of ternary nitride alloy $A_xB_{1-x}N$ is related to the lattice constants of its binary components AN and BN and the mole fraction x by the Vegard law:

$$a(A_xB_{1-x}N) = x \cdot a(AN) + (1-x) \cdot a(BN) \tag{2.3}$$

$$c(A_xB_{1-x}N) = x \cdot c(AN) + (1-x) \cdot c(BN) \tag{2.4}$$

The band gaps of a nitride alloy can be roughly estimated by the Vegard law, but an accurate calculation requires consideration of its nonlinear relation to the band gaps of the binary materials in the alloy, which is referred to as the bowing effect

$$E_g(A_xB_{1-x}N) = x \cdot E_g(AN) + (1-x) \cdot E_g(BN) - b \cdot x \cdot (1-x) \tag{2.5}$$

where b is the bowing coefficient, the relevant term of which introduces the nonlinear effect. For AlGaN, the presently reported typical value for the bowing coefficient is 1.0 eV. The band gap of In-contained alloys remains inconclusive since the quality of InN and In-rich AlInN and InGaN is not good enough and it is difficult to reduce the background electron concentration. The bowing constants for AlInN and InGaN were reported to be 5.4 and 2.5 eV (Piprek 2007), respectively, based on an InN band gap of 1.97 eV or 3.4 eV and 1.4 eV (Piprek 2007), respectively, based on an InN band gap of 0.77 eV.

When the indium content is less than 2%, the band gap of the quaternary alloy AlGaInN was reported to decline in a linear manner with increasing indium content. Monroy et al. described AlGaInN band gap by an empirical formula (Monroy et al. 2003).

$$\begin{aligned} E_g(Al_xIn_yGa_{1-x-y}N = xE_g(AlN) + (1-x-y)E_g(GaN) + yE_g(InN) \\ - bAl_x(1-x) - bIn_y(1-y) \end{aligned} \tag{2.6}$$

According to photoluminescence characterization results of AlGaInN grown by molecular beam epitaxy (MBE), the indium content-related bowing constant b_{In} is 2.5 eV with E_g(InN) = 1.9 eV, b_{Al} = 1 eV.

2.1.3 Band Offsets at Heterointerfaces

The heterojunction formed between a binary nitride and its alloy is a type I heterojunction, that is, the material with wider band gap at the heterointerface has a higher conduction band edge and lower valence band edge than the material with narrower band gap. An empirical band offset ratio for nitride heterojunctions is such that the conduction band and valence band offsets of AlN/GaN account for 73% and 27%, respectively, of the band gap difference between AlN and GaN, and the conduction band and valence band offsets of GaN/InN are 57% and 43%, respectively, of the band gap difference between GaN and InN. A good agreement is often found between the empirical band offset ratio and the theoretical calculation and experimental results of the band offsets, whether pure binary nitride heterojunctions or AlGaN or InGaN alloy heterojunctions (Takahashi et al. 2007).

Note that test results indicate the influence of growth sequence and polarization effect on conduction and valence band offsets between binary nitrides leads to the dispersion of the experimental data. For instance, using X-ray photoelectron spectroscopy, Martin et al. (1996) measured a valence band offset of 0.93 ± 0.25 eV for InN/GaN (top-down) and 0.59 ± 0.24 eV for GaN/InN grown on c-plane sapphire. However, in the same report both AlN/GaN and GaN/AlN valence band offsets were about 0.6 ± 0.2 eV. Therefore, the immature quality and undetermined energy band parameters of InN may have an influence on the measured band offsets of In-contained nitride heterojunctions.

2.2 Electron Velocity–Field Relationship and Low-Field Mobilities of Nitrides

For nitride materials and electronic devices, the electron transport properties such as the low-field mobility and velocity–field relationship are of great importance. For GaN, the dominant scattering mechanisms affecting the directional electron drift in an applied electric field are the ionized impurity scattering and the lattice vibration scattering (including the polar optical phonon scattering, the longitudinal acoustic phonon scattering, the piezoelectric scattering, etc.).

2.2.1 Electron Velocity–Field Relationship for GaN

The low-field electron mobility of GaN can be determined by the Hall measurements, showing an increase followed by a decrease with rising temperature (Dhar and Ghosh 1999) (Figure 2.4), which is caused by the weakened ionized impurity scattering and enhanced lattice vibration scattering at higher temperature. The dislocations, doping, and compensation also have an impact on the low-field mobility. At room temperature, the low-field electron mobility for GaN with an n-type background doping concentration of 10^{16}–10^{17} cm^{-3} is about 200–1000 cm^2/(V·s). The mobility dispersion is strongly correlated to the GaN dislocation density: the higher the dislocation density in GaN, the lower mobility it has.

The electron velocity–field relationship for GaN is often theoretically investigated by the Monte Carlo method to establish the analytical model by fitting the data, and can be obtained experimentally from the measured *I–V* characteristics of specific samples using high-voltage pulse signals. Shown in Figure 2.5 are the theoretical electron velocity–field curves of GaN with a background ionized impurity concentration of 10^{17} cm^{-3} obtained from the full band Monte Carlo simulation (Kolnik et al. 1995).

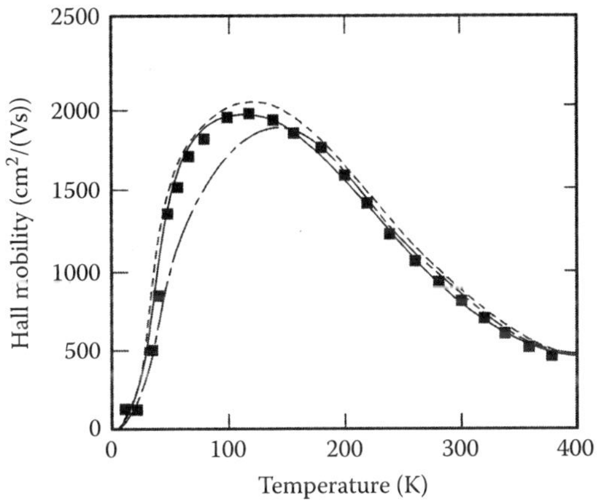

Figure 2.4 Temperature dependence of electron Hall mobility for GaN with a background ionized impurity concentration of 10^{17} cm^{-3}. (Reproduced from Dhar, S. and Ghosh, S., *J. App. Phys.*, 86, 2668, 1999.)

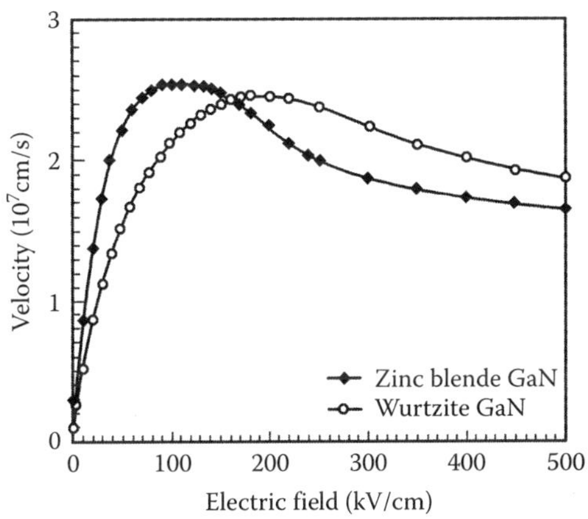

Figure 2.5 Electron drift velocity as a function of applied electric field for GaN with a background ionized impurity concentration of 10^{17} cm^{-3}. (Reproduced from Kolnik, J. et al., *J. Appl. Phys.*, 78, 1033, 1995.)

It can be seen that the GaN electron drift velocity rises with increasing strength of the applied electric field and reaches the peak value (~2.5×10^7 cm/s) at a relatively strong electric field (~180 kV/cm for wurtzite GaN); further increase in electric field yields a negative differential resistance zone where the drift velocity gradually drops until reaching the saturation drift velocity (~1.5×10^7 cm/s) in a strong electric field.

Owing to high breakdown fields (>3×10^6 V/cm for GaN, >1.2×10^7 V/cm for AlN), GaN and AlN can withstand very strong electric fields. In a low electric field, all electrons are located at the bottom of the conduction band (Γ valley) with nearly linear velocity–field dependence, and the slope is the low-field mobility. With the increase of the electric field, electrons obtain more energy and gain the opportunity to move to the satellite valley of higher energy and greater effective mass, thus a general decline of mobility. However in the case of GaN, the quite large electron effective mass at the Γ valley, the strong polar optical phonon scattering at room temperature and the relatively large energy separation between the Γ and satellite valleys make large-scale electron transfer between valleys possible only in high electric fields, at which moment the electron drift velocity reaches peak. Once the electrons enter the satellite valley, the drift motion is weakened due to less kinetic energy, smaller velocity and greater effective mass, thus forming negative differential resistance characteristics; with further increase in the electric field, electrons keep moving into the satellite valley with the electron energy still increasing but at a lower rate due to greater effective mass and the fact that the new intervalley deformation potential scattering becomes the effective scattering mechanism. The two effects combine to form a broad negative differential resistance region so as for electron velocity to saturate. Similar curves have been obtained by many theoretical calculations only with some difference in the peak velocity and corresponding electric field and the saturation velocity depending on different assumptions and given conditions for calculation.

AlN and InN are similar to GaN in the velocity–field relationship and the formation mechanism, but differ in the low-field mobility, peak velocity, and saturation drift velocity (the values are given in Section 2.4).

2.2.2 Analytical Model of Low-Field Electron Mobility and Velocity–Field Relationship for GaN and AlGaN

The analytical model of low-field electron mobility and velocity–field relationship for nitrides and nitride alloys has been derived from numerous theoretical calculations and experimental results. In 2001, Farahmand et al. proposed a unified model of low-field electron mobility and velocity–field dependence (hereinafter referred to as the FMCT model) for GaN, AlN, InN, AlGaN, and InGaN with recommended model parameters for each material by fitting the Monte Carlo calculation results (Farahmand et al. 2001). Covering the transport properties of almost all nitride semiconductors and their ternary alloys, this model has been adopted by the nitride material model libraries of the commercial device simulation software ATLAS@Silvaco and found wide applications. Based on this model, our group established a sophisticated analytical model of low-field electron mobility and velocity–field relationship for full aluminum composition AlGaN (including GaN) considering the alloy composition, temperature dependence of low-field and high-field parameters, and alloy disorder effect in order to accurately predict the characteristics of the terahertz (THz) GaN Gunn diodes that operate based on the negative resistance characteristics of the velocity–field relationship (Yang et al. 2011a).

In the FMCT model, the low-field electron mobility of GaN as a function of the doping concentration and the temperature satisfies

$$\mu_0(\mathrm{GaN}) = \mu_{\min}\left(\frac{T}{300}\right)^{\beta_1} + \frac{(\mu_{\max} - \mu_{\min})(T/300)^{\beta_2}}{1 + \left(N/(N_{\mathrm{ref}}(T/300)^{\beta_3})\right)^{\alpha(T/300)^{\beta_4}}} \tag{2.7}$$

Table 2.2 Model Parameters of Low-Field Electron Mobility for GaN

PARAMETER	μ_{min} [cm²/(V·s)]	μ_{max} [cm²/(V·s)]	α	β_1	β_2	β_3	β_4
Value	295.0	1460.7	0.66	−1.02	−3.84	3.02	0.81

Source: Yang, L.-A., et al., *IEEE Trans. Electron Devices*, 58(4), 1076, 2011a. With permission.

Table 2.3 Fitting Parameters for $f_Z(x)$

λ^x	*a*	*b*	*c*
λ_0^x	-8.699×10^{-2}	6.662×10^{-1}	2.813×10^{-1}
λ_1^x	9.834×10^{-1}	1.226	1.362
λ_2^x	3.286×10^{-2}	-2.503×10^{-1}	-3.849×10^{-1}
λ_3^x	-1.628×10^{-2}	2.433×10^{-2}	2.277×10^{-2}

Source: Yang, L.-A., et al., *IEEE Trans. Electron Devices*, 58(4), 1076, 2011a. With permission.

where:

N is the total doping concentration

$N_{ref} = 1 \times 10^{17}$ cm^{-3}

μ_{min}, μ_{max}, α, and β_1–β_4 are the fitting parameters whose values are listed in Table 2.2

The applicable doping concentration for our model is 1×10^{16} to 1×10^{18} cm^{-3}.

The variation of the low-field electron mobility for AlGaN with the aluminum composition x is related to AlGaN material parameters such as the effective mass of the center valley electrons. By introducing coefficient $f_Z(x)$ to consider the nonlinear relation between the mobility and the aluminum composition as follows, the low-field electron mobility for AlGaN can be obtained from that for GaN:

$$\mu_0(\mathrm{Al}_x\mathrm{Ga}_{1-x}\mathrm{N}) = f_Z(x)\mu_0(\mathrm{GaN}) \tag{2.8}$$

$$f_Z(x) = \frac{1}{(1 + ax + bx^2 + cx^3)} \tag{2.9}$$

where a, b, and c are the fitting parameters denoted generally by λ^x, which is related to the doping concentration N via

$$\lambda^x = \lambda_0^x\left[\lambda_1^x + \lambda_2^x\left(\frac{N}{N_{ref}}\right) + \lambda_3^x\left(\frac{N}{N_{ref}}\right)^2\right] \tag{2.10}$$

Values for the parameters in Equation 2.10 are listed in Table 2.3.

As the aluminum composition x for AlGaN have an impact both on electron effective mass and on alloy disorder scattering induced by the random distribution

Table 2.4 Electron Velocity–Field Model Parameters for GaN

PARAMETER	v_{sat} (10^7 cm/s)	F_C (kV/cm)	n_1	n_2	a
Value	1.9064	220.8936	7.2044	0.7857	6.1973

Source: Yang, L.-A., et al., *IEEE Trans. Electron Devices*, 58(4), 1076, 2011a. With permission.

of aluminum and gallium atoms, Equation 2.8 needs to be corrected by introducing the random alloy factor f^{a1} into

$$\mu_0(\text{alloy}) = f^{a1}\mu_0(\text{Al}_x\text{Ga}_{1-x}\text{N}) \tag{2.11}$$

$$f^{a1} = 1 - p(f_1^{a1}x - f_1^{a1}x^2) \tag{2.12}$$

where $p = (U_{\text{alloy}}/\Delta E_C)^2$ is the alloy disorder intensity factor (U_{alloy} is alloy disorder potential and ΔE_C is conduction band offset between GaN and AlN) valued 0–1 to regulate the alloy disorder effect. From the Monte Carlo calculation of the velocity–field characteristics of $\text{Al}_{0.5}\text{Ga}_{0.5}\text{N}$ with the strongest alloy disorder scattering, we can obtain $f_1^{a1} = 3.393$.

The FMCT model of high-field electron transport velocity for GaN is expressed as

$$v = \frac{\mu_0 F + v_{\text{sat}}(F/F_C)^{n1}}{1 + a(F/F_C)^{n2} + (F/F_C)^{n1}} \tag{2.13}$$

where:

μ_0 is the low-field mobility
F is the electric field
v_{sat} the saturation drift velocity

The critical electric field F_C and the parameters a, n_1, and n_2 need to be obtained by fitting the Monte Carlo calculation results of the velocity–field characteristics for GaN, whose values are listed in Table 2.4.

As shown is Figure 2.6, this model can perfectly reproduce the features of electron velocity–field curve of GaN but suffers from the major defect that the effect of temperature on high-field parameters is not included. We propose the high-field electron transport model (YHT model) for GaN by fitting the Monte Carlo simulation curve of the electron velocity–field characteristics of GaN at temperatures 300–600 K. Denoting the fitting parameters v_{sat}, F_C, a, n_1, and n_2 as λ^F, λ^F is related to temperature T via

$$\lambda^F = \lambda_0^F\left[\lambda_1^F + \lambda_2^F\left(\frac{T}{300}\right) + \lambda_3^F\left(\frac{T}{300}\right)^2\right] \tag{2.14}$$

Parameter values in the equation are given in Table 2.5.

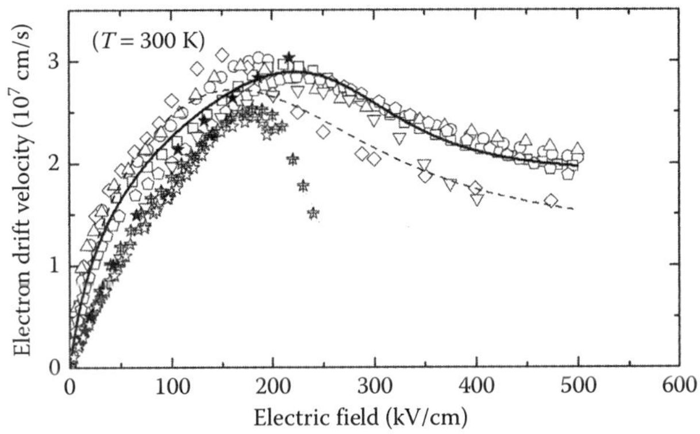

Figure 2.6 GaN electron velocity–field curves at 300 K. The hollow signs are the Monte Carlo simulation data (Albrecht et al. 1998, Bertazzi et al. 2009, Djeffal et al. 2009, Farahmand et al. 2001, Reklaitis and Reggiani 2004, Tomita et al. 2007), the star signs are experimental data (Barker et al. 2002, 2005, Liberis et al. 2006), and the solid line and dotted line are the fitting curves based on the analytical models. The solid line data is from the YHT model and the dotted line data is from Schwierz (2005). (Data from Yang, L.-A. et al., *IEEE Trans. Electron Devices*, 58, 1076, 2011a. With permission.)

Table 2.5 Fitting Parameters for Equation 2.14

λ^F	v_{sat} (cm/s)	F_C (V/cm)	n_1	n_2	a
λ_0^F	1.907×10^7	2.209×10^5	7.144	0.783	5.362
λ_1^F	1.777	1.220	2.108	2.437	3.302
λ_2^F	−0.983	−0.420	−1.643	−2.318	−3.102
λ_3^F	0.206	0.200	0.535	0.881	0.800

Source: Yang, L.-A., et al., *IEEE Trans. Electron Devices*, 58(4), 1076, 2011a. With permission.

It can be seen in Figure 2.7 that the predicted velocity–field characteristics by the YHT model agree with those from experimental and theoretical data, that is, the increase in temperature leads to a decrease in the peak drift velocity and saturation drift velocity but an increase in the threshold electric field; whereas the calculated velocity–field characteristic curves from the FMCT model fail to correctly describe the electron velocity–field characteristics at corresponding temperatures.

On the basis of electron velocity–field characteristics of GaN with temperature effect in consideration (Equations 2.13 and 2.14), the YHT model for AlGaN electron velocity–field relationship considers first the relation between the high-field fitting parameters and the aluminum composition. In this relation, denoting v_{sat}, F_C, a, n_1, and n_2 as λ^{Fx}, λ^{Fx} is related to aluminum composition via

$$\lambda^{Fx} = \lambda_0^{Fx}\left[1 + \lambda_1^{Fx}x + \lambda_2^{Fx}x^2 + \lambda_3^{Fx}x^3\right] \tag{2.15}$$

The parameter values for (2.15) are listed in Table 2.6.

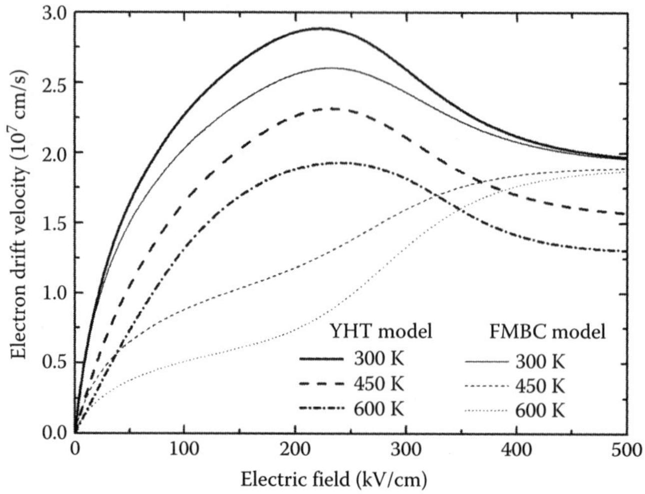

Figure 2.7 Calculated GaN electron velocity–field characteristics at different temperatures by the YHT model (Yang et al. 2011a) and the FMCT model (Farahmand et al. 2001). (Courtesy of Qingyang Yao.)

Table 2.6 Fitting Parameters for Equation 2.15

λ^{Fx}	v_{sat} (cm/s)	F_C (V/cm)	n_1	n_2	a
λ_0^{Fx}	The same as those for (2.14)				
λ_1^F	3.557×10^{-1}	5.129×10^{-1}	9.497×10^{-1}	7.188×10^{-2}	7.975×10^{-1}
λ_2^F	-2.198×10^{-1}	5.365×10^{-1}	-2.125	4.660×10^{-2}	-0.349
λ_3^F	0	0	2.743	0	0

Source: Yang, L.-A., et al., *IEEE Trans. Electron Devices*, 58(4), 1076, 2011a. With permission.

Then, the relation of high-field fitting parameter to the random alloy factor f^{a2} in the AlGaN electron velocity–field model is considered. In this relation, denoting v_{sat}, F_C, a, n_1, and n_2 as λ^{Fa}, λ^{Fa} is related to λ^{Fx} and f^{a2} in Equation 2.15 via

$$\lambda^{Fa} = f^{a2}\lambda^{Fx} \tag{2.16}$$

$$f^{a2} = 1 - \left[p_a \left(f_1^{a2} x - f_1^{a2} x^2 \right) \right] \tag{2.17}$$

The parameter values are shown in Table 2.7, where p is the alloy disorder intensity factor in Equation 2.12, and p_{FC} and p_{n2} are

$$p_{FC} = (-3.457 \times 10^{-3})\exp(6.168p) + (1.617 \times 10^{-4})\exp(9.704p) \tag{2.18}$$

$$p_{n2} = (-1.494 \times 10^{-2})\exp(2.009p) + (3.826 \times 10^{-6})\exp(12.580p) \tag{2.19}$$

Table 2.7 Fitting Parameters for Equations 2.16 and 2.17

λ^F	v_{sat} (cm/s)	F_C (V/cm)	n_1	n_2	a
f_1^{a2}	1.372	−2.234	-1.607×10^{-1}	−2.171	2.694
p_a	p	p_{FC}	p	p_{n2}	p

Source: Yang, L.-A., et al., *IEEE Trans. Electron Devices*, 58(4), 1076, 2011a. With permission.

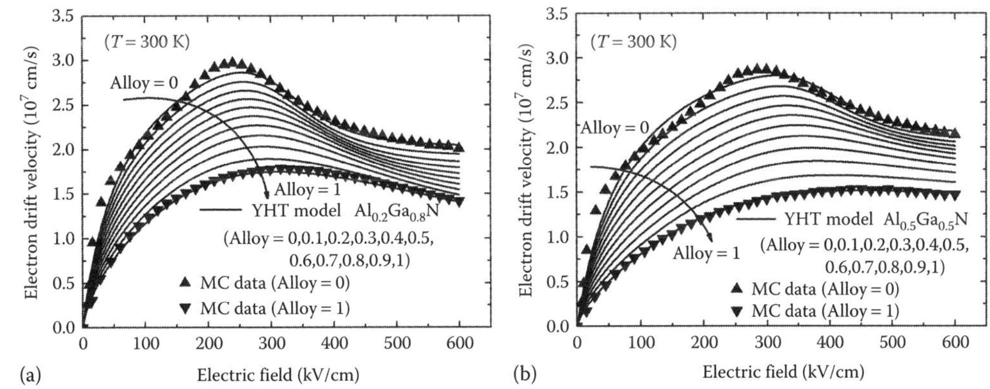

Figure 2.8 Electron velocity–field curves of AlGaN under different alloy disorder intensity factors. The Monte Carlo simulation data are from Farahmand et al. (2001). (a) $Al_{0.2}GaN$ and (b) $Al_{0.5}GaN$. (Courtesy of Qingyang Yao.)

As shown in Figure 2.8, a gradual increase of the alloy disorder intensity factor p from 0 ($U_{alloy} = 0$) to 1 ($U_{alloy} = \Delta E_C$) yields an AlGaN electron velocity–field curve with a changing alloy disorder effect from the weakest to the strongest, which agrees very well with the Monte Carlo simulation.

The YHT model proves an electron velocity–field model very close to the actual material physical mechanism for full aluminum composition AlGaN owing to its inclusion of the dependence of low-field and high-field parameters on the temperature, alloy composition and alloy disorder effect on the basis of the FMCT model (Yang et al. 2011a). Fully compatible with the commercial device simulator ATLAS@Silvaco, this model has been verified in GaN-based negative resistance device simulations (Yang et al. 2011b).

2.3 Polarization Effect of Nitrides

2.3.1 Polarity

Both wurtzite and zinc-blende structure crystals are noncentrosymmetrical crystals with polar axes. Wurtzite structure nitrides have only one polar axis, namely the *c*-axis. The crystal is formed by the stacking of III-N bilayers in different sequences along the two opposite directions ([0001] and $[000\bar{1}]$) parallel to the *c*-axis (Figure 2.9). For GaN, if the Ga–N bond parallel to the *c*-axis shows nitrogen atom at the top,

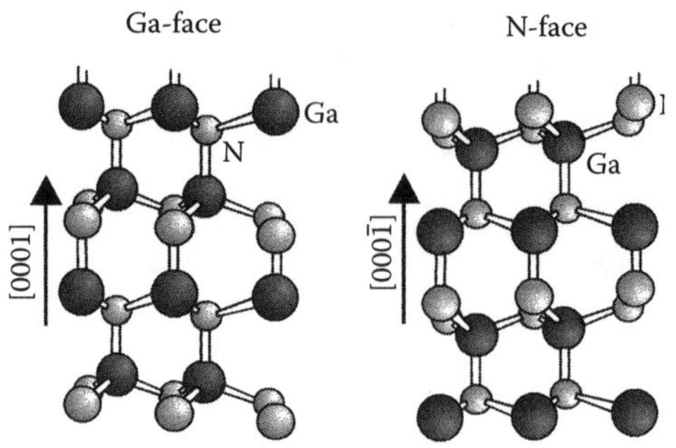

Figure 2.9 Schematic structure of hexagonal wurtzite GaN with different polarities. (Reproduced from Ambacher, O. et al., *J. Appl. Phys.*, 85, 3222, 1999.)

Ga-face polarity forms and GaN crystal growth is along the [0001] direction. Whereas if gallium atom of the same bond is at the top, N-face polarity forms and growth is along the [000$\bar{1}$] direction. Note that the polarity is irrelevant with whether the surface of GaN is terminated by gallium or nitrogen. Different polarity differ greatly in physical and chemical properties such as reaction to acid and alkali, surface adsorption, Schottky barrier (Stutzmann et al. 2001), and band offsets (Martin et al. 1996) at heterointerfaces, and therefore the two polarity are not equivalent.

Few direct methods of precision are available for predicting the material polarity in nitride thin-film epitaxial growth, which requires experimental techniques such as the convergent beam electron diffraction, chemical etching and circular polarization spin photoelectric effect. High quality MOCVD-grown nitride films with smooth surfaces are generally of Ga-face polarity, while the rough-surfaced ones may have Ga- or N-face polarity. Mixed polarity may also exist on relatively low-quality epitaxial films. High quality N-face polarity films are generally grown by MBE, but in recent years MOCVD has been employed on vicinal substrates (such as c-plane sapphire and SiC with misorientation toward a-plane or m-plane) to successfully grow high quality smooth surface N-polar GaN. According to current reports, surface polarity may vary with substrates, nucleation layers, growth conditions, and growth techniques.

2.3.2 Spontaneous and Piezoelectric Polarization

Both wurtzite and zinc-blende are noncentrosymmetric compound structures, thus having the piezoelectric effect. With applied stress, the lattice deformation induces a separation of the centers of positive and negative charges in the crystal thus forming dipole moments, the accumulation of which gives rise to polar charges on the crystal surface, hence the piezoelectric polarization. Furthermore, since wurtzites have poorer crystal symmetry than zinc-blendes, their positive and negative charge centers do not coincide even without stress, thereby inducing the spontaneous polarization along the polar axis.

The polarization effect can be described by the polarization strength P, the spatial variation of which induces the polarization bound charge. Assuming the z direction is perpendicular to the interface between two material of different polarization strengths, the interface location is $z = z_0$, and $\sigma_{pol} = P(z_0^-) - P(z_0^+)$ is the polar sheet charge induced by the change of P, by integrating the Gauss equation across the interface we have (Ridley 2004)

$$\varepsilon(z_0^-)\varepsilon_0 F(z_0^-) + \sigma_{pol} = \varepsilon(z_0^+)\varepsilon_0 F(z_0^+) \quad (2.20)$$

where:

F is the electric field
ε is the relative dielectric constant
ε_0 is the vacuum permittivity

Equation 2.20 applies to abrupt interfaces. For gradual interfaces, the polar charge is the bulk charge, whose density is

$$\rho_{pol}(z) = -\nabla P(z) \quad (2.21)$$

If both the opposite surfaces of a uniformly polarized crystal film are adjacent to non-polar materials (say the substrate on one side, and the air on the other), polar charges $\pm\sigma_{pol}$ of opposite electrical property and same magnitude will induce a built-in electric field in the crystal. In actual semiconductor crystals, the carriers induced by background doping ionization or the extrinsic charges absorbed on the surface have a shielding effect on polarization-induced field, which makes it difficult to observe the polarization strength itself by experiment. Instead, what is observed is its change, such as the variation of polarization strength with temperature in pyroelectricity and with the applied stress under the piezoelectric effect. Semiconductor heterostructures have internal interfaces, which facilitate the study and utilization of the polarization effect.

Nitride semiconductors have strong spontaneous polarization due to the strong polarity of chemical bonds between Group III and nitrogen atoms. The polarization strengths (Table 2.8) were calculated based on the modern polar theory by Bernardini et al. (1997), while the piezoelectric polarization strength is related to the crystal strain ε. In the case of spontaneous polarization in nitride heterojunction without strain, the spontaneous polarization occurs along the $[000\bar{1}]$ direction with polarization-induced positive interface charges for the Ga face samples and negative ones for the N-face samples; while with strain the piezoelectric polarization along the c-axis with the strained layer under tensile/compressive strain is in the same/opposite direction with respect to the spontaneous polarization (Figure 2.10). Theoretical calculations and experimental measurements show that the polarization effect induces strong built-in electric fields of the order of MV/cm and polar-bound charges of a density up to 10^{13} cm^{-2} in nitride heterostructures, significantly modulating the energy band structure and influencing the distribution of free carriers.

Table 2.8 Polarization-Related Physical Quantities

	LATTICE CONSTANT *a* (Å)	e_{33} (C/m^2)	e_{31} (C/m^2)	C_{13} (GPa)	C_{33} (GPa)	P_{SP} (C/m^2)
AlN	3.112	1.55[a]	−0.58[a]	127	382	−0.081
		1.46[b]	−0.60[b]	108	373	−0.090
		1.50	−0.53			
GaN	3.189	1[c]	−0.36[c]	100	392	−0.029
		0.44[d]	−0.22[d]	103	405	−0.034
		0.65[e]	−0.33[e]			
		0.73[b]	−0.49[b]			
		0.67	−0.34			
InN	3.548	0.43[f]	−0.22[f]	94	200	−0.032
		0.97[b]	−0.57[b]	92	224	−0.042
		0.81	−0.41			
ZnO		1.32[*]	−0.57[*]			
GaAs		0.093[*]	−0.185[*]			

Sources: Bernardini, F. et al., *Phys. Rev. B* (*Condensed Matter*) 56(16), 10024, 1997; Shur, M.S. et al., Pyroelectric and piezoelectric properties of GaN-based materials, *Materials Research Society Symposium Proceedings*, 1999; Yu, E.T. et al., Spontaneous and piezoelectric polarization effects in III-V nitride heterostructures, *26th Conference on the Physics and Chemistry of Semiconductor Interfaces*, January 17–21, San Diego, CA, 1999.

Note: The piezoelectric coefficients indicated by * and letters are from Shur et al. (1999) using the following measurements and calculation methods: *a* is measured by acoustic surface wave; *b* is obtained from first principle calculations; *c* is estimated from the electromechanical coupling coefficient measurements; *d* is extracted from experimental low-field mobility data in combination with the AlGaN/GaN 2DEG piezoelectric scattering mechanism analysis; *e* and *f* are from the analysis of the estimated optical phonon frequency. The lattice constants and elastic constants are from Yu et al. (1999).

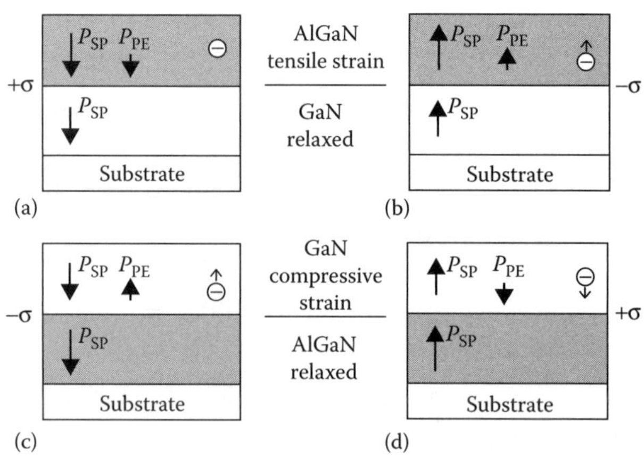

Figure 2.10 Spontaneous and piezoelectric polarization in the nitride heterostructures. The top-down structures are (a) Ga-face AlGaN/GaN, (b) N-face AlGaN/GaN, (c) Ga-face GaN/AlGaN, and (d) N-face GaN/AlGaN. (Reproduced from Ambacher, O. et al., *J. Appl. Phys.*, 85, 3222, 1999.)

2.3.3 Piezoelectric and Spontaneous Polarization Strengths for Nitride Alloys

Since most nitride electronic devices are heterostructures containing nitride alloys, the piezoelectric and spontaneous polarization strengths of the nitride alloy are parameters of great significance. In the case of biaxial strain in the c-plane under moderate stress, for instance the pseudomorphic growth of the nitride alloy $A_xB_{1-x}N$ on the nitride buffer layer, the piezoelectric polarization strength (P_{PE}) along the *c*-axis is in linear relation to the strain ε induced by the relative change of the lattice constant *a* by

$$P_{PE} = 2\varepsilon\left(e_{31} - e_{33}\frac{C_{13}}{C_{33}}\right), \quad \varepsilon = \frac{a_{buffer} - a(x)}{a(x)}, \tag{2.22}$$

where:

e_{ij} and C_{ij} are the piezoelectric and elastic coefficients of $A_xB_{1-x}N$, respectively, which can be obtained by a linear interpolation of corresponding physical quantities of the binary materials AN and BN according to their mole fractions

$a(x)$ is the lattice constant of $A_xB_{1-x}N$

a_{buffer} the lattice constant of the buffer layer

The spontaneous polarization (P_{SP}) of alloy $A_xB_{1-x}N$ is also linearly related to alloy composition via

$$P_{SP}(A_xB_{1-x}N) = P_{SP}(AN)x + P_{SP}(BN)(1-x) \tag{2.23}$$

Equations 2.22 and 2.23 give a linear model of the polarization effect for nitride alloys. Ambacher et al. investigated further into the polarization of AlGa(In)N and presented the nonlinear relation of the spontaneous polarization to the alloy composition and of the piezoelectric polarization under large stress to the strain (Ambacher et al. 2002), that is, the nonlinear model of polarization effect in nitride alloy materials. With the AlGaN alloy on the GaN substrate as an example, take the alloy composition *x* as argument and we get

$$P_{PE}(Al_xGa_{1-x}N/GaN) = [-0.0525x + 0.0282x(1-x)]\ C/m^2 \tag{2.24}$$

$$P_{SP}(Al_xGa_{1-x}N) = -0.090x - 0.034(1-x) + 0.021x(1-x)\ C/m^2 \tag{2.25}$$

Another nonlinear piezoelectric polarization strength model employs in-plane strain as the argument. For a binary semiconductor with an in-plane strain η_1, its nonlinear piezoelectric polarization strength as a function of η_1 is

$$
\begin{aligned}
P_{PE}^{AlN} &= -1.808\eta_1 + 5.624\eta_1^2 \quad \text{when } \eta_1 < 0 \\
P_{PE}^{AlN} &= -1.808\eta_1 - 7.888\eta_1^2 \quad \text{when } \eta_1 > 0 \\
P_{PE}^{GaN} &= -0.918\eta_1 + 9.541\eta_1^2 \\
P_{PE}^{InN} &= -1.373\eta_1 + 7.559\eta_1^2
\end{aligned} \tag{2.26}
$$

Assuming an AlGaN lattice constant $a(x)$ and an in-plane strain $\eta_1(x)$, we have

$$\eta_1(x) = \frac{a^{GaN} - a(x)}{a(x)} \tag{2.27}$$

and the AlGaN piezoelectric polarization strength as a function of $\eta_1(x)$ is

$$P_{PE}(Al_xGa_{1-x}N, \eta_1) = xP_{PE}(AlN, \eta_1) + (1-x)P_{PE}(GaN, \eta_1) \tag{2.28}$$

The calculation formulas (Equations 2.29 through 2.32) for the spontaneous and piezoelectric polarization strength of the quaternary alloy AlInGaN were also given in Piprek (2007).

$$
\begin{aligned}
P_{PE}(Al_xIn_yGa_{1-x-y}N, \eta_1) = {} & xP_{PE}(AlN, \eta_1) + yP_{PE}(InN, \eta_1) \\
& + (1-x-y)P_{PE}(GaN, \eta_1)
\end{aligned} \tag{2.29}
$$

$$\eta_1(Al_xIn_yGa_{1-x-y}N) = \frac{x(a^{GaN} - a^{AlN}) + y(a^{GaN} - a^{InN})}{xa^{AlN} + ya^{InN} + (1-x-y)a^{GaN}} \tag{2.30}$$

$$
\begin{aligned}
P_{SP}(Al_xIn_yGa_{1-x-y}N) = {} & xP_{SP}(AlN) + yP_{SP}(InN) + (1-x-y)P_{SP}(GaN) \\
& + b_{AlGaN}x(1-x-y) + b_{InGaN}y(1-x-y) \\
& + b_{AlInN}xy + b_{AlInGaN}xy(1-x-y)
\end{aligned} \tag{2.31}
$$

$$
\begin{aligned}
b_{AlInGaN} = {} & 27P_{SP}(Al_{1/3}In_{1/3}Ga_{1/3}N) - 9(b_{AlGaN} + b_{InGaN} + b_{AlInN}) \\
& - 3(P_{SP}(AlN) + P_{SP}(GaN) + P_{SP}(InN))
\end{aligned} \tag{2.32}
$$

2.3.4 Polarization Relaxation Mechanisms

The piezoelectric polarization discussed above is of the material under complete strain. If certain stress relaxation mechanism arises in the alloy layer $A_xB_{1-x}N$ grown on relatively thick buffer layers of binary nitride semiconductors, a strain relaxation is induced, greatly weakening the piezoelectric effect. According to the elastic strain relaxation theory, the relaxation degree of the $A_xB_{1-x}N$ internal stress is a function of x and the alloy layer thickness d (Figure 2.11). When d for the alloy layer of a given x

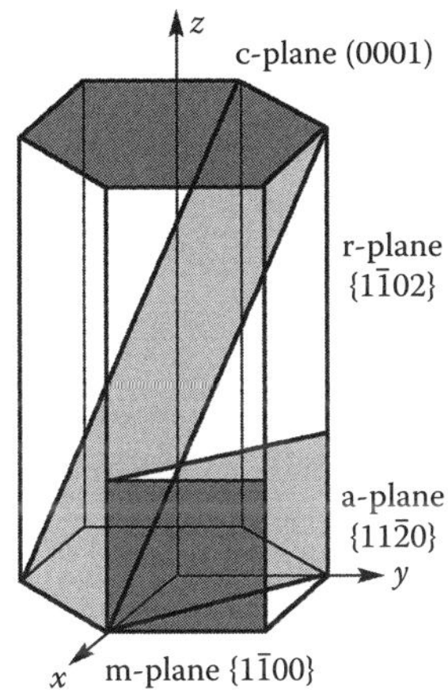

Figure 2.11 Directions of lattice planes of wurtzite GaN in relation to the polar axis. (Reproduced from Moram, M.A. and Vickers, M.E., *Rep. Prog. Phys.*, 72, 036502, 40 pp., 2009.)

exceeds the so-called critical thickness d_{crit} or x for that of a given d_{crit} exceeds a certain value (Bykhovski et al. 1998), strain relaxation arises. The polarization strength is higher with a small d_{crit} and a large x. Strain relaxation degree can be calculated indirectly by C–V measurements. Strain relaxation is usually assumed having no effect on spontaneous polarization.

If no abrupt change on the atomic scale but certain interdiffusion within a relatively large range of thickness occurs at the heterointerface, the gradient of spontaneous polarization is also reduced with weaker polarization effect. If mixed polarity, that is, the random distribution of polarity inversion domains, exists in the epitaxial film, the total polarization effect may disappear.

In polar materials, the polarization-induced internal electric field is often partly shielded by the oppositely charged defects and the absorbed charges from the ambience between the buffer layer and the substrate or at the surface and the heterointerface (Piprek 2007), some of them could work as various surface states or interface states (Shur et al. 1999). Both the trapping effect of fast and slow states and the change of the piezoelectric and spontaneous polarizations may occur under an applied high-frequency alternating-current field, affecting device performance. The major countermeasure is the growth of a passivation layer on the surface of the heterostructure to stabilize the strain and change the distribution of surface and interface states, also suppressing the change of polarization to some extent.

2.3.5 Polar and Nonpolar/Semipolar Nitrides

For nitride semiconductors, the (0001) plane (Ga polar) and (000$\bar{1}$) plane (N polar) are generally referred to as the c-plane. The nitride epitaxial films are polar materials if the film surface is c-plane. As shown in Figure 2.11, if the surface of the nitride film is the {1$\bar{1}$00} plane (m-plane) or the {11$\bar{2}$0} plane (a-plane) parallel to the polar axis (*c*-axis), no polarization effect exists along the growth direction, and the nitride is nonpolar. If the film surface is neither parallel nor normal to the *c*-axis, the material is semipolar, for instance the $\{1\bar{1}02\}$ plane (r-plane).

Polarization induces tilted energy bands in polar nitride heterostructures, which is in favor of the formation of high-density two-dimensional electron gases in electronic devices, but the resulting spatial separation of electrons and holes in photoelectric heterostructures reduces the overlap of wave functions of the two carriers, thus decreased luminescence efficiency and a luminescence wavelength redshift occur. This is referred to as the quantum-confined Stark effect (QCSE) (Chichibu et al. 1998, Miller et al. 1984). As the polarization of nonpolar (semipolar) nitrides vanishes (diminishes) along the growth direction, theoretically the QCSE can be eliminated (weakened), and therefore many researches have been devoted to their use in photoelectric materials and devices, together with some reports on their applications in enhancement-mode field effect transistors (Fujiwara et al. 2009).

Nonpolar and semipolar materials can be prepared by epitaxy on the specific crystal planes of a foreign substrate, for example, the a-plane GaN grown on r-plane sapphire or a/m-plane GaN on a/m-plane SiC. Due to the significant anisotropy (e.g., different lattice constants of a-plane materials along the *c*-axis and *m*-axis) of the nonpolar and semipolar material, stacking faults are an important type of epitaxial defects along with dislocations. In recent years, the crystalline quality and surface smoothness of heteroepitaxially grown nonpolar and semipolar materials have been significantly improved close to those of the polar nitrides through growth process optimization and the adoption of the epitaxial lateral overgrowth (ELOG) and such techniques as the porous interlayer (see Section 5.2 for details). Homoepitaxial high-quality nonpolar nitrides are also achievable on nonpolar GaN substrates obtained by side cleavage of several millimeter-thick polar GaN grown by hydride vapor phase epitaxy (HVPE).

2.4 Doping and Other Properties of Nitrides

N- and p-type nitrides are generally prepared by silicon doping and magnesium doping, respectively, which can be performed simultaneously in epitaxial growth, known as *in situ* doping. Recently, success in selective doping by ion implantation combined with high-temperature annealing has also been reported. Silicon is a shallow donor in GaN with the ionization energy of 0.012–0.02 eV, and an ionized electron concentration of the order of 1×10^{20} cm^{-3} can be realized. While the acceptor magnesium has a large ionization energy (in the vicinity of 0.14–0.21 eV) which is influenced by residual impurities, so a well-ionized hole concentration can be higher than 1×10^{18} cm^{-3}. Without intentional doping, the grown nitride epitaxial thin films often contain residual impurities such as hydrogen, oxygen, carbon atoms, and nitrogen vacancies. There are many sources of hydrogen in the nitride growth (e.g., the hydrogen atoms in the methyl (CH_3) and ethyl (C_2H_5) from the MO source of gallium and aluminum in MOCVD, but mainly those from the decomposed carrier gas H_2 at high temperature), and the hydrogen concentration in as-grown nitride thin films is considerably high. The adverse effect of hydrogen on nitride materials lies primarily in the forming of neutral complex with magnesium [$Mg^- + H^+ \rightarrow (Mg - H)^0$], bringing difficulty in magnesium-acceptor ionization and inducing high resistance, which is one of the major hindrances to p-type doping of GaN. Present methods for the "dehydrogenation" of magnesium acceptor are annealing in N_2 ambience or the use of low-energy electron beam irradiation (LEEBI), and the effective hole ionization concentration can reach 1×10^{18} cm^{-3}. Among other residual impurities, oxygen is a shallow donor having significant impact on the background carrier concentration of the material. The oxygen comes generally from the gaseous precursor and carrier gas in MOCVD growth and some from the sapphire substrate (Al_2O_3), for example, in GaN growth the oxygen in the sapphire substrate can diffuse into the GaN layer. Carbon is an amphoteric impurity for GaN, a fact confirmed by both theoretical calculations and experiments. Carbon is one of the major residual impurities for MOCVD-grown GaN, generally from the MO source precursor. A close connection between carbon and the yellow luminescence of GaN was observed by experiments (Feng et al. 2003). N vacancy (V_N) is a native point defect in nitrides. Theoretical calculation indicates that V_N serves as a shallow donor with the ionization energy of about 40 meV, and experiments show that the film resistance increases with rising flow ratio of N/Ga source in the growth.

Given in Table 2.9 are the parameters for nitride material properties including the impurity ionization energies, dielectric properties, electronic transport properties, heat diffusivity, and thermal expansion, all being of great importance in the study of nitride materials and electronic devices.

2.5 Characterization Methods of Nitride Films

The research on materials of a definite crystal structure generally consists of two parts: one is the material growth and the other is the material measurement and analysis, also called the material characterization. Characterization is a very important part of the research on materials. Only by a series of characterizations is it possible to accurately determine the corresponding material properties, and the feedback of the characterization results to material growth helps adjusting the growth processes and improving material properties. Accordingly, material growth and characterization are inseparable. This section introduces the common techniques for the property measurement and analysis of GaN-based materials.

2.5.1 High-Resolution X-Ray Diffraction

High-resolution X-ray diffraction (HRXRD) is a very important characterization technique for current research on materials, with semiconductor single crystals and different low dimensional semiconductor heterostructures as its major subjects.

Table 2.9 Parameters for Nitrides

	GaN	AlN	InN
Donor ionization energy (eV)	Si (substituting Ga): 0.012–0.02 V_N (N vacancy): 0.03; 0.1 C (substituting Ga): 0.11–0.14 O (substituting N): 0.03	Si (substituting Al): 1 V_N (N vacancy): 0.17; 0.5; 0.8–1 C (substituting Al): 0.2	V_N (N vacancy): 0.04–0.05 (Tansley and Egan 1992)
Acceptor ionization energy (eV)	V_{Ga} (Ga vacancy): 0.14 Mg (substituting Ga): 0.14–0.21 C (substituting N): 0.89	V_{Al} (Al vacancy): 0.5 Mg (substituting Al): 0.1 C (substituting N): 0.4	
Mass density ρ (g/cm³)	6.15	3.23	6.81
Static dielectric constant ε_s	8.9	8.5	15.3
High-frequency dielectric constant ε_h	4.6	5.35	8.4
Infrared refractive index	2.3	2.15	2.9
Optical phonon energy (meV)	91.2	99.2	73
Breakdown electric field (10^6 V/cm)	~3	~12	—
Room temperature electron mobility μ_n [cm²/(V·s)]	≤1000	≤1000	≤3200
Room temperature hole mobility μ_p [cm²/(V·s)]	≤200	14	—
Electron saturation velocity (10^7 cm/s)	2–2.7	2.2	2.5
Thermal conductivity κ [W/(cm·°C)]	2.0	2.85	0.45
Thermal expansion coefficient (K⁻¹)	$\frac{\Delta a}{a}=5.59\times10^{-6}$ $\frac{\Delta c}{c}=3.17\times10^{-6}$	$\frac{\Delta a}{a}=4.15\times10^{-6}$ $\frac{\Delta c}{c}=5.27\times10^{-6}$	$\frac{\Delta a}{a}=3.8\times10^{-6}$ $\frac{\Delta c}{c}=2.9\times10^{-6}$

Probing the internal structure of crystals by using their diffraction of the X-ray, HRXRD has the advantages of high efficiency and accuracy with no damage or contamination to the sample. Its basic principle is Bragg's law

$$2d\sin\theta = n\lambda \tag{2.33}$$

where:

- λ is the X-ray wavelength
- n is the series of the diffraction peak
- d and θ are the interplanar spacing and Bragg reflection angle, respectively

In the measurements, diffraction occurs only in lattice planes that satisfy the Bragg equation.

The optical path of Bruker D8 discover HRXRD is shown in Figure 2.12. Its basic working principles are described as follows. The X-ray generated by electron beam bombardment on the Cu target in the X-ray tube passes the Göbel mirror and the

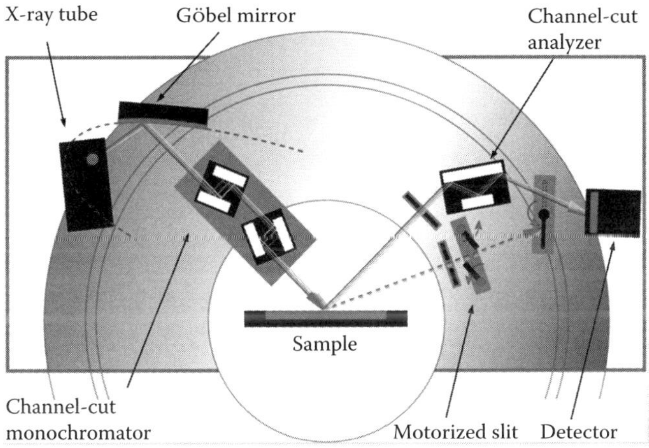

Figure 2.12 HRXRD optical system. (Courtesy of Bruker [Beijing, China] Scientific Technology Co., Ltd.)

four-bounce monochromator and arrives at the sample. The beams from the X-ray tube converged by the Göbel mirror still has quite poor parallelism and includes components of various wavelengths, so the four-bounce monochromator is employed to obtain parallel and monochromatic X-ray beams. When the X-ray hits the sample, diffraction will occur if the diffraction conditions are met, and the diffracted beams will be received by the detector after passing through the triple-axis crystal or the variable slit. The use of variable slit offers a diffraction pattern of higher intensity but with lower angular resolution and sometimes failure to recognize two diffraction peaks very close to each other, while the triple-axis crystal provides better resolution but the maximum light intensity received by the detector is decimated. Therefore, the choice of the optical path depends on specific cases.

The measurements with XRD involve two typical scan types:

1. $2\theta-\omega$ scan: ω and 2θ rotate around the same axis with an angular ratio of 1:2. When 2θ satisfies the Bragg diffraction condition, the diffraction peak of the corresponding lattice plane emerges. It is referred to as $2\theta-\omega$ scan if the x-axis of the measured diffraction spectra is 2θ, and as $\omega-2\theta$ scan if the x-axis is ω.
2. ω scan: after obtaining the diffraction peak position of the measured plane, the detector (2θ) is kept stationary and the sample is rotated about the ω-axis to collect diffraction information, and the obtained curves are known as the rocking curves.

$2\theta-\omega$ curves are mainly adopted for analysis of material composition, strain, and so on of GaN and AlGaN/GaN; while the rocking curves for evaluation of the crystalline quality, and its full width at half magnitude (FWHM) can be used to calculate the dislocation density. The $2\theta-\omega$ scan curves of the (002) plane of the AlGaN/GaN heterostructure sample given in Figure 2.13 show the diffraction peaks of the GaN buffer

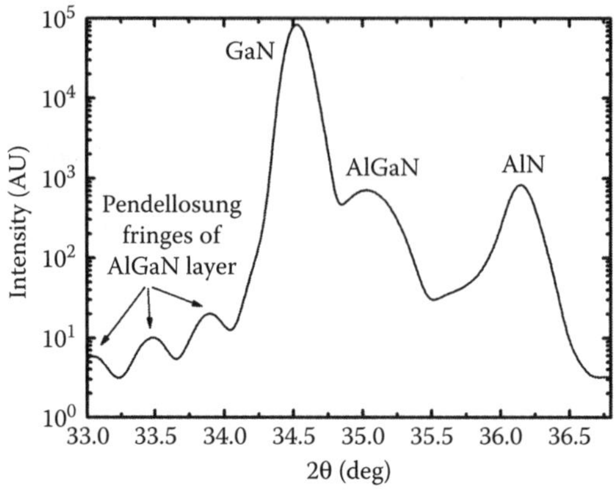

Figure 2.13 (002) 2θ – ω scan curve for AlGaN/GaN heterojunction sample. (Courtesy of Dr. Huantao Duan.)

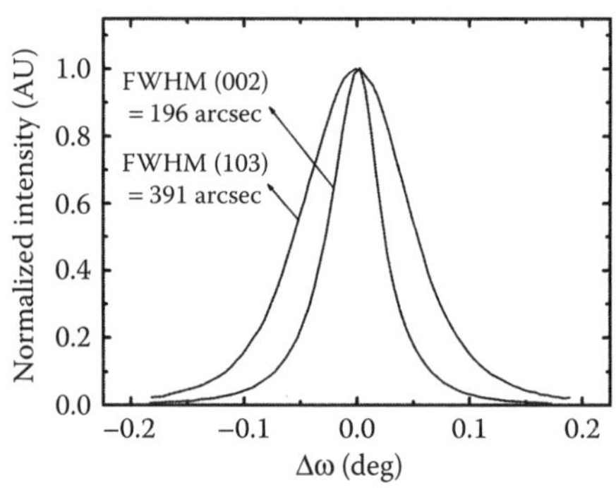

Figure 2.14 (002) and (103) rocking curves for the epitaxial GaN monolayer. FWHM is the full width at half maximum. (After Zhang, J.-F. et al., *Acta Phys. Sin.*, 60, 611, 2011.)

layer, AlGaN barrier layer, and AlN nucleation layer, respectively. The *c*-axis lattice constant of each layer can be estimated from its peak position. The appearance of the interference peak of AlGaN layer indicates a uniform AlGaN thickness and smooth AlGaN/GaN interface, and its period along the angular axis depends on the AlGaN layer thickness, that is, the smaller the thickness is, the longer the period will be. The rocking curves of the (002) and (103) reflections for epitaxial GaN monolayer are shown in Figure 2.14 with both peaks placed at 0° for convenient comparison of the broadening of the diffraction peaks. FWHM values for the curves are given in the figure.

The heteroepitaxial c-plane GaN film exhibits a mosaic structure consisting of multiple columnar subgrains parallel to the growth direction with a height approximately equal to the film thickness, and the subgrain boundaries are composed of screw, edge, and mixed dislocations. The threading dislocation line is oriented along the [0001] direction with the Burgers vector $\boldsymbol{b}$ for the edge, screw, and mixed dislocations being 1/3 $\langle 11\bar{2}0\rangle$, $\langle 0001\rangle$, and 1/3 $\langle 11\bar{2}3\rangle$, respectively.

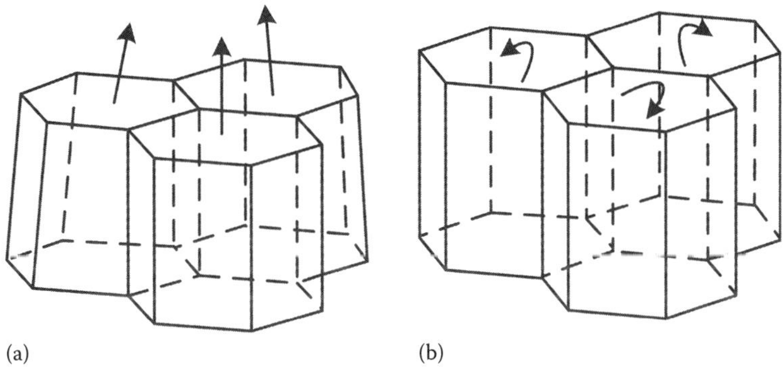

Figure 2.15 Mosaic structure of the hexagonal GaN film. (a) Tilt and (b) twist.

The lattice deformation resulting from screw dislocations causes the direction change of the normal of the basal (001) plane in the hexagonal unit cell (often called tilt as shown in Figure 2.15a), and therefore the rocking curve at the $\{00l\}$ planes (e.g., the (002) plane) can reflect the screw dislocation density. According to the invisibility criterion for the X-ray diffraction of the close-packed hexagonal structure, the diffraction of the $(hkil)$ plane is invisible if $h+2k$ is an integral multiple of 3 with l being odd. Consequently, in the $\{00l\}$ planes the (002), (004), (006), and so on can be detected by diffraction, whereas the (001) plane cannot be detected. For c-plane nitride films, the $\{00l\}$ diffraction is geometrically symmetric, and the incident and diffracted beams are at an equal angle to the sample surface normal and on the same plane it. In asymmetric diffractions, the incident and diffracted beams are at different angles to the sample surface normal, and the crystal planes with such diffraction are at an angle to the sample surface (c-plane), for example, the (104) plane. For mosaic structures, skew symmetric diffraction will also occur if the incident and diffracted beams are at an equal angle to the sample surface normal but not on the same plane with it. For c-plane GaN, crystal planes showing skew symmetric diffraction are the (102), (104), (302), and so on.

Edge dislocation-induced lattice deformation leads to changes mainly in the direction of the side plane $(1\bar{1}0)$ of the hexagonal unit cell (generally referred to as twist, as shown in Figure 2.15b). However, it is very difficult to directly characterize the $(1\bar{1}0)$ plane perpendicular to the surface of c-plane nitride films. So the skew symmetric rocking curves of the lattice planes with diffraction vectors containing the a-axis components (e.g., the [102] and [302] planes) are employed for an indirect characterization of the edge dislocation density. The dislocation density reflected by the XRD measured twist for GaN is the sum of edge dislocations and mixed dislocations with edge-type components, and the further discrimination of the two requires the transmission electron microscopy (TEM).

Typically, GaN has only a small portion of screw dislocations (less than 10% of the total dislocation density) and many mixed and edge dislocations whose proportions vary greatly. The dislocation density can be estimated by (2.34) and (2.35). For randomly distributed dislocations, there is

$$\rho = \frac{\beta^2}{9b^2} \tag{2.34}$$

where:

β is the half width of the rocking curve
b is Burgers vector length of the dislocation

For films whose dislocations are mostly located at the grain boundaries, there is

$$\rho = \frac{\beta}{2.1bd_0} \tag{2.35}$$

where d_0 is the lateral coherence length (Moram and Vickers 2009).

For multilayer epitaxial materials, the strain, composition, and crystalline quality of each layer can also be measured by the reciprocal-space mapping obtained by the HRXRD (Moram and Vickers 2009, Xu 2007). According to solid-state physics, the lattice planes in the direct space have an one-to-one correspondence to the reciprocal-lattice vector or reciprocal-lattice points in the reciprocal space: for any set of lattice planes $\{h_1h_2h_3\}$, there exists a reciprocal-lattice vector $\boldsymbol{k}_h = h_1\boldsymbol{b}_1 + h_2\boldsymbol{b}_2 + h_3\boldsymbol{b}_3$ in the reciprocal space, with $\boldsymbol{k}_h$ oriented in the normal direction of $\{h_1h_2h_3\}$ and $|\boldsymbol{k}_h|$ directly proportional to the reciprocal of the $\{h_1h_2h_3\}$ interplanar distance. The reciprocal space mapping measures the X-ray diffraction intensity distribution near the reciprocal-lattice points, or the positions and shapes of the diffraction patterns of the reciprocal-lattice spots. It can be drawn in a special way like continuous measurement alternating $2\theta - \omega$ scans and ω scans. The reciprocal-lattice vectors $\boldsymbol{k}_x$ and $\boldsymbol{k}_y$ are in common use as the two reference axes of the reciprocal space mapping, corresponding to the a-axis and c-axis of the direct lattice respectively for wurtzite crystals, and therefore the measured reciprocal-lattice point positions reflect the lattice constants a and c. As shown in Figure 2.16, the two reference axes can be converted into $2\theta - \omega$ ($\omega - 2\theta$) scan (along the line from the reciprocal space origin to the reciprocal-lattice point being measured, or along the direction of reciprocal-lattice vector $\boldsymbol{k}_h$) and ω scan (along the circular arc with $\boldsymbol{k}_h$ as the radius) by coordinate transformation. Consequently, the strain and defects of the epitaxial material reflected by both scan types can be seen from the shape of the lattice points in the reciprocal space mapping. In the reciprocal space mapping of multilayer epitaxial materials, the diffraction patterns of different materials at the same reciprocal-lattice point are separated from each other with their relative positions dependent on composition and strain.

2.5.2 Atomic Force Microscopy

Atomic force microscopy (AFM) is mainly used to investigate the atomic-level microscopic surface morphologies and has extensive applications in the characterization of GaN morphology. Compared with the conventional scanning electron microscopy

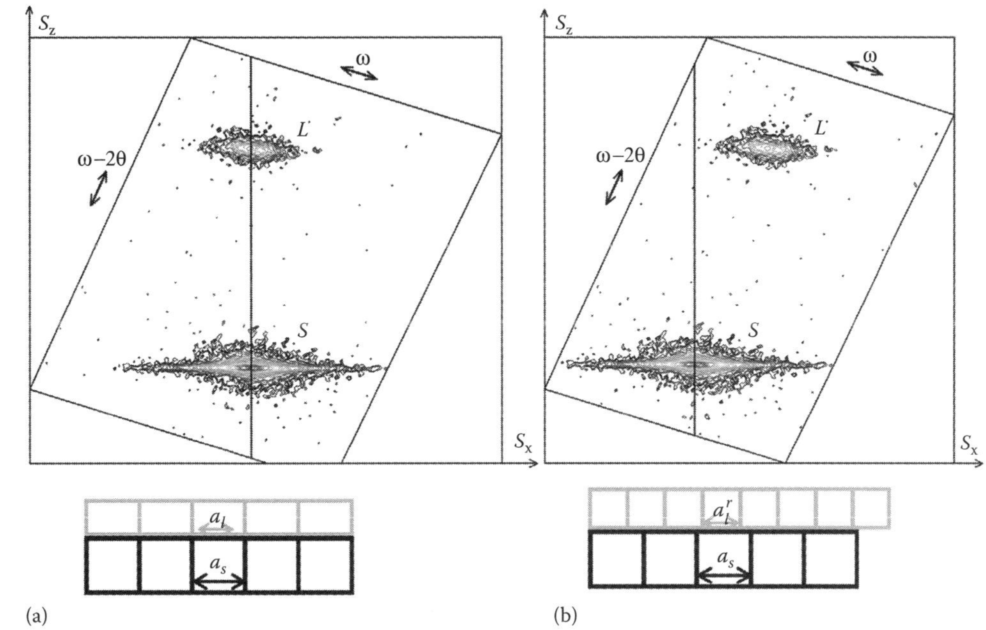

Figure 2.16 Reciprocal space mapping of the AlGaN layer (spot L) on top of the thick GaN buffer layer (spot S). The vertical line through the GaN spot center, that is, $k_x = C$ (constant), gives the GaN lattice constant a_s. (a) AlGaN in complete strain, with its spot center also on the vertical line $k_x = C$, that is, its lattice constants $a_l = a_s$; (b) AlGaN in strain relaxation, with its spot center coordinate $k_x > C$, that is, its lattice constants $a_l^r < a_s$. (Reproduced from Moram, M.A. and Vickers, M.E., *Rep. Prog. Phys.*, 72, 036502, 40 pp., 2009.)

(SEM), AFM offers a three-dimensional plot of the direct space at better lateral resolution (0.1–0.2 nm) and vertical resolution (0.01 nm) and with large depth of field and contrast. It uses an elastic cantilever with one end fixed on the scanner and an AFM tip at the other end to detect the surface morphology or other surface properties of the sample. As the tip scans, the interaction force (attractive or repulsive) between the tip and sample induces a deflection of the cantilever. When a laser beam strikes the back of it, the cantilever reflects the laser beam onto a photodetector, which converts the deflection to photoelectric signal since the intensity difference of the laser received by different quadrants of the detector is proportional to the cantilever deflection. An image of the sample surface morphology is obtained by measuring the variation of detector voltage versus the scan positions, as illustrated in the schematic diagram in Figure 2.17. The AFM images of two- and three-dimensional surface morphologies of GaN are presented in Figure 2.18.

2.5.3 *Scanning Electron Microscopy*

Invented in 1965, SEM uses the imaging of mainly secondary electron signal to observe the surface morphology. SEM consists of the electron gun, the lens system, the electron collection system, the scanning coils, and the cathode ray tube (CRT). The electron

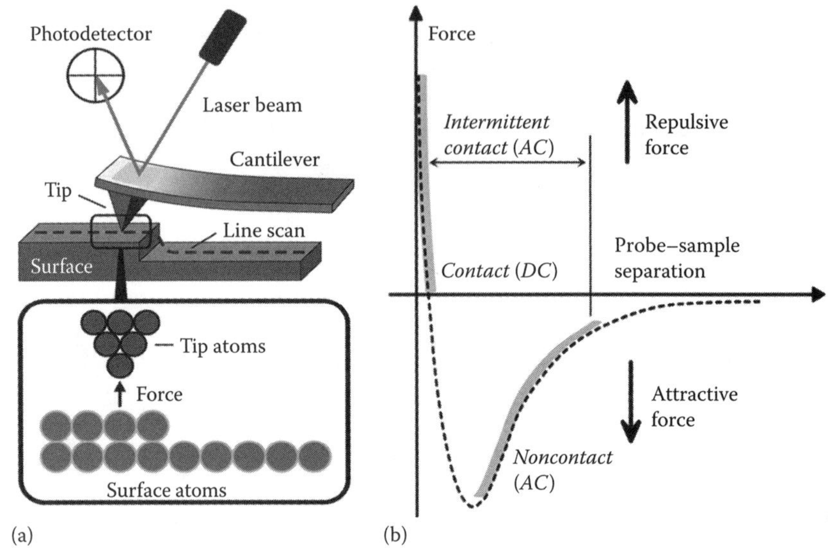

Figure 2.17 Principles and working modes of AFM measurements. (Courtesy of Keysight Technologies [Beijing, China] Co., Ltd.) (a) Principles of measurements. (b) Acting force vs. separation between the tip and the sample and the related working modes.

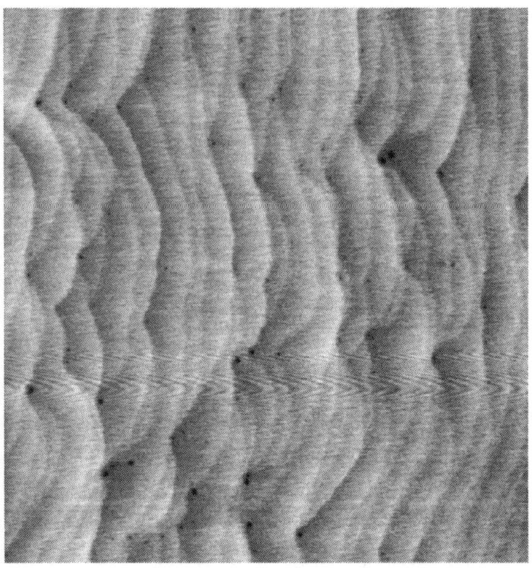

Figure 2.18 AFM images of two- and three-dimensional surface morphologies of GaN. The scan area is $2 \times 2\ \mu m^2$, and the root mean square (RMS) roughness is 0.185 nm. Distinct and straight atomic steps indicate good material quality, and the dark spots are the surface terminations of the threading defects. (Courtesy of Dr. Xiaowei Zhou.)

beams produced by the electron gun, focused by the electromagnetic transmission system and converged by the aperture, are driven by a two-dimensional scan voltage generated by a digital scan generator to perform lateral and longitudinal two-dimensional scans for sample excitation. When the electron beam interacts with a solid sample, not only will secondary electrons, Auger electrons, and back-scattered electrons be produced, but X-rays and bremsstrahlung can also result with cathode luminescence (CL) excited. Among them secondary electrons can produce a magnified image of the surface morphology which is formed point by point chronologically during the scan.

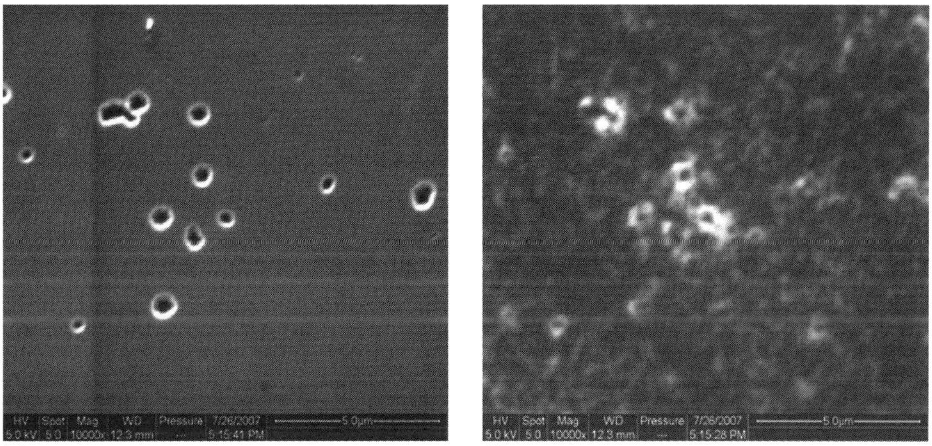

(a) (b)

Figure 2.19 SEM surface morphology and CL intensity distribution in the same region of the GaN film with defect pits. (a) SEM surface morphology. (b) CL intensity distribution in the same region. (After Gao, Z. et al., *Cailiao Yanjiu Xuebao/Chin. J. Mater. Res.*, 22, 657, 2008.)

SEM is much similar to the optical microscopy, except for its use of electron beams to replace light beams as well as a different method for imaging. As compared with the light microscopy, SEM offers a far greater magnification and depth of field owing to the far shorter wavelength of the electron than that of the photon. The advantage of SEM lies in its short measurement period and much larger measurable area than AFM. Meanwhile, many current SEM systems comprise the CL module. Note that CL is a very effective method for characterizing dislocations (Figure 2.19).

2.5.4 Transmission Electron Microscopy

TEM (Xu 2007) is similar in principles to optical microscopy with a series of lenses for magnification, but with an advantageous resolution up to 0.15 nm. If combined with electron energy loss measurements and the light/X-ray detection, it is referred to as analytical transmission electron microscopy (TEM). The TEM has a vertical columnar structure consisting of the electron-optical unit, the electronic control unit, and the vacuum unit. The electron-optical unit functions as the core and the other two as the auxiliaries.

In TEM, visible light beams are replaced by electron beams, and optical glass lenses by electromagnetic lenses. Each lens's name corresponds to that of the optical microscope lens for easier understanding of their similar functions. The TEM beam path diagram is shown in Figure 2.20, where AB is the object, OO' is the objective lens, F is the focus of the objective lens, LF is the focal length, the plane passing through the focus and normal to the optical axis XX' is called the back focal plane, A' is the image of the object point A, B' is the image of B, and the $A'B'$ plane perpendicular to the optical axis XX' is called the image plane.

Owing to its ability to reflect the evolution of threading dislocation and to distinguish various types of dislocations by changing the diffraction vector, TEM is very

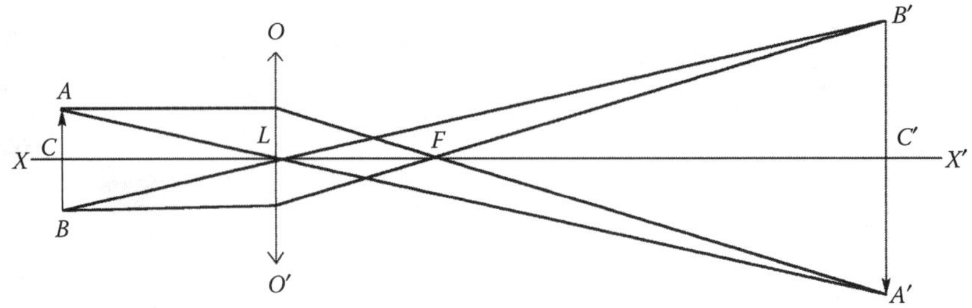

Figure 2.20 Schematic diagram of TEM. (Reproduced from Figure 6.2 of Chapter 6 written by Jiajun Qian in Xu, Z., *Measurement and Analysis of Semiconductors*, 2nd edn., [in Chinese], Science Press, Beijing, China, 2007.)

effective tool for analyzing the dislocation formation mechanism in GaN, evaluating the dislocation density, and determining the layered structure. It is currently one of the most intuitive and effective methods for dislocation analysis. A weak link in TEM applications has been the sample preparation, which is realized usually by machine grinding and polishing and ion thinning and requires good skills.

In the characterization of threading dislocations in GaN by cross-sectional TEM images, the dislocation type is determined by the invisibility criterion in the TEM diffraction contrast analysis. In elastically isotropic materials, for the edge dislocations, both $\boldsymbol{g}{\cdot}\boldsymbol{b} \times \boldsymbol{u} = 0$ and $\boldsymbol{g}{\cdot}\boldsymbol{b} = 0$ must be satisfied for contrast to disappear; whereas for the screw dislocations, only $\boldsymbol{g}{\cdot}\boldsymbol{b} = 0$ is required. Here, $\boldsymbol{g}$ is the diffraction vector, $\boldsymbol{b}$ is the Burgers vector, and $\boldsymbol{u}$ is the dislocation line spatial orientation. Therefore, one can observe the dislocations having screw components with $\boldsymbol{g} = [0002]$ and those having edge components with $\boldsymbol{g} = [11\bar{2}0]$, while the mixed dislocations appear in both cases (Figure 2.21).

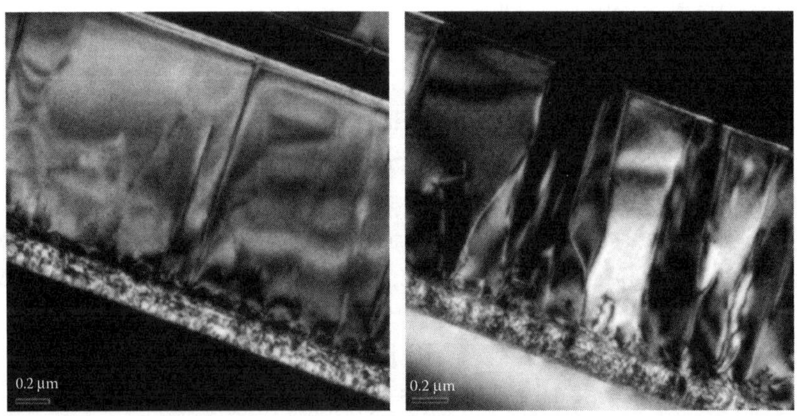

(a) (b)

Figure 2.21 Cross-sectional TEM images of an epitaxial film of AlGaN/GaN heterojunction on sapphire substrate near the $[1\bar{1}00]$ zone axis. Bottom left is the substrate, and upper right is the film surface. (a) g = [0002] and (b) g = [11–20]. (Courtesy of Xinxiu Ou.)

2.5.5 Photoluminescence Spectra

The photoluminescence spectra (PL spectra) are the luminescence of semiconductors when excited by illumination with higher energy than the band gap, and serve as one of the common methods for the detection of semiconductor band structure and defect luminescence. The physical process of semiconductor photoluminescence can be divided into three steps. The first step is light absorption, in which electron–hole pairs are generated through light excitation to form nonequilibrium carriers. Since intrinsic absorption typically occurs when the photon energy is greater than the band gap E_g, a large light absorption coefficient is required to ensure effective generation of electron–hole pairs. The second step is the relaxation and diffusion of photon-induced nonequilibrium carriers, in which spatial diffusion and energy transfer of the carriers may result. Generally, the vast majority of carriers will be relaxed to the energy band edge before recombination. The third step is the luminescence induced by the electron–hole radiative recombination.

The common radiative recombination processes in semiconductors are illustrated in Figure 2.22: (a) interband transition corresponding to the recombination of conduction band electron e and valence band hole h, covering direct transition (e–h) and nondirect optical transition (e–h^p) with phonons; (b) radiative recombination transition through localized impurity levels in band gap, namely the radiative recombination transition between the energy levels of the conduction/valence band and impurity center in band gap (e–A^0, D^0–h, e–D^+, h–A^-, etc.), where e–A^0 is the transition of the electrons from the conduction band edge to the neutral acceptor level, D^0–h is the transition of electrons from the neutral donor level to the valence band edge, e–D^+ is the transition of the electrons from the conduction band edge to the ionized donor level, and h–A^- is the transition of the valence band holes to the ionized acceptors; and (c) radiative recombination transition between donor–acceptor pairs, that is, the transition of electrons from the donor level to the acceptor level.

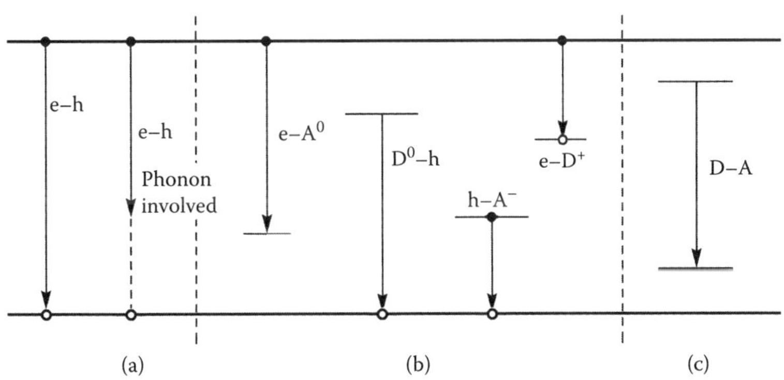

Figure 2.22 Schematic of common radiative recombination processes in semiconductors. These processes are (a) interband transition, (b) transition through localized impurity levels in band gap, and (c) transition between donor–acceptor pairs. (Reproduced from Figure 3.5 of Chapter 3 written by Desheng Jiang in Xu, Z., *Measurement and Analysis of Semiconductors*, 2nd edn., [in Chinese], Science Press, Beijing, China, 2007.)

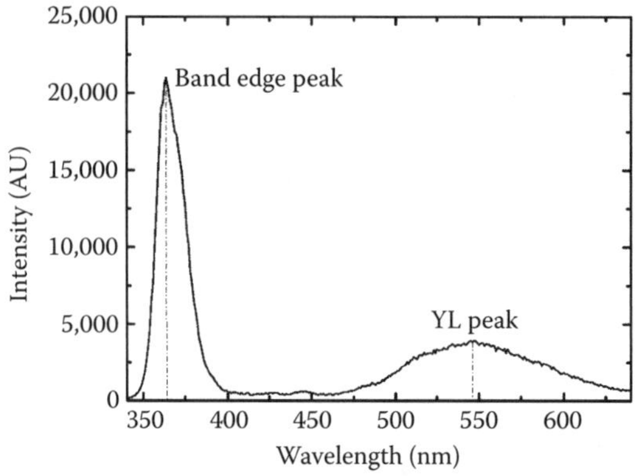

Figure 2.23 PL spectra of epitaxial GaN film at room temperature. The band edge peak position is 365 nm corresponding to a photon energy of 3.4 eV, that is, the band gap of GaN. This peak is generated by the luminescence due to the direct transition of conduction band electrons to the valence band. The peak position 550 nm corresponds to the photon energy of about 2.25 eV, which is the typical yellow luminescence of GaN caused by the radiative recombination between donors and acceptors. (Courtesy of Chuankai Yang.)

The PL spectra are a graphical presentation of the optical signals of different energy emitted by the optically excited semiconductor (Figure 2.23). One can determine the PL peak properties and evaluate the crystalline quality and the density and type of point defects, and so on, according to the position and shape of the PL peak and the variation of the PL peak with optical excitation intensity as well as with ambient conditions such as temperature and pressure.

2.5.6 Capacitance–Voltage (C–V) *Measurements*

For the pn junction, Schottky barrier and MIS structures, the width of the space-charge region and the number of charges vary with the applied voltage across the junction, therefore the barrier capacitance C is related to the applied voltage V. It is this behavior that is exploited in C–V measurements to determine the charge concentration and distribution of the lightly doped side of the pn junction or the semiconductor side of the Schottky junction. This fast and convenient method finds wide applications in charge concentration measurement of semiconductor materials.

For the Schottky junction formed by the metal and noncompensated n-type semiconductor, assuming a donor concentration N_D and a junction area A, in case of depletion layer approximation and complete ionization of impurity, the corresponding carrier concentration N_{CV} can be obtained by C–V measurements via

$$N_{CV} = N_D = -\frac{2}{e\varepsilon A^2}\frac{1}{d(1/C^2)/dV} \tag{2.36}$$

where:

e is the elementary charge quantity
ε is the dielectric constant of the semiconductor

The carrier concentration can be determined by the fact that the free carrier concentration is equal to the donor concentration. In semiconductors and heterostructures with greatly varied doping concentrations, especially in the presence of two-dimensional electron gases, theoretical research indicates that N_{CV} is approximately equal to the free carrier concentration in the material and satisfies charge conservation (Ambacher et al. 1999), that is

$$N_{CV}(z_{CV}) \cong n(z) \quad (2.37)$$

$$z_{CV} = \frac{\varepsilon A}{C} = z \quad (2.38)$$

$$n_s = \int_{-\infty}^{\infty} N_{CV}(z_{CV})dz_{CV} = \int_{-\infty}^{\infty} n(z)dz \quad (2.39)$$

where:

z is the depth from the surface of the Schottky junction

z_{CV} is the width of the depletion region

Therefore, *C–V* measurements can give the heterostructure carrier concentration as a function of the depth, that is, the *C–V* carrier profile, from which the position and distribution of 2DEGs and the sheet electron density can be readily obtained.

The room temperature *C–V* curve of a Schottky circular diode (the inner circle is the Schottky contact, and the exterior ring is the ohmic contact) prepared from the AlGaN/GaN heterostructure is shown in Figure 2.24a. According to the depletion conditions, the curve is divided into four regions. Region 0 reflects the entry of electrons into the AlGaN barrier layer in the forward biased Schottky junction. Region 1 is the 2DEG accumulation terrace in case of depletion under a negative bias

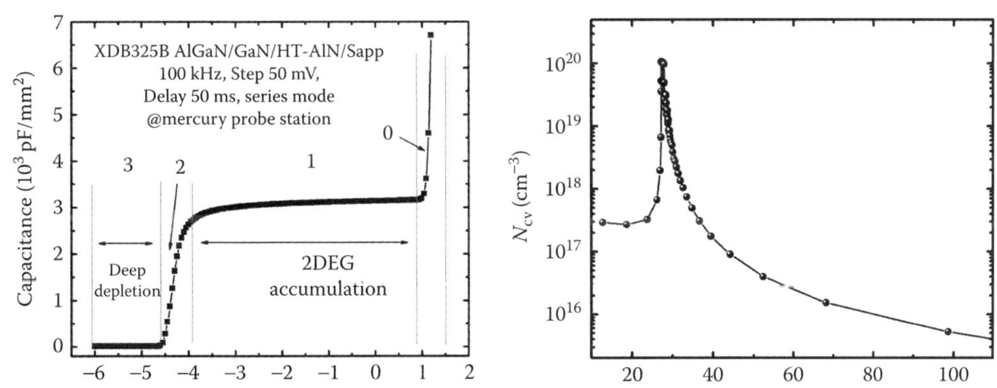

Figure 2.24 Room temperature *C–V* characteristics of the Schottky diode prepared on the AlGaN/GaN heterostructure. (a) *C–V* curve measured at 100 kHz in series signal mode. (b) Corresponding *C–V* carrier concentration N_{CV} as a function of depth. (Courtesy of Zhongfen Zhang.)

and indicates the presence of 2DEGs. The flatter the terrace, the better 2DEG confinement. Region 2 is the 2DEG depletion region. The steeper the curve, the better 2DEG confinement. Region 3 indicates the depletion of electrons in GaN layer after the 2DEG in the channel is depleted. The smaller the depletion capacitance, the small the background carrier concentration is. Shown in Figure 2.24b is the calculated *C–V* carrier profile from the *C–V* curve in Figure 2.24a, giving a good reflection of the 2DEG distribution and the background carrier concentration.

2.5.7 Van der Pauw Measurements

Using the van der Pauw measurements, one can calculate the semiconductor electrical properties such as the sheet resistance R_{SH} and the carrier mobility. Similar to the four-probe method, the Van De Pauw method requires the fabrication of four contact points on the semiconductor film surface as near to the wafer edge as possible to form good ohmic contacts. The schematic of van der Pauw measurements is shown in Figure 2.25. By applying electric current on two neighboring ohmic contact points and measuring the voltage across the other two points, the characteristic resistances R_A and R_B can be determined. The four ohmic contact points are numbered 1, 2, 3, and 4 in the figure, and the line from 1 to 3 should be as normal as possible to the line from 2 to 4.

With R_A and R_B measured, from the van der Pauw theory follows

$$R_{SH} = \frac{\pi}{\ln 2} \frac{R_A + R_B}{2} f_P \tag{2.40}$$

where:

f_p is the van der Pauw correction factor
R_{SH} is the sheet resistance

The difference between R_A and R_B is a measure of the resistivity inhomogeneity of semiconductor materials. To reduce the measurement error, one can successively

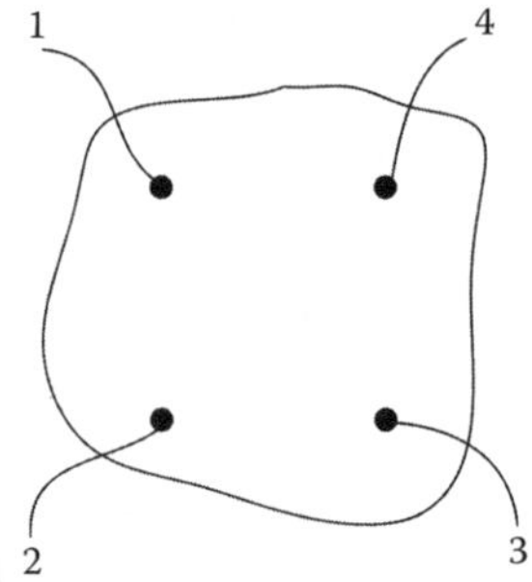

Figure 2.25 Schematic of van der Pauw measurements.

switch the entrance and exit of the applied current and measure the voltage for each time, and then average the calculate sheet resistance values.

After obtaining the sheet resistance value, the sample is placed in a magnetic field perpendicular to its surface and applied a current, then the Hall voltage V_H is measured using the Hall effect, and the carrier mobility μ_H and the sheet density n_{sheet} are obtained by

$$\mu_H = \frac{V_H}{R_{SH}BI} \tag{2.41}$$

$$n_{sheet} = \frac{1}{R_{SH}e\mu_H} \tag{2.42}$$

where:

B and I are the magnitudes of the magnetic field and the electric current, respectively
e is the elementary charge

If the thickness of the semiconductor film is known, the bulk carrier density can be determined. Generally, the sheet carrier density obtained by the Hall measurements is slightly higher than the C–V measurements, mainly because the Schottky barrier in the sample has depleted part of the carriers in the C–V measurements.

Shown in Figure 2.26 are the variable temperature Hall data for a nearly lattice-matched InAlN/GaN heterostructure, showing mainly the change in the 2DEG electrical properties.

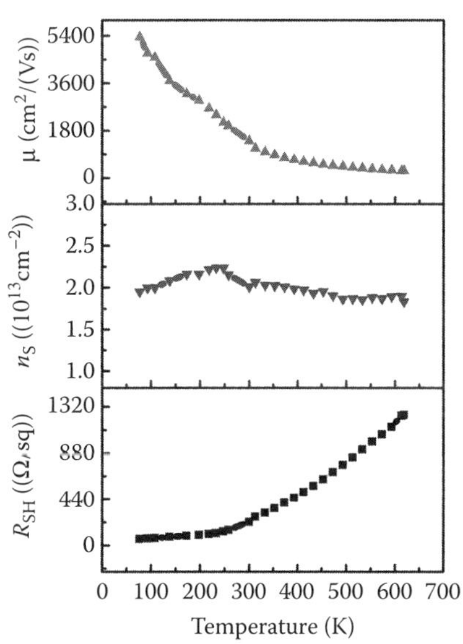

Figure 2.26 Temperature-dependent Hall data for a nearly lattice-matched InAlN/GaN heterostructure. (After Xue, J.S. et al., *J. Cryst. Growth*, 314, 359, 2011.)

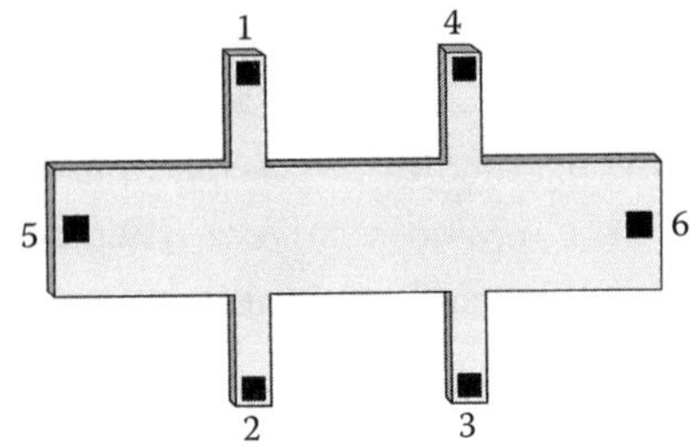

Figure 2.27 Schematics of Hall bars with six electrodes and feasible test connections.

2.5.8 Analysis of 2DEG Transport Properties with SdH Oscillation by Hall Bar Measurements

The standard pattern for Hall effect measurements is the Hall bar (Figure 2.27), which is more intuitive in measurement principles and enjoys a flexible number of electrodes for different measurements, thus applicable in more complicated measurements, such as the Hall measurements with applied gate voltage.

Taking the six-electrode Hall bar (Figure 2.27) as example, with a constant current I applied to 5 and 6, the voltage V_ρ is measured across (2, 3) or (1, 4), and the Hall voltage V_H is measured across (1, 2) or (3, 4). Here, w and d are the width and thickness of the sample, respectively, and l is the distance between 3 and 5 (typically taking $l/w > 3$). Then the resistivity ρ and the Hall coefficient R_H can be expressed as

$$\rho = \frac{V_\rho w d}{lI} \tag{2.43}$$

$$R_H = \frac{V_H d}{BI} \tag{2.44}$$

Applying a strong magnetic field perpendicular to the surface (Z direction assumed) of a 2DEG-containing heterostructure sample, the 2DEG already quantized in the Z direction is further quantized parallel to the sample surface (XY-plane) under the influence of the magnetic field. Each former Z-direction subband divided into a series of evenly distributed discrete energy levels (without considering the magnetic field-induced electron energy along the Z-direction), that is, the Landau levels E_n:

$$E_n = E_i + \left(n + \frac{1}{2}\right)\hbar\omega_C, \quad \omega_C = eB/m^*, n, i = 0,1,2,3\ldots \tag{2.45}$$

where:

E_i is the bottom energy of the 2DEG subband in the Z-direction

$\hbar\omega_C$ is the interspacing of Landau levels

$\hbar$, e, m^* are the reduced Planck constant, the elementary charge, and the electron effective mass, respectively

the values of n and i are independent from each other

The Landau level is analogous in form to the one-dimensional harmonic oscillator energy, and ω_C is the oscillation frequency.

Named after its finder Shubnikov and de Hass, SdH oscillation is a magnetoresistance oscillation induced by the quantum effect of the electrons in materials under a continuously changing magnetic field, which actually indicates the change in the density of states of Landau level at the Fermi surface (Shubnikov and de Haas 1930). The Landau levels are periodically aligned with the Fermi level in a continuously changing magnetic field, and the magnetoresistance peaks when the Landau level is in alignment with the Fermi level, thus a periodic oscillation in the magnetoresistance forms with a oscillation period of

$$\Delta\left(\frac{1}{B}\right)=\frac{1}{B_{n+1}}-\frac{1}{B_n}=\frac{e\hbar}{\left(E_F-E_i\right)m^*} \tag{2.46}$$

where $(E_F - E_i)$ is the difference between the Fermi surface and the subband bottom energies.

Low temperature and a strong magnetic field are generally required for the SdH oscillation to be observed. For semiconductor heterostructures with high mobility 2DEGs, the SdH effect is an effective method for measuring the 2DEG properties. Without considering the impact of the spin splitting, the SdH oscillation of the magnetoresistance can be expressed by (Coleridge et al. 1989)

$$\frac{\Delta R_{XX}}{R_0}=4\frac{X}{\sinh(X)}\exp\left(-\frac{\pi}{\omega_C\tau_q}\right)\cos\left(\frac{2\pi(E_F-E_i)}{\hbar\omega_C}-\pi\right) \tag{2.47}$$

where:

R_0 is the resistance at zero magnetic field

$X = 2\pi^2 k_B T/(\hbar\omega_C)$ is a temperature-related term

k_B and T are the Boltzmann constant and temperature respectively

τ_q is the quantum scattering time

The cosine term in the above equation reflects the periodic change resulting from the subband bottom passing through the Fermi level. For 2DEGs, $(E_F - E_i)$ is inversely proportional to m^* as

$$E_F - E_i = \frac{n_i\pi\hbar^2}{m^*} \tag{2.48}$$

where n_i is the electron sheet density of each 2DEG subband, and thus

$$n_i = \frac{2ef_i}{h} \tag{2.49}$$

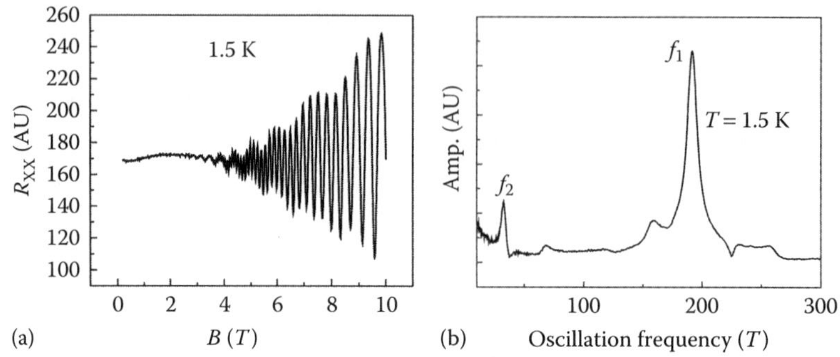

Figure 2.28 Oscillation of 2DEG SdH magnetoresistance for AlGaN/GaN heterostructure showing the two-subband occupation properties of the 2DEGs. (a) Increase of magnetoresistance with magnetic field showing the characteristic double-cycle beat frequency oscillation. (b) Fourier transform curve of the oscillation curve in (a), with the oscillation frequency and amplitude of the two subbands given. (Courtesy of Dr. Ning Tang, Peking University, Beijing, China.)

where $f_i = 1/(\Delta(1/B))$. It follows that the 2DEG SdH oscillation period $1/f_i$ is in fact dependent only on the carrier concentration n_i and irrelevant to m^*. At enough low temperatures, if the 2DEGs occupy multiple subbands (e.g., two subbands with i = 0, 1), then the multiperiod beat frequency SdH oscillation will arise (Figure 2.28a); while if only one subband (i = 0) is taken, then the SdH oscillation has only one period. The Fourier transform can be applied to the SdH oscillation waveforms in the analysis of the oscillation frequency and amplitude corresponding to each subband to estimate the 2DEG subband occupation properties such as the number and interspacing of subbands (Figure 2.28b). For different subbands, the quantum scattering time τ and the quantum mobility of electrons $\mu_{q,i} = e\tau_{q,i}/m^*$ are different.

References

Albrecht, J. D., R. P. Wang, P. P. Ruden, M. Farahmand, and K. F. Brennan. 1998. Electron transport characteristics of GaN for high temperature device modeling. *Journal of Applied Physics* 83(9):4777.

Ambacher, O., J. Majewski, C. Miskys, A. Link, M. Hermann, M. Eickhoff, M. Stutzmann et al. 2002. Pyroelectric properties of Al(In)GaN/GaN hetero- and quantum well structures. *Journal of Physics: Condensed Matter* 14(13):3399–3434.

Ambacher, O., J. Smart, J. R. Shealy, N. G. Weimann, K. Chu, M. Murphy, W. J. Schaff et al. 1999. Two-dimensional electron gases induced by spontaneous and piezoelectric polarization charges in N- and Ga-face AlGaN/GaN heterostructures. *Journal of Applied Physics* 85(6):3222–3233.

Barker, J. M., R. Akis, T. J. Thornton, D. K. Ferry, and S. M. Goodnick. 2002. High field transport studies of GaN. *International Workshop on Physics of Light-Matter Coupling in Nitrides*, September 26–29, 2001, Rome, Italy.

Barker, J. M., D. K. Ferry, D. D. Koleske, and R. J. Shul. 2005. Bulk GaN and AlGaN/GaN heterostructure drift velocity measurements and comparison to theoretical models. *Journal of Applied Physics* 97(6):63705-1.

Bernardini, F., V. Fiorentini, and D. Vanderbilt. 1997. Spontaneous polarization and piezoelectric constants of III-V nitrides. *Physical Review B (Condensed Matter)* 56(16):10024–10027.

Bertazzi, F., M. Moresco, and E. Bellotti. 2009. Theory of high field carrier transport and impact ionization in wurtzite GaN. Part I: A full band Monte Carlo model. *Journal of Applied Physics* 106(6):063718 (12 pp.).

Bougrov, V., M. E. Levinshtein, S. L. Rumyantsev, and A. Zubrilov. 2001. *Properties of Advanced SemiconductorMaterials GaN, AlN, InN, BN, SiC, SiGe.* Rumyantsev, S. L., Levinshtein, M. E., and Shur, M. S. (eds.), New York: John Wiley & Sons.

Bykhovski, J. H., R. Gaska, and M. S. Shur. 1998. Piezoelectric doping and elastic strain relaxation in AlGaN–GaN heterostructure field effect transistors. *Applied Physics Letters* 73(24):3577–3577.

Chichibu, S. F., A. C. Abare, M. S. Minsky, S. Keller, S. B. Fleischer, J. E. Bowers, E. Hu, U. K. Mishra, L. A. Coldren, S. P. DenBaars, and T. Sota. 1998. Effective band gap inhomogeneity and piezoelectric field in InGaN/GaN multiquantum well structures. *Applied Physics Letters* 73(14):2006–2008.

Coleridge, P. T., R. Stoner, and R. Fletcher. 1989. Low-field transport coefficients in GaAs/Ga1-xAlxAs heterostructures. *Physical Review B (Condensed Matter)* 39(2):1120–1124.

Dhar, S. and S. Ghosh. 1999. Low field electron mobility in GaN. *Journal of Applied Physics* 86(5):2668–2676.

Djeffal, F., N. Lakhdar, M. Meguellati, and A. Benhaya. 2009. Particle swarm optimization versus genetic algorithms to study the electron mobility in wurtzite GaN-based devices. *Solid-State Electronics* 53(9):988–992.

Farahmand, M., C. Garetto, E. Bellotti, K. F. Brennan, M. Goano, E. Ghillino, G. Ghione, J. D. Albrecht, and P. P. Ruden. 2001. Monte Carlo simulation of electron transport in the III-nitride Wurtzite phase materials system: Binaries and ternaries. *IEEE Transactions on Electron Devices* 48(3):535–542.

Feng, Q., D. Meng, and H. Yue. 2003. Influence on Yellow-band emission of undoped GaN induced by structural defects. *Acta Photonica Sinica* 32(11):1340–1342.

Fujiwara, T., S. Rajan, S. Keller, M. Higashiwaki, J. S. Speck, S. P. Denbaars, and U. K. Mishra. 2009. Enhancement-mode m-plane AlGaN/GaN heterojunction field-effect transistors. *Applied Physics Express* 2(1):0110011–0110012.

Gao, Z., H. Duan, Y. Hao, P. Li, and J. Zhang. 2008. Formation and optical properties of the large V-shaped surface pits in GaN thin film. *Cailiao Yanjiu Xuebao/Chinese Journal of Materials Research* 22(6):657–663.

Guo, Q. and A. Yoshida. 1994. Temperature dependence of band gap change in InN and AlN. *Japanese Journal of Applied Physics* 33(part 1, 5A):2453–2456.

Holt, D. B. and B. G. Yacobi. 2007. *Extended Defects in Semiconductors.* New York: Cambridge University Press.

Kolnik, J., I. H. Oguzman, K. F. Brennan, R. Wang, P. P. Ruden, and Y. Wang. 1995. Electronic transport studies of bulk zincblende and wurtzite phases of GaN based on an ensemble Monte Carlo calculation including a full zone band structure. *Journal of Applied Physics* 78(2):1033.

Leszcynski, M., I. Grzegory, and M. Bockowski. 1993. X-ray examination of GaN single crystals grown at high hydrostatic pressure. *Journal of Crystal Growth* 126(4):601–604.

Levinshtein, M. E., S. L. Rumyantsev, and M. S. Shur. 2001. *Properties of Advanced Semiconductor Materials GaN, AlN, InN, BN,SiC, SiGe* (1st edn.). New York: John Wiley & Sons.

Liberis, J., M. Ramonas, O. Kiprijanovic, A. Matulionis, N. Goel, J. Simon, K. Wang, H. Xing, and D. Jena. 2006. Hot phonons in Si-doped GaN. *Applied Physics Letters* 89(20):202117-1.

Martin, G., A. Botchkarev, A. Rockett, and H. Morkoc. 1996. Valence-band discontinuities of wurtzite GaN, AlN, and InN heterojunctions measured by X-ray photoemission spectroscopy. *Applied Physics Letters* 68(18):2541–2543.

Miller, D. A. B., D. S. Chemla, T. C. Damen, A. C. Gossard, W. Wiegmann, T. H. Wood, and C. A. Burrus. 1984. Band-edge electroabsorption in quantum well structures: The quantum-confined Stark effect. *Physical Review Letters* 53(22):2173–2176.

Monroy, E., N. Gogneau, F. Enjalbert, F. Fossard, D. Jalabert, E. Bellet-Amalric, L. S. Dang, and B. Daudin. 2003. Molecular-beam epitaxial growth and characterization of quaternary III-nitride compounds. *Journal of Applied Physics* 94(5):3121–3127.

Moram, M. A. and M. E. Vickers. 2009. X-ray diffraction of III-nitrides. *Reports on Progress in Physics* 72(3):036502 (40 pp.).

Piprek, J., ed. 2007. *Nitride Semiconductor Devices: Principles and Simulation*. Wiley-VCH Verlag GmbH & Co. KGaA.

Reklaitis, A. and L. Reggiani. 2004. Monte carlo study of hot-carrier transport in bulk wurtzite GaN and modeling of a near-terahertz impact avalanche transit time diode. *Journal of Applied Physics* 95(12):7925–7935.

Ridley, B. K. 2004. Analytical models for polarization-induced carriers. *Semiconductor Science and Technology* 19(3):446–450.

Schwierz, F. 2005. An electron mobility model for wurtzite GaN. *Solid-State Electronics* 49(6):889–895.

Shubnikov, L. W. and W. J. de Haas. 1930. A new phenomenon in the change of resistance in a magnetic field of single crystals of bismuth. *Nature* 126:500.

Shur, M. S., A. D. Bykhovski, and R. Gaska. 1999. Pyroelectric and piezoelectric properties of GaN-based materials, *Materials Research Society Symposium Proceedings*, Vol. 537.

Siklitsky, V. New semiconductor materials. Characteristics and properties, Electronic archive.

Stutzmann, M., O. Ambacher, M. Eickhoff, U. Karrer, A. Lima Pimenta, R. Neuberger, J. Schalwig, R. Dimitrov, P. J. Schuck, and R. D. Grober. 2001. Playing with polarity. *4th International Conference on Nitride Semiconductors*, July 16–20, Denver, CO.

Suzuki, M, T. Uenoyama, and A. Yanase. 1995. First-principles calculations of effective-mass parameters of AlN and GaN. *Physical Review B* 52(11):8132–8139.

Takahashi, K., A Yoshikawa, and A. Sandhu. 2007. *Wide Bandgap Semiconductors*. Berlin, Germany: Springer-Verlag.

Tansley, T. L. and Egan R. J. 1992. Point-defect energies in the nitrides of aluminum, gallium, and indium. *Physical Review B* 45:10942–10950.

Tomita, Y., H. Ikegami, and H. I. Fujishiro. 2007. Monte Carlo study of high-field electron transport characteristics in AlGaN/GaN heterostructure considering dislocation scattering. *Physica Status Solidi C* 4(7):2695–2699.

Wu, J., W. Walukiewicz, W. Shan, K. M. Yu, J. W. Ager III, S. X. Li, E. E. Haller, H. Lu, and W. J. Schaff. 2003. Temperature dependence of the fundamental band gap of InN. *Journal of Applied Physics* 94(7):4457–4460.

Xu, Z., ed. 2007. *Measurement and Analysis of Semiconductors* (2nd edn.) (in Chinese). Beijing, China: Science Press.

Xue, J. S., Y. Hao, X. W. Zhou, J. C. Zhang, C. K. Yang, X. X. Ou, L. Y. Shi, H. Wang, L. A. Yang, and J. F. Zhang. 2011. High quality InAlN/GaN heterostructures grown on sapphire by pulsed metal organic chemical vapor deposition. *Journal of Crystal Growth* 314(1):359–364.

Yang, L.-A., Y. Hao, Q. Yao, and J. Zhang. 2011a. Improved negative differential mobility model of GaN and AlGaN for a terahertz Gunn diode. *IEEE Transactions on Electron Devices* 58(4):1076–1083.

Yang, L.-A., W. Mao, Y. Hao, and Q. Yao. 2011b. Temperature effect on the submicron AlGaN/GaN Gunn diodes for terahertz frequency. *Journal of Applied Physics* 109(2):024503 (6 pp.).

Yu, E. T., X. Z. Dang, P. M. Asbeck, S. S. Lau, and G. J. Sullivan. 1999. Spontaneous and piezoelectric polarization effects in III-V nitride heterostructures. *26th Conference on the Physics and Chemistry of Semiconductor Interfaces*, January 17–21, San Diego, CA.

Zhang, J.-F., P.-Y. Wang, J.-S. Xue, Y.-B. Zhou, J.-C. Zhang, and Y. Hao. 2011. High electron mobility lattice-matched InAlN/GaN materials. *Acta Physica Sinica* 60(11):611–616.

3

Heteroepitaxial Growth and Defect Properties of Nitrides

Due to the absence of natural native substrates for nitrides, nitride single crystals need to be grown artificially. Nitride crystal growth can be divided into the bulk crystal growth and the thin film epitaxy.

Nitride bulk crystals are very difficult to grow. Taking GaN as example, due to its high melting point (over 2500°C) and a decomposition point of about 900°C, that is, a very high equilibrium nitrogen pressure is required for GaN to exist at its melting point, it is virtually impossible to grow single crystal GaN by using the standard preparation processes for single crystal silicon.

Among the extensively investigated bulk crystal growth techniques are the ammonothermal method, the high nitrogen pressure solution (HNPS) method, and the Na flux method. The ammonothermal method makes use of the chemical reaction of the active supercritical ammonia with metals to grow GaN, AlN, or BN crystals. Based mostly on chemical reaction, the growth can be performed at a relatively low temperature (400°C–500°C) and pressure (200–300 MPa). The HNPS method involves the reaction between the liquid state gallium and nitrogen atoms to generate GaN at 1400°C–1700°C in very high pressure (1–2 GPa) N_2 ambient, which is to suppress the decomposition of GaN crystals. In the Na flux method, N_2 is dissolved into the molten mixture of Na and Ga at 600°C–800°C and 5–10 MPa for continuous precipitation of Na to prepare GaN crystals. However, all the above three methods have a very low growth rate with a typical growth time of 100–200 h. In recent years, major breakthroughs have been seen in the large size single crystal GaN preparation by the ammonothermal method with reported GaN substrates of a diameter of 2 in., a thickness of 4 mm, a dislocation density of about 5×10^4 cm^{-2}, and a background carrier density of about 10^{16} cm^{-3}, while the GaN crystals by HNPS method and Na flux method are still limited in size (typically 10 mm × 10 mm). Presently, the widely used GaN single crystal substrates are mostly grown by the hydride vapor phase epitaxy (HVPE), where a thick (from hundreds of micrometers to several millimeters) GaN film is heteroepitaxially grown on the substrate such as sapphire, which is then lift off and polished to obtain the GaN crystal. A diameter of 3–4 in. is now available with a typical dislocation density of 10^5–10^7 cm^{-2}. In this book, the HVPE is introduced as an epitaxial technique.

Heteroepitaxy of single crystal films on other substrates is currently the mainstream technology for preparing nitride semiconductors and is relatively stable and mature especially for GaN and GaN heterostructures. The epitaxial growth is closely related to the type of the material, the epitaxy method, the type and size of the substrate, and so on involving many aspects of material science.

3.1 Nitride Epitaxial Growth Techniques

By crystal epitaxy is meant the deposition of single crystal films of a specific crystalline orientation (in the same direction as or at a slight angle to the substrate orientation) on the surface of single crystal substrates. As the grown films look like an outgrown extension of the original crystal, thereby is termed epilayer. There are two kinds of GaN crystal thin film epitaxy: the homoepitaxy and heteroepitaxy. For homoepitaxy, only a small fraction of the substrates uses bulk GaN crystals, and the majority adopts heteroepitaxial GaN films grown by metalorganic chemical vapor deposition (MOCVD) or HVPE. For heteroepitaxy, the typical substrates are sapphire, SiC, Si, $LiAlO_2$, and diamond.

Which epitaxial method to use is the primary concern in material growth. For the homoepitaxial or heteroepitaxial growth of GaN, there are three common techniques: MOCVD, molecular beam epitaxy (MBE) and HVPE.

3.1.1 MOCVD

MOCVD, or MOVPE (metalorganic vapor phase epitaxy), is an important and the most widely adopted method for nitride semiconductor preparation by the industrial community. MOCVD performs vapor phase epitaxy using group III and II organic compounds and V- and VI-hydrides as the sources to grow single crystal thin films of III–V and II–VI compound semiconductors by thermal decomposition reaction on the substrate. MOCVD has the following advantages:

1. Wide applications in the growth of almost all compound semiconductor films.
2. Suitability for the growth of various heterostructures.
3. Capability of ultrathin epilayer growth with steep interface transitions and ease in the preparation of steep-interface heterostructures or multilayer compounds with different components by adjusting the flow and type of the precursors.
4. Easy growth control. The metal organic molecules are generally in the form of liquid, and it is easy to regulate the amount of metal organic molecules by accurate control of the carrier gas flow. Meanwhile, the composition of the formed compounds and the growth rate can be easily controlled by adjusting the flows of gaseous precursors.
5. High material purity.
6. Good uniformity of large area epilayers.
7. Mass production capability.
8. Easy doping.

These advantages make MOCVD one of the most popular and effective methods for semiconductor growth at present.

Nitride crystal growth is generally performed at atmospheric pressure or a low pressure (in order of magnitude of 10^3–10^4 Pa) in cold- or hot-wall reactors. For GaN growth, trimethylgallium (TMGa) and the ammonia (NH_3) carried by H_2 are often simultaneously sent into the reaction chamber, and the reactant gases are transported to and mixed at and above the surface of the high temperature substrate (e.g., sapphire substrate with a temperature of 800°C–1200°C) for the following chemical reaction:

$$Ga(CH_3)_3 \text{ (gas)} + NH_3 \text{ (gas)} \rightarrow GaN \text{ (solid)} + 3CH_4 \text{ (gas)} \quad (3.1)$$

Resulting GaN molecules deposit on the substrate surface to form the epitaxial film.

3.1.2 MBE

MBE performs epitaxy by the deposition of molecular or atomic beams from an ultrahigh vacuum system on a heated crystal substrate surface. The beams are generally generated in heated Knudsen cells and maintained in a quasi-equilibrium state, thus invariable beam composition and intensity. The ejected beams from the Knudsen cells are regulated by the aperture and the gate and directed through a straight path to the substrate surface, where they condense and grow under thermodynamically controlled conditions.

The low growth temperature and growth rate of MBE enable accurate control of the epilayer thickness with atomic level smoothness for the growth surface or interface, and if Knudsen cells with proper gates are added, it is possible to introduce molecular beams of various types. Such features are much favorable for the growth of ultrathin semiconductor films and complex structures with very steep interfaces and accurate control of the thickness, doping concentration, and components. The ultrahigh vacuum capability of MBE allows the addition of various *in situ* monitoring and characterization devices, such as the mass spectrometer, the Auger analyzer, the ion bombardment device, various electron microscopes as well as diffractometers, and the film thickness analyzer. These devices can provide important information on the crystallinity, composition, and layered structure of the deposited film for the control and change of growth conditions, greatly enhancing the controllability of the MBE system.

The nitrogen source in GaN heteroepitaxy is mainly the atomic nitrogen beams from the exciting of N_2 by the RF plasma, and the group III source is mostly the group III atomic beams from the thermal evaporation of high purity metals. Being electroneutral, the nitrogen and group III atomic beams move toward the substrate relying not on the electric field but on the pressure gradient-induced diffusion in the vacuum chamber, thus a low atomic beam kinetic energy of about 1 eV upon its arrival at the substrate surface, thereby no damage to the surface.

3.1.3 HVPE

HVPE is the earliest method for GaN epilayer growth. In this epitaxial process gallium is chloridized and transported. In the reaction boat in the source zone, the following reaction occurs at 800°C:

$$2HCl\ (gas) + 2Ga\ (liquid) \rightarrow 2GaCl\ (gas) + H_2\ (gas) \tag{3.2}$$

Then in the main growth chamber, GaN is formed on the substrate surface heated to a temperature of about 1000°C–1050°C by

$$GaCl\ (gas) + NH_3\ (gas) \rightarrow GaN\ (solid) + HCl\ (gas) + H_2\ (gas) \tag{3.3}$$

A prominent characteristic of HVPE is its high growth rate (tens to hundreds of μm/h), making it an excellent candidate for the growth of thick GaN layers. Free-standing bulk GaN crystals of a 3-in. diameter are currently available by this approach, that is, growing by heteroepitaxy a 0.5–1 mm GaN layer on the substrate, then removing the substrate by laser lift-off, and finally performing surface polishing. A major disadvantage of HVPE is the high point defect density in the grown GaN crystal, typically in the vicinity of 10^{16}–10^{17} cm^{-3}, higher than that by MOCVD and MBE.

The HVPE combined with the epitaxial lateral overgrowth (ELOG) in MOCVD can yield thick (up to 500 μm) low-defect GaN layers and high-quality GaN wafers.

3.1.4 Comparison between MOCVD, MBE, and HVPE

Based on the above introduction, the advantages and disadvantages of the three epitaxial growth techniques are summarized in Table 3.1.

Table 3.1 Comparison between MOCVD, MBE, and HVPE

GROWTH TECHNIQUE	ADVANTAGES	DISADVANTAGES
MOCVD	Atomic level interface *In situ* thickness monitoring High growth rate Ultrahigh film quality High output Moderate cost	Lack of *in situ* characterization Large NH_3 consumption Requirement of after-growth activation of p-type Mg doping due to the Mg-H complex
MBE	Atomic level interface *In situ* characterization High-purity growth Hydrogen-free ambient Availability of plasma- or laser-assisted growth	Need for ultrahigh vacuum Low growth rate (1–1.5 um/H) Low output High cost
HVPE	Simple growth technique Ultrahigh growth rate High film quality Quasi-bulk GaN crystals	Nonsmooth interface Hydrogen ambient

In terms of GaN crystal quality, MBE is the best, followed by MOCVD, with HVPE being the lowest; while in terms of growth rate, HVPE is the highest, followed by MOCVD, MBE being the lowest.

It is often the case that all the three techniques are combined in practical use, for example, the HVPE- and MOCVD-grown GaN epifilm used as the substrate for MBE homoepitaxy, or the HVPE-grown GaN epifilm as the substrate for MOCVD homoepitaxy.

For the growth of AlGaN/heterojunction and GaN quantum well materials, both MOCVD and MBE offer desirable results. Owing to its moderate growth rate, high crystal quality, simple equipment, and good process repeatability for mass production, MOCVD is currently the mainstream technique for the growth of GaN and GaN heterostructures.

3.2 Basic Epitaxial Growth Modes and Choice of Epitaxial Substrate for Nitrides

For the MOCVD growth, a macroscopic description is given in the chemical equation (Equation 3.1) of the molecular formation and deposition mechanism of GaN crystals, but in a microscopic view the growth process is much more complicated. The decomposition of the source and other vapor phase reactions at high temperature form the precursors for film growth and byproducts on the substrate surface. The growth precursors are transported to and adsorbed by the growth surface, and diffuse toward the growth sites with lower energies. Then constituent atoms combine with each other through surface chemical reaction and are incorporated into the films, while the surface reaction byproducts are desorbed from the surface and finally carried out of the reaction chamber by the gas flow. The above microscopic and dynamic processes of MOCVD only describe the formation of epifilms, and are not sufficient to explain the formation of the epitaxial material morphologies such as the layer or island structures or the origin of lattice defects, which can be interpreted by the basic modes of heteroepitaxial growth.

3.2.1 Basic Epitaxial Growth Modes

Heteroepitaxy can be divided into three general growth modes: the Frank–van der Merwe mode (two-dimensional [2D] layer-by-layer growth), the Volmer–Weber mode (three-dimensional [3D] island nucleation of deposited atoms on the surface), and the Stranski–Krastanow mode (initial layer-by-layer growth of a few monolayers followed by 3D island nucleation). The schematics of the three modes are shown in Figure 3.1.

According to the fundamental theory of heteroepitaxial growth, when the sum of the surface energy and interface energy of the deposited material is far less than the substrate surface energy, the deposited material is strongly inclined to cover the substrate surface completely (Frank–van der Merwe mode); in the opposite case, when the sum of the surface energy and interface energy of the deposited material is far greater than the substrate surface energy, in order to lower the surface energy for a minimal surface area of the deposited material, 3D islands of the deposited material

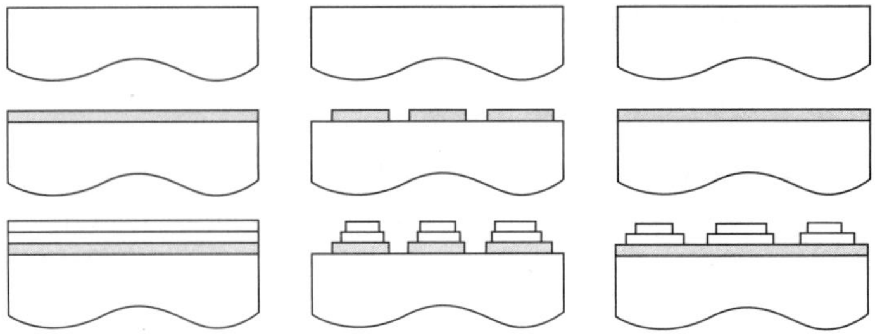

Figure 3.1 Basic heteroepitaxial growth modes. (a) Frank–van der Merwe, (b) Volmer–Weber, and (c) Stranski–Krastanov.

are formed on the substrate surface (Volmer–Weber mode). The most complex case occurs if the sum of surface energy and interface energy of the deposited material is slightly higher or lower than the substrate surface energy, then the epitaxial growth is dependent significantly on the lattice match between the substrate and the epilayer. The heteroepitaxial GaN growth on sapphire or SiC and the fabrication of heterostructures on GaN epilayers fall into this category.

For an imperfect lattice match, the accumulation of lattice strain in the growth of the epilayer leads to an increase in the total energy of the material system. If the thickness and strain energy of the epilayer is large enough, a rearrangement of the system structure (i.e., reconstruction) occurs to release part of the strain energy, and this thickness is referred to as the critical thickness. Reconstruction happens in the form of the generation of misfit dislocations, the formation of 3D islands, and so on.

In nitride epitaxy, the growth modes are related to lattice match as follows.

1. *Frank–van der Merwe mode*: The growth with a very small lattice mismatch and very large critical thickness, or with a smaller grown layer thickness than the critical thickness, follows the Frank–van der Merwe mode, in which semiconductors of different lattice constants can grow pseudomorphically on the substrate (i.e., no dislocation at the interface). In this case, strain occurs in the grown layer for the in-plane lattice constant to match the lattice constant of the substrate. Misfit dislocations are generated at the interface if the grown layer thickness exceeds the critical thickness. AlGaN/GaN heterostructures formed by growing AlGaN on the GaN epilayer is an example of this mode.
2. *Stranski–Krastanov mode*: If the epilayer strain behaves as compressive strain and with a proper range of lattice mismatch (2%–10%), a 2D thin epilayer (called wetting layer) will result, followed by a release of accumulated lattice strain energy by the formation of 3D islands, that is, the Stranski–Krastanov mode. These 3D islands are formed largely by the decomposition of the epilayer whose strain reaches the critical strain.
3. *Volmer–Weber mode*: For a lattice mismatch greater than 10%, the 3D islands are formed directly on the substrate without any wetting layer, namely the Volmer–Weber mode. With a lattice mismatch of 16% between GaN and

sapphire, GaN growth on sapphire falls into this mode. In growing relatively thick GaN, 3D islands form and continuously grow on the substrate until adjacent islands coalesce leading to a 2D-like growth mode, whereas the crystal surface still exhibits the characteristics of 3D islands.

3.2.2 Choice of Substrates

In order to reduce the lattice mismatch between the substrate and epitaxial material, it is generally required that the substrate should have a similar crystal structure and lattice constant to the epitaxial material, and of stable physical and chemical properties, high crystalline quality, low cost as well as the availability of bulk size and a desirable resistivity.

The lattice mismatch rate r between the epitaxial material and the substrate is calculated by

$$r = \frac{a_{\text{sub}} - a_{\text{epi}}}{a_{\text{sub}}} \times 100\% \tag{3.4}$$

where a_{sub} and a_{epi} are the lattice constants for the epitaxial material and the substrate, respectively. A negative r indicates the presence of compressive strain in the crystal while a positive one the tensile strain. The lattice parameters of GaN and substrate candidates for GaN heteroepitaxy are listed in Table 3.2.

The most popular substrates for GaN heteroepitaxy are sapphire, SiC, and Si. It can be seen in Table 3.2 that the lattice mismatch between sapphire and GaN is very

Table 3.2 Fundamental Parameters of GaN and Substrate Candidates

		LATTICE PROPERTIES		
SUBSTRATE	CRYSTAL STRUCTURE	LATTICE CONSTANT a (nm)	LATTICE MISMATCH TO GaN (%)	THERMAL EXPANSION COEFFICIENT (10^{-6}/K)
GaN	Hexagonal (wurtzite)	$a = 0.3189$ $c = 0.5185$	0	5.59 3.17
AlN	Hexagonal (wurtzite)	$a = 0.3112$ $c = 0.4982$	$-2.5\langle 0001 \rangle$	4.2 5.3
InN	Hexagonal (wurtzite)	$a = 0.3548$ $c = 0.5760$	$10\langle 0001 \rangle$	
α-Al_2O_3 (sapphire)	Hexagonal (wurtzite)	$a = 0.4758$ $c = 1.2991$	$-16.1\langle 0001 \rangle$	7.5 8.5
6H-SiC	Hexagonal (wurtzite)	$a = 0.3081$ $c = 1.5092$	$3.5\langle 0001 \rangle$	4.2 4.68
4H-SiC	Hexagonal (wurtzite)	$a = 0.3073$ $c = 1.0053$	$3.8\langle 0001 \rangle$	
ZnO	Hexagonal (wurtzite)	$a = 0.3252$ $c = 0.5213$	$-1.9\langle 0001 \rangle$	2.9 4.75
Si	Cubic (diamond)	$a = 0.5431$	$16.9\langle 0001 \rangle$	2.59

large. Note that if we substitute the lattice constants of GaN and a-axis sapphire in Table 3.2 into Equation 3.4, the calculated lattice mismatch is –33%, which is much greater than the actual value because the epitaxial relation between GaN and (0001) Al_2O_3 (sapphire) substrate is (0001) GaN//(0001) Sapp and $[01\bar{1}0]$ GaN//$[11\bar{2}0]$ Sapp, that is, there is an in-plane difference of 30° between the (0001) reference systems of two crystals, and thus the GaN/sapphire lattice mismatch is

$$r = \frac{a_{\text{sapp}}/\sqrt{3} - a_{\text{GaN}}}{a_{\text{sapp}}/\sqrt{3}} \times 100\% = -16.09\% \tag{3.5}$$

The lattice mismatch between SiC and GaN is visibly much smaller with lattice mismatches of 4H-SiC and 6H-SiC to GaN being 3.8% and 3.5%, respectively. Experiments also show that it is much easier to perform GaN heteroepitaxial growth on SiC substrates.

Although the lattice mismatch between the sapphire substrate and GaN is far larger than that of SiC to GaN, high-quality GaN crystal films can be obtained on both sapphire and SiC. Because sapphire is much cheaper than SiC and enjoys large size substrate and high crystal quality, many researches on the growth of GaN films and AlGaN/GaN heterojunctions are based on the sapphire substrates. The obvious advantages of SiC in its small lattice mismatch to GaN and a far greater thermal conductivity than sapphire makes it an ideal substrate material for high-power nitride devices.

Silicon remains the mainstream of semiconductor technology. Nitride epitaxy on silicon substrates is of a strategic significance for both the development of nitrides and the integration of compound semiconductors with Si. However, it is better to choose the (111) silicon (in-plane triple symmetry) rather than the typical (100) silicon for GaN epitaxy on Si. Even so, the lattice mismatch and thermal mismatch between silicon and GaN are still as high as 16.9% and 56%, respectively, thus a very large strain that makes it difficult to grow high-quality and crack-free GaN and GaN heterojunctions on silicon. Currently, strain release and control has been realized by using the transition layer, the interlayer, *in situ* passivation, and so on (Amano et al. 1986, Nakamura 1991) for the epitaxy, and high-performance nitride materials and microwave power devices on silicon have been obtained.

3.3 Two-Step MOCVD Growth of Nitrides

The lattice mismatch and thermal mismatch are always present in heteroepitaxy between the epitaxial material (e.g., GaN) and the substrate (e.g., sapphire). In high temperature epitaxy of GaN films directly on the sapphire substrate, owing to the large lattice mismatch between GaN and the sapphire substrate, theoretically the critical thickness for misfit dislocations to occur is far smaller than the thickness of an atomic monolayer, and therefore it is impossible for an intact atomic layer to form in the initial growth. Besides, the strong binding energy of Ga-N bond and great mismatch between the substrate and the epilayer lead to difficult diffusion of adsorbed GaN atoms on the surface, thus a Volmer–Weber growth mode. In the early GaN growth techniques it was difficult to obtain crack-free films with smooth surface, and moreover the epitaxial films usually had very high background carrier concentration.

In order to deal with the mismatch and high background carrier concentration, Amano et al. proposed in 1986 a two-step growth method: the initial growth of the AlN nucleation layer at low temperature followed by the GaN growth at high temperature, which greatly improved the quality of GaN epilayers (Amano et al. 1986). Nakamura et al. also reported this technique, who found that low temperature GaN nucleation layers can play the same role, and that the thickness of the nucleation layer has significant influence on both GaN morphology and crystal quality (Nakamura 1991). The two-step growth has become the mainstream approach to MOCVD growth of nitride materials at present.

3.3.1 Two-Step Growth Process

As shown in Figure 3.2, the major procedures of two-step GaN growth on sapphire substrates are:

1. *High temperature substrate cleaning*: Flows of hydrogen into the reaction chamber at about 1000°C can remove the contaminants and form microstep structures on the surface of sapphire substrate to improve GaN crystalline quality.
2. *Substrate surface nitrification*: Nitrification of the sapphire surface by passing NH_3 into the reaction chamber at low or high temperature prior to the growth of the GaN nucleation layer can not only facilitate the formation of nucleation centers and increase the adhesion of the nucleation layer to the substrate, but also improve the GaN surface morphology. Nonetheless, long-time nitrification may lead to a transformation of polycrystalline nitrided layers into monocrystalline ones, thereby forming relatively high-density surface bulges.
3. *Nucleation layer growth*: The common nucleation layers are low temperature (500°C–650°C) GaN or low-temperature AlN. Nucleation layers grown at

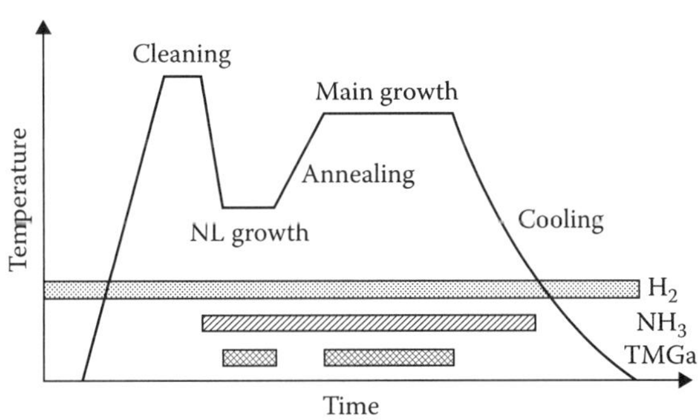

Figure 3.2 Two-step growth process of GaN. (Reproduced from Figure 7.41 in Lu, D. and Duan, S., *Fundamentals and Applications of Metalorganic Compound Vapor Epitaxy* [in Chinese], Science Press, Beijing, *China,* 2009.)

low temperature have continuous and rather smooth surfaces but with many defects and mixed crystal phases (such as cubic and hexagonal phases). The significance of the nucleation layer lies in its ability to convert the initial 3D GaN growth into the 2D layer growth.

4. *Annealing and surface reconstruction of nucleation layer*: When the low temperature growth of the nucleation layer is done, the temperature elevation to the epilayer growth temperature has a high temperature annealing effect on the nucleation layer. The surface morphology of the nucleation layer depend strongly on the temperature and time of annealing as well as on the heating rate. Experiments indicate that a short annealing time is in favor of the formation of high-quality GaN epilayers, whereas long-time annealing etches the nucleation layer (due to the absence of TMGa or TEGa) and reduces its thickness. Therefore, the optimization of high temperature annealing is also a key procedure for obtaining high-quality GaN.
5. *High temperature growth and cooling of GaN epilayer*: The growth temperature of GaN epilayers has great impact on GaN quality. The high bond energy of GaN requires a high growth temperature for the atoms to migrate on the growth surface. High-quality GaN epilayers can be obtained only at high temperature, typical 1000°C–1100°C for growing GaN bulk epilayer on AlN or GaN nucleation layer. The high N_2 partial pressure needed to suppress GaN decomposition at high growth temperature yields a high V/III ratio. Post-growth cooling in NH_3 ambient is often required due to the GaN decomposition temperature in the vicinity of 900°C.

3.3.2 Growth Mode Evolution in Two-Step Growth of GaN on Sapphire

As shown in Figure 3.3, there exists a gradual evolution of the growth mode and the defect behavior in two-step GaN growth on sapphire. After the formation of a 10–100 nm thick low temperature nucleation layer (Figure 3.3a), the heating and annealing induces recrystallization on the nucleation layer surface, thus weakening its grain boundary characteristics and blocking the boundary dislocations to prepare the ground for the subsequent GaN epilayer growth. The GaN epilayer starts to grow at high temperature (1000°C–1150°C) with the initial thin layer offering high temperature nucleation (Figure 3.3b) on the low temperature nucleation layer. When the epilayer thickness increases to about 50 nm, close-packed columnar crystals are formed and continue to grow along a specific orientation (Figure 3.3c). Subsequently 3D islands arise with the continuous growth of GaN (Figure 3.3d). When the epilayer thickness reaches 200–300 nm, neighboring islands begin to coalesce owing to the lateral growth of 3D islands (Figure 3.3e). With further growth, complete coalescence of islands is achieved (Figure 3.3f) with dislocations forming at the front of coalescence, and from this moment GaN continues to grow in the 2D layer mode.

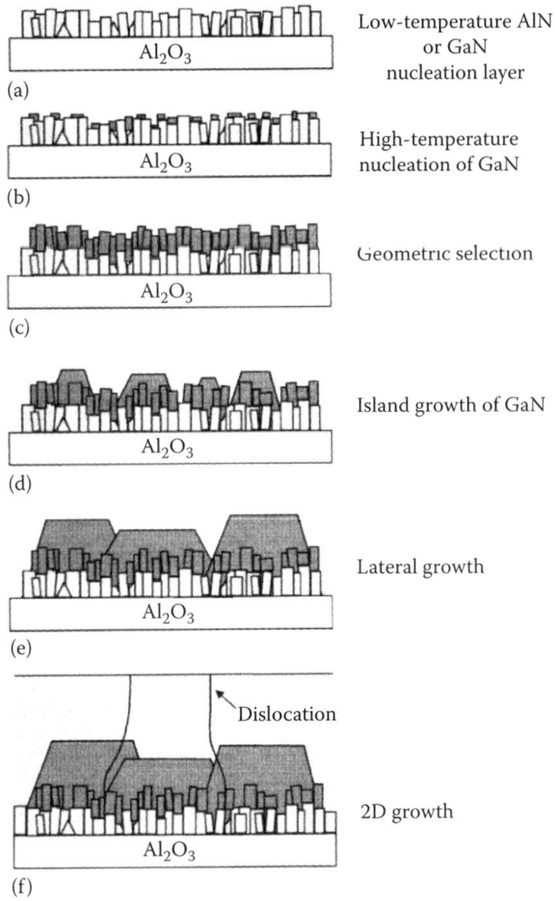

Figure 3.3 (a–f) Growth process of GaN on sapphire. (Reproduced from Ambacher, O., *J. Phys. D: Appl. Phys.*, 31, 2653–2710, 1998.)

3.4 Nucleation Layer Optimization in Nitride Epitaxy

The nucleation layer in the two-step nitride growth method have three functions: (1) to provide the nucleation centers in the same orientation as the substrate; (2) to release the mismatch stress caused by lattice mismatch as well as thermal stress induced by the expansion coefficient mismatch between GaN and the substrate; (3) to offer a smooth nucleation surface for further epilayer growth and to reduce the contact angle of the initial nucleation, helping the island-like crystal grains to coalesce into a planar layer at a relatively small thickness, thus transforming into 2D growth.

Both the growth conditions and thickness of the nucleation layer have remarkable impact on the crystalline quality and defect properties of nitride epitaxial films, and thus its growth optimization is of great significance. The optimization of growth conditions involves series of experiments. In the following, the methods for improving the crystalline quality of epilayer are introduced.

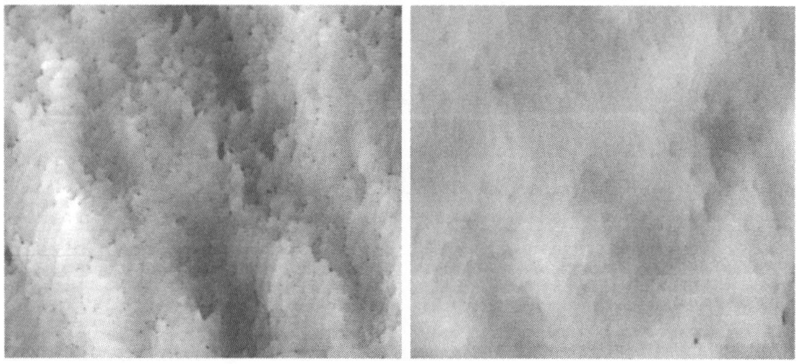

(a) (b)

Figure 3.4 Influence of low temperature GaN nucleation layers with different growth rates on the AFM surface morphology of 1.5 μm GaN epilayer. (a) Low growth rate nucleation layer, RMS (surface RMS roughness) = 1.5 nm; (b) high growth rate nucleation layer, RMS = 0.4 nm. (After Duan, H. et al., *J. Semiconduct.*, 30, 105002, 3 pp., 2009b.)

3.4.1 Low-Temperature Nucleation Layer

For low temperature GaN nucleation layer on sapphire with identical growth conditions (reaction chamber pressure, V/III ratio and temperature, etc.) and thickness but for growth rates, the sample with the nucleation layer of the higher growth rate enjoys a lower dislocation density and smoother surface (Figure 3.4) (Duan et al. 2009b). Since edge dislocations arise from islands coalescence, it is suggested that the nucleation layer of a high growth rate can develop large-size and low-density nucleation islands with coalescence interface of small area, which is in favor of the decrease in the dislocation density. Further analysis suggests that the nucleation layer with large-size low-density nucleation islands plays a positive role in the release of residual stress in the GaN epilayer, which also help improve crystal quality. In terms of electrical properties, samples with the high growth rate nucleation layers have high background electron concentrations and electron mobilities, which are attributed to the reduced trapping of background carriers by the fewer edge dislocations that act as deep acceptors.

The low temperature GaN nucleation layer growth is generally followed by a temperature rise procedure equivalent to a variable temperature annealing, but a specific constant temperature annealing technique can also be introduced to improve the crystalline quality at an optimal annealing temperature possibly between the growth temperatures for the nucleation layer and the GaN epilayer (Figure 3.5). It is analyzed that better island orientation uniformity and easier formation of large islands by neighboring islands during nucleation are achieved due to the recrystallization of nucleation layer after specific constant temperature annealing. Large islands have fewer coalescence interface, thus less coalescence-induced dislocations and better crystal quality.

Low-temperature AlN nucleation layers on sapphire can yield GaN epilayers of high crystalline quality, but due to the strong adhesion and poor mobility of the

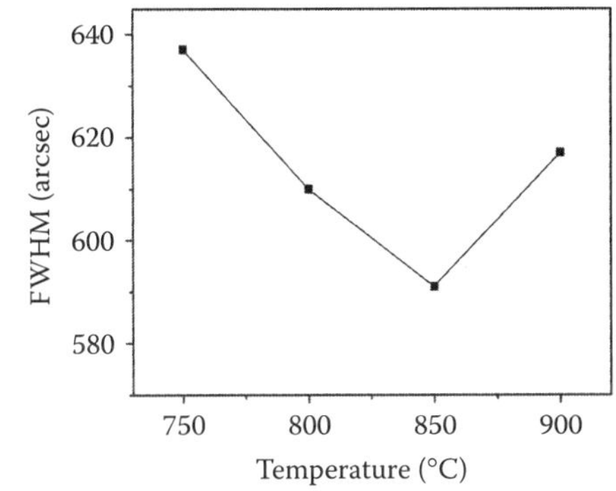

Figure 3.5 FWHM of GaN (002) rocking curves versus annealing temperature of the low temperature GaN nucleation layer. (Courtesy of Dr. Xiaowei Zhou.)

Al atoms in the growth, the lateral growth of AlN is slow and high-density 3D islands are formed, making it difficult for the AlN nucleation layer to completely cover the sapphire substrate. As a result, in the high temperature growth of GaN epilayer, the oxygen atoms in the sapphire substrate diffuse upward into GaN as shallow donors and a buried charge layer forms at the interface between the low-temperature AlN nucleation layer and the high temperature GaN epilayer, giving rise to a leakage path, which is confirmed by mesa isolation leakage, *C–V* and secondary ion mass spectrometry (SIMS) measurements.

3.4.2 High-Temperature AlN Nucleation Layers

The optimized high-temperature AlN nucleation layer not only yields high-quality GaN epilayers but also effectively prevents electric leakage. According to our experiments (Duan 2011), the AlGaN/GaN heterojunction grown on the high-temperature AlN nucleation layer with optimized growth thickness, V/III ratio and temperature has a Hall mobility of 1549 $cm^2/(Vs)$ with the background electron concentration-related *C–V* carrier concentration in the depth of GaN reduced to about 10^{13} cm^{-3}, thus achieving high resistivity GaN epilayers. With the increase of the thickness of the high-temperature AlN nucleation layer within the range of 30–150 nm, the GaN background electron concentration gradually decrease while the electron mobility rises together with improved material surface morphology and crystal quality, a trend basically saturating at a nucleation layer thickness of 100 nm. The optimal V/III ratio is about 800 for high-temperature AlN nucleation layers, and further increase in the V/III ratio induces poorer GaN epilayer morphology and increased defects, which is considered to be the consequence of the enhanced pre-reaction and decreased surface mobility of Al atoms on the sapphire substrate. The optimal growth temperature for the high-temperature AlN nucleation layer is about 1000°C (Figures 3.6 and 3.7). It is suggested that the lower temperature decreases the surface migration of Al atoms

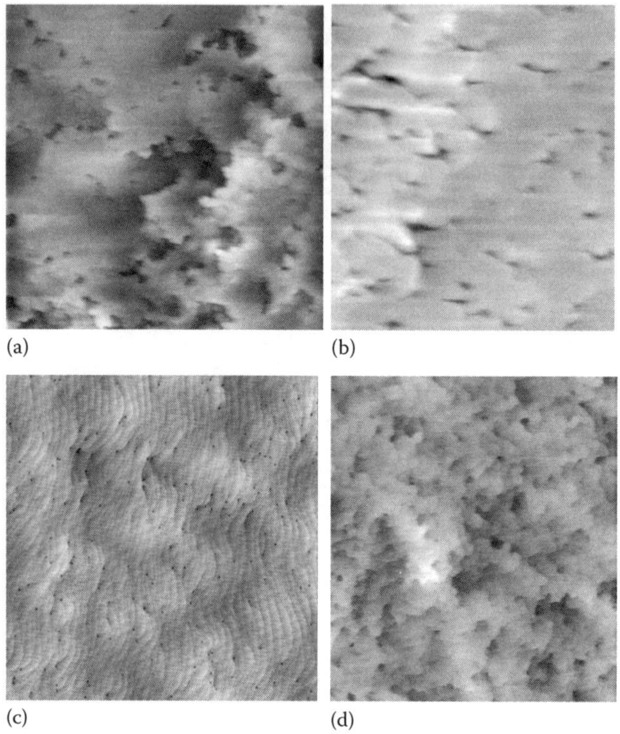

Figure 3.6 AFM images of GaN surface morphology at different growth temperatures of the high-temperature AlN nucleation layer: (a) $T = 800°C$, (b) $T = 900°C$, (c) $T = 1000°C$, and (d) $T = 1100°C$. (Courtesy of Dr. Huantao Duan.)

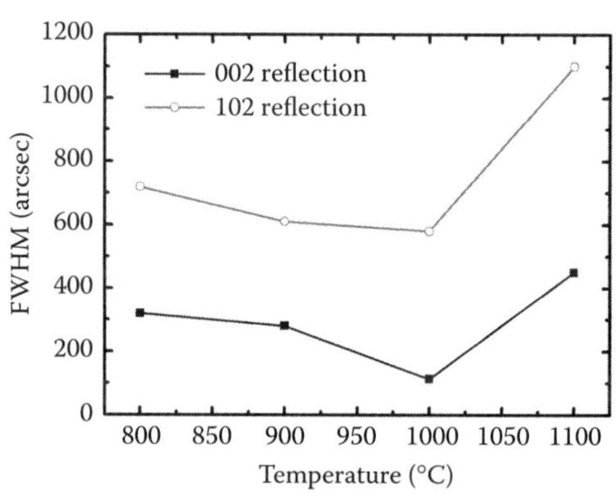

Figure 3.7 GaN crystalline quality versus growth temperature of the high-temperature AlN nucleation layer. (Courtesy of Dr. Huantao Duan.)

and the density of AlN nucleation islands, causing difficulty in island coalescence in subsequent GaN growth, thereby a surface morphology with incomplete coalescence and a low crystal quality (Figure 3.6a and b). While too high a temperature intensifies the pre-reaction between TMA and NH_3, leading to poor AlN quality and in turn affecting the GaN quality (Figure 3.6d).

3.4.3 *High-Temperature AlN Nucleation Layer Grown by Alternating Supply of Ammonia*

Relatively large strain exists in continuously grown high-temperature AlN nucleation layers on sapphire substrates, leading to rather great residual stress in GaN, which has negative impact on device reliability. The residual stress in GaN can be greatly reduced by growing the AlN nucleation layer by the method of alternating supply of ammonia (Figure 3.8) (Duan et al. 2009a). According to the XRD $2\theta - \omega$ scan curves of GaN, the *c*-axis GaN lattice constant is 0.5187 nm, which is close to the ideal GaN lattice constant. The crystalline quality (Figure 3.9) and surface smoothness of the material are also improved with the Hall mobility of the AlGaN/GaN heterojunction rising to 1700 $cm^2/(Vs)$ along with a further decrease of GaN background electron concentration.

It is suggested that the alternating supply of ammonia in AlN growth increases the surface migration time of Al atoms with suppressed pre-reaction between TMA and NH_3 due to the time-sharing transport of TMA and NH_3 and enhances coalescence of AlN nucleation islands and 2D growth, yielding AlN nucleation layers with relatively good crystal quality and smaller strain. Owing to the small lattice mismatch between AlN and GaN, GaN grown further on the better AlN nucleation layer can achieve relatively good crystal quality, surface quality, and very small residual stress. High crystal quality coupled with smooth surface and interface is the key to the high electrical performance of AlGaN/GaN heterostructures.

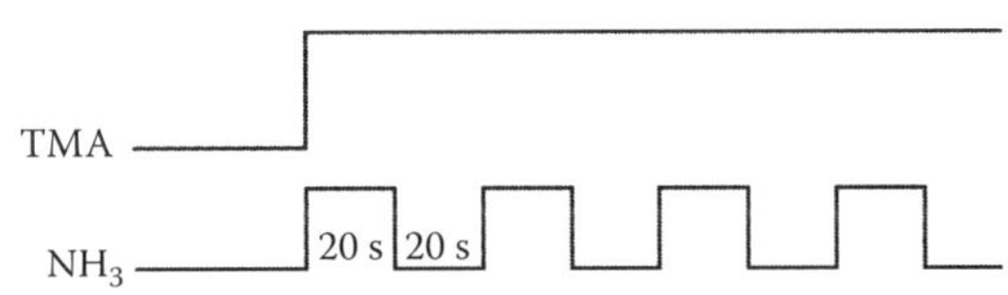

Figure 3.8 Schematic of AlN growth by alternating supply of ammonia. (After Duan, H. et al., *J. Semiconduct.*, 30, 093001, 4 pp., 2009a.)

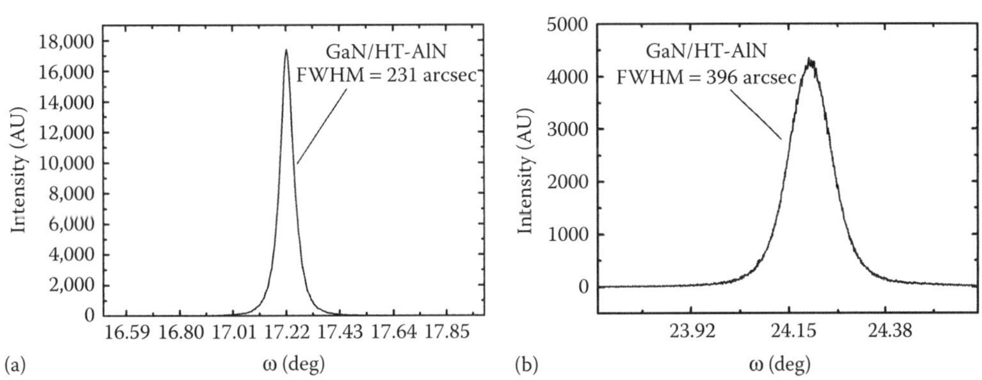

Figure 3.9 (a) (002) and (b) (102) rocking curves of GaN epitaxial layer on high-temperature AlN nucleation layer grown by alternating supply of ammonia. (After Duan, H. et al., *J. Semiconduct.*, 30, 093001, 4 pp., 2009a.)

3.5 Influence of Growth Conditions of Nitride Epilayer on Material Quality

The growth conditions (such as temperature, V/III ratio, growth rate, and reaction chamber pressure) and the thickness of the nitride epilayer have significant influence on its crystalline quality and defect properties. An ideal GaN epilayer has a low background carrier concentration, a high mobility, and a low defect density. MOCVD-grown GaN is known to typically exhibit the unintentional n-type doping, and for high-quality GaN the ideal background carrier concentration should be less than 10^{16} cm^{-3}. Extensive reports suggest that the major sources of background carriers in GaN are the oxygen atoms, unintentional impurities, and nitrogen vacancies.

The V/III ratio influences the stoichiometric ratio for GaN, that is, the concentrations of vacancies and interstitial atoms in the material. These intrinsic defects themselves are also electrically active, among which nitrogen vacancy shows a single donor level and gallium vacancy shows three acceptor levels. When the concentration of a vacancy in the crystal increases, the probability for impurities to occupy this vacancy rises. Therefore, the V/III ratio has an impact on impurity incorporation and the electrical and optical properties. For GaN, oxygen and carbon may occupy the nitrogen vacancies while silicon and germanium are inclined to take the sites of gallium. The V/III ratio also has certain effect on the GaN residual stress.

A high ammonia gas partial pressure in the MOCVD reaction chamber can effectively suppress the formation of nitrogen vacancies, and therefore a very high V/III ratio is required in high temperature GaN growth. Of course, the need for a high V/III ratio is also owing to the low decomposition efficiency of ammonia. Nonetheless, a very high V/III ratio (>5000) may intensify the pre-reaction between the ammonia gas and the MO source of gallium, thus a poor growth rate and surface morphology.

In the initial growth of the high temperature GaN epilayer, the V/III ratio is pivotal to the growth mode. A low V/III ratio is favorable for the surface diffusion of gallium atoms with nucleation islands tending to grow laterally, facilitating the coalescence of nucleation islands and the shift into the 2D growth mode; while at a high V/III ratio, increased nitrogen atoms make the surface diffusion length of gallium atoms shorter with nucleation islands inclined to grow longitudinally in a 3D growth mode, thus increasing the surface roughness and the time for nucleation islands to coalesce. The optimal V/III ratio for 2D growth ranges between 1000 and 5000 and is dependent on the reaction chamber design.

The influences of growth temperature, pressure, and thickness on the epilayer quality are discussed based on the experimental results obtained from our self-developed vertical low-pressure MOCVD. The data may vary greatly on different MOCVD systems but should be in a similar pattern.

The surface morphologies of GaN epilayers of the same thickness grown on low temperature GaN nucleation layers at different temperatures (950°C–1100°C) are observed by optical microscopy (not shown here) and show that the optimal growth temperature of GaN epilayers is around 1050°C. If temperature is much higher, hexagonal defects of a very high density arise on the surface. The hexagonal defect density drops with decreasing temperature, and when the temperature drops to about 1050°C, the hexagonal surface morphology disappears, replaced by a mirror surface. When the temperature is below 1000°C, dark spots are observed on the surface. The possible reasons for the severely worsened surface quality of GaN materials at high growth temperatures are as follows. On the one hand, intensified GaN decomposition at high temperature causes deterioration of the surface morphology of the low temperature nucleation layer even under the protection from NH_3, negatively influencing the subsequent nucleation, growth mode, stress, and defect evolution of high temperature GaN epilayer, thus poor surface morphologies. On the other hand, very high temperature may induce the too small density and overly large size of the nucleation islands and cause incomplete coalescence in high temperature GaN growth.

In the experiments on the influence of growth pressure, ~1.1 μm thick GaN epilayers were grown at 500; 5,000; 10,000; and 20,000 Pa, respectively (Ni et al. 2009). Generally, the GaN film surface morphologies observed by AFM gradually coarsen with increasing reaction chamber pressure (Figure 3.10). When the growth pressure increases from 2500 to 5000 Pa, atomic steps with increasing width were observed on the GaN surface. The surfaces of GaN films grown at higher pressure

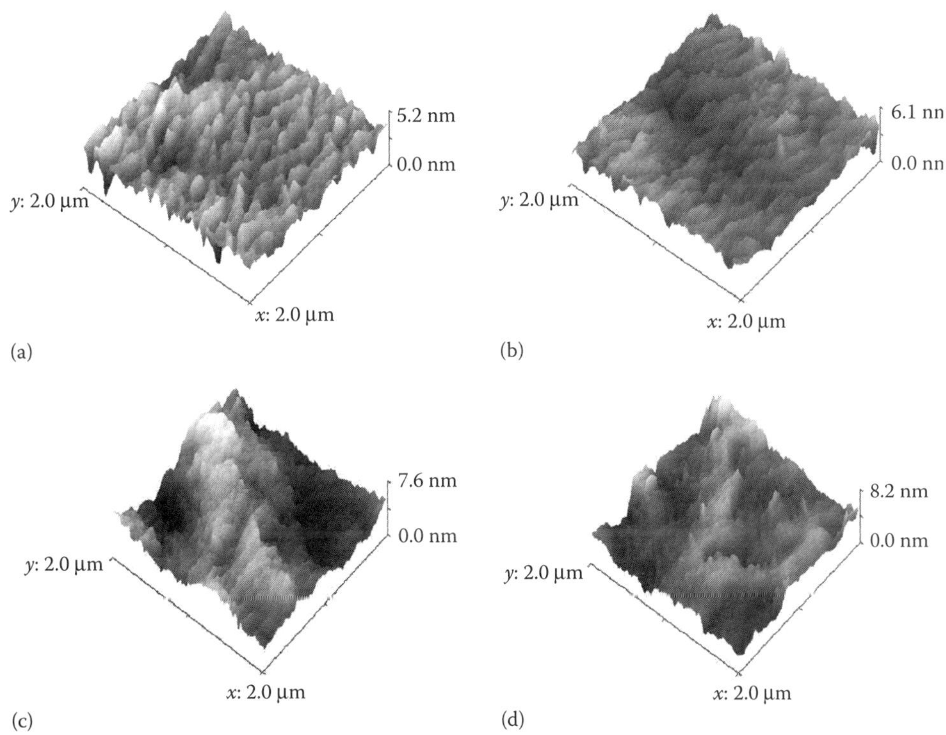

Figure 3.10 AFM images of GaN film surfaces. (a) 2,500 Pa, (b) 5,000 Pa, (c) 10,000 Pa, and (d) 20,000 Pa. (After Ni, J.Y. et al., *Chin. Sci. Bull.*, 54, 2595–2598, 2009.)

deteriorated dramatically with blurred atomic steps, and large hillocks and valleys were observed. It is widely observed that low pressures are favorable for the lateral growth of high temperature GaN islands while high pressures are in favor of the vertical growth. Therefore, the initial nucleation islands of high temperature GaN grown at high pressures have small densities and large sizes, which makes it difficult for them to coalesce, thus rather rough surfaces of the GaN epilayers. Nevertheless, the fewer dislocations generated in the coalescence of the low-density large-size nucleation islands are favorable for crystal quality improvement, with the FWHM values of XRD rocking curve declining with rising growth pressure. Hall effect measurements showed a high background electron concentration of the GaN epilayer grown at high pressure, possibly due to the low density of dislocations that can capture free carriers.

GaN epilayer thickness also bears an influence on its crystalline quality and 2D film integrity. For GaN epilayers of different thicknesses grown on like nucleation layers, it is found that the 0.8 μm GaN has an apparently better surface morphology than the 0.4 μm GaN. Complete coalescence is already achieved at the GaN thickness of 0.8 μm. As the thickness increases, the surface smoothness of GaN improves. However, when the GaN thickness exceeds 1.2 μm, no obvious change in GaN surface morphology is observed. The variation of FWHM of XRD rocking curve with GaN thickness follows similar pattern, indicating that the GaN dislocation density decreases fast with the increase in thickness when the thicknesses are small but the decrease slows down when the GaN thickness exceeds 1.2 μm. In general, the GaN dislocations are reduced by dislocation blocking and inclining. Without interlayers or other blocking mechanisms, dislocation inclination is dominant. The initial high temperature GaN growth focuses on the nucleation islands, mainly in the 3D growth mode. As the nucleation islands grow larger, the 3D nucleation islands gradually coalesce into a 2D film. During the shift from 3D to 2D, some dislocations have an inclined extension and fail to reach the surface of the film, thereby an effective suppression of dislocation density to some extent. When the GaN grows thick enough, the growth changes into a quasi-2D mode in which the probability of dislocation inclination and the annihilation rate decline and ultimately the GaN crystal quality stabilizes.

In a microscopic view, the growth process of the GaN epilayer starts with the high temperature GaN nucleation on the low temperature GaN nucleation layer, followed by the lateral growth and coalescence of the 3D nucleation islands, after which the GaN grows in the 2D layer mode (Figure 3.3e). This process can be performed under the same process conditions all along or under different conditions in different stages of growth. With GaN epilayer growth divided into the initial growth of the coalescence layer and the subsequent growth of the 2D layers (Zhou 2010), reducing the V/III ratio in the coalescence layer growth helps to decrease the dislocation density in the material. It is proposed that the higher lateral growth rate under a low V/III ratio in the initial growth of high temperature GaN can increase the lateral size of

3D nucleation islands and decrease their density. So the coalescence between islands and the transition into the quasi-2D growth mode speed up with a decreased dislocation density at the coalescence sites, which can improve the quality of GaN growth. Nevertheless, there exists an optimal V/III ratio for the coalescence layer, and too low a V/III ratio may induce excessive gallium sources in the reactants, in which case a lot of dark spot defects on the sample surface will be observed by microscopy.

We further employed the low-temperature AlN followed by pulsed MOCVD growth of AlN to grow a multiple AlN nucleation layer (Zhou 2010), and divide the GaN epilayer growth into the three stages of GaN I, GaN II, and GaN III, with a total GaN thickness of 2 μm. GaN I is the initial high temperature GaN growth with a low V/III ratio (1700) for the nucleation islands to coalesce rapidly; GaN II adopts a high V/III ratio (4000) to reduce nitrogen vacancies thus improving the GaN growth quality; and GaN III uses a low V/III ratio (1700) to improve GaN surface smoothness. As-grown GaN epitaxial materials show distinct and nearly parallel atomic steps on the surface as shown in the $2 \times 2\ \mu m^2$ AFM image (Figure 3.11) with an RMS roughness of only 0.185 nm, an FWHM value of 348 arcsec for the (102) and of merely 70 arcsec for the (002) XRD rocking curves (Figure 3.12), a high-level crystal quality for GaN films on sapphire substrates.

3.6 Defect Microstructures of Nitride Single Crystal Thin Films

GaN heteroepitaxial thin films have many defect microstructures. The reported defects in heteroepitaxial GaN grown by MOCVD are shown schematically in Figure 3.13, which fall into the following types:

1. *Point defects*: Substitutional sites, interstitials, atomic vacancy, and complexes (Pearton et al. 1999).

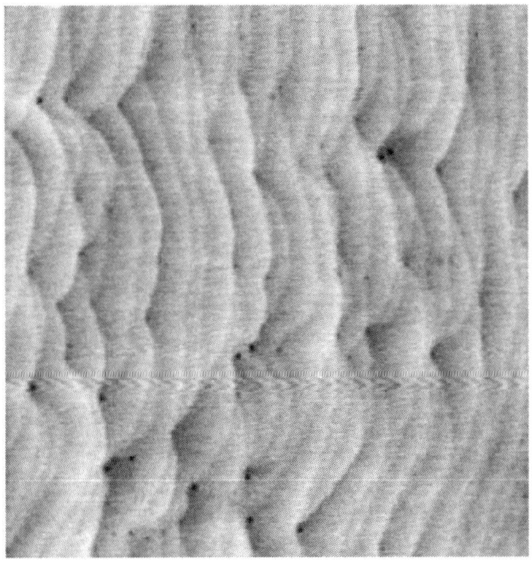

Figure 3.11 AFM surface morphology of GaN grown on multiple AlN nucleation layer with variable V/III ratios. (Courtesy of Dr. Xiaowei Zhou.)

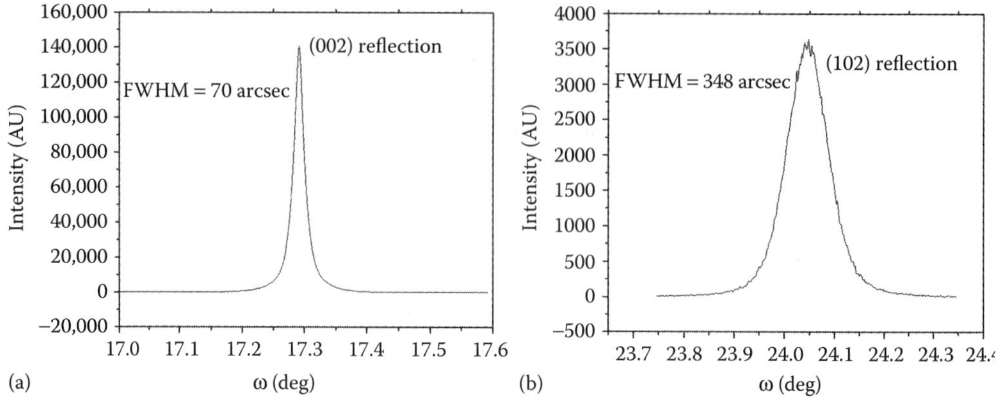

Figure 3.12 XRD rocking curves of GaN on multiple AlN nucleation layer with a three-stage growth. (a) (002) reflection and (b) (102) reflection. (Courtesy of Dr. Xiaowei Zhou.)

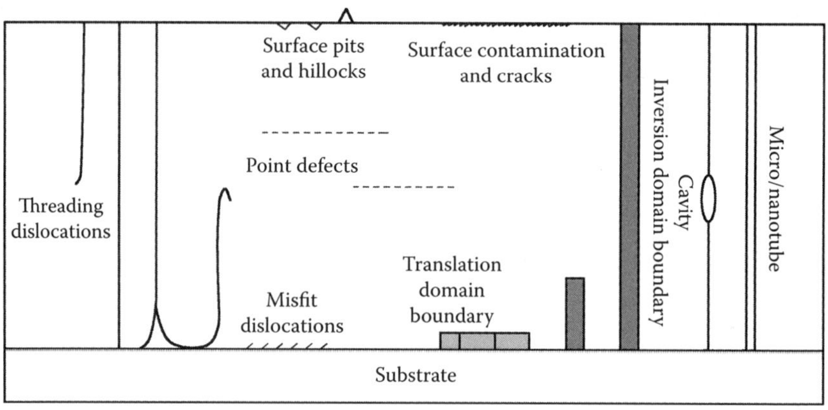

Figure 3.13 Schematic of defects existing in GaN heteroepitaxial films grown by MOCVD. (Courtesy of Dr. Zhiyuan Gao.)

2. *Line defects*:
   a. *Threading dislocations* (Bai et al. 2005): Screw dislocations with $\boldsymbol{b} = \langle 0001 \rangle$, edge dislocations with $\boldsymbol{b} = 1/3\langle 11\bar{2}0 \rangle$, and mixed dislocations with $\boldsymbol{b} = 1/3\langle 11\bar{2}3 \rangle$.
   b. *Misfit dislocations* (Ruterana and Nouet 2001).
3. *Planar defect*:
   a. *Phase boundary*:
      i. *Inversion domain boundary* (IDB) (Wu et al. 1996, Rouviere et al. 1997).
      ii. *Translation domain boundary* (TDB): stacking fault (SF), antiphase boundary (APB) (Rouviere et al. 1997), and double-positioning boundary (DPB) (only observed in epilayers on SiC substrates [Smith et al. 1995, Sverdlov et al. 1995]).
      iii. *Twin* (only observed in epilayers on SiC substrates) (Liliental-Weber et al. 1995).
   b. *Grain boundary*: low-angle grain boundary (LAGB) (Qian et al. 1995).
4. *Bulk defects*: Deposits, cracks, and voids.

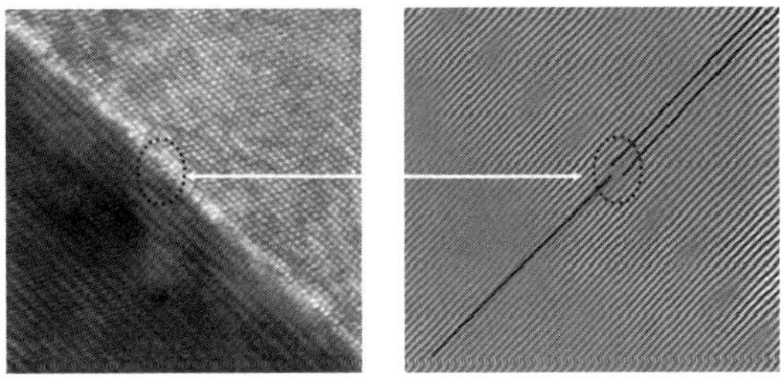

Figure 3.14 Characterization of misfit dislocations at sapphire/nucleation layer interface. (a) HRTEM lattice image at sapphire/nucleation layer interface; (b) Fourier transformed image with a clearer view of the misfit dislocations. (Courtesy of Dr. Zhiyuan Gao.)

In the following, the microstructural characteristics of these defects are introduced in the spatial bottom-up order of material growth.

3.6.1 Microstructures at the Substrate/Nucleation Layer Interface—Misfit Dislocations

Misfit dislocations abound at the substrate/nucleation layer interface. Shown in Figure 3.14a is the image of the crystal lattices at the interface between the sapphire substrate and GaN nucleation layer observed by high resolution transmission electron microscope (HRTEM). For a more distinct picture of the lattice defects, a Fourier transform is performed on this image to remove the diffraction spots from the planes parallel to the interface, followed by an inverse Fourier transform to yield the image in Figure 3.14b, which shows clearly the lattice mismatches at the interface with nearly one in seven to eight atomic layers. The dislocation line is normal to the plane of the paper, that is, parallel to the substrate surface.

Misfit dislocations, as the name suggests, stem from lattice mismatch. The abundant misfit dislocations at the substrate/nucleation layer interface indicate that the nucleation layer as the sacrificial layer has released the heteroepitaxy-induced lattice mismatch strain, thus a guarantee of the quality of subsequent high temperature GaN epilayers. This is one of the major functions of the nucleation layer.

3.6.2 Microstructures in the Nucleation Layer: Stacking Faults, Local Cubic Phases, and Antiphase Boundaries

A large number of translation domain boundaries exist inside the nucleation layer, and for samples with sapphire substrates, these planar defects are primarily SF (Wu et al. 1996). A typical HRTEM lattice image of such planar defects is shown in Figure 3.15a. Similarly, the image after Fourier transform in Figure 3.15b shows more distinctly the disordered arrangement at the interface, with the stacking sequence of atoms translating from wurtzite ABABAB... into zinc-blende ABCABC... The SF plane separates the wurtzite hexagonal phases from the zince-blende cubic phases. Samples with 6H-SiC substrates also contain antiphase boundaries (Rouviere et al.

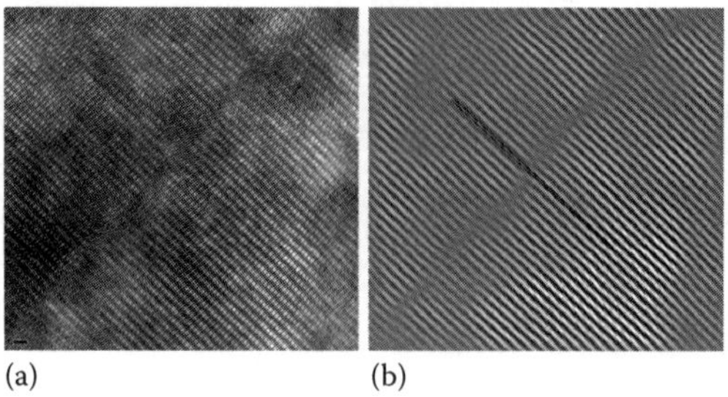

Figure 3.15 Characterization of SF in the GaN nucleation layer. (a) HRTEM lattice image of SF in the GaN nucleation layer; (b) Fourier transformed image with a clearer view of the sequence of atomic planes. (Courtesy of Dr. Zhiyuan Gao.)

1997), double-positioning boundaries (Smith et al. 1995, Sverdlov et al. 1995) and twins (Liliental-Weber et al. 1995), and the formation of these planar defects has much to do with the substrate surface condition. Compared to SF, these defects are rare and have little influence on the material properties.

Although containing many crystal grains and different phases, the low temperature nucleation layer still obtains near uniformly oriented GaN layers owing to the fact that the transformation between the cubic phase to hexagonal phase requires only an introduction of a SF, with which the wurtzite structure can grow readily on the {111} plane of the zince-blende structure. However, it is also due to this fact that the cubic phases in the nucleation layer sometimes extends into the GaN layer to form local cubic phases. Part of the cubic phases terminates within the layer and some extend up to the surface. Local cubic phases are often observed in Mg-rich samples with surface terminations in the shape of triangular pit (Figure 3.16), which is in agreement with the symmetry of the zinc-blende structure. Total energy calculations indicate that disordered growth sequence results readily from the preferential combination of magnesium with nitrogen (Kitamura et al. 1995), thereby a higher probability of the cubic phase formation.

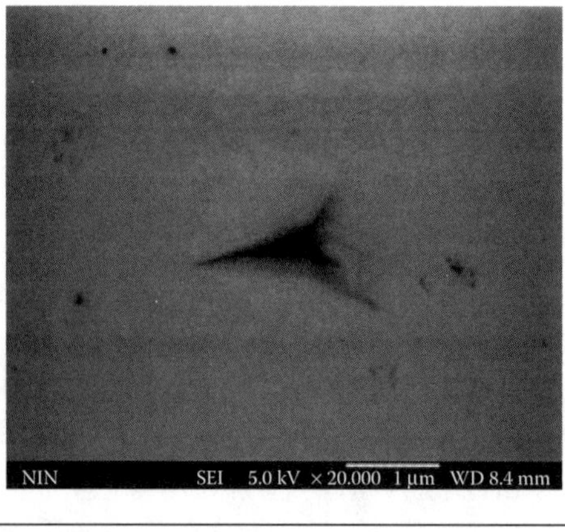

Figure 3.16 SEM image of surface morphology of Mg-doped GaN. The local cubic phase propagates to the crystal surface to form a triangular pit. (After Gao, Z. et al., *Mater. Rev.*, 23, 1–5, 2009.)

Electron backscatter diffraction (EBSD) is an effective tool for detecting the distribution of the crystal orientation and phase domains. Different from TEM, it uses the diffraction patterns produced by the backscattered electrons and reflects mainly the surface layer structure. For the cubic phase extending through to the surface or for relatively thin GaN crystals, this technique can be used to characterize the cubic phase in GaN. Shown in Figure 3.17a is the distribution of the different phases over an area of 30 μm × 30 μm on a 1 μm thick GaN epilayer, with an average grain size of 0.5 μm in the dotted region, which has a zinc-blende structure as indicated by its corresponding EBSD patterns and polar diagram in Figure 3.17b. The EBSD patterns and polar diagram for the large area background region as shown in Figure 3.17c suggest a wurtzite structure.

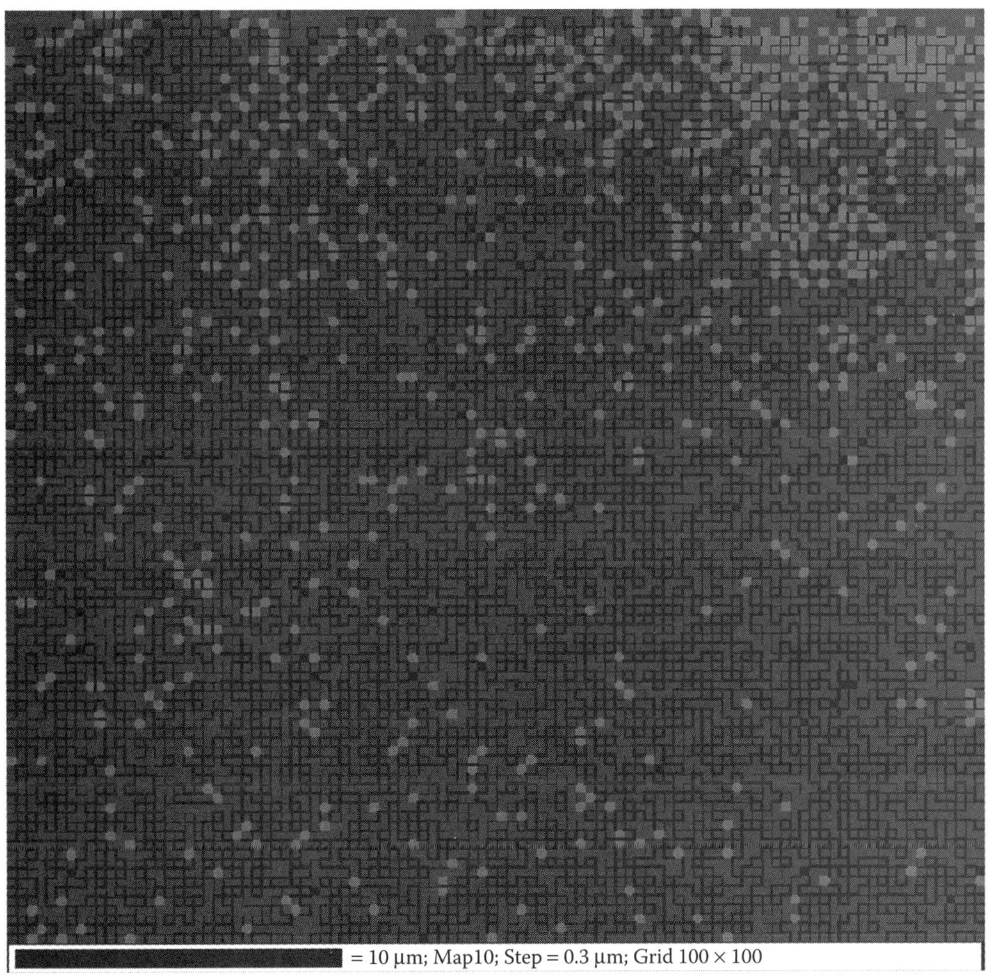

(a)

Figure 3.17 Characterization of different GaN crystal phases. (a) Distribution of different phases over an area of 30 μm × 30 μm on a GaN epilayer of a thickness of 1 μm. *(Continued)*

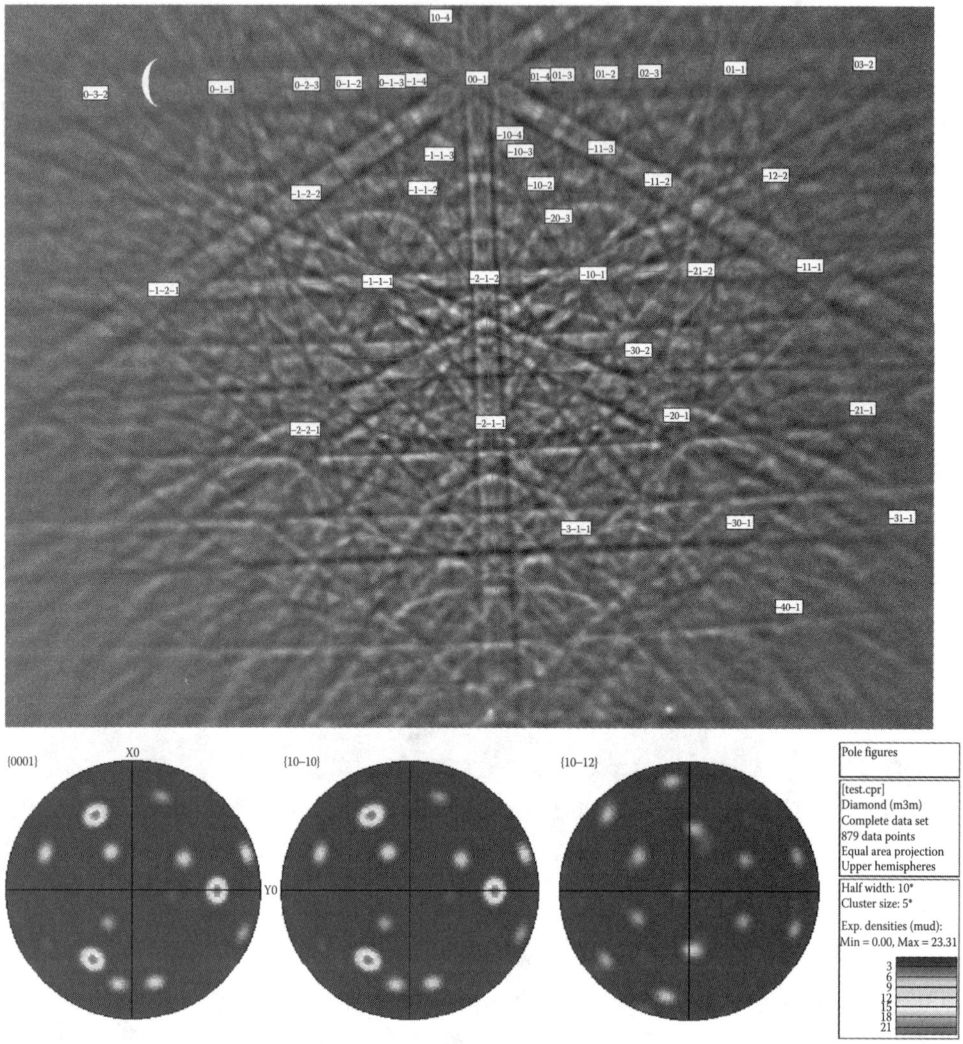

(b)

Figure 3.17 (Continued) Characterization of different GaN crystal phases. (b) EBSD patterns and the polar diagram of the dotted region showing a zinc-blende structure. *(Continued)*

The function of nitrification and the nucleation layer is also reflected in the polarity choice (Rouviere et al. 1997). Under the MOCVD growth condition with high V/III ratio, since the N-face is stable at low temperature, the nitrification of the sapphire substrate renders the initial growth planes N-face polarity, and the subsequently grown high temperature GaN layers with proper control of the nucleation layer growth time are typically of Ga-face polarity. Improper nitrification or nucleation layer process control may induce local inversion domain boundaries with opposite polarity to the ambient in the high temperature GaN layer. Some inversion domain boundaries terminate inside the layer, while some terminate at the surface typically forming hillocks as shown in the AFM image in Figure 3.18.

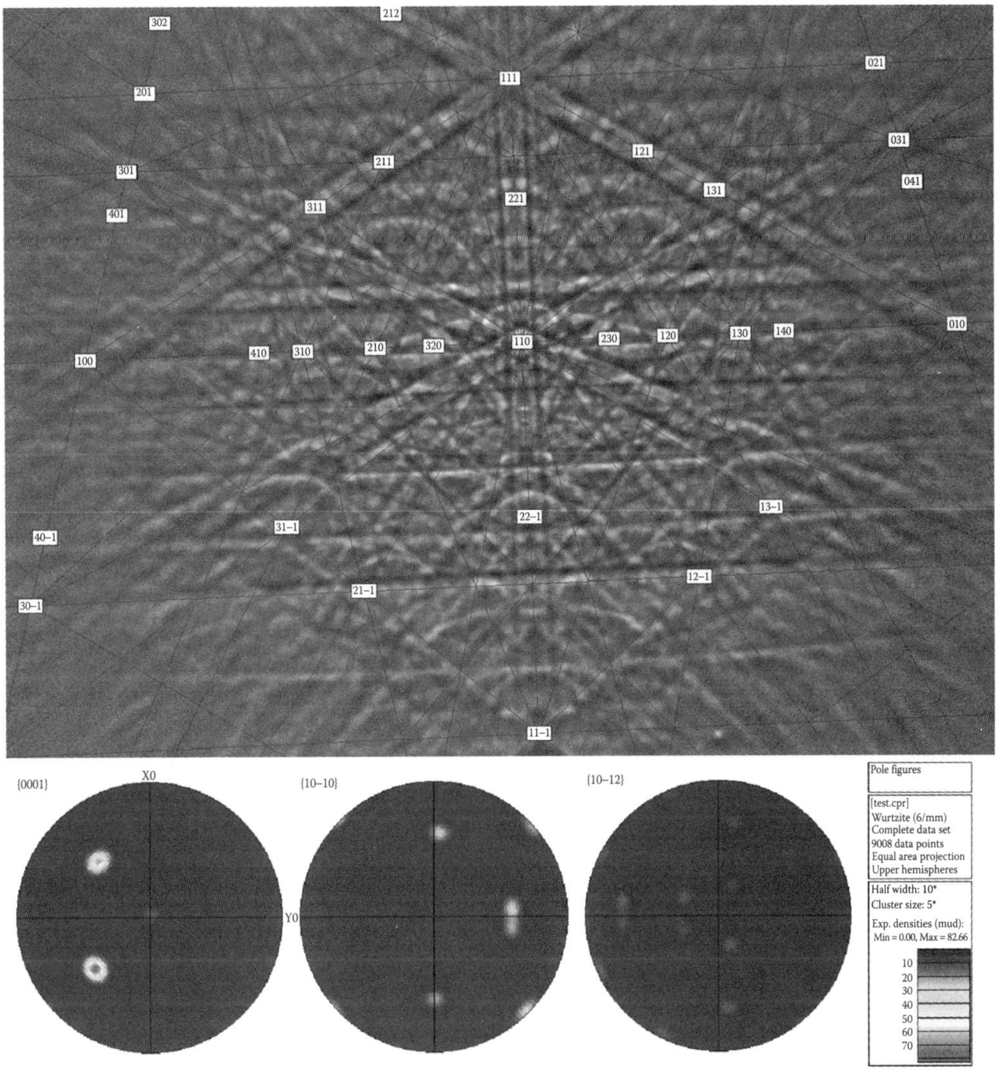

(c)

Figure 3.17 (Continued) Characterization of different GaN crystal phases. (c) EBSD patterns and polar diagram of the background region showing a wurtzite structure. (After Gao, Z. et al., *Mater. Rev.*, 23, 1–5, 2009.)

The difference between the IDB and TDB is that the crystal structure of the former on the defect side can be deduced to the opposite side by both reversion and translation while that of the latter only by translation. Accurate determination of the inversion domain boundaries requires the convergent beam electron diffraction (CBED) method, by which the electron beams are converged to a very small spot for the sample and its diffraction patterns to appear together so as to determine the crystal orientations of different adjacent regions. Experiments show that the inversion domain boundaries can also be exposed by wet etching.

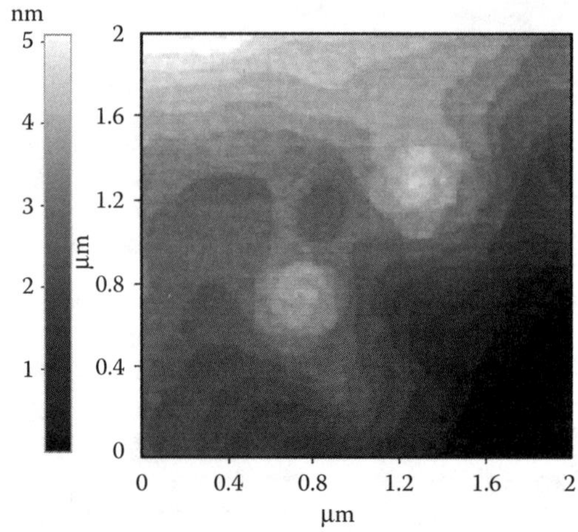

Figure 3.18 2 µm × 2 µm AFM image of surface morphology of GaN with hillocks formed on the surface by the IDBs. (After Gao, Z. et al., *Mater. Rev.*, 23, 1–5, 2009.)

3.6.3 Microstructures in High-Temperature GaN Layer: Low-Angle Grain Boundaries, Threading Dislocations, and Point Defects

Nitrification and the nucleation layer also help determine the crystal orientation. The solid phase recrystallization of the amorphous low temperature nucleation layers by annealing at high temperature prior to the growth of GaN layers gives rise to the columnar crystals. The initial GaN growth is the nucleation of islands on these columnar structures. These islands grow and coalesce in subsequent growth and the resulting columnar subgrains maintain the tiny orientation difference present in the initial growth. In some cases, the subgrains fail to form an integral unit by coalescence, thus the low-angle grain boundaries between the grains.

The cross-sectional TEM image can reveal the subgrains. Shown in Figure 3.19 is the cross-sectional two-beam TEM image with a dark field on the left and a bright field on the right, and the difference in diffraction contrast indicates that they are two different crystal grains with the dislocation right on the juncture of the

Figure 3.19 Cross-sectional two-beam TEM image of GaN epilayer with $g = [10\bar{1}1]$. (Courtesy of Dr. Zhiyuan Gao.)

two. The rugged GaN surface is caused by ion thinning with the original sample surface bombarded off by the ion beam. The LAGB can also be characterized by wet etching.

There are two types of dislocations in heteroepitaxial GaN: the aforementioned misfit dislocations parallel to the substrate surface and the threading dislocations perpendicular to the substrate surface. Of all defects, threading dislocations are the most widely distributed in GaN epilayers and have the greatest influence on the material and device performance.

Threading dislocations stem from the coalescence between the high temperature GaN nucleation islands grown on the nucleation layer, that is, the low angle boundary formed by coalescence of the columnar subgrains with tiny orientation difference are composed of substantial threading dislocations. This can be illustrated by Figure 3.19, in which contrast of dislocation arises right at the juncture between two subgrains.

The dislocation Burgers vector is conventionally measured by two-beam TEM with several proper reflections chosen for imaging. The firs imaging uses a reflection g1 for the dislocation image to be invisible and the second uses a reflection $g2$ to make the dislocations contrastless, and then we have $b = g1 \wedge g2$. Heteroepitaxial GaN contains three types of threading dislocations: the edge dislocations (b = 1/3 $\langle 11\bar{2}0 \rangle$), edge dislocations ($b$ = $\langle 0001 \rangle$), and mixed dislocations (b = 1/3 $\langle 11\bar{2}3 \rangle$). For edge dislocations, although the eight-atom core structure has the least energy, there is also evidence of the existence of four- or seven-atom core structures. The case is more complicated for screw dislocations as their core structures vary greatly and irregularly from the open-core (micro/nanopipes) to closed-core structure, although its balanced structure is the closed-core one.

Point defects are another major type of defects. The native point defects in GaN have six forms: (a) the nitrogen vacancy V_N, (b) the gallium vacancy V_{Ga}, (c) the antisite nitrogen (namely gallium-site nitrogen) N_{ant}, (d) the antisite gallium (namely nitrogen-site gallium) Ga_{ant}, (e) the interstitial nitrogen N_{int}, and (f) the interstitial gallium Ga_{int}. In p-type GaN, V_N has the least formation energy, and in n-type GaN V_{Ga} has the least formation energy. The low energy originates from that the vacancies are charged, thereby greatly reduced formation energy. In contrast to V_N and V_{Ga}, both n-type and p-type antisite defects have a large formation energy, and thus are less likely to form.

The impurities introduced by unintentional doping of MOCVD-grown GaN are primarily silicon, carbon, oxygen, and hydrogen. Silicon comes from the Si-contaminated growth material, and if the reaction chamber walls are made of materials containing silicon such as quartz glass, it is also likely for silicon to be released during the high temperature growth; carbon comes from the MO source, and oxygen from the MOCVD-grown gaseous precursors and the substrate, the carrier gas, and so on; hydrogen has many sources. In p-type materials, hydrogen is a donor which combines the acceptor magnesium to form a complex. Thus in order to activate the acceptor, a post-growth annealing is often performed to get rid of hydrogen in the material.

The yellow luminescence and high background carrier concentration of GaN are closely related to the substitutional doping and complexes formed by these impurities.

3.6.4 Cracks and Deposits

The formation of cracks in wafer growth largely limits the quality and performance of the heterostructure-based devices, and a good understanding of cracking behaviors is necessary for growth process optimization and the design and choice of the device structure. Basically, the formation of cracks is due to the presence of strain, and the strain is released by crack formation. Strain may arise both from the difference in the inherent material properties and from the growth process. The major origin of cracks is the strain arising from the lattice mismatch between the substrate and the epitaxial film, and their different thermal expansion coefficients in the growth process.

The nucleation layer releases largely the lattice mismatch-induced strain, but a new form of strain is introduced thereafter (Etzkorn and Clarke 2004): the tensile strain generated by coalescence of neighboring islands grown in the initial growth of the high temperature GaN layers on the nucleation layer due to the orientation difference between these islands. One way for such strain to be released is the generation of substantial dislocations along the grain boundary, and the other is the formation of cracks. There are generally few dislocations at the sites where cracks are present.

Figure 3.20 shows the SEM image of cracks on the surface of GaN grown on sapphire. The hexagonal pattern formed by the intersection of the cracks reflects the sixfold rotation symmetry of the GaN crystal structure along the *c*-axis and shows the fact that the GaN cleavage planes are the $\{10\bar{1}0\}$ planes. However, deposits can be also observed besides the cracks in Figure 3.20, especially with many cracks running across the large-size deposits.

Different from the bulk crystal growth, the vapor phase growth technique of MOCVD offers atomic level smoothness for epitaxial materials, where large-size bulk defects such as deposits are rarely seen. But in case of insufficient reaction chamber vacuum or improper substrate surface cleaning, large-size deposits containing carbon

Figure 3.20 SEM image of surface cracks of GaN on sapphire substrate showing intersecting cracks at an angle of 60° with several deposits right on the cracks. (After Gao, Z. et al., *Mater. Rev.*, 23, 1–5, 2009.)

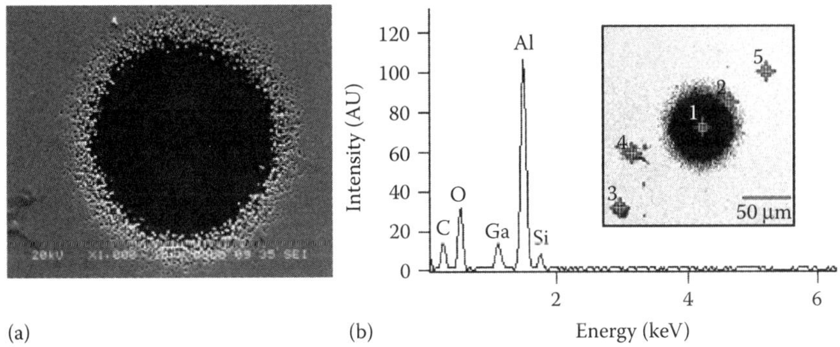

Figure 3.21 A deposit in the Si-doped GaN film grown on sapphire. (a) SEM image of the deposit. (b) Energy spectrum at the deposit center (Spot 1). (After Gao, Z. et al., *Mater. Rev.*, 23, 1–5, 2009.)

and oxygen may exist in the epitaxial wafer. Such deposits are not necessarily contained in the crystal, and sometimes can be seen on the surface of the epitaxial film because no or few GaN crystals exist at the sites with deposits where nucleation and growth may come to a stop in the vicinity.

Deposit-induced surface pits are of large dimensions and typically in irregular shapes. The SEM image of a deposit in Si-doped GaN grown on sapphire is shown in Figure 3.21a. The deposit is approximately circular and 100 μm in diameter. A composition analysis of the deposit in Figure 3.21a is made using SEM electronic probes, and the energy spectrum obtained at the deposit center is shown in Figure 3.21b. Different from the energy spectrum of the defect-free region, the energy spectrum of the deposit center shows a small gallium content and no sign of nitrogen while a very high Al peak with relatively high peaks of oxygen and carbon. This suggests that the deposit consists mainly of carbon and oxygen and that the Al peak originates from the Al_2O_3 substrate that is exposed, where GaN growth stops due to the existence of deposit.

The relation between cracks and deposits in the surface morphology (Figure 3.20) suggests that deposits can promote crack initiation: the tensile strain concentrates more around the deposit in the film growth and due to the relatively weak compressive strength of the deposit, plastic deformation of neighboring crystal lattices makes it easy for cracks to form in the vicinity of the deposit, which then spread out and intersect into a net. Almost all the deposits in Figure 3.20 are unexceptionally located on the cracks, from which it can be seen that the cracks in GaN can be effectively reduced simply by reducing or eliminating the deposits.

References

Amano, H., N. Sawaki, I. Akasaki, and Y. Toyoda. 1986. Metalorganic vapor phase epitaxial growth of a high quality GaN film using an aln buffer layer. *Applied Physics Letters* 48(5):353–355. doi: 10.1063/1.96549.

Ambacher, O. 1998. Growth and applications of group III-nitrides. *Journal of Physics D: Applied Physics* 31(20):2653–2710. doi: 10.1088/0022-3727/31/20/001.

Bai, J., T. Wang, P. J. Parbrook, K. B. Lee, and A. G. Cullis. 2005. A study of dislocations in AlN and GaN films grown on sapphire substrates. *Journal of Crystal Growth* 282:290–296.

Duan, H., Y. Hao, and J. Zhang. 2009a. Effect of a high temperature AlN buffer layer grown by initially alternating supply of ammonia on AlGaN/GaN heterostuctures. *Journal of Semiconductors* 30(9):093001 (4 pp.). doi: 10.1088/1674-4926/30/9/093001.

Duan, H., Y. Hao, and J. Zhang. 2009b. Effect of nucleation layer morphology on crystal quality, surface morphology and electrical properties of AlGaN/GaN heterostructures. *Journal of Semiconductors* 30(10):105002 (3 pp.). doi: 10.1088/1674-4926/30/10/105002.

Duan, H. T. 2011. Growth of AlGaN/GaN on high temperature AlN nucleation layer and new double heterostructure. PhD Thesis, Xidian University, Xi'an, China.

Etzkorn, E. V. and D. R. Clarke. 2004. Cracking of GaN films. *International Journal of High Speed Electronics and Systems* 14(1):63–81. doi: 10.1142/s0129156404002247.

Gao, Z., Y. Hao, and J. Zhang. 2009. Effect of structural defects in GaN epitaxial layer on its surface morphology. *Materials Review* 23(4):1–5.

Kitamura, S., K. Hiramastsu, and N. Sawaki. 1995. Fabrication of GaN hexagonal pyramids on dot-patterned GaN/sapphire substrates via selective metalorganic vapor phase epitaxy. *Japanese Journal of Applied Physics* 34:L1184–1186.

Liliental-Weber, Z., H. Sohn, N. Newman, and J. Washburn. 1995. Electron microscopy characterization of GaN films grown by molecular-beam epitaxy on sapphire and SiC. *Journal of Vacuum Science and Technology B: Microelectronics and Nanometer Structures* 13(4):1578–1581. doi: 10.1116/1.588190.

Lu, D. and S. Duan. 2009. *Fundamentals and Applications of Metalorganic Compound Vapor Epitaxy* (in Chinese). Beijing, China: Science Press.

Nakamura, S. 1991. GaN growth using GaN buffer layer. *Japanese Journal of Applied Physics, Part 2: Letters* 30(10 A):L1705–L1707.

Ni, J. Y., Y. Hao, J. C. Zhang, and L. A. Yang. 2009. Effect of reactor pressure on the growth rate and structural properties of GaN films. *Chinese Science Bulletin* 54(15):2595–2598. doi: 10.1007/s11434-009-0300-6.

Pearton, S. J., J. C. Zolper, R. J. Shul, and F. Ren. 1999. GaN: Processing, defects, and devices. *Journal of Applied Physics* 86(1):1–78. doi: 10.1063/1.371145.

Qian, W., M. Skowronski, M. De Graef, K. Doverspike, L. B. Rowland, and D. K. Gaskill. 1995. Microstructural characterization of GaN films grown on sapphire by organometallic vapor phase epitaxy. *Applied Physics Letters* 66(10):1252–1254.

Rouviere, J. L., M. Arlery, B. Daudin, G. Feuillet, and O. Briot. 1997. Transmission electron microscopy structural characterization of GaN layers grown on (0001) sapphire. *Materials Science and Engineering B* 50:61–71.

Ruterana, P. and G. Nouet. 2001. Atomic structure of extended defects in wurtzite GaN epitaxial layers. *Physica Status Solidi (b)* 227(1):177–228.

Smith, D. J., D. Chandrasekhar, B. Sverdlov, A. Botchkarev, A. Salvador, and H. Morkoc. 1995. Characterization of structural defects in wurtzite GaN grown on 6H SiC using plasma-enhanced molecular beam epitaxy. *Applied Physics Letters* 67(13):1830–1832.

Sverdlov, B. N., G. A. Martin, H. Morkoc, and D. J. Smith. 1995. Formation of threading defects in GaN wurtzite films grown on nonisomorphic substrates. *Applied Physics Letters* 67(14):2063–2065. doi: 10.1063/1.115079.

Wu, X. H., L. M. Brown, D. Kapolnek, S. Keller, B. Keller, S. P. DenBaars, and J. S. Speck. 1996. Defect Structure of metal organic chemical vapor deposition grown epitaxial (0001) GaN/Al2O3. *Journal of Applied Physics* 80(6):3228–3237.

Zhou, X. W. 2010. Growth of high Al fraction AlGaN and GaN semiconductor materials. PhD Thesis, Xidian University, Xi'an, China.

4

Electrical Properties and Related Mechanisms of GaN Heterostructures Used in High Electron Mobility Transistors

The high electron mobility transistor (HEMT) gets its name by the adoption of the conducting channel formed by a high mobility two-dimensional electron gas (2DEG). Such devices are fabricated from heterostructures whose conduction band edge forms band offsets and quantum wells at the heterointerface, and the electrons distributed in the quantum wells become the 2DEG that can move freely along the heterointerface but is confined perpendicular to the interface by the quantum wells. The source and the drain of the HEMT are to form ohmic contacts to the 2DEG for electric current to arise from the 2DEG transport along the heterointerface, while the Schottky barrier gate controls the on and off of the 2DEG channel by gate voltage. Therefore, such a device is essentially a heterojunction field effect transistor (HFET).

In the analysis of electrical properties of heterojunction material and HEMT device performance, a 2DEG is typically viewed as a thin layer of electrons with a thickness approaching zero, a sheet electron concentration of n_{s2D} (in units of cm^{-2}), and a mobility μ. The 2DEG sheet conductivity $G = e{\cdot}n_{s2D}{\cdot}\mu$ is in units of S (siemens), where e is the elementary charge. Thus, the improvement of 2DEG conductivity for greater current driving ability of the HEMT requires increasing the 2DEG density and mobility, which are also employed together with the sheet resistance ($R_{SH} = 1/G$) as the major physical parameters for the electrical property evaluation of heterojunctions.

The 2DEG is a quantized system, whose conduction varies greatly from that of the bulk electrons. For GaN-based heterostructures, the 2DEG density and mobility are closely related to the layered structure, polarization, and quality of the heterojunction.

4.1 2DEGs in GaN-Based Heterostructures

4.1.1 Formation Mechanism of 2DEGs in GaN-Based Heterostructures

In the conventional AlGaAs/GaAs material system, 2DEG electrons stem mainly from the donor ionization in AlGaAs and GaAs. However, for AlGaN/GaN heterostructures, 2DEGs of sheet densities of the order of 10^{13} cm^{-2} are available even without intentional doping. Thus, the origin of the 2DEGs in AlGaN/GaN heterostructures is intriguing. If the electrons originate completely from the ionization of background n-type doping, such a high 2DEG density requires a nitride film thickness at least of 10–100 μm with a typical background doping concentration of 10^{15}–10^{16} cm^{-3}, whereas the AlGaN/GaN heterojunction has a typical thickness of only 1–3 μm, so the bulk donor could merely supply a small fraction of electrons for the 2DEGs.

The 2DEGs in the AlGaN/GaN heterostructures are generally referred to as polarization-induced 2DEGs owing to the very strong built-in electric fields formed in the heterojunctions due to the polarization effect. The built-in field modulates the band structure of the nitride heterojunction and makes the quantum well at the GaN side of the heterointerface deep and narrow, favorable for attracting free electrons into the well to form the 2DEG. But the sum of all the polar charges at the heterojunction surface and heterointerfaces is 0, which means that they cannot provide electrons for the 2DEGs.

Taking $Al_{0.27}Ga_{0.73}N$/GaN for example (Zhang and Hao 2006), if only net polar charge presents at the surface and the heterointerface and the external electric field is 0, the polarization-induced internal electric field can yield a band bending up to the band gap of AlGaN (GaN) within a depth of 10 nm from the AlGaN surface (the GaN/substrate interface) into the heterojunction. Opposite to this unrealistic finding, the polarization-induced field is partly shielded under various mechanisms in the physical picture of electric–static equilibrium of the actual nitride heterojunction. Currently, nitride heterojunctions are grown mostly by epitaxy techniques, featuring high density lattice defects in the nucleation layer near substrate. Since the nucleation layer usually has a thickness of tens of nanometers, and the polarization effect is very sensitive to the lattice construction, the density of polar charge here cannot reach the theoretical value. Usually, substantial charged defects and impurities are accumulated in this region, and they also have a cancelling effect on the polar charge. At the surface, the surface states and the external charge adsorbed cancel the polar charge. At the internal interfaces the polar charges cause the sudden changes of the electric field, and quantum wells of conduction band or valence band are often formed with electrons or holes accumulated to screen the polarization-induced field (Figure 4.1).

Smorchkova et al. (1999) observed a critical thickness of AlGaN layer for the formation of the 2DEG in AlGaN/GaN heterostructure samples of different AlGaN layer thicknesses, where the 2DEG density increases rapidly with further increase in AlGaN thickness and then saturates. Accordingly, they proposed a model in which the 2DEG originates mainly from the ionization of donor-like AlGaN surface states. As Figure 4.2 shows, as the AlGaN layer thickness increases before the formation of the 2DEG, the polarization field raises the surface potential ($e\Phi_S$, the height of the conduction band edge E_C with respect to the Fermi level E_F at AlGaN surface), and a surface donor-like state with the energy level E_{DS} below the conduction band rises accordingly. When E_{DS} aligns with the Fermi level, the 2DEG emerges as electrons are released by surface donor state ionization with a reduced AlGaN built-in field. As the AlGaN layer grows thicker (assuming the absence of strain relaxation),

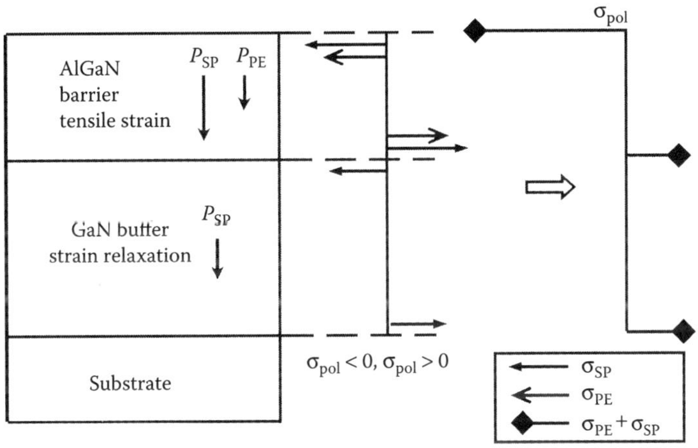

Figure 4.1 Polarization effect in Ga-face AlGaN/GaN heterostructures. (a) Strain and polarization strength orientation of each layer. (b) Polar charge distribution on each interface. (c) Equivalent polar charge distribution. The suffix SP stands for the spontaneous polarization, *PZ* for the piezoelectric polarization, and pol for the polarization. The types and relative magnitudes of the polar charges in (b) and (c) are shown with different arrows with their real positions at the abrupt heterointerface.

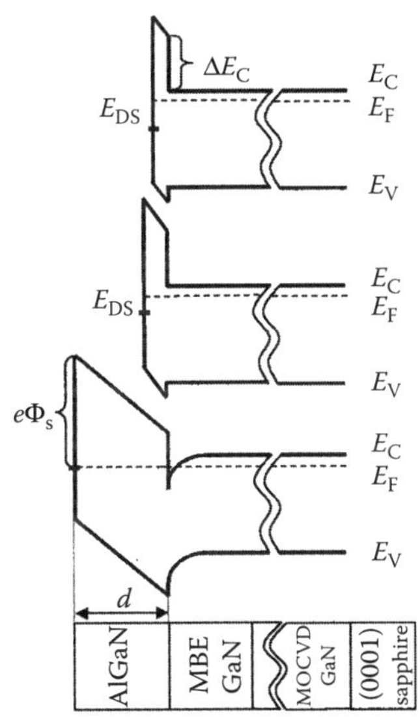

Figure 4.2 Energy band of AlGaN/GaN heterojunction as a function of AlGaN layer thickness. The donor-like surface state is at the level E_{DS}. (Reproduced from Smorchkova, I.P. et al., *J. Appl. Phys.*, 86, 4520–4526, 1999.)

the 2DEG density tends to saturate approaching the density of the positive polar charge at the AlGaN/GaN interface with E_{DS} kept aligning with the Fermi level (the surface potential $e\Phi_S$ remaining basically constant and equal to E_C-E_{DS}). This model interprets the variation of the 2DEG density with AlGaN thickness, and proposes the contribution of surface charges to the 2DEG formation where unintentional doping fails to provide enough electrons for the 2DEGs.

However, fitting of the experimental data by this model yields a donor-like surface state at 1.42 eV below the conduction band edge. If there is so high a surface potential after the 2DEG is formed in the AlGaN/GaN heterojunction, the surface barrier height (on bare AlGaN) should be greater than most reported Schottky barrier heights on AlGaN surface (Al content 0.1–0.4), which means that once the Schottky barrier is formed at the AlGaN surface, the 2DEG density under the barrier will increase. However, experiments show that the Schottky barrier in the AlGaN/GaN heterojunction can generally deplete part of the 2DEG. Therefore, the potential of exposed AlGaN surface should be lower than the typical Schottky barrier height, and a new interpretation of the critical thickness for the 2DEG formation is needed.

Koley and Spencer (2001) studied the surface potentials of GaN epilayers with different doping and $Al_{0.35}Ga_{0.65}N$/GaN heterostructures having different AlGaN layer thickness by scanning Kelvin probe microscopy (SKPM). They found that the surface potential of GaN is greater than the energy difference of E_C to E_F inside GaN and decreases with increasing n-type doping of GaN, but the surface potential of $Al_{0.35}Ga_{0.65}N$/GaN decreases with the increase in AlGaN layer thickness (completely opposite to the surface potential change illustrated by the model of surface donor state ionization as shown in Figure 4.2). So it is proposed the presence of negatively charged acceptor-like states at the GaN surface and in the AlGaN barrier layer. Based on this model, the acceptor-like states are negatively charged (or occupied by electrons) when the AlGaN layer of AlGaN/GaN heterojunction is thin, so the surface potential is high and no 2DEG is present. As the AlGaN layer thickness increases, the acceptor-like states release electrons and the surface potential is lowered, and so the 2DEG emerges and grows in density resulting from both the accumulation of electron released from the acceptor-like states and the weakening of the surface depletion effect. Current collapse of AlGaN/GaN HEMTs also proves the existence on the AlGaN surface of native acceptor states that can trap electrons, and the trapping/release of electrons has notable influence on the surface potential, which is responsible for the so-called virtual gate effect.

Therefore, the variation of 2DEG density and surface potential of AlGaN/GaN heterostructure with AlGaN content and thickness can be attributed to the co-work of the surface state and the polarization effect.

4.1.2 Sheet Electron Concentration of 2DEGs in GaN-Based Heterostructures

As shown in Figure 4.3, suppose the AlGaN barrier layer of the AlGaN/GaN heterostructure is of thickness d and relative dielectric constant $\varepsilon(x)$, composed of the doped AlGaN layer (of doping concentration N_D), and an unintentionally doped spacer of thickness d_i. The $Al_xGa_{1-x}N$ surface potential can be approximated by the Schottky barrier height $e\Phi_b(x)$ formed by the metal–semiconductor junction of Ni at the AlGaN surface, and this potential barrier may vary with the Al content x of AlGaN, that is,

$$e\phi_b(x) = (1.3x + 0.84)\ \text{eV} \tag{4.1}$$

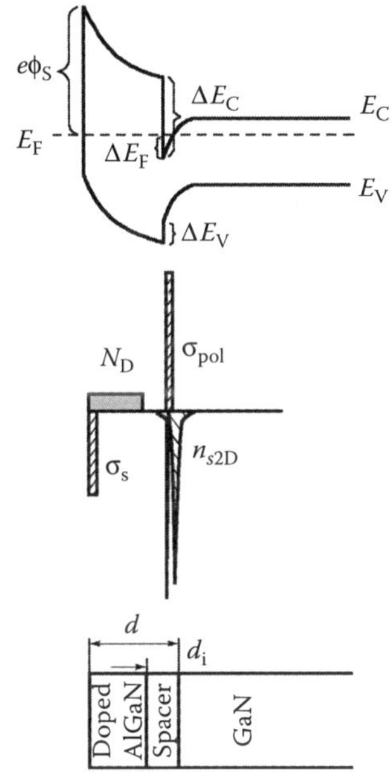

Figure 4.3 Energy band structure and charge distribution of AlGaN/GaN heterostructure.

From the electrostatic analysis of the heterostructure, we have (Delagebeaudeuf and Linh 1982)

$$n_{s2D} = \frac{\sigma_{pol}(x)}{e} - \frac{\varepsilon_0\varepsilon(x)}{de^2}[e\phi_b(x) + \Delta E_F - \Delta E_C(x)] + \frac{N_D(d - d_i)^2}{(2d)} \tag{4.2}$$

where:

$\sigma_{pol}(x)$ is the polar sheet charge at the AlGaN/GaN interface
e is the elementary charge
ΔE_F is the energy difference between the Fermi level and the bottom of the GaN side quantum well

The conduction band offset at AlGaN/GaN interface is $\Delta E_C(x) = 0.7\Delta E_g$, where ΔE_g is the band gap difference between AlGaN and GaN.

ΔE_F and n_{s2D} in Equation 4.2 have the following relation, in which the relation between the ground state E_0 and n_{s2D} are obtained by modifying the expression of the ground state in the triangular well approximation

$$\Delta E_F = E_0 + \frac{\pi\hbar^2}{m^*} n_{s2D}, \quad E_0 = \left(\frac{9\pi\hbar}{8\sqrt{8m^*}} \frac{e^2}{\varepsilon_0\varepsilon_{GaN}} n_{s2D} \right)^{2/3} \tag{4.3}$$

where m^* is the effective mass of electron in the quantum well.

In an actual AlGaN/GaN heterostructure, a thin AlN layer is usually inserted at the heterointerface to form the AlGaN/AlN/GaN heterostructure so as to improve

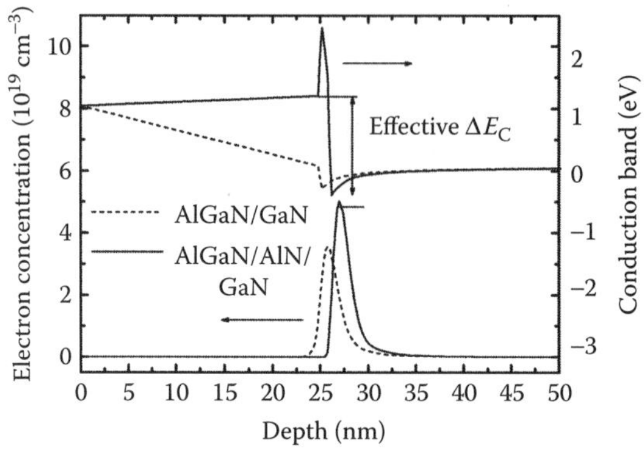

Figure 4.4 Conduction band profile and electron density distribution of AlGaN/GaN and AlGaN/AlN/GaN heterostructures. (After Zhang, J.-F. et al., *Chin. Phys.*, 15, 1060–1066, 2006.)

the material quality and electrical properties. In solving n_{s2D} for this material, the magnifying effect of the AlN interlayer on the effective conduction band offset $\Delta E_{C,\text{eff}}$ (see Figure 4.4) should be considered, and the difference between the dielectric constants of AlGaN and AlN can be neglected for simplicity, and then Equation 4.2 can be rewritten as (Gonschorek et al. 2008)

$$n_{s2D} = \frac{\sigma_{\text{AlGaN}} \cdot d}{e(d + d_{\text{AlN}})} - \frac{\varepsilon_0 \varepsilon(x)}{e^2(d + d_{\text{AlN}})} \times [e\phi_b(x) + \Delta E_F - \Delta E_{C,\text{eff}}] + \frac{N_D(d - d_i)^2}{(2d)} \tag{4.4}$$

$$\Delta E_{C,\text{eff}} = \Delta E_{C,\text{AlN/GaN}} - \Delta E_{C,\text{AlGaN/AlN}}(x) + \frac{e\sigma_{\text{AlN}} \cdot d_{\text{AlN}}}{\varepsilon_0 \varepsilon(x)} \tag{4.5}$$

where the signs d, $\varepsilon(x)$, $e\Phi_b(x)$, ΔE_F, N_D, and d_i have the same meaning as those in Equation 4.2; σ_{AlGaN} and σ_{AlN} are the polar charge densities at the AlGaN/GaN and AlN/GaN interfaces; $\Delta E_{C,\text{AlN/GaN}}$ and $\Delta E_{C,\text{AlGaN/AlN}}(x)$ are the conduction band offsets (note that both take positive values) at the AlN/GaN and AlGaN/AlN interfaces; and d_{AlN} is the AlN interlayer thickness, respectively.

The addition of a GaN cap layer on top of the surface of AlGaN/GaN heterostructure to form a GaN/AlGaN/GaN heterostructure is also a common practice to improve material and electrical properties. The introduction of the GaN cap layer raises the effective barrier height $e\Phi_{b,\text{eff}}$ of the AlGaN layer. For an unintentionally doped GaN layer of thickness d_{GaN} and a relative dielectric constant ε_{GaN}, when the structural parameters and doping of AlGaN layer are unchanged, Equation 4.2 needs to be rewritten for n_{s2D} solution as (Yu et al. 1999)

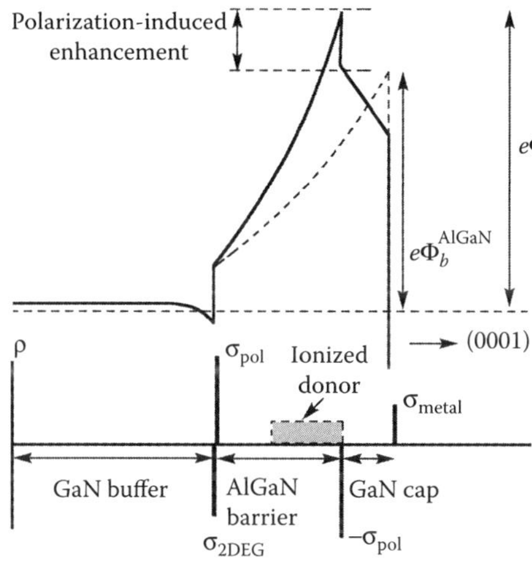

Figure 4.5 Band structure and charge distribution of GaN/AlGaN/GaN heterostructure. (a) Changed shape of the conduction band edge, showing an increased effective barrier height induced by the negative polar sheet charge between the GaN cap and AlGaN. (b) Charge distribution. (After Yu, E.T. et al., *J. Vac. Sci. Technol.*, B, 17, 1742–1749, 1999.)

$$n_{s2\mathrm{D}} = \frac{\begin{array}{c}\sigma_{\mathrm{pol}}(x) - (\varepsilon_0 \varepsilon(x)/d)\left(\phi_b^{\mathrm{GaN}} + \Delta E_F/e\right) + \\ eN_D(d-d_i)^2/(2d) + (\varepsilon(x)/\varepsilon_{\mathrm{GaN}})(d_{\mathrm{GaN}}/d)eN_D(d-d_i)\end{array}}{e[1 + (\varepsilon(x)/\varepsilon_{\mathrm{GaN}})(d_{\mathrm{GaN}}/d)]} \tag{4.6}$$

$$\phi_{b,\,\mathrm{eff}} = \frac{\Delta E_{\mathrm{C}}}{e} + \phi_b^{\mathrm{GaN}} + \frac{ed_{\mathrm{GaN}}}{\varepsilon_0 \varepsilon_{\mathrm{GaN}}}[n_{s2\mathrm{D}} - N_D(d-d_i)] \tag{4.7}$$

Please refer to the note in Figure 4.5 for the doping- and thickness-relevant parameters.

4.2 One-Dimensional Quantum Effect Self-Consistent Solution for Distribution of Conduction Band and Charge in GaN-Based Heterostructures

For semiconductor heterostructures with 2DEGs, the accurate theoretical analysis of the 2DEG quantization properties such as the subband levels and wave function distribution requires numerical calculation of the energy band and charge distribution of the heterostructure along the direction the quantum effect occurs. Since the energy band can be found by solving the Poisson equation and the 2DEG energy level and wave function by solving the Schrödinger equation, a self-consistent solution of the coupled equation of the two gives the distribution of both conduction band and charge at one time.

Tan et al. presented a detailed investigation of the mathematical and physical models and numerical methods of such one-dimensional self-consistent solution

(Tan et al. 1990), and developed a calculation program. This program applies not only to the GaAs heterostructures and the GaN-based heterostructures with polarization effect, but also to user-defined new materials. However, as far as the electrons in the heterostructure materials are concerned, this program considers only the 2DEG, no bulk electrons considered. Based on this model, we establish a one-dimensional self-consistent solution model with both 2DEG and bulk electrons in consideration (Zhang et al. 2006) and with little increase in the algorithm complexity, which is more practical for actual heterostructures.

4.2.1 *Physical Models of One-Dimensional Self-Consistent Solution of Schrödinger–Poisson Equations*

The quantum confinement of electrons by the potential wells formed by the conduction band edge at the heterointerface can be described by the Schrödinger equation in the effective mass approximation as (Tan et al. 1990)

$$-\frac{\hbar^2}{2m^*}\frac{d^2}{dz^2}\psi_k(z)+E_C(z)\psi_k(z)=E_k\psi_k(z) \tag{4.8}$$

where:

m^* is the electron effective mass
$\hbar$ is the reduced Planck constant
E_k and Ψ_k are eigenenergy and wave function of the 2DEG, respectively

The conduction band edge can be expressed as $E_C(z) = -eV(z) + \Delta E_C(z)$, where $V(z)$ is the electrostatic potential and ΔE_C is the conduction band offset ($\Delta E_C = 0.7\Delta E_g$ is typically chosen for the GaN material systems).

The sheet electron concentration n_{s2D} and spatial distribution $n_{2D}(z)$ of the 2DEG are

$$n_{s2D}=\sum_k n_k=\sum_k \frac{m^* k_B T}{\pi\hbar^2}\ln\left[1+\exp\left(\frac{E_F-E_k}{k_B T}\right)\right] \tag{4.9}$$

$$n_{2D}(z)=\sum_k n_k\psi_k^*(z)\psi_k(z) \tag{4.10}$$

where:

E_F is the Fermi level
k_B is the Boltzmann constant
T is the temperature

The eigen wave functions satisfy $\int\psi^*_k(z)\,\psi_k(z)dz = 1$, $k = 0, 1, 2 \ldots$

The solution of the conduction band edge by the charge distribution is described by the Poisson equation:

$$\frac{d}{dz}\left[\varepsilon_0\varepsilon(z)\frac{d}{dz}V(z)\right]=-\rho(z)$$
$$\rho(z)=e\left[N_D^+(z)-n_{2D}(z)-n_{3D}(z)\right] \tag{4.11}$$

where:

ε_0 is the dielectric permittivity of vacuum
$\varepsilon(z)$ is the static dielectric constant
$\rho(z)$ is the sum of all charges

The ionized impurity concentration is

$$N_D^+(z)=\frac{N_D(z)}{1+2\exp[(E_F-E_C(z)+E_d)/(k_BT)]} \tag{4.12}$$

where E_d is the ionization energy of the donor impurity, typically taking the value 20–40 meV for GaN.

The electrons consist of the two-dimensional electrons and the bulk electrons whose density obeys the Fermi distribution:

$$n_{3D}(z)=N_CF_{1/2}\left(\frac{E_F-E_C(z)}{k_BT}\right) \tag{4.13}$$

$$F_{1/2}(\xi)=\int_0^\infty\frac{x^{1/2}}{1+e^{x-\xi}}dx \tag{4.14}$$

where:

N_C is the conduction band effective density of states
$F_{1/2}(\xi)$ is the Fermi integral of order 1/2

The two-dimensional electrons and bulk electrons are discriminated by their energies according to a natural criteria: if two neighboring eigenenergy levels obtained from solving the Schrödinger equation have an energy separation smaller than k_BT, the lower level is still regarded as a discrete energy level of two-dimensional electron while the upper as the lower limit of the bulk electron energy.

4.2.2 Numerical Algorithm for the Model of One-Dimensional Self Consistent Solution of Schrödinger–Poisson Equations

The one-dimensional self-consistent solution of Schrödinger–Poisson equations boils down mathematically to the problem of boundary values of the partial differential equations, which requires solution by iterative method. Suppose the heterostructure is discretized perpendicular to the heterointerface into a nonuniform mesh, and the grids at the spatial positions with drastically changing electric field or charge density such as nearby the heterointerface should be set dense, and the relatively sparse grids be set for the other places so as to shorten the computation time while maintaining the accuracy. The mesh size should change gradually in space to reduce the calculation error.

Assuming the position of the i_{th} mesh point is z_i (i = 1, 2, ..., n) and the spacing between points z_i and z_{i+1} is h_i, for the i_{th} point after the discretization of the Schrödinger equation we have

$$-\frac{\hbar^2}{2}\left(\frac{2(\psi_{i+1}-\psi_i)}{m^*_{i+1/2}h_i(h_i+h_{i-1})}-\frac{2(\psi_i-\psi_{i-1})}{m^*_{i-1/2}h_{i-1}(h_i+h_{i-1})}\right)+E_{Ci}\psi_i=\lambda\psi_i \tag{4.15}$$

that is, $\sum_{j=1}^{n} A_{ij}\psi_j=\lambda\psi_i$, where

$$A_{ij}=\begin{cases}-\dfrac{\hbar^2}{2}\dfrac{2}{m^*_{i+1/2}h_i(h_i+h_{i-1})} & j=i+1\\ -\dfrac{\hbar^2}{2}\dfrac{2}{m^*_{i-1/2}h_{i-1}(h_i+h_{i-1})} & j=i-1\\ -A_{ii+1}-A_{ii-1}+E_{Ci} & j=i\\ 0 \quad \text{other cases} & \end{cases} \tag{4.16}$$

The half-integer index here implies the location midway between two adjacent grid points.

Thus solving the Schrödinger equation turns into the problem of the eigenvalue of the coefficient matrix A with its eigenvalue λ being the eigenenergy E_k(k = 0, 1, 2, ..., n) and the eigenfunction Ψ_k being the corresponding wave function. The boundary conditions are $\Psi(0) = 0$ (i.e., $\Psi_1 = 0$) and $\Psi(z_{max}) = 0$ (i.e., $\Psi_n = 0$).

The Poisson equation can be solved by Newton's method, that is, substituting into the equation the assumed initial value of the physical quantity (potential V) to be found and then solving repeatedly for the variation (δV) of this physical quantity to achieve convergence. It follows from the properties of the above self-consistent model that the electron concentration is a function of the potential $V(z)$. It is relatively easy to represent the δV-induced change in bulk electron concentration δn_{3D} by Newton's method, but the relation of δV to the change in the two-dimensional electron concentration δn_{2D} needs to be expressed as follows (according to the nondegenerate perturbation theory),

$$\delta n_{2D}[V]=\sum_k(\psi^*_k\psi_k)\frac{m^*}{\pi\hbar^2}\frac{1}{1+\exp((E_k-E_F)/k_BT)}\langle\psi_k\,|\,e\delta V\,|\,\psi_k\rangle \tag{4.17}$$

Thus the Poisson equation is transformed into

$$-\left[\frac{d}{dz}\varepsilon\varepsilon_0\frac{d}{dz}V+e(N_D{}^{+}[V]-n_{2D}[V]-n_{3D}[V])\right]$$
$$=\left[\begin{array}{l}\dfrac{d}{dz}\left(\varepsilon\varepsilon_0\dfrac{d}{dz}\right)+e\left(\dfrac{\delta N_D{}^{+}[V]}{\delta V}-\dfrac{\delta n_{3D}[V]}{\delta V}\right)\\ -\dfrac{e^2m^*}{\pi\hbar^2}\displaystyle\sum_k|\psi_k^*\psi_k|\dfrac{1}{1+\exp((E_k-E_F)/(k_BT))}\end{array}\right]\delta V \tag{4.18}$$

The term on the left-hand side of Equation 4.18 gives the error of the previous potential solution V_{old} substituting into the Poisson equation at an iterative step, and the potential variation δV can be found by using this equation, and then the new potential solution $V_{\text{new}} = V_{\text{old}} + \delta V$ is solved. Repeated iteration is performed until δV approaches 0 to converge.

Defining

$$C_{ij}=\begin{cases}\dfrac{2\varepsilon_{i+1/2}\varepsilon_0}{h_i(h_i+h_{i-1})} & j=i+1\\ \dfrac{2\varepsilon_{i-1/2}\varepsilon_0}{h_{i-1}(h_i+h_{i-1})} & j=i-1\\ -C_{ii+1}-C_{ii-1} & j=i\\ 0 & \text{other cases}\end{cases} \tag{4.19}$$

for the i_{th} mesh point, the Poisson equation (Equation 4.18) can be discretized as

$$-\left[\sum_{j=1}^{n}C_{ij}V_j+e\left(N_{Di}^{+}-n_{2Di}-n_{3Di}\right)\right]$$
$$=C_{ii+1}\delta V_{i+1}+C_{ii-1}\delta V_{i-1} \tag{4.20}$$
$$+\left[C_{ii}+e\left(N_D^{+'}-n_{3D}'\right)-\frac{e^2m_i^*}{\pi\hbar^2}\sum_k|\psi_{ki}^*\psi_{ki}|\frac{1}{1+\exp((E_k-E_F)/k_BT)}\right]\delta V_i$$

In this way, the Poisson equation is reduced to an equation of $Cx = b$ whose matrix H is a tridiagonal matrix. The typical boundary conditions are a constant V_1 at heterojunction surface and $(V_n - V_{n-1})/h_{n-1} = 0$ at the bottom, that is, as for the AlGaN/GaN heterostructures, both AlGaN surface potential and electric field at the bottom of GaN being given.

The self-consistent solution consists of two major steps. An initial solution of the potential/energy band and the carrier distribution is found out without considering the quantum effect after inputting the heterojunction structure and material parameters, and this initial solution is then used in the self-consistent iterations of the Schrödinger equation and the Poisson equation to obtain the final solution. The first-step solution is relatively simple. An initial charge distribution is assumed according to the doping of the heterostructure and substituted into the Poisson equation to determine the potential and the conduction band, thus obtaining a new charge distribution, after which iterations are performed by Newton's method to solve for the variation of potentials until convergence is achieved. In the second-step solution, the solution obtained from the first step is used as the initial solution, and the conduction band is substituted into the Schrödinger equation to solve for the new distribution of electron concentrations, which is substituted into the Poisson equation for the potential variation to obtain the new energy band, which is then substituted into the Schrödinger equation again, and so on until convergence. The merit of such solution procedures is that the first-step calculation offers a rather accurate initial solution for the second step, which matters a lot for the convergence of Newton's method. If no quantum effect arises in the considered structure, it can be observed from the solution from the first step. The convergence criterion is that the maximum potential difference between two successive iterations should be less than a constant value (e.g., 0.001 V in the first step, and $10^{-5}\cdot(k_BT)/e$ in the second), and this value should be properly adjusted in cases where convergence is difficult to achieve, such as at low temperature. Taking the $Al_{0.27}Ga_{0.73}N$/GaN heterostructure with a background doping of 1×10^{15} cm^{-3} and a 24 nm barrier layer as an example, the first- and second-step solutions are shown in Figure 4.6.

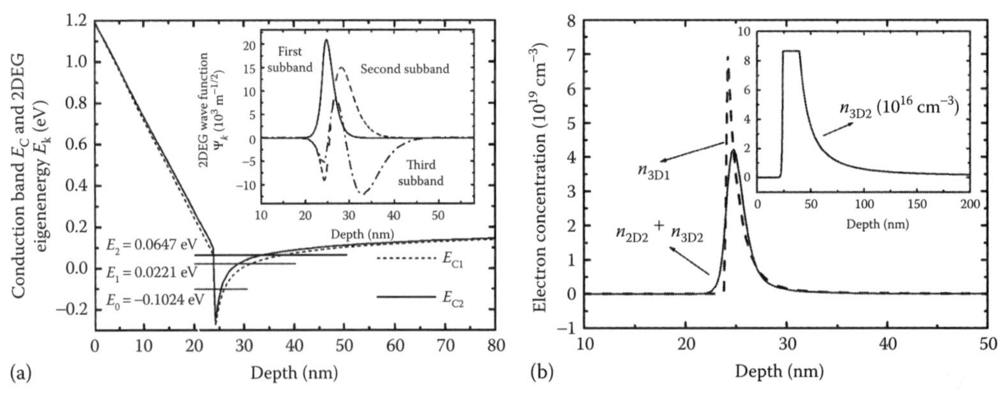

Figure 4.6 Initial solution without considering quantum effect and self-consistent solution taking quantum effect into consideration. In (a) E_{C1} is the initial solution and E_{C2} the self-consistent solution. The inset is the normalized 2DEG wave function of the self-consistent solution. In (b) n_{3D1} is the initial solution, and n_{2D2} and n_{3D2} are the self-consistent solutions.

4.2.3 Application of One-Dimensional Quantum Effect Self-Consistent Solution in GaN-Based Heterostructures

For GaN-based heterostructures, the subband structure and electron distribution of the 2DEG are significantly affected by the polarization effect.

Taking $Al_{0.25}Ga_{0.75}N$(30 nm)/GaN heterostructre as an example, we suppose that the $Al_{0.25}Ga_{0.75}N$ barrier layer is composed from surface to bottom of a 25 nm Si-doped layer ($N_D = 10^{18}$ cm^{-3}) and a 5 nm spacer layer, and there is an unintentional n-type doping of 10^{15} cm^{-3} for the whole structure. 2DEG densities are calculated using Equations 4.2 and 4.3. For $e\Phi_{ch} \approx 0$ and the polar charge at AlGaN/GaN heterointerface $\sigma_{pol} = 0$, the 2DEGs will be totally depleted when $e\Phi_s$ is higher than about 0.95 eV. According to this result, we take 0.4 eV (Figure 4.7a) and 1.2 eV (Figure 4.7b) for the surface potentials (so as to consider two cases of the heterostructre with different surface state densities), and it can be observed that in the absence of the polarization effect, there is still a tiny amount of 2DEG in the former while neither quantum well nor 2DEG is formed in the latter, and that the depletion region caused by high surface potential not only covers the AlGaN layer, but also enter into the depth of GaN. Whereas with polarization effect in consideration, the polar charge density at the $Al_{0.25}Ga_{0.75}N$/GaN interface is $\sigma_{pol} = 1.2 \times 10^{13}$ cm^{-2}, resulting in the formation of a quantum well under the 1.2 eV surface potential and a deeper quantum well under the 0.4 eV surface potential; on the other hand, regardless of surface potentials, the difference between the total sheet electron concentration (obtained from the integration of n_t, i.e., the sum of 2DEG and bulk electron, on the whole structure) with and without polarization effect into account is indeed about 1.2×10^{13} cm^{-2}.

Since the polarization field can modulate energy bands and charge distribution, the layered structure of the heterojunction can be adjusted to optimize the quantum wells and potential barriers and corresponding carrier distributions, which is referred to as the polarization engineering of GaN-based heterostructures. A major application of

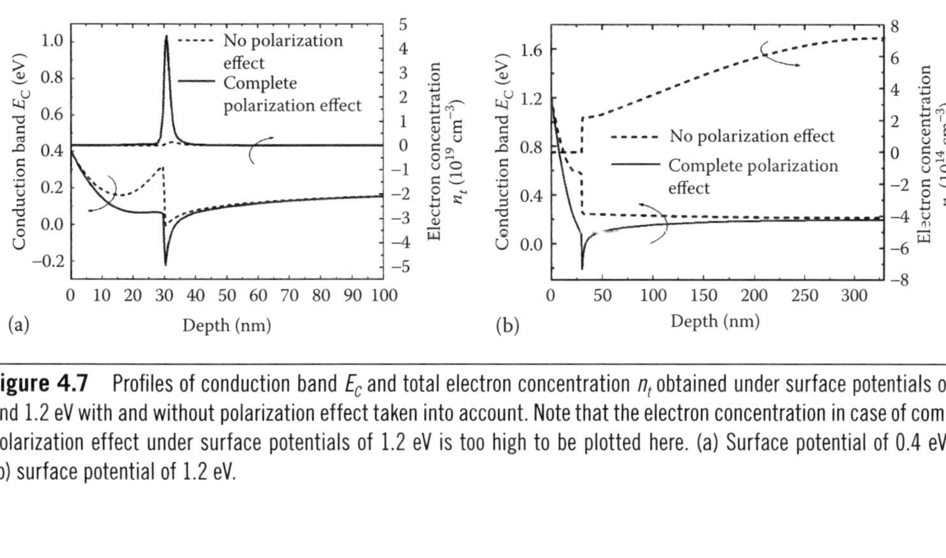

Figure 4.7 Profiles of conduction band E_C and total electron concentration n_t obtained under surface potentials of 0.4 and 1.2 eV with and without polarization effect taken into account. Note that the electron concentration in case of complete polarization effect under surface potentials of 1.2 eV is too high to be plotted here. (a) Surface potential of 0.4 eV and (b) surface potential of 1.2 eV.

the one-dimensional quantum effect self-consistent solution is the polarization engineering of GaN-based heterostructures, which is very helpful in the study of GaN-based multiple heterostructure materials.

Inserting a thin layer of AlN about 1 nm thick between AlGaN and GaN is a widely adopted optimization technique for AlGaN/GaN heterostructure materials. Under polarization effect, this interlayer improves the effective conduction band offset between the AlGaN barrier layer and the GaN channel layer (Figure 4.4) favorable for deeper potential wells and better 2DEG confinement. Theoretical calculations considering both two-dimensional electrons and bulk electrons and experimental investigation (Zhang et al. 2006) show that in case of relatively large variation in the doping concentration of the barrier layer and ambient temperature, the AlN interlayer has the additional benefit of maintaining stable quantum properties of 2DEG and suppressing the thermally activated parallel conduction outside the channel.

The samples used in theoretical investigation include an $Al_{0.25}Ga_{0.75}N$(30 nm)/GaN heterostructure and an $Al_{0.25}Ga_{0.75}N$(30 nm)/AlN(1 nm)/GaN heterostructure, both with a donor density $N_D = 5 \times 10^{18}$ cm^{-3} within 25 nm from the top of AlGaN. Defining the equivalent sheet density of bulk electrons as $n_{s3\mathrm{D}} = \int n_{3\mathrm{D}}(z)dz$ and the total sheet electron concentration in each sample as $n_{st} = n_{s2\mathrm{D}} + n_{s3\mathrm{D}}$. For a barrier layer of thickness d, two ratios of the sheet electron densities are defined as

$$r_{2\mathrm{D}} = \int_0^d \frac{n_{2\mathrm{D}}(z)dz}{n_{s2\mathrm{D}}} \qquad r_t = \int_0^d \frac{n_t(z)dz}{n_{st}} \tag{4.21}$$

Both $r_{2\mathrm{D}}$ and r_t can reflect the ratio of electrons in barrier layer to those in the whole structure with the former for pure 2DEG and the latter for both 2DEG and bulk electrons.

With the temperature dependence of the band gap and the lattice constant included in the calculation model, the calculated results at 50–500 K for the two structures are shown in Figures 4.8 and 4.9, respectively. The temperature dependence of ΔE_F (defined in Equation 4.2) and the 2DEG eigenenergy is shown in Figure 4.8.

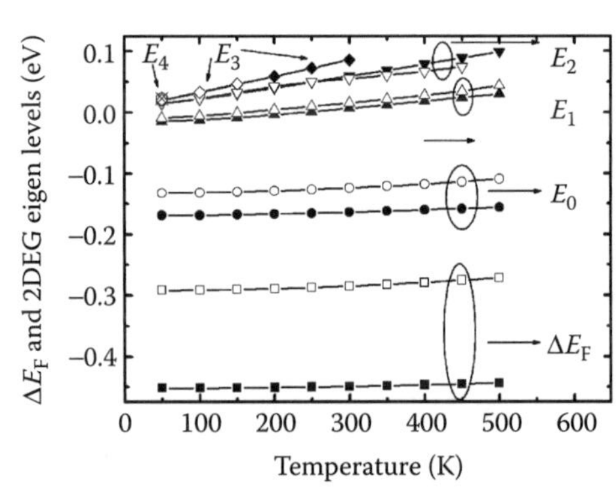

Figure 4.8 ΔE_F and E_i for the two structures as a function of temperature T. The hollow (solid) symbol line stands for AlGaN/GaN (AlGaN/AlN/GaN) structure in this figure and the next three figures. (After Zhang, J.-F. et al., *Chin. Phys.*, 15, 1060–1066, 2006.)

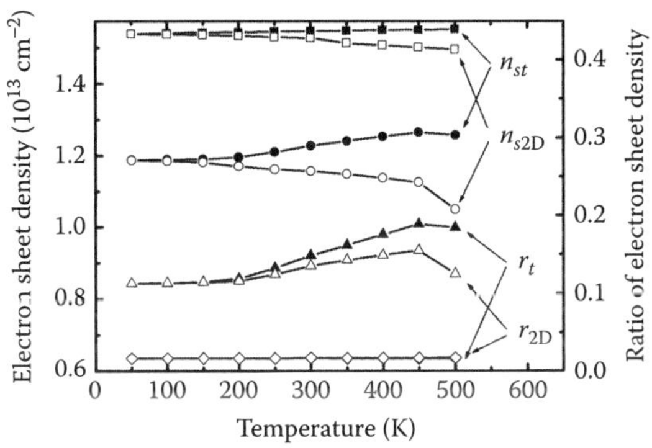

Figure 4.9 n_{st} and n_{s2D}, and r_{2D} and r_t for the two structures as a function of temperature *T*. (After Zhang, J.-F. et al., *Chin. Phys.*, 15, 1060–1066, 2006.)

The temperature dependence of the electron density ratios r_{2D} and r_t and the electron densities n_{st} and n_{s2D} is given in Figure 4.9. As the temperature rises, the reduced band gap results in a smaller ΔE_C and shallower quantum well at the heterointerface (as shown by the variation of ΔE_F in Figure 4.8), hence decreasing the 2DEG concentration, a trend that can be alleviated by the increase in lattice expansion-induced strain. The bulk electron concentration increases as a whole with increasing temperature since N_C is directly proportional to $T^{2/3}$. Thus n_{st} increases with temperature. However, the 2DEGs in the two structures are different in stability as the temperature rises.

As shown in Figure 4.9, the difference between n_{st} and n_{s2D} reflects the proportion of bulk electrons in the electron system of the whole structure while the difference between the electron density ratios r_{2D} and r_t demonstrates the proportion of bulk electrons in the barrier layer. Both differences increase with temperature, suggesting the thermal activation of the bulk electrons. It can be found by an analysis of both Figures 4.8 and 4.9 that the 2DEG eigenenergy levels of the AlGaN/GaN structure (without AlN interlayer) reduce from 5 at 50 K to 2 at 500 K, showing the gradual conversion from the 2DEG into bulk electrons with rising temperature, but the mechanisms of the change in the number of energy levels are not the same at different temperatures. At 50–200 K, the reduction in the number of 2DEG levels is due primarily to the shallower quantum well. At 200 K bulk electrons begin to arise in the AlGaN layer, hence the difference between r_{2D} and r_t. Increasingly significant thermal excitation of bulk electrons is observed as the temperature rises. At 200–450 K, smaller ΔE_C and the shallower quantum well lead to increased 2DEG penetration into the barrier layer and larger r_{2D} while the bulk electron thermal activation causes larger difference between r_{2D} and r_t. At 450–500 K, the number of 2DEG energy levels reduces from 3 to 2, leading to an apparent decrease in n_{s2D} and r_{2D} as well as greater proportion of bulk electron conductions in the barrier layer. For AlGaN/AlN/GaN structure, in comparison, the number of 2DEG eigenenergy levels changes only at temperatures above 300 K from 4 to 3 with r_{2D} and r_t always less than 2% in the considered temperature range.

It can be concluded from the above analysis that the thermal activation of the bulk electrons in the AlGaN/GaN heterostructure without AlN interlayer has much greater impact than in the heterostructure with AlN interlayer. The magnitude of and difference between r_{2D} and r_t of the former structure show larger 2DEG penetration into the barrier layer and easier thermal activation of the parallel bulk conduction in the barrier layer. Therefore, the electron system of the latter structure has stronger two-dimensional properties and 2DEG stability in the considered range of temperature. The temperature dependence of the *C–V* carrier profiles (Figures 4.10 and 4.11) agrees with the calculations.

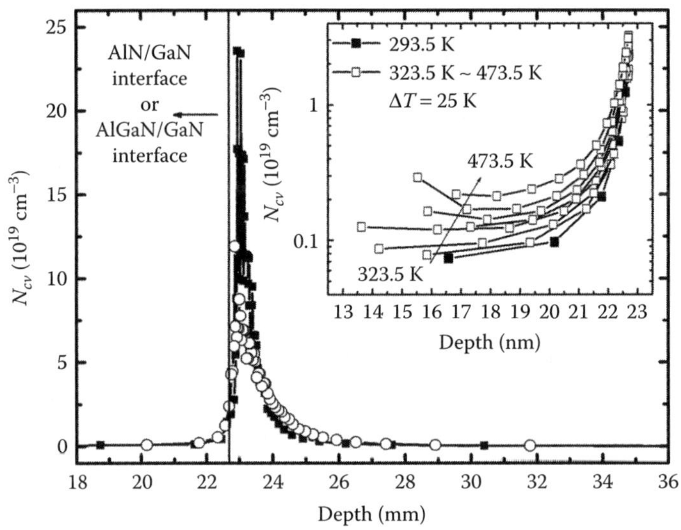

Figure 4.10 *C–V* carrier concentration N_{CV} profiles at 293.5 K of two diodes fabricated on AlGaN/GaN heterostructures with and without AlN interlayer, respectively. The heterointerface is at the depth of 22.7 nm. The inset shows the temperature-dependent electron distribution of the barrier layer in the structure without AlN interlayer. (After Zhang, J.-F. et al., *Chin. Phys.*, 15, 1060–1066, 2006.)

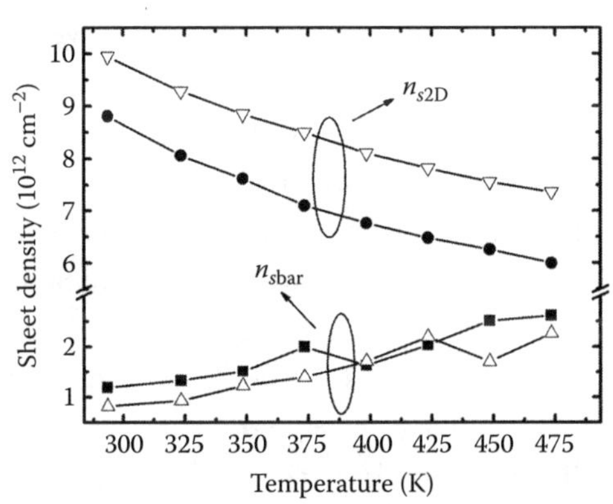

Figure 4.11 Temperature-dependent sheet electron densities obtained from the integration of N_{CV} profiles, where n_{s2D} is obtained by integrating from 22.7 to 26 nm, and n_{sbar} by integrating from the depth at which carriers can be detected in the barrier layer to 22.7 nm. (After Zhang, J.-F. et al., *Chin. Phys.*, 15, 1060–1066, 2006.)

4.3 Analysis of 2DEG Low-Field Mobility for GaN-Based Heterostructures by Analytical Model

In the low-field mobility model of bulk electrons, the major parameters influencing the mobility are the doping concentration and the temperature. The 2DEG is confined in spatial distribution by the quantum effect, and thus influenced by very different scattering mechanisms from those working on the bulk electrons. In heterostructures utilizing the 2DEG as channels, the channel layers are generally undoped with as low a background ionized impurity concentration as possible to raise the mobility; the steepness and uniformity of the heterointerfaces are controlled in growth to decrease the interface roughness (IFR) scattering; and the penetration of the 2DEG into the barrier layer is reduced by structure optimization to decrease the alloy disorder (ADO) scattering. Compared with bulk electrons, the 2DEG has more concentrated distribution, higher densities, and better screening of various scattering mechanisms. Therefore, the 2DEG mobility depends not only on doping and temperature, but also on the 2DEG electron density and the specific structure of the heterostructure.

For a simple analysis of the influencing factors of 2DEG mobility for GaN-based heterostructures, an analytical model of the 2DEG low-field mobility needs to be introduced.

4.3.1 Analytical Model of 2DEG Low-Field Mobility of GaN-Based Heterostructures

According to the Matheissen's rule, the 2DEG mobility is related to the mobilities limited by various scattering mechanisms as follows

$$\mu^{-1} = \sum_i \mu^{-1}{}_i \tag{4.22}$$

In the momentum relaxation approximation, the component mobility μ_i satisfies

$$\mu_i = \frac{e\tau_i}{m^*} \tag{4.23}$$

where τ_i is the momentum relaxation time, and m^* is the electron effective mass. The 2DEG in nitride heterostructures generally has quite high a density and populates mainly the first subband with the lowest energy, so all the scattering mechanisms considered adopt the model based on 2D degenerate statistics in single subband approximation to find the momentum relaxation rate ($1/\tau_i$ for the i_{th} scattering mechanism).

In the analytical model of the momentum relaxation rate, the spatial distribution of 2DEG requires approximation, and the Hartree–Fork wave function of the ground state for GaN-based heterostructures is approximated by the analytical Fang–Howard variational wave function (Davies 1998)

$$\psi(z) = \begin{cases} 0 & z < 0 \\ \sqrt{\frac{b^3}{2}} z \exp\left(-\frac{bz}{2}\right) & z \geq 0 \end{cases} \tag{4.24}$$

where b is a variational parameter, and for the 2DEG system energy to be the lowest its value should be

$$b = \left(\frac{33m^* e^2 n_{s2\mathrm{D}}}{8\hbar^2 \varepsilon_0 \varepsilon_s} \right)^{1/3} \tag{4.25}$$

The distance of the wave function centroid from the heterointerface is

$$z_0 = \int_0^\infty z |\psi(z)|^2 \, dz = \frac{3}{b} \tag{4.26}$$

In calculating the scattering probability, the Fang–Howard variational wave function elicits two form factors (let k be the wave vector), which are expressed as

$$F(k) = \eta^3 = \left(\frac{b}{(b+k)} \right)^3, \quad G(k) = \frac{1}{8}(2\eta^3 + 3\eta^2 + 3\eta) \tag{4.27}$$

If the 2DEG spatial distribution has a thickness reduced to nearly 0, η as well as the form factors $F(k)$ and $G(k)$ will approach unity.

The typical scattering mechanisms are the acoustic deformation-potential (DP) scattering, the piezoelectric (PE) scattering, the POL scattering, the ADO scattering, the IFR scattering, the dislocation (DIS) scattering, and the remote modulation doping (MD) scattering. The ADO scattering and the remote MD scattering are related to the properties of the barrier layer, and the IFR scattering is related to the heterointerface properties. The material parameters employed in other scattering mechanisms are only related to the semiconductor in the channel layer such as GaN. Using the Fang–Howard variational wave function one can obtain the momentum relaxation rates of these mechanisms or corresponding mobilities (by substituting the momentum relaxation rate into Equation 4.23) as follows.

The acoustic DP scattering is expressed as

$$\frac{1}{\langle \tau_{\mathrm{DP}} \rangle} = \frac{3m^* a_c^2 k_B T}{16\rho v_s^2 \hbar^3} b \tag{4.28}$$

where a_c is the acoustics DP, ρ is the GaN mass density, and v_s is the sound velocity.

The PE scattering is expressed as

$$\frac{1}{\tau_{\mathrm{PE}}} = \frac{e^2 M^2 k_B T m^*}{4\pi\varepsilon_0 \varepsilon_s \hbar^3 k_F^3} \int_0^{2k_F} \frac{F(q)\, q^3}{(q + q_{\mathrm{TF}} F(q))^2 \sqrt{1 - (q/(2k_F))^2}} dq \tag{4.29}$$

where:

M^2 is the electromechanical coupling coefficient
ε_s is the static dielectric constant of GaN
k_F is the Fermi wave vector
n_{s2D} is the 2DEG sheet density

Suppose the scattering occurs mostly in the vicinity of the Fermi surface, the difference of the wave vectors between the initial state and the final state in scattering satisfies

$$q = 2k_F \sin\left(\frac{\theta}{2}\right), \quad k_F = \sqrt{2\pi n_{s2D}}, \quad \theta \in (0, \pi) \tag{4.30}$$

The Thomas–Fermi wave vector q_{TF} reflects the screening length of 2DEG:

$$q_{TF} = \frac{m^* e^2}{(2\pi\varepsilon_0\varepsilon_s\hbar^2)} \tag{4.31}$$

The polar optical phonon (POL) scattering is expressed as

$$\frac{1}{\langle \tau_{POL} \rangle} = \frac{e^2 \omega_{POL} m^* N_B(T) G(k_o)}{2\varepsilon^* k_o \hbar^2 P_{POL}(y)} \tag{4.32}$$

where:

$\varepsilon^* = \varepsilon_0 / (1/\varepsilon_h - 1/\varepsilon_s)$ is defined with ε_h being the high frequency dielectric constant of GaN
$\hbar\omega_{POL}$ = the POL energy
$k_o = \sqrt{2m^*(\hbar\omega_{POL})/\hbar^2}$ is the POL wave vector
$N_B(T) = 1/(\exp[\hbar\omega_{POL}/(k_B T)] - 1)$ is the Bose–Einstein distribution function
$P_{POL}(y) = 1 + (1 - e^{-y})/y$ is with a dimensionless variable $y = \pi\hbar^2 n_{s2D}/(m^* k_B T)$

The ADO scattering is expressed as

$$\frac{1}{\langle \tau_{ADO} \rangle} = \frac{m^* \Omega(x)(V_A - V_B)^2 x(1-x)}{\hbar^3} \times \frac{\kappa_b P_b^2}{2} \tag{4.33}$$

where $\Omega(x)$ is the volume of the unit cell of the alloy $A_xB_{1-x}N$ (e.g., $Al_xGa_{1-x}N$) related to the lattice constants $a_0(x)$ and $c_0(x)$ of $A_xB_{1-x}N$.

$$\Omega(x) = \frac{\sqrt{3}}{2} c_0(x) \cdot a_0^2(x) \tag{4.34}$$

In actual AlGaN/GaN heterostructures the lattice constant $a(x)$ is greater than $a_0(x)$ while $c(x)$ is less than $c_0(x)$ since the AlGaN layer is generally in tensile strain. $(V_A - V_B)$

is the alloy scattering potential resulting from the replacement of gallium atoms by aluminum atoms in the AlGaN barrier layer.

To take into account the penetration of wave functions into the barrier layer, one should use the modified Fang–Howard variational function

$$\psi(z)=\begin{cases} N_1\exp\left(\dfrac{\kappa_b z}{2}\right) & z<0 \\ N_2(z+z_0)\exp\left(\dfrac{-bz}{2}\right) & z\geq 0 \end{cases} \tag{4.35}$$

$$\kappa_b = 2\sqrt{\frac{2m^*\Delta E_C(x)}{\hbar^2}} \tag{4.36}$$

$$N_2=\sqrt{\frac{b^3}{2(1+bz_0+0.5b^2z_0{}^2(1+b/\kappa_b))}} \qquad N_1=N_2z_0 \tag{4.37}$$

Here, the distance of the wave function centroid from the heterointerface is $z_0 = 2/(b+\kappa_b m_B/m_A)$, and m_A and m_B are the electron effective masses of the potential barrier and the quantum well, respectively.

Thus, the probability for the electrons in the quantum well to penetrate into the potential barrier is

$$P_b=\frac{N_2{}^2z_0{}^2}{\kappa_b} \tag{4.38}$$

The IFR scattering is expressed as

$$\frac{1}{\langle\tau_{\text{IFR}}\rangle}=\frac{\Delta^2L^2e^4m^*}{2(\varepsilon_0\varepsilon_s)^2\hbar^3}\left(\frac{1}{2}n_{s2\text{D}}\right)^2\int_0^1\frac{u^4\exp\left(-k_F{}^2L^2u^2\right)}{[u+G(q)q_{\text{TF}}/(2k_F)]^2\sqrt{1-u^2}}du \tag{4.39}$$

where:

Δ is the root mean square (RMS) roughness height
L is the correlation length

They are usually combined to show the interfacial roughness. The variable $u = q/(2k_F)$ is used, where k_F is the Fermi wave vector, and the definition of the wave vector q can be found in Equation 4.30.

The DIS scattering (Jena et al. 2000) is expressed as

$$\frac{1}{\tau_{\text{DIS}}}=\frac{N_{\text{DIS}}m^*e^4f_{\text{DIS}}{}^2}{\hbar^3\varepsilon_0^2\varepsilon_s^2c_0^2(0)}\cdot\frac{1}{4\pi k_F{}^4}\int_0^1\frac{du}{[u+q_{\text{TF}}/(2k_F)]^2\sqrt{1-u^2}} \tag{4.40}$$

where the DIS sheet density N_{DIS} and the 2DEG sheet density n_{s2D} are in the unit of cm^{-2}, f_{DIS} is the occupancy of the states in the energy band gap introduced by the DISs and $u = q/(2k_F)$ is used for simplicity.

Suppose the distances between both boundaries of the modulation-doped part and the AlGaN/GaN heterointerface are d_1 and d_2, respectively, with $d_1 > d_2$ and under complete ionization, the remote MD scattering is expressed as

$$\frac{1}{\langle\tau_{MD}\rangle} = N_D \frac{m^*}{4\pi\hbar^3 k_F{}^3}\left(\frac{e^2}{2\varepsilon_0\varepsilon_s}\right)^2 \int_0^{2k_F} \frac{F(q)^2[\exp(-2qd_2)-\exp(-2qd_1)]q}{[q+q_{TF}G(q)]^2\sqrt{1-(q/(2k_F))^2}}dq \tag{4.41}$$

where N_D is the MD concentration.

The 2DEG total mobility can be calculated by substituting the mobilities of the above scattering mechanisms into Equation 4.22. The typical values for the material parameters used in the mobility calculation are listed in Table 4.1.

4.3.2 *Influence of Al Content on 2DEG Mobility in AlGaN/GaN Heterostructures*

The low-field mobility of the 2DEG in GaN-based heterostructures varies with the temperature, the doping of the structure, the 2DEG density, and the Al content and the thickness of the AlGaN barrier layer.

Miyoshi et al. reported the experimental data on the Hall electron concentration and mobility of different Al content AlGaN/GaN heterostructures on sapphire

Table 4.1 Parameters Used in Mobility Calculation

Electron effective masses of GaN and AlGaN (m_0 is the static electron mass)	$m^* = 0.2\ m_0$ $m_x^* = (0.2 + 0.2\ x)\ m_0$
Acoustics deformation potential of GaN (eV)	$a_C = 9.10$
GaN mass density (g/cm^3)	$\rho = 6.15$
Longitudinal acoustic phonon velocity of GaN (cm s^{-1})	$v_s = 8.00 \times 10^5$
Static and high frequency dielectric constants of GaN	$\varepsilon_s = 8.90$, $\varepsilon_h = 5.35$
Electromechanical coupling coefficient for GaN	$M^2 = 0.039$
POL energy for GaN (meV)	$\hbar\omega_{POL} = 91.2$
Lattice constant of $Al_xGa_{1-x}N$ (nm)	$a_0(x) = (0.3189 + 0.0077\ x)$ $c_0(x) = (0.5186 + 0.0204\ x)$
Alloy scattering potential of AlGaN (eV)	$V_A - V_B = 1.80$
AlGaN/GaN conduction band offset	$\Delta E_C(x) = 0.75(E_g(\text{AlGaN}) - E_g(\text{GaN}))$
Thomas–Fermi wavevector of GaN (m^{-1})	$q_{TF} = 8.4994 \times 10^8$
Dislocation filling factor of GaN	$f_{DIS} = 0.3$

Sources: Knap, W. et al., *Appl. Phys. Lett.*, 70(16), 2123–2123, 1997; Leung, K. et al., *Appl. Phys. Lett.*, 74(17), 2495–2497, 1999; Levinshtein, M.E. et al., *Properties of Advanced Semiconductor Materials GaN, AlN, InN, BN, SiC, SiGe*, 1st ed, John Wiley & Sons, New York, 2001; Zanato, D. et al., *Semiconduct. Sci. Technol.* 19(3), 427–432, 2004; After Zhang, J.-F. et al., *Sci. China Ser. F (Inform. Sci.)*, 51(6), 177–186, 2008.

substrate at 77 K and room temperature (Miyoshi et al. 2005). The AlGaN barrier layer is modulation doped with a DIS density of GaN template $N_{DIS} = 3 \times 10^9$ cm^{-2}. With the Al content x ranging from 0.16 to 0.42, the 2DEG sheet density increases nearly linearly from about 7×10^{12} cm^{-2} to approximately 1.7×10^{13} cm^{-2} at 77 K, and the whole curve is slightly higher at room temperature; the mobility decreases from ~6600 cm^2/(Vs) to ~2000 cm^2/(Vs) at 77 K and from ~1400 cm^2/(Vs) down to ~900 cm^2/(Vs) at room temperature. These data are analyzed by the mobility models in Section 4.3.1. A linear fitting of the $n_{s2D} \sim x$ curve measured at 77 K is performed, and the fitted data are substituted into the mobility model to calculate the mobilities for various scattering mechanisms (Zhang et al. 2008).

Shown in Figure 4.12 are the 2DEG mobility profile and the mobility profiles of various scattering mechanisms, including all the mechanisms introduced in Section 4.3.1. Both DIS scattering and remote MD scattering are coulomb, and they are weak at both 77 K and room temperature due to the greatly screened coulomb force under the quite high 2DEG density in the considered samples. At 77 K, lattice vibration scattering is relatively weak, involving the acoustic DP scattering, the PE scattering, and the POL scattering (with a mobility of the order of 10^7 cm^2/(Vs), which is not given in Figure 4.12a). The decrease of 2DEG mobility with increasing Al content is governed primarily by the IFR scattering and the ADO scattering, and at a relatively high Al content it is the IFR scattering that has a dominant impact on the decrease in 2DEG mobility. The other scattering mechanisms exert certain influence on the 2DEG mobility at a relatively low Al content.

At room temperature (Figure 4.12b), the lattice vibration-related scattering mechanisms are significantly intensified with the POL scattering as the dominating mechanism while other scattering mechanisms are not temperature-dependent. The mobilities due to the acoustic DP scattering and the PE scattering are proportional to T^{-1}, whereas the mobility due to the POL scattering decreases nearly exponentially with temperature. The large POL energy of GaN (91.2 meV)

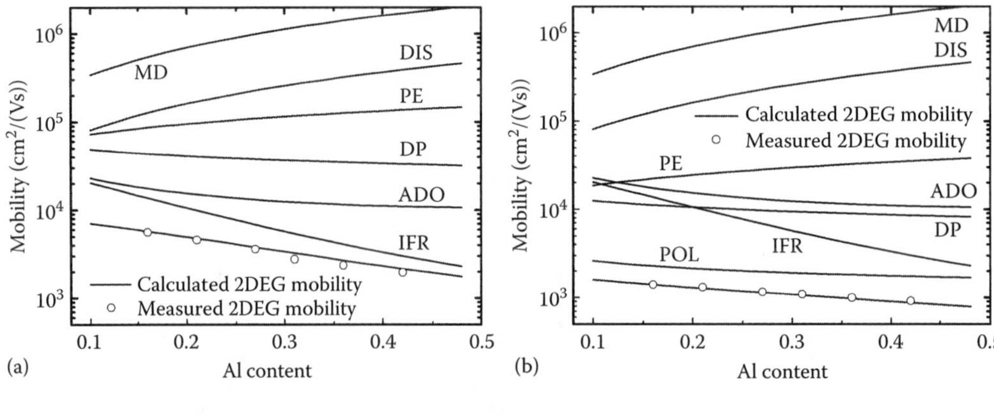

Figure 4.12 Measured 2DEG mobility and the calculated 2DEG mobility and component mobilities of the individual scattering mechanisms as a function of Al content x in the barrier layer for AlGaN/GaN heterostructures at 77 K and room temperature. (a) 77 K and (b) room temperature. (After Zhang, J.-F. et al., *Sci. China Ser. F* [*Inform. Sci.*], 51, 177–186, 2008.)

strengthens the scattering further. So the magnitude and of 2DEG mobility at room temperature and its decrease with x is mainly determined by the POL scattering and the IFR scattering. The former dominates in the considered x range, and the latter is also important at high x.

The strength of every scattering process changes with the Al content at given temperature, though none of the scattering mechanisms except the ADO scattering is relevant to x according to the models of their momentum relaxation rates. So the changes of the scattering strength are caused by the change of density and spatial extension of the 2DEG as a result of increased polarization effect when the Al content in the AlGaN/GaN structure increases.

The remote MD scattering is a kind of coulomb scattering, and the ionized donor ions as the scattering centers are in the doping layer with the spacer layer between it and the 2DEG. When the donor density is constant and n_{s2D} increases, the screening effect of the electrons on the columbic force is stronger and the scattering effect is weaker, so the mobility increases. The DIS scattering is also columbic and the scattering center is the charged DIS lines composed of the suspended bonds, and the line charge density is $e f_{DIS}/c_0(0)$, so the corresponding mobility also increases with n_{s2D}.

The mobility due to acoustic DP scattering is proportional to $n_{s2D}^{-1/3}$, so it decreases with n_{s2D}. The mobility limited by PE scattering is a complicated function of n_{s2D}, and it increases in the involved n_{s2D} range. The mobility limited by POL scattering decreases almost exponentially with n_{s2D}.

The ADO scattering originates from the effect of the randomly fluctuating aluminum and gallium atom potential in the AlGaN material on the penetration part of the 2DEG wave function into the barrier layer. When x increases, on one hand the ADO extent in AlGaN (proportional to $x(1-x)$) changes and reaches a maximum point at $x = 0.5$, and on the other hand, the energy barrier height at the AlGaN/GaN hetero-interface increases as well as 2DEG density. As Figure 4.13 shows, in the AlGaN/GaN heterostructures with increasing x while constant n_{s2D}, for $x \leq 0.5$, the scattering

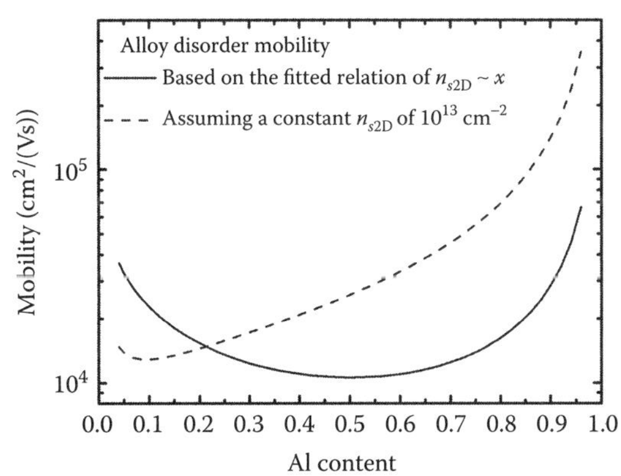

Figure 4.13 The mobility due to ADO scattering versus Al content x. (Solid line: n_{s2D} increases with x; dotted line: $n_{s2D} = 10^{13}$ cm^{-2}.) (After Zhang, J.-F. et al., *Sci. China Ser. F* [*Inform. Sci.*], 51, 177–186, 2008.)

is strengthened only at small x because of the more disordered III-group atom distribution, but weakened when x is larger as a result of the weaker 2DEG penetration caused by the higher energy barrier at heterointerface. For $x \geq 0.5$, both the even weaker 2DEG penetration and the less disordered alloy atom distribution weaken the scattering further. As for the AlGaN/GaN heterostructures with the 2DEG naturally changing its density with x, the mobility reflects the interplay of the above mechanisms, and the dominant factor changes from the increase of n_{s2D} to that of x at the bottom of the mobility curve.

The IFR scattering is very sensitive to n_{s2D}. The heterointerface is a region where the semiconductor composition changes in several atomic layers, and the irregularity emerging in this change leads to the macroscopically "rough" heterointerface and scatters the electrons which transport along the interface. When n_{s2D} increases, the 2DEG electron distribution shifts closer to the AlGaN/GaN heterointerface and the electron transport becomes more sensitive to the IFR, so the IFR scattering is stronger and the mobility is lower. Note that when calculating the IFR mobility in Figure 4.12, an assumption that the AlGaN/GaN IFR increases with x must be made, or the fast decrease of the IFR mobility with x cannot be fully explained. The assumption is indirectly supported by the atomic force microscopic pictures of the surface morphology of the samples. When the Al content of AlGaN/GaN structure increases, the surface becomes rougher, and this is attributed to the increasing strain in the AlGaN layer (Miyoshi et al. 2005). The impact of the strain could also cause the roughening of AlGaN/GaN interfaces. Since the IFR parameters (Δ and L in Equation 4.39) take empirical values (typically $\Delta \approx 0.3$ nm, $L = 1.0$–3.0 nm) but cannot be definitely solved, a single value of IFR mobility may correspond to many possible (Δ, L) values. But we propose in this case the strain accumulation from the bottom to the top of the AlGaN layer (along the growth direction) should lead to a rougher surface than the AlGaN/GaN interface, which is different from the analysis of Miyoshi et al. (2005).

It is found from the above analysis that the decrease of 2DEG mobility with Al content in AlGaN/GaN heterostructures is mainly determined at 77 K by the IFR scattering and the ADO scattering, and at room temperature mainly by the POL scattering and the IFR scattering. This gives a possible explanation under single subband approximation of 2DEG, but we won't exclude the possibilities such as the crystal quality degradation of AlGaN layer with higher Al content decreasing the 2DEG mobility or the working of multi-subband scattering in the 2DEG.

4.3.3 Transport Properties of 2DEGs in Lattice-Matched InAlN/GaN and InAlN/AlN/GaN Structures

The early lattice-matched InAlN/GaN structures generally has a comparatively low electron mobility (70–500 cm^2/[Vs]) at room temperature (Gonschorek et al. 2006), and its temperature-dependent Hall mobility often exhibits the bulk electron behavior (increasing to maximum and then dropping with decreasing temperature)

(Xie et al. 2007), with rare reports of 2DEG behavior (increasing and saturating with decreasing temperature). InAlN/AlN/GaN structure formed by inserting a thin AlN interlayer at the InAlN/GaN interface can obtain the 2DEG transport behavior and an electron mobility of 812–1510 cm^2/(Vs) at the same level with that of the AlGaN/GaN structure at room temperature (Gonschorek et al. 2006, Wang et al. 2011). The 2DEG density also has remarkable impact on mobility, and the reported electron densities for InAlN/GaN and InAlN/AlN/GaN are sparse (0.6–4.23 × 10^{13} cm^{-2}) (Katz et al. 2004, Wang et al. 2011). All this makes it difficult to analyze the transport properties of InAlN/GaN structure and the mechanism of improvement in InAlN/GaN electron mobility by the AlN interlayer.

We grow high quality lattice-matched InAlN/GaN and InAlN/AlN/GaN samples by pulsed metalorganic chemical vapor deposition (PMOCVD), and make a quantitative analysis of the 2DEG low-field mobility (Wang et al. 2011). InAlN/GaN and InAlN/AlN/GaN structures have a room temperature Hall mobility of 949 cm^2/(Vs) and 1437 cm^2/(Vs), respectively, and the temperature dependence of the Hall mobility showing the typical 2DEG behavior. The mobilities due to various scattering mechanisms and the final 2DEG mobility are calculated by substituting the parameters in Table 4.2 into the theoretical model given in Section 4.3.1, in which the ADO scattering is not considered for the InAlN/AlN/GaN sample and the remote MD scattering is absent for both samples. The calculated values of the 2DEG mobility for both InAlN/GaN and InAlN/AlN/GaN samples agree well with the measured data at temperatures ranging from 77 to 300 K, showing the fitness of the adopted model and parameters.

A comparison between Figure 4.14a and 4.14b shows that InAlN/AlN/GaN structure has significantly higher 2DEG mobility than InAlN/GaN structure, which can be explained by the competition of three mechanisms. First, the AlN interlayer deepens the quantum well in GaN layer and increases the 2DEG density, resulting in weaker DIS scattering and stronger lattice vibration-related scatterings and also stronger IFR scattering in case of invariable IFR, hence a slight decrease in the 2DEG

Table 4.2 Model Parameters Used in the Calculation

PARAMETERS	InAlN/GaN	InAlN/AlN/GaN
2DEG density (cm^{-2})	1.650×10^{13}	1.750×10^{13}
IFR (nm)	0.432	0.245
Correlation length (nm)	1.500	1.500
ADO scattering potential (eV)	3.848	
Dislocation density (cm^{-2})	8.869×10^{0}	
Electron effective mass m_x^* of InAlN	$(0.400 - 0.330x)m_0$ (m_0 is the electron rest mass)	
Band gap of InAlN	$6.130(1 - x) + 0.626\,x - 5.400\,x(1 - x)$	
Lattice constant of InAlN	$a(x) = 3.548x + 3.112(1 - x)$ $c(x) = 5.760x + 4.982(1 - x)$	

Source: Wang, P.-Y. et al., *Acta Phys. Sinica* 60(11), 117304 (6 pp.), 2011.

Note: The 2DEG density is the measured value, the dislocation density is estimated based on X-ray diffraction experimental data, and the ADO scattering potential assumed equaling the conduction band offset between AlN and InN, and the In content $x = 0.18$.

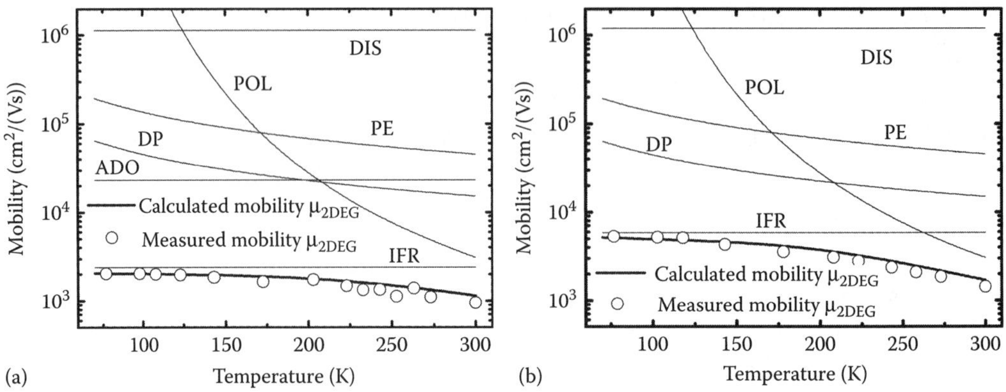

Figure 4.14 Temperature dependence of 2DEG mobility μ_{2DEG} and component mobilities due to the individual scattering mechanisms for InAlN/GaN and InAlN/AlN/GaN samples. (a) InAlN/GaN and (b) InAlN/AlN/GaN. (After Wang, P.-Y. et al., *Acta Phys. Sinica* 60, 117304, 6 pp., 2011.)

mobility of InAlN/AlN/GaN structure. Second, the AlN interlayer suppresses the ADO scattering, thereby increasing the 2DEG mobility, which is also an important mechanism in the 2DEG mobility improvement by the introduction of the AlN interlayer into conventional AlGaN/GaN structure. Third, the impact of the IFR scattering is notably weaker in InAlN/AlN/GaN sample than in InAlN/GaN sample with the IFR parameter dropping from 0.432 to 0.245 nm. Since both samples are grown under the same growth conditions except for the AlN interlayer, the difference in the IFR suggests it is the AlN interlayer that reduces the interface microfluctuation affecting 2DEG transport, thus a smoother interface and weaker IFR scattering. Consequently, the increase in the 2DEG mobility of InAlN/AlN/GaN sample indicates that the elimination of the ADO scattering and the suppression of the IFR scattering play a dominant role among the three AlN interlayer-induced changes.

The 2DEG density is another important influencing factor of the 2DEG mobility. Currently, the reported data on electron densities for InAlN/GaN and InAlN/AlN/GaN samples are rather dispersed owing to differences in InAlN barrier layer thicknesses, alloy compositions, growth systems, and growth conditions. The dependence of the 2DEG mobility on 2DEG density for InAlN/GaN and InAlN/AlN/GaN structures (In content $x = 0.18$) at room temperature is calculated for an analysis of how the 2DEG density influences the mobility. In view of the difference in crystal quality and interface condition of the actual samples, the DIS scattering is left out of account while retaining other scattering mechanisms for simplicity, and a series of IFR values are taken for the IFR scattering, and thus the resulting 2DEG mobility can be regarded as the theoretical upper limit of the actual mobility. The calculation results and experimental data are shown in Figure 4.15.

In Figure 4.15a, the experimental data of both the room temperature mobility and the electron density for InAlN/GaN samples are scattered. According to the reports on the temperature-dependent Hall characteristics, the samples with 2DEG transport behavior have higher mobilities (Katz et al. 2004), while the samples showing bulk

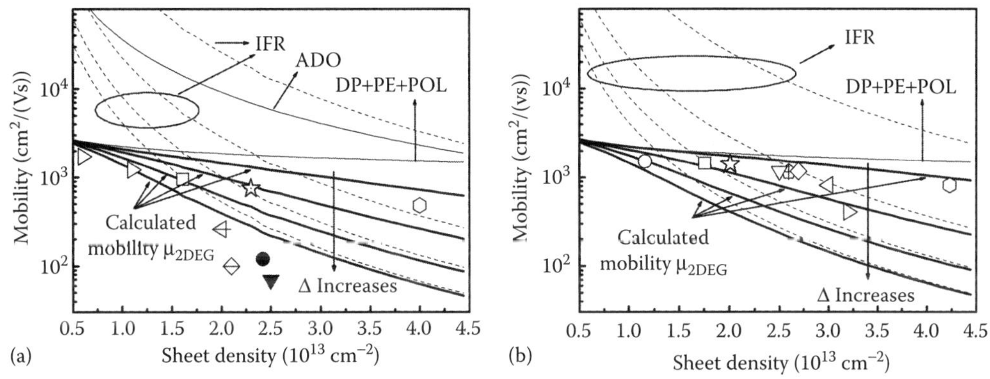

Figure 4.15 2DEG mobility μ_{2DEG} and component mobilities limited by the individual scattering mechanisms as a function of the electron sheet density for InAlN/GaN and InAlN/AlN/GaN structures. The IFR $\Delta = 0.1$, 0.3, 0.5, 0.7 nm (all with a correlation length of 1.5 nm) are chosen for the IFR scattering to form a family of mobility curves to consider the IFR dispersion of different samples, and the corresponding 2DEG mobilities μ_{2DEG} also form a family of curves. The solid and hollow symbols are the experimental data. (a) InAlN/GaN and (b) InAlN/AlN/GaN. (After Wang, P.-Y. et al., *Acta Phys. Sinica* 60, 117304, 6 pp., 2011.)

electron transport characteristics (the low temperature mobilities are even lower than the room temperature mobilities) have lower mobilities (Kuzmik et al. 2007). The experimental data of the room-temperature mobilities for InAlN/AlN/GaN samples are all relatively high (except for the early report by Dadgar et al. (2004), in which the sample was grown on the Si substrate, thus a low mobility). In terms of the effect of scattering mechanisms, no significant change in lattice vibration-related mobilities is observed with increasing 2DEG density, while the ADO scattering and the IFR scattering are remarkably enhanced owing to the shorter distance of the 2DEG to the heterointerface and the greater proportion of penetration into the alloy barrier layer. It can also be found from the distribution of the experimental data that the mobility data of the InAlN/GaN samples with 2DEG transport behavior are distributed widely on the calculated μ_{2DEG} curves with the IFR ranging from 0.1 to 0.7 nm, whereas almost all the mobility data of InAlN/AlN/GaN samples are in the close vicinity of the calculated μ_{2DEG} curves with an IFR of only 0.1 nm, a demonstration of the improvement in InAlN/GaN interface by the AlN interlayer.

In summary, the significant increase in the electron mobility of InAlN/AlN/GaN structure with respect to that of InAlN/GaN structure is due primarily to the fact that the AlN interlayer not only effectively suppresses the ADO scattering, but also improves interface smoothness and weakens the IFR scattering.

References

Dadgar, A., F. Schulze, J. Blasing, and A. Diez. 2004. High-sheet-charge–carrier-density Al In *N*/Ga *N* field-effect transistors on Si(111). *Applied Physics Letters* 85(22):5400–5402.

Davies, J. H. 1998. *The Physics of Low-Dimensional Semiconductors, 1st ed.* Cambridge, UK: Cambridge University Press.

Delagebeaudeuf, D. and N. T. Linh. 1982. Metal-(n) AlGaAs-GaAs two-dimensional electron gas fet. *IEEE Transactions on Electron Devices* ED-29(6):955–960.

Gonschorek, M., J. F. Carlin, E. Feltin, M. A. Py, and N. Grandjean. 2006. High electron mobility lattice-matched AlInN/GaN field-effect transistor heterostructures. *Applied Physics Letters* 89(6):062106. doi: 10.1063/1.2335390.

Gonschorek, M., J. F. Carlin, E. Feltin, M. A. Py, N. Grandjean, V. Darakchieva, B. Monemar, M. Lorenz, and G. Ramm. 2008. Two-dimensional electron gas density in Al1-xInxN/AlN/GaN heterostructures ($0.03 \leq x \leq 0.23$). *Journal of Applied Physics* 103(9):093714-1. doi: 10.1063/1.2917290.

Jena, D., A. C. Gossard, and U. K. Mishra. 2000. Dislocation scattering in a two-dimensional electron gas. *Applied Physics Letters* 76(13):1707–1709. doi: 10.1063/1.126143.

Katz, O., D. Mistele, B. Meyler, G. Bahir, and J. Salzman. 2004. InAlN/GaN heterostructure field-effect transistor DC and small-signal characteristics. *Electronics Letters* 40(20):1304–1305. doi: 10.1049/el:20045980.

Knap, W., S. Contreras, H. Alause, C. Skierbiszewski, J. Camassel, M. Dyakonov, J. L. Robert et al. 1997. Cyclotron resonance and quantum Hall effect studies of the two-dimensional electron gas confined at the GaN/AlGaN interface. *Applied Physics Letters* 70(16):2123–2123. doi: 10.1063/1.118967.

Koley, G. and M. G. Spencer. 2001. Surface potential measurements on GaN and AlGaN/GaN heterostructures by scanning Kelvin probe microscopy. *Journal of Applied Physics* 90(1):337–344. doi: 10.1063/1.1371941.

Kuzmik, J., J. F. Carlin, M. Gonschorek, A. Kostopoulos, G. Konstantinidis, G. Pozzovivo, S. Golka et al. 2007. Gate-lag and drain-lag effects in (GaN)/InAlN/GaN and InAlN/AlN/GaN HEMTs. *Physica Status Solidi A* 204(6):2019–2022. doi: 10.1002/pssa.200674707.

Leung, K., A. F. Wright, and E. B. Stechel. 1999. Charge accumulation at a threading edge dislocation in gallium nitride. *Applied Physics Letters* 74(17):2495–2497. doi: 10.1063/1.123018.

Levinshtein, M. E., S. L. Rumyantsev, and M. S. Shur. 2001. *Properties of Advanced Semiconductor Materials GaN, AlN, InN, BN, SiC, SiGe,* 1st ed. New York: John Wiley & Sons.

Miyoshi, M., T. Egawa, and H. Ishikawa. 2005. Structural characterization of strained AlGaN layers in different Al content AlGaN/GaN heterostructures and its effect on two-dimensional electron transport properties. *Journal of Vacuum Science and Technology B (Microelectronics and Nanometer Structures)* 23(4):1527–1531. doi: 10.1116/1.1993619.

Smorchkova, I. P., C. R. Elsass, J. P. Ibbetson, R. Vetury, B. Heying, P. Fini, E. Haus, S. P. DenBaars, J. S. Speck, and U. K. Mishra. 1999. Polarization-induced charge and electron mobility in AlGaN/GaN heterostructures grown by plasma-assisted molecular-beam epitaxy. *Journal of Applied Physics* 86(8):4520–4526. doi: 10.1063/1.371396.

Tan, I. H., G. L. Snider, L. D. Chang, and E. L. Hu. 1990. A self-consistent solution of Schrodinger-Poisson equations using a nonuniform mesh. *Journal of Applied Physics* 68(8):4071–4076. doi: 10.1063/1.346245.

Wang, P.-Y., J.-F. Zhang, J.-S. Xue, Y.-B. Zhou, J.-C. Zhang, and Y. Hao. 2011. Transport properties of two-dimensional electron gas in lattice-matched InAIN/GaN and InAIN/AIN/GaN materials. *Acta Physica Sinica* 60(11):117304 (6 pp.).

Xie, J., X. Ni, M. Wu, J. H. Leach, U. Ozgur, and H. Morkoc. 2007. High electron mobility in nearly lattice-matched AlInN/AlN/GaN heterostructure field effect transistors. *Applied Physics Letters* 91(13):1–3. doi: 10.1063/1.2794419.

Yu, E. T., X. Z. Dang, P. M. Asbeck, S. S. Lau, and G. J. Sullivan. 1999. Spontaneous and piezoelectric polarization effects in III-V nitride heterostructures. *Journal of Vacuum Science & Technology B* 17(4):1742–1749. doi: 10.1116/1.590818.

Zanato, D., S. Gokden, N. Balkan, B. K. Ridley, and W. J. Schaff. 2004. The effect of interface-roughness and dislocation scattering on low temperature mobility of 2D electron gas in GaN/AlGaN. *Semiconductor Science and Technology* 19(3):427–432. doi: 10.1088/0268-1242/19/3/024.

Zhang, J.-F. and Y. Hao. 2006. GaN-based heterostructures: Electric-static equilibrium and boundary conditions. *Chinese Physics* 15(10):2402–2406. doi: 10.1088/1009-1963/15/10/036.

Zhang, J.-F., Y. Hao, J.-C. Zhang, and J.-Y. Ni. 2008. The mobility of two-dimensional electron gas in AIGaN/GaN heterostructures with varied Al content. *Science in China Series F (Information Science)* 51(6):177–186. doi: 10.1007/s11432-008-0056-7.

Zhang, J.-F., C. Wang, J.-C. Zhang, and Y. Hao. 2006. Effects of donor density and temperature on electron systems in AlGaN/AlN/GaN and AlGaN/GaN structures. *Chinese Physics* 15(5):1060–1066. doi: 10.1088/1009-1963/15/5/032.

5

Growth and Optimization of AlGaN/GaN Heterostructures

The AlGaN/GaN heterostructure is the mainstream structure of GaN HEMTs. Based on a type I heterojunction band structure similar to that of AlGaAs/GaAs heterostructure, the AlGaN/GaN heterostructure has a two-dimensional electron gas (2DEG) at the GaN side. Whereas AlGaN/GaN heterostructure as a whole has a rather wide band gap and a high breakdown voltage; the large conduction band offset at the AlGaN/GaN interface, together with its strong piezoelectric and spontaneous polarization, is favorable for the formation of deep and narrow quantum wells and the accumulation of high density 2DEGs. The 2DEG possesses, owing to its distribution and transport characteristics, remarkably higher mobility and saturation velocity. Therefore the AlGaN/GaN heterostructure is an ideal candidate for microwave power devices (Figure 5.1).

The crystalline quality of the AlGaN/GaN heterostructure has an impact on its electrical properties, both of which are closely related to the layered structure and growth technique of the material. For the optimization of the electrical properties of the AlGaN/GaN heterostructure, an in-depth investigation of the structure and growing techniques of the AlGaN/GaN heterostructure is needed.

5.1 Layered Structure of the AlGaN/GaN Heterostructure

The early AlGaN/GaN heterostructure has only two layers above the substrate and the nucleation layer: the GaN buffer layer (also acting as the channel layer) and the AlGaN barrier layer. The key to the optimization of such heterostructures is the improvement of the crystalline quality of the layers, obtaining smooth surface and interface as well as improving the electrical properties. The GaN buffer layer has a typical thickness of 1 ~ 3 μm, and a high quality buffer layer requires:

1. *High crystalline quality or few epitaxial defects.* The heteroepitaxy of III-nitrides renders a mosaic structure of the grown thin films, thus high density extended defects (majorly threading dislocations), which affect surface and interface qualities as well as the crystalline quality of the alloy barrier layer. Moreover, the edge dislocations in the extended defects can introduce into the band gap deep levels as acceptor traps, and their charged dangling bonds and the strain field around the dislocations have a scattering effect on the 2DEG transport

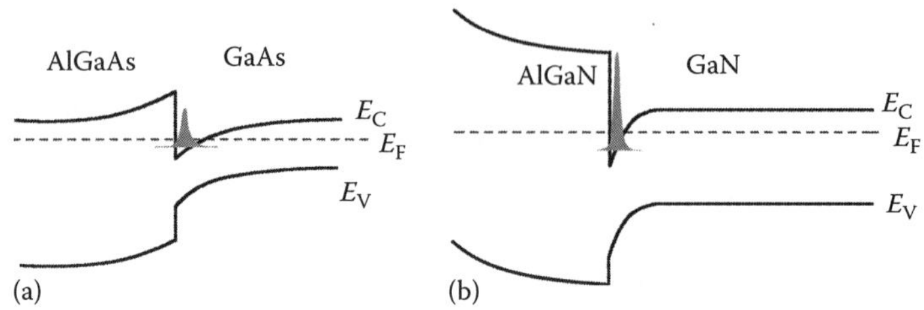

Figure 5.1 Band structures of (a) AlGaAs/GaAs and (b) AlGaN/GaN heterostructures. The gray shadow shows the 2DEG distribution.

along the heterointerface. The screw dislocations, electrically neutral as they are, may form a leakage path down through from the surface. Therefore, the buffer layer defect density should be reduced as much as possible.

2. *A high resistivity*. A high resistivity buffer layer can offer a smaller off-state leakage current, better subthreshold characteristics, a higher on/off ratio, and a higher breakdown voltage. Theoretically, GaN has very good high resistivity characteristics with an intrinsic carrier density of the order of 10^{-11} cm^{-3} at room temperature, but unintentionally doped GaN usually has a background n-type doping of the order of $10^{15} \sim 10^{17}$ cm^{-3}, primarily owing to the shallow donor impurities of V_N, silicon and oxygen, and so on. Therefore, a reduction of shallow donor impurities or an introduction of compensating acceptor doping in the buffer layer is required for achieving high resistivity characteristics. On the other hand, there generally exists a region of high n-type doping density (up to 10^{18} cm^{-3}) near the substrate in the GaN buffer layer grown on sapphire, thus forming a leakage path, referred to as the buried charge layer. It is formed by the out diffusion of oxygen element from the sapphire substrate in the high temperature growth due to incomplete covering of the nucleation layer over the substrate. So the elimination of this buried charge layer is also needed to obtain a high resistivity. Another method for growing high resistivity buffer layers is to introduce an AlGaN back barrier buffer layer with a wider band gap while maintaining the crystalline quality of the whole heterostructure.

As for the AlGaN barrier layer, an increase in its Al content or thickness leads to a higher 2DEG density but stronger tensile strain in AlGaN. Too strong a strain may induce a relaxation in the heterostructure, during which the stress is released in the form of emerging defects such as dislocations or even cracks, thus a significantly weakened piezoelectric effect. According to the experimental data reported by Ambacher et al., strain relaxation arises with Al content greater than 0.38 for the AlGaN barrier layer of a thickness of ~30 nm (Ambacher et al. 2000). The strain relaxation point of actual heterostructures varies with the growth system and growth conditions, and is usually observed by changed surface morphology, crystalline quality, and electrical properties. Apart from its influence on the defect density and surface morphology of the barrier layer and a significant drop in 2DEG density, strain relaxation also induces a lower 2DEG mobility, which may be explained by the increased interface roughness scattering due to the impact of strain relaxation on heterointerface smoothness. Besides the limit in the Al content and thickness of the AlGaN barrier layer imposed by strain relaxation, the AlGaN/GaN HEMT reliability study reveals that the inverse piezoelectric effect at high electrical field that strongly influence the device reliability is also related to the initial strain of the AlGaN layer.

Therefore, for a high electron concentration in the channel without invoking strain relaxation, the AlGaN barrier layer in the HEMT heterostructure has a typical Al content of 0.15 ~ 0.3 and thickness of 10 ~ 30 nm (an enhancement-mode HEMT may have even a smaller thickness), and a common practice is the introduction of an ~1 nm thick AlN interlayer at the AlGaN/GaN interface. Under the polarization effect, this interlayer can increase both the channel electron density and mobility as well as improve the heterointerface.

The surface morphology, microstructure, and surface condition of the GaN-based heterostructure also have considerable influence on the device processes and device reliability. If the surface of AlGaN barrier layer acts directly as the surface of HEMT structures, the easily oxidized aluminum may cause degradation of the heterostructure characteristics over time, an effect particularly significant in the AlGaN/GaN heterostructure with a relatively high Al content and decreased AlGaN crystalline quality. A thin GaN cap layer can be introduced on top of AlGaN to improve the heterostructure surface and the material properties in general.

The above optimizations of the growth and structure of the AlGaN/GaN heterojunction are summarized in Figure 5.2.

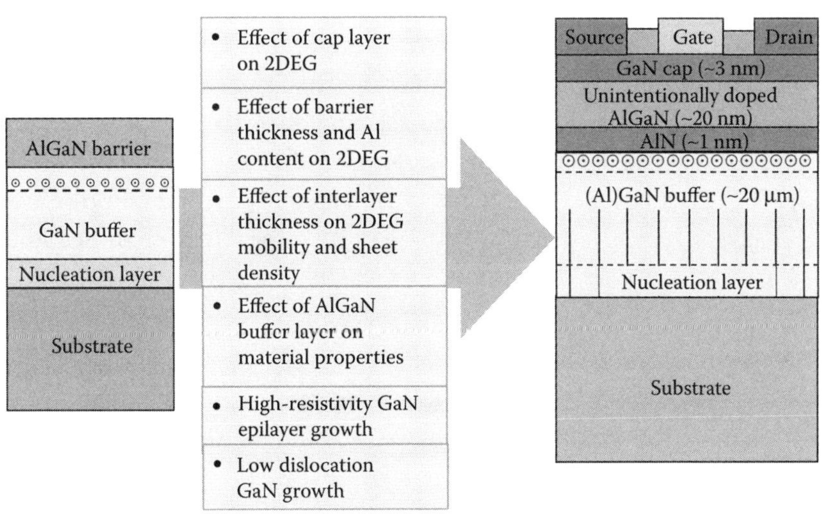

Figure 5.2 Major optimizations in growth and structure of an AlGaN/GaN heterojunction.

5.2 Growth Methods of Low-Defect Density Nitrides

In the heteroepitaxial growth of GaN films, the substrate surface treatment and nucleation layer optimization are requisite for the reduction of the high density dislocations (of the order of $10^9 \sim 10^{10}$ cm^{-3}) and residual stress in the epilayer induced by the lattice and thermal mismatch between the substrate and GaN so as to improve the GaN crystalline quality. An effective method for further decreasing the density of extended defects in epitaxial nitrides, the epitaxial lateral overgrowth (ELOG) can reduce the dislocation density of the epitaxial GaN on sapphire by 2–4 orders of magnitude to 10^6 cm^{-3} (Kato et al. 1994), and is becoming a major growth technique of low-defect density nitrides.

The implementation of ELOG is described as follows:

1. The conventional growth of the GaN template of a thickness of a few μm.
2. The growth of the mask such as SiO_2 or SiN and the exposure of GaN beneath the mask by etching a periodic array of windows (typically in the shape of stripes or hexagons).
3. The growth of GaN in selected regions under proper growth conditions. Since GaN has far greater nucleation energy on the mask surface than on GaN, the initial nucleation and growth occurs only in the unmasked areas (window regions) and the lateral growth initiates along with the vertical growth when GaN grows thicker than the mask.
4. The lateral overgrowth of GaN in neighboring windows till coalescing into a continuous GaN layer.

The successful lateral overgrowth coalescence (fast coalescence and smooth material surface after coalescence) depends heavily on the shape and position of the window regions (such as the ratio of the window width to the mask width and the orientation of the stripes) and epitaxial growth conditions (such as the V/III ratio and temperature). As shown in Figure 5.3a, the lateral growth rate of GaN is higher with the stripe-shaped window oriented in the $[1\bar{1}00]$ direction, the facet of GaN side walls changes gradually from $(1\bar{1}01)$ into $(11\bar{2}0)$ in the selective growth with a rectangular cross section of ELOG layers, and the thin film formed after the lateral coalescence grows along the [0001] orientation. If the window is in the $[11\bar{2}0]$ orientation (Figure 5.3b), the ELOG layers typically have a triangular or trapezoidal cross section (higher

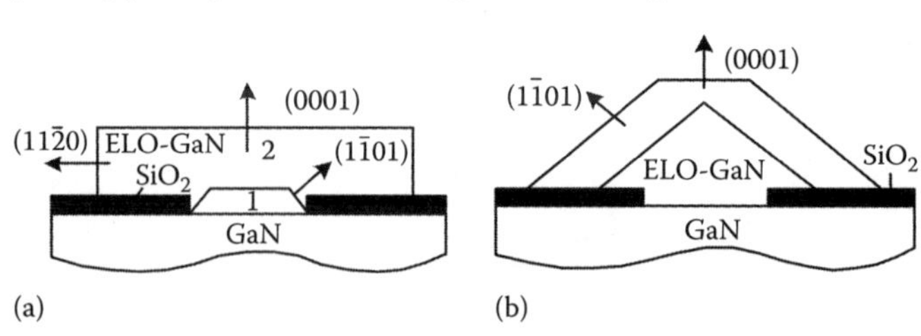

Figure 5.3 Schematic of GaN ELOG. The stripe-shaped window is in the orientation of (a) $[1\bar{1}00]$ and (b) $[11\bar{2}0]$.

vertical growth rate than lateral one) with $(1\bar{1}01)$ sidewalls and there are still high ridges on the surface of the resulting thin film even after lateral epitaxial coalescence.

ELOG can reduce the dislocation density in the GaN layer because on one hand the mask blocks the underlying extended dislocations, and on the other hand annihilation of dislocations occurs in the formation of closed dislocation loop due to dislocation bending in the lateral overgrowth.

Low-defect density GaN epitaxy can be achieved with simple ELOG processes. However, only the lateral grown part and coalescence region on top of the mask enjoys a reduced dislocation density while the region on top of the window still has a high dislocation density, thus unfavorable for the fabrication of large size devices. What's more, it is generally time-consuming, and unintentional doping or contamination is easily introduced during the mask deposition and secondary growth.

Based on ELOG, the patterned substrate (PS) epitaxy and porous interlayer technique were developed. A PS is produced by etching the substrate surface to form periodic micro/nanometer patterns of various shapes, including convex (columnar, pyramid, spherical, etc.) and concave structures as well as etch pits. In GaN epitaxy on PSs, the defect density can be remarkably reduced by controlling the lateral and vertical growth rate in a similar growth mechanism to ELOG while avoiding the unintentional contamination in ELOG since the growth is on the whole a single step process. The introduction of PSs to light emitting diodes (LEDs) also increases the luminescence efficiency. PSs have found wide uses in optoelectronic devices but comparatively less applications in electronic materials and devices.

The porous interlayer technique refers to the growth of low-defect density GaN by ELOG on the porous mask with nanometer holes, and the mask is formed by the chemical reaction between the reactant gas of nitride epitaxy (e.g., NH_3) and the deposited thin layer of foreign material on top of the surface of the GaN template. Porous interlayers are mainly of two types: the in situ grown SiN layers and the metal nitride interlayers formed by nitrification of some deposited metal.

In metalorganic chemical vapor deposition (MOCVD), the in situ growth of SiN interlayers can be performed by simultaneous inflows of SiH_4 and NH_3. The good capability of continuous film formation of SiN makes the coalescence of GaN rather difficult unless a porous structure with enough dense holes is formed by controlling the SiN interlayer to be very thin. In comparison to GaN grown without the SiN interlayer under the same MOCVD process conditions, the GaN material we obtained using the in situ SiN interlayer has an obviously smaller full width at half magnitude (FWHM) of XRD rocking curve. The (002) FWHM decreases from 300 to 220 arcsec and (102) FWHM from 800 to 198 arcsec, showing a significant decline in the dislocation density. Nevertheless, the silicon donor doping of GaN in the growth of SiN has an adverse impact on the application of in situ SiN interlayers in electronic devices.

As for metal nitride interlayers, we obtained low-defect density nonpolar a-plane GaN by using a TiN interlayer (Xu et al. 2011). Moram et al. (2007) reported that 30 min nitrification of a 5 nm thick titanium metal film resulted in a thin porous TiN film with relatively uniform and close-packed holes and a greater spacing than the size of the hole, a desirable candidate for the mask in ELOG. Titanium of varied thicknesses was adopted to grow the TiN films by nitrification on nonpolar a-plane GaN templates of the same quality in the MOCVD reaction chamber, and then a second a-plane GaN layer of the same thickness was grown on all the samples. The HRXRD and TEM measurements show the optimum titanium thickness is 10 nm (Figure 5.4). At a titanium thickness of 5 nm, an apparent improvement of the surface morphology is already observed (as shown in Figure 5.5) with greatly reduced dislocation pit density and in-plane anisotropy-induced surface roughness, and the RMS roughness of the $5 \times 5\ \mu m^2$ AFM images decreases from 5.73 nm (no TiN interlayer) to 0.594 nm. The cross section TEM image of the TiN interlayer using 10 nm titanium is shown in Figure 5.6. A significant blocking of the two major extended defects (dislocations and stacking faults) in nonpolar a-GaN is observed at the TiN interlayer, and the bending of the majority of the defects penetrating through the interlayer facilitates their annihilation. The TiN interlayer significantly reduces the defect density of the material above it. The FWHM values of $(11\bar{2}0)$ rocking curves for a-plane GaN reported

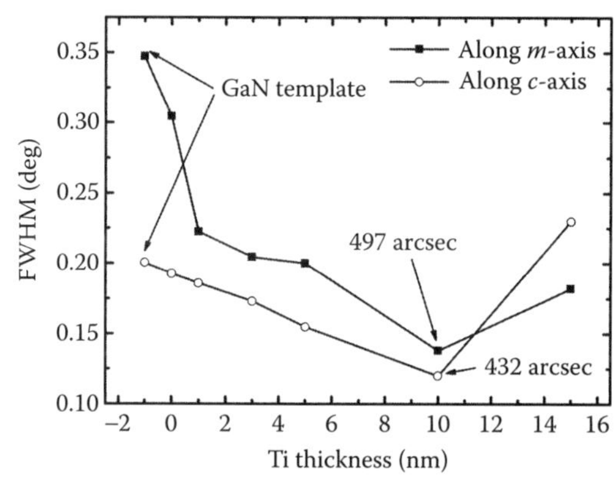

Figure 5.4 Measured FWHM of $(11\bar{2}0)$ rocking curves of a-plane GaN grown with TiN interlayer using titanium of varied thicknesses. (After Xu, S.R. et al., *J. Cryst. Growth*, 327, 94–97, 2011.)

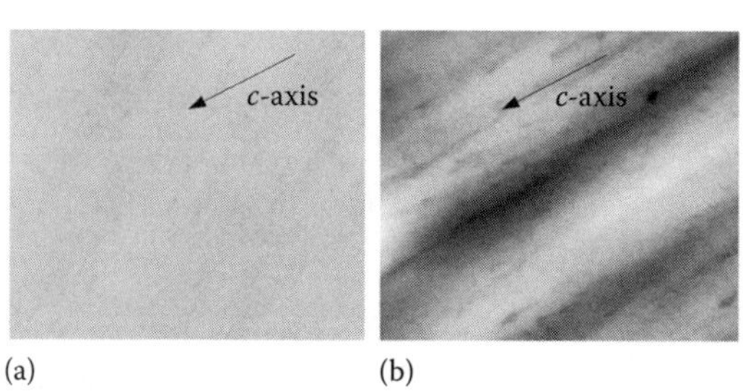

Figure 5.5 5 μm × 5 μm AFM surface images of a-plane GaN. (a) With TiN interlayer using 5 nm titanium and (b) without TiN interlayer. (After Xu, S.R. et al., *J. Cryst. Growth*, 327, 94–97, 2011.)

Figure 5.6 Cross-sectional TEM image of a-plane GaN grown with TiN interlayer using 10 nm titanium. The dotted line shows the position of the TiN interlayer. (a) $g = [11\bar{2}0]$, showing the dislocations. (b) $g = [1\bar{1}00]$, showing the stacking faults. (After Xu, S.R. et al., *J. Cryst. Growth,* 327, 94–97, 2011.)

Table 5.1 Comparison of XRD Data for Nonpolar a-Plane GaN

	XRD $(11\bar{2}0)$ PLANE ROCKING CURVE FWHM (arcsec)		
GROWTH TECHNIQUE	ALONG *c*-AXIS	ALONG *m*-AXIS	SOURCE OF DATA
SiN ELOG	669	1111	Huang et al. (2008)
In-situ SiN interlayer	936	1080	Chakraborty et al. (2006)
ELOG+flow modulation epitaxy (FME)	459	1062	Huang et al. (2008)
TiN interlayer formed on 10 nm titanium	432	497	Xu (2010)

Source: Courtesy of Dr. Shengrui Xu.

during the same time period are listed in Table 5.1. Since the TiN interlayer leads to very fast coalescence of GaN in ELOG, the repeated use of the interlayer technique can further improve the material quality.

5.3 Growth of Low-Defect GaN Buffer Layers on Vicinal Substrates

Besides ELOG and the PS epitaxy and porous interlayer technique based on the similar principles, the growth of GaN on vicinal substrates can also reduce the dislocation density of GaN and improve the material characteristics of AlGaN/GaN heterostructures without the need for extra processing in the growth.

We analyzed the impact of the misorientation angle of the substrate on the crystalline quality and electrical properties of AlGaN/GaN heterostructures on (0001) sapphire substrates (Xu et al. 2009), and found that both the (002) and (102) XRD rocking curve FWHMs of GaN decrease with increasing misorientation angles (0°–0.5°) whether toward a-plane or toward m-plane, indicating a reduction in the screw and edge dislocation densities as the misorientation increases (Figure 5.7). Better quality of GaN is obtained on the substrate with a miscut toward m-plane than that miscut toward a-plane. The strain in GaN films on substrates of varied misorientations was investigated by XRD 2θ-ω scan. It was found that in all the samples GaN is under in-plane compressive strain. The smaller in-plane compressive strain in

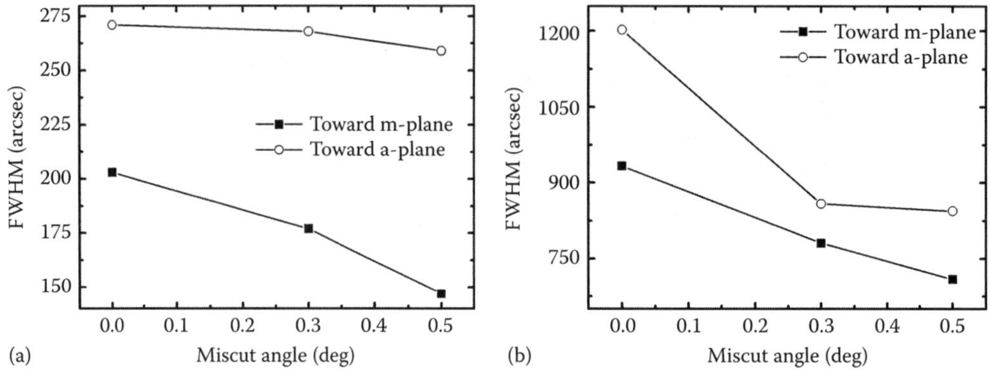

Figure 5.7 FWHM of XRD rocking curves of GaN on vicinal sapphire substrates as a function of the miscut angle. (a) FWHM of (002) XRD rocking curves and (b) FWHM of (102) XRD rocking curves. (Courtesy of Xinxiu Ou.)

Table 5.2 Electrical Properties of GaN-Based Heterostructure Samples on Miscut (0001) Sapphire Substrates

SAMPLE	SUBSTRATE CHARACTERISTICS	300 K MOBILITY (CM^2/[Vs])	77 K MOBILITY (CM^2/[Vs])	300 K 2DEG DENSITY ($\times 10^{13}$/CM^2)	77 K 2DEG DENSITY ($\times 10^{13}$/CM^2)
Ia	0.5° toward a-plane	1419	6106	1.49	1.35
Ib	0.3° toward a-plane	1277	5270	1.58	1.42
Ic	0.0° toward a-plane	1215	4294	1.5	1.36
IIa	0.5° toward m-plane	1406	6397	1.19	1.08
IIb	0.3° toward m-plane	1353	5980	1.21	1.12
IIc	0.0° toward m-plane	1305	4852	1.25	1.13

Source: Courtesy of Zhihao Xu.

the GaN films on vicinal substrates than that on conventional substrates indicates a good release of residual stress in GaN on the vicinal substrates. At identical miscut angles, the GaN films on the substrates miscut toward m-plane have smaller in-plane compressive strain, thus a better GaN quality.

The results of variable temperature Hall measurement (Table 5.2) of the heterostructure samples show that the carrier mobility increases with larger miscut angles of the substrate in both misorientations (toward a-plane and toward m-plane), and that the carrier mobilities rise faster with dropping temperature in the samples with greater miscut angles. The heterojunctions on the substrates miscut toward m-plane have slightly lower Hall electron concentrations and generally a higher Hall mobility. According to the AFM surface morphology measurements (Figure 5.8), the use of vicinal substrate brings obviously wider atomic steps at the growth surface, which hints the promotion of the step-flow mode of crystal growth. It is the very effect that results in a reduction of threading dislocations in the crystal, and an improvement in the electrical properties of the AlGaN/GaN heterostructure.

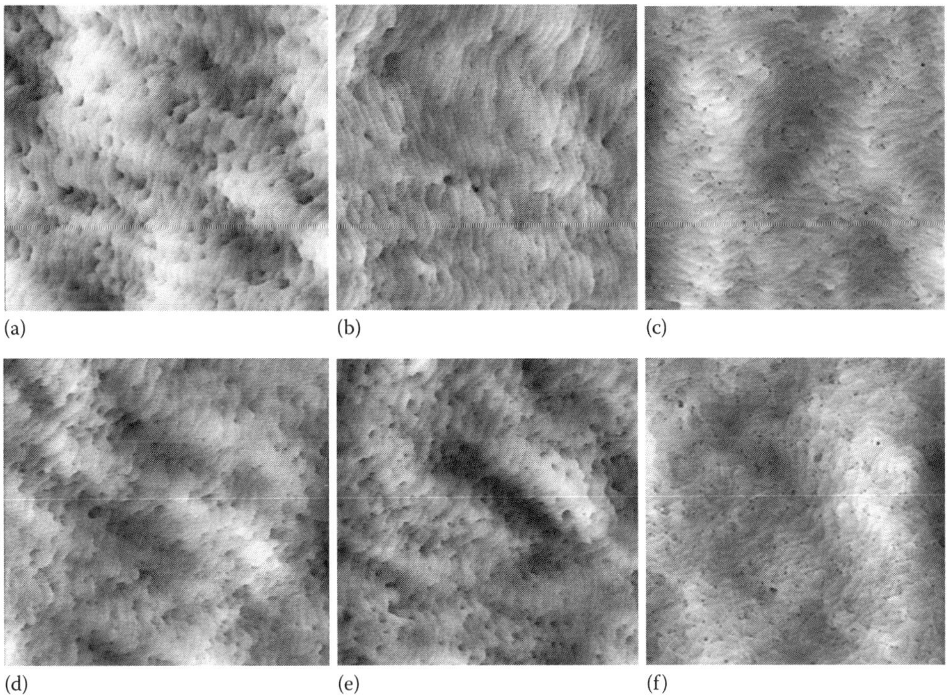

Figure 5.8 5 × 5 μm² AFM surface images of GaN films on various vicinal sapphire substrates. (a) 0.5° toward m-plane, RMS = 0.76 nm, (b) 0.3° toward m-plane, RMS = 0.34 nm, (c) 0.0° toward m-plane, RMS = 0.32 nm, (d) 0.5° toward a-plane, RMS = 0.82 nm, (e) 0.3° toward a-plane, RMS = 0.72 nm, and (f) 0.0° toward a-plane, RMS = 0.39 nm. (Courtesy of Xinxiu Ou and Zhihao Xu.)

5.4 GaN Homoepitaxy

Homoepitaxy can significantly improve the crystalline quality of GaN because no lattice mismatch and thermal mismatch between the substrate and the epitaxial material are involved. Currently, free-standing GaN films up to 300 μm thick have been obtained by HVPE, sufficient for use as the substrate for further homoepitaxy of GaN. The thick GaN film grown by HVPE also requires first a GaN template heteroepitaxially grown by MOCVD, on which the film is grown and then the foreign substrate is removed with the thick GaN film left as the substrate for homoepitaxy.

Simple as it is, homoepitaxy of GaN films on monocrystal GaN substrates involves a few major problems:

1. The existence of polar charges and substantial surface terminations of threading dislocations on the GaN surface makes it easy for the monocrystal GaN substrate surface to attract impurities such as carbon and oxygen that bind easily with gallium and aluminum. This may lead to the formation of a current leakage layer at the interface between the substrate and the epilayer if the surface contamination is not completely removed in homoepitaxy.

2. The difference in the growth methods of monocrystal substrates and epilayers (for instance, the monocrystal GaN substrate by HVPE while the epitaxial GaN film by MOCVD) causes the difference in surface morphology and dislocation density between the homoepitaxially grown GaN epilayer and the substrate.
3. The formation of V-shape etch pits by proper etch of the dislocation terminations on the monocrystal GaN substrate surface prior to homoepitaxy and the use of dislocation inclination and annihilation behaviors during homoepitaxy can yield a far smaller dislocation density in the epilayer than in the substrate.

A simple discussion of the experimental results is presented below of the HVPE-grown thick GaN films on the MOCVD-grown GaN templates and the MOCVD epitaxial GaN on HVPE grown GaN templates.

5.4.1 Growth of GaN by HVPE on Vicinal Substrates

Thick GaN films grown by HVPE are usually grown on on-axis c-plane sapphire substrates with few reports on vicinal sapphire substrates. Based on our findings in the GaN dislocation annihilation induced by vicinal substrates (see Section 8.3 for details), a comparison is made between the HVPE-grown thick GaN films on the on-axis c-plane sapphire substrate and on the sapphire substrate miscut 0.5° toward m-plane. The samples were prepared by growing an AlN nucleation layer and a 1.5 μm thick GaN by MOCVD on the two types of sapphire substrates followed by the growth of 30 μm thick GaN by HVPE.

The AFM surface images of GaN on the vicinal substrate and the on-axis substrate before and after HVPE over a scan area of 2×2 μm^2 are shown in Figure 5.9. It can be seen that the dark spots corresponding to the surface terminations of extended defects at the GaN surface are significantly reduced after HVPE, indicating a remarkable decrease in the dislocation density. The wide and flat surface steps of a width several times the width of the surface steps of the GaN templates, suggesting the formation of larger steps by coalescence of adjacent steps with a larger size of the crystal grains and a decreased dislocation density. A close look at the image also reveals that most of surface area of the HVPE-grown thick GaN film on the epitaxially grown GaN based on the vicinal substrate is very smooth with defects clustering in a few local areas, which agrees with the clustering of dislocations in GaN on vicinal substrates (see Section 8.3 for detail), indicating more decrease of dislocation density on vicinal substrates.

The surface morphologies over a larger area of the thick GaN films grown by HVPE on the vicinal substrate and on-axis substrate are shown in Figure 5.10, in which it can be seen that the film grown on the vicinal substrate is very smooth over a relatively large area, while many large pits exists on the surface of the film grown on the on-axis substrate after HVPE. The arising of such large defects suggests a

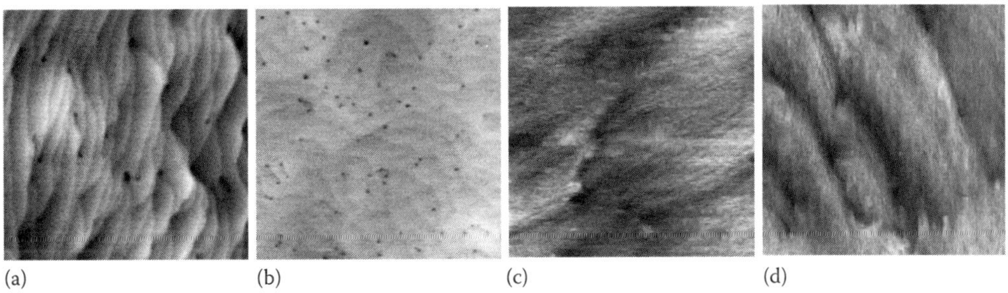

Figure 5.9 2 × 2 μm² AFM surface images of the GaN film samples grown on (a) vicinal substrate and (b) on-axis substrate before HVPE, and of those grown on (c) vicinal substrate and (d) on-axis substrate after HVPE. (Courtesy of Xinxiu Ou.)

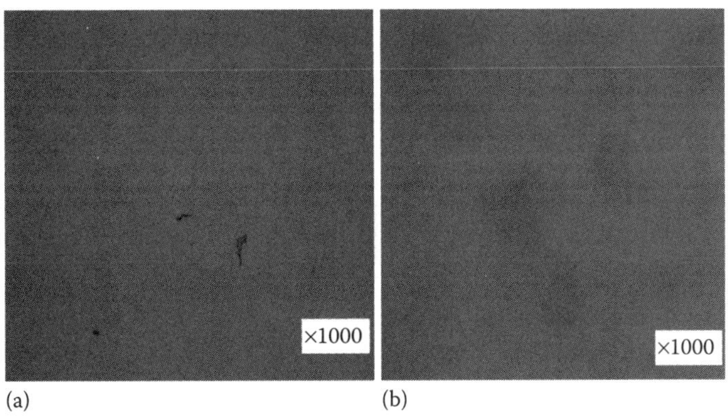

Figure 5.10 1000-times magnified surface morphologies of the thick GaN films grown after HVPE on (a) vicinal substrate and (b) on-axis substrate observed by an optical microscope. (Courtesy of Xinxiu Ou.)

relatively high initial defect density or large residual stress, which is verified by the XRD rocking curve measurements.

The (002) and (102) XRD rocking curves of the thick GaN films grown by HVPE on the vicinal substrate and the on-axis substrate are shown in Figure 5.11, from which one can see that GaN grown on the vicinal substrates both before and after HVPE have a lower defect density than those on the on-axis substrates. Further examination indicates that GaN grown on vicinal substrates exhibits a concentrated dislocation annihilation behavior, which may arise at several different depths in the film. The thicker the HVPE-grown GaN film is, the more regions there are where concentrated annihilation of dislocations occurs, resulting in greatly reduced dislocations that can penetrate to the GaN surface, thus a high quality of the thick GaN films.

5.4.2 Epitaxy of GaN by MOCVD on HVPE-Grown GaN Templates

Presently, the GaN homoepitaxy is performed by growing first in the HVPE system thick GaN films, which is polished as the substrate, and then put into the MOCVD system to carry out the homoepitaxy of thin GaN films. Surface treatment prior to

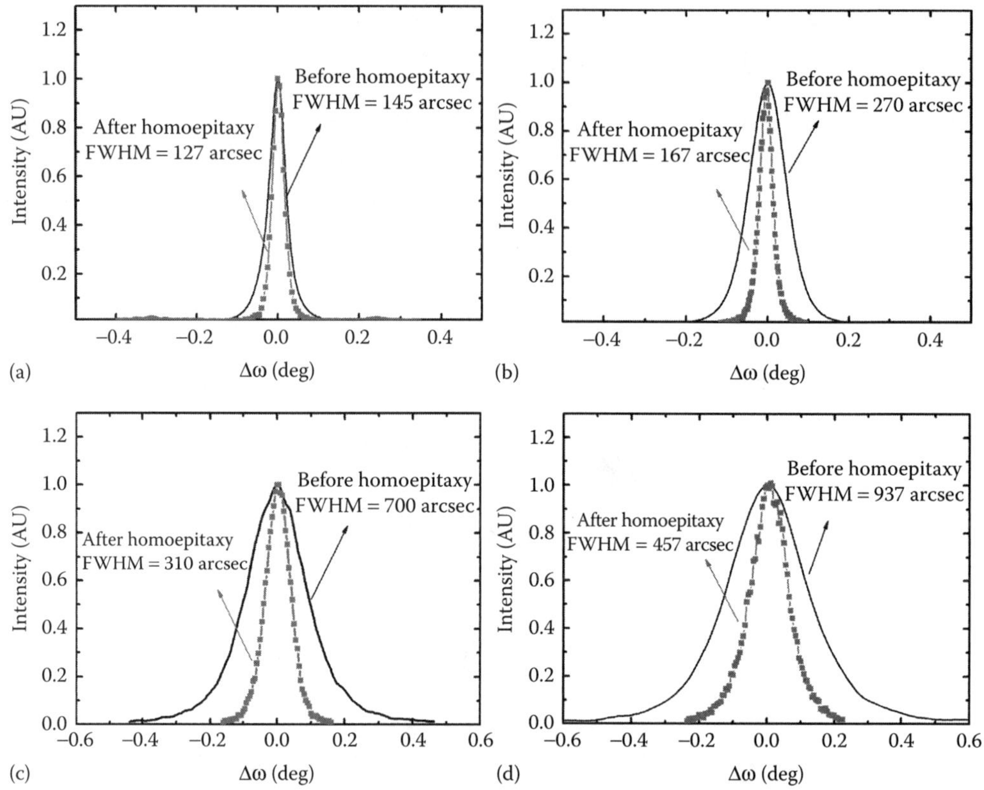

Figure 5.11 (002) and (102) XRD rocking curves of thick GaN films grown before and after HVPE. (a) On vicinal substrate, (002) plane; (b) on on-axis substrate, (002) plane; (c) on vicinal substrate, (102) plane; and (d) on on-axis substrate, (102) plane. (Courtesy of Xinxiu Ou.)

the secondary MOCVD growth is essential due to the serious contamination of the HVPE-grown GaN substrate caused by the polishing and the exposure in the air.

We divided a polished 2-in. GaN film grown by HVPE on sapphire into four samples, marked as I, II, III, and IV, respectively. Sample I was ultrasonically cleaned with de-ionized water; sample II was ultrasonically cleaned with acetone, ethanol, and de-ionized water; sample III was immersed in HF for 5 s followed by ultrasonic cleaning with acetone, ethanol, and de-ionized water; and sample IV was immersed in HF for 10 s followed by ultrasonic cleaning with acetone, ethanol, and de-ionized water. The secondary epitaxy of the four samples were performed in the same reaction chamber where high temperature surface treatment is conducted with NH_3 and H_2 flux, and then 1.4 μm-thick GaN was grown.

The FWHMs of (002) and (102) XRD rocking curves of the GaN samples before and after secondary growth with different surface treatments are shown in Figure 5.12. It can be found that the homoepitaxially grown GaN films are far superior in quality to those HVPE-grown GaN substrates with a greatly reduced screw and edge dislocation densities. Moreover, the quality of the epitaxial GaN varies greatly after different surface treatments. The four samples had basically the same crystalline quality

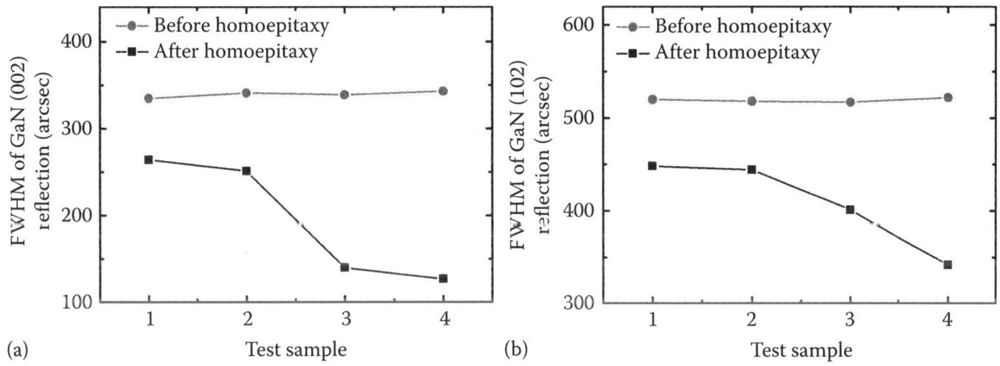

Figure 5.12 FWHM of XRD rocking curves of the four samples before and after the secondary growth with different surface treatments. (a) (002) plane and (b) (102) plane. (Courtesy of Xinxiu Ou.)

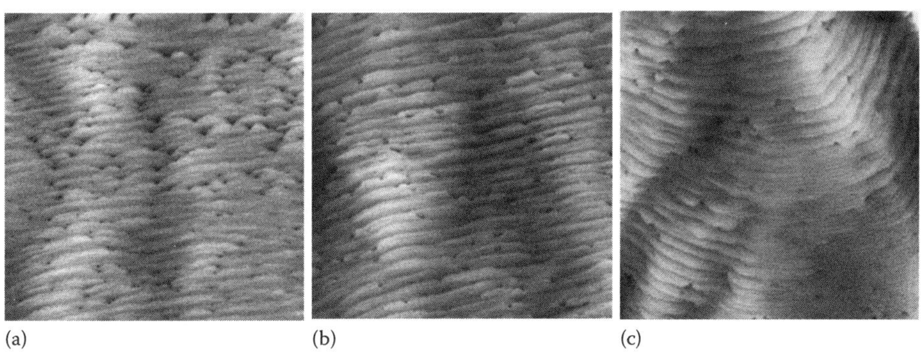

Figure 5.13 AFM images of GaN surfaces before and after secondary growth. (a) Surface morphology of the GaN substrate before surface treatment (RMS = 0.417 nm). (b) Surface morphology of sample I after secondary growth of GaN (RMS = 0.675 nm). (c) Surface morphology of sample IV after secondary growth of GaN (RMS = 0.803 nm). (Courtesy of Xinxiu Ou.)

before the secondary epitaxial growth, whereas the different surface treatment techniques lead to much different qualities of the epitaxially grown GaN films afterward. The best crystalline quality was observed in sample IV which was treated first in HF 10 s followed by ultrasonic cleaning with acetone, ethanol, and de-ionized water, with a decreasing (002) XRD FWHM from 343 to 127 arcsec and a dropping (102) FWHM from 522 to 342 arcsec.

For an in-depth analysis of the effect of surface treatment on the subsequent GaN epitaxial growth, the surface morphologies of the substrates before and after the surface treatment and the GaN films after the secondary growth were investigated by atomic force microscopy (AFM) and SEM, as shown in Figures 5.13 and 5.14. According to the AFM images, the density of the surface defects (dark spots) of the GaN film was reduced with wider surface steps and greater surface roughness after the secondary growth. It is attributed that homoepitaxy plays a fundamental role in the improvement of the crystal quality of epitaxial film owing to the elimination of the lattice- and thermal-mismatch, and that the step coalescence at the growth front also

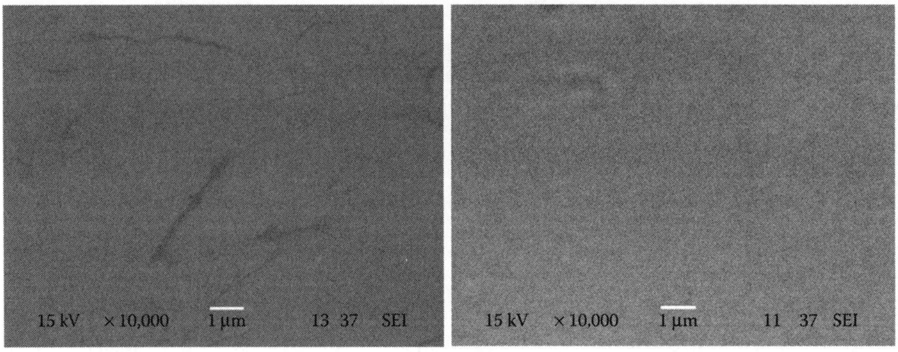

Figure 5.14 SEM images of thick GaN film used as substrates after different surface treatments. (a) Sample I and (b) Sample IV. (Courtesy of Xinxiu Ou.)

promotes the dislocation bending and annihilation. However, the AFM images fail to provide explanations for the reason why surface treatments have remarkable impact on crystalline quality, which is revealed by the SEM images.

According to Figure 5.14, sample IV has a very clean surface, whereas large size deposits are clearly visible on the surface of sample I that was simply treated by ultrasonic cleaning with de-ionized water. The composition analysis (Figure 5.15) of the deposits in Figure 5.14a by using the energy dispersive spectroscopy (EDS) of SEM finds a very high oxygen and carbon concentration at the deposit sites, suggesting that the residual deposits are mainly inclusion compounds containing carbon and oxygen. Consequently, the cleaning effect of different surface treatments of the substrate is one of the major reasons for the difference in crystalline quality of homoepitaxial GaN films. In the presence of deposits, it is difficult for gallium and nitrogen atoms to nucleate around the deposits and thus the deposits are located along the crystal grain boundaries, where the stress accumulates to a certain extent and then is released to produce the dislocations. Therefore, secondary epitaxial GaN grown on the poorly cleaned sample surfaces has low crystalline quality. In contrast, samples III and IV were treated by the same technique except for the HF cleaning time with longer time for the latter, thus its better crystalline quality. The possible reason for this is that moderate etching of the dislocation sites by HF on the GaN substrate surface gives rise to V-shape etch pits, and many dislocations extending from the substrate are annihilated during the lateral growth of GaN in the vicinity of the V pits in MOCVD, hence yielding GaN epilayers of a lower dislocation density.

5.5 High-Resistivity GaN Epitaxy Techniques

5.5.1 Characterization of GaN Buffer Layer Leakage

The leakage current of GaN buffer layer can be comprehensively characterized by *C–V*, mesa leakage, and secondary ion mass spectrometry (SIMS) measurements.

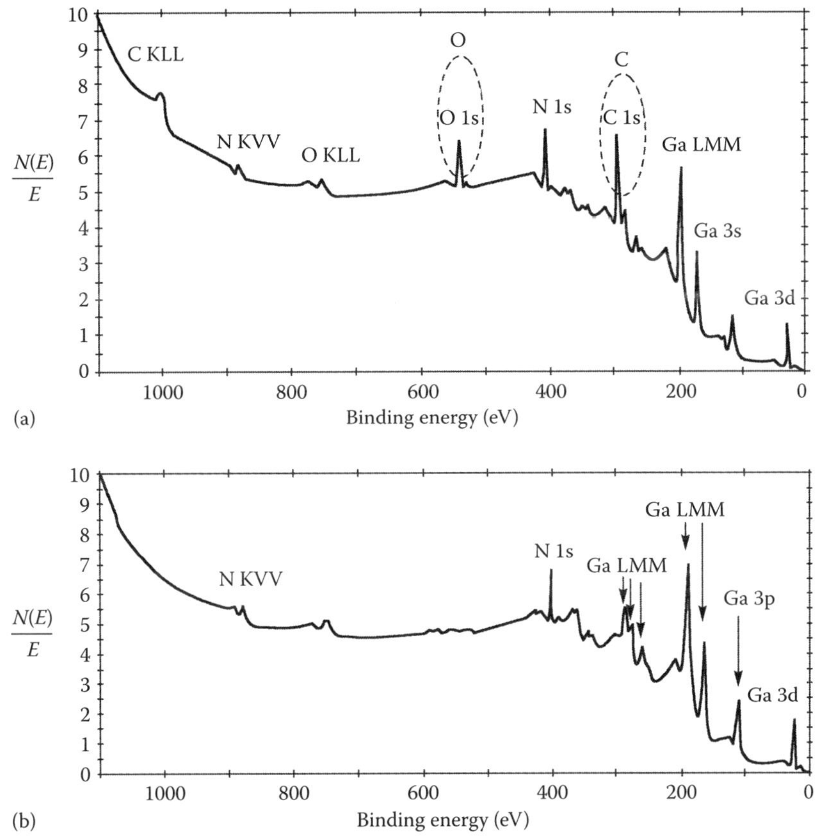

Figure 5.15 Energy spectra of the surface of the thick GaN film used as substrates after different surface treatments. (a) Energy spectrum at the deposit sites at the surface of sample I. (b) Energy spectrum of the surface of sample IV. (Courtesy of Xinxiu Ou.)

The measured *C–V* curve of the AlGaN/GaN heterostructure is shown in Figure 5.16. Region 1 is the 2DEG depletion region and the measured capacitance is mainly that of the AlGaN barrier; in region 2 the GaN buffer layer begins to deplete upon the depletion of the 2DEG; region 3 indicates depletion deep into the GaN buffer layer and the measured capacitance can be used to determine the presence of leakage. The capacitance of region 3 C_{min} is usually very small owing to the low carrier density in GaN buffer layer, and ideally there is $C_{min} \approx 0$. For a GaN buffer layer with leakage, C_{min} could be relatively large, and the greater value suggests more serious leakage. Moreover, the *C–V* curve can be transformed into the carrier density profile to facilitate the analysis of the background doping distribution at different depths in the GaN epilayer.

The mesa leakage measurements are a device-level characterization technique, in which the distribution of the buffer layer resistivity over the whole epitaxial wafer is calculated based on the measured *I-V* properties between any two adjacent mesas after mesa isolation etch (into the GaN buffer layer) and ohmic contact fabrication of the whole epitaxial wafer is performed. The resistivity is usually calculated at 50 V because

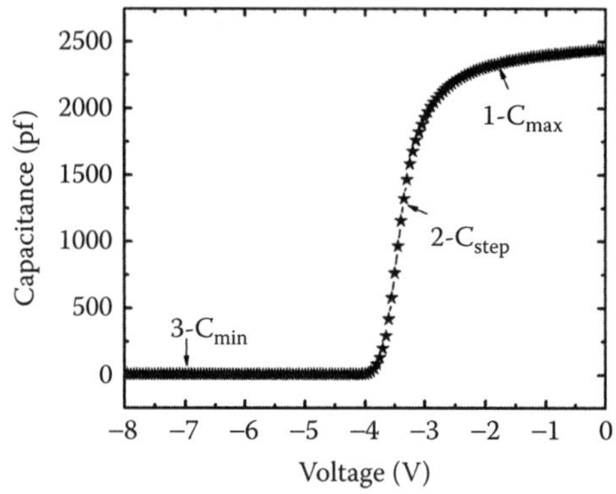

Figure 5.16 *C–V* curve of AlGaN/GaN heterostructure. (Courtesy of Chuankai Yang.)

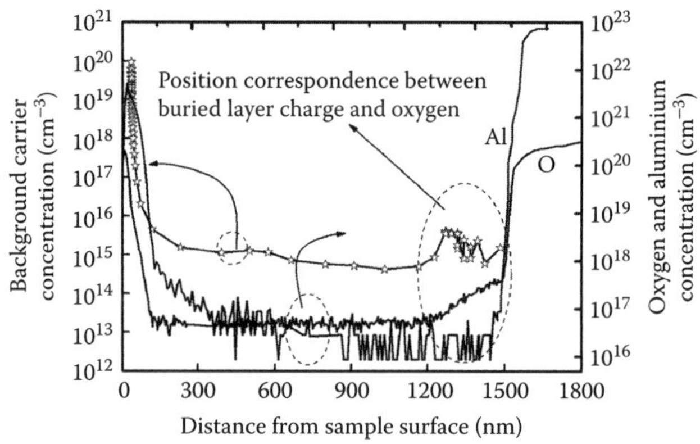

Figure 5.17 *C–V* carrier profile and SIMS measurements results of AlGaN/GaN sample with a leaky buffer layer. (Courtesy of Hao Wang.)

at a high bias voltage the electrons captured by traps can also participate in electric conduction and the calculated result can be closer to the real resistivity of the GaN buffer layer in GaN HEMT devices which operate at a relatively high drain-source voltage.

SIMS measurements can be employed in the accurate analysis of the densities of such impurities as carbon, oxygen, and silicon that are capable of inducing the background doping and the estimate of the leakage behavior.

Shown in Figure 5.17 is a comparison between the *C–V* carrier profile and the distribution of SIMS measured oxygen and aluminum concentrations as a function of the distance from the surface of the sample grown on sapphire using high-temperature AlN nucleation layer with a leaky buffer layer. Judging from the SIMS measured oxygen and aluminum concentrations, the interface between the high-temperature AlN nucleation layer and GaN buffer layer is 1.5 μm from the sample surface with an extremely high concentration of oxygen in the high-temperature AlN nucleation layer. No intentional oxygen doping is introduced in the growth of AlN nucleation layer, so the oxygen comes from sapphire substrate whose surface is partially decomposed in high temperature MOCVD process. From this interface up to the surface, the oxygen concentration reduces but is still higher than the background carrier concentration, suggesting the diffusion of part of the oxygen atoms in the nucleation layer into the GaN layer. These oxygen impurities form donor levels in the GaN buffer layer, introduce the background carriers, and concentrate within a small region to give rise to the buried charge layer. This agrees with the *C–V* carrier profile. The oxygen concentration in Figure 5.17 is as high as 5×10^{16} cm^{-3}, while the background carrier concentration is of the order of 1×10^{16} cm^{-3}, which is attributed to the presence of carbon impurity as acceptors and the edge dislocations that can trap the background electrons, thus a lower background electron concentration than that of oxygen.

It can be seen that the leakage of the buried charge layer cannot be diminished by the growth of relatively thick buffer layer because the spacing between the 2DEG conducting channel and the buried charge layer can hardly be increased. Due to the fact that the buried charge layer is formed essentially by the out-diffusion of oxygen from the sapphire substrate in the high temperature buffer layer growth, a longer time growth would lead to the continuous diffusion of oxygen deeper into the buffer layer, therefore the buried charge layer become thicker and leakage still remain.

5.5.2 Influence of Dislocations on Oxygen Diffusion from Substrate

In high temperature growth of GaN, the oxygen in substrate diffuses gradually upward via its substitution of nitrogen to form point defects, and the threading dislocations can also act as the diffusion path. Then the question arises whether the screw dislocations or edge dislocations are at work here.

We mark the sample in Figure 5.17 as sample A, which is known to have leakage induced by the buried charge layer. Sample B, which has no leakage, was grown using the same nucleation layer, epitaxial layer structure, and growth technique but with a sapphire substrate from another manufacturer and a different polishing quality from sample A. The root reason for the different leakage behaviors of the two samples is the difference in the substrate quality, which in turn leads to the difference in the nucleation layer coverage and in epitaxial layer crystalline quality under identical growth conditions. The dislocations were evaluated from the XRD rocking curve measurement results (Table 5.3), which show that the leaky sample A has a high screw dislocation density and a low edge dislocation density, suggesting the screw dislocations may be the dominant path for the upward oxygen diffusion from the substrate. The size of the pits (dark spots) as the surface terminations of the dislocations in nitrides suggests a relatively wider leakage path that can be formed by the screw dislocations than that by the edge dislocations, which also indicates that oxygen impurities diffuse more readily upward along the screw dislocations.

This conclusion is supported by another experiment. A 2-in. epitaxial wafer with leakage occurring at its center but not at the edge due to nonuniform growth is examined, and the presence of the buried charge layer caused by oxygen diffusion from

Table 5.3 Measured Data by XRD, AFM, and *C–V* of the Two Samples with Different Leakage Behaviors due to Different Substrates

	AFM	XRD FWHM (arcsec)		*C–V*
SAMPLE	RMS (nm)	(002)	(102)	DEPLETION CAPACITANCE (pf)
A	0.273	369	743	55
B	0.328	297	689	6

Source: Courtesy of Chuankai Yang.

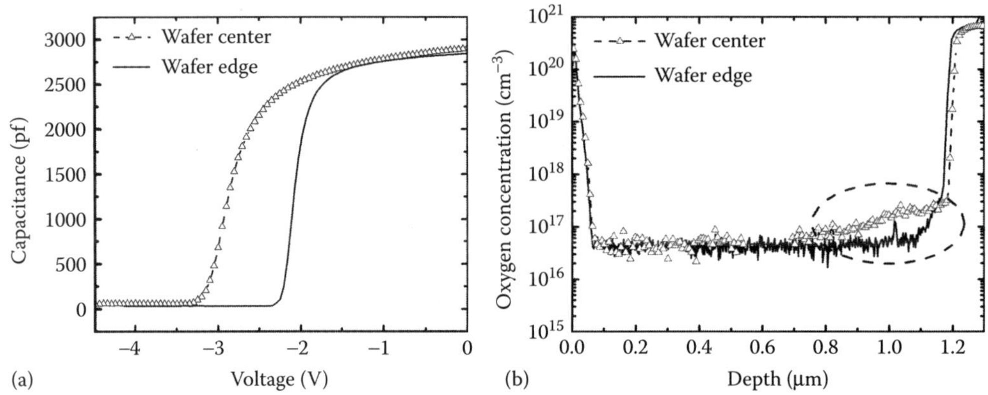

Figure 5.18 Comparison of *C–V* characteristics and SIMS measured oxygen profiles of sample B at the center and the edge of the epitaxial wafer. (a) *C–V* characteristics and (b) oxygen profiles. (Courtesy of Chuankai Yang.)

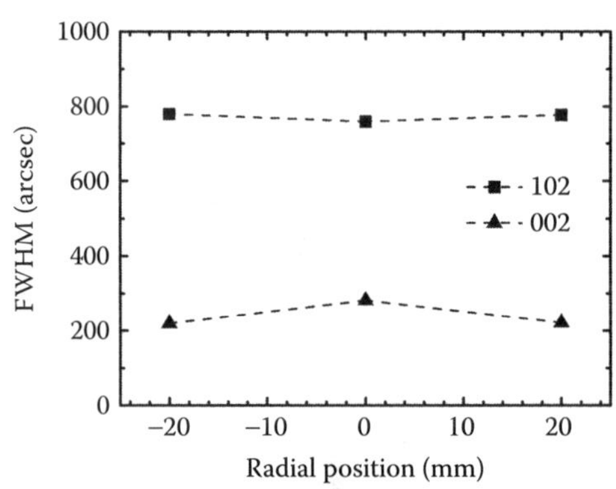

Figure 5.19 FWHM of XRD rocking curves of sample B along the diameter of the 2-in. epitaxial wafer. (Courtesy of Chuankai Yang.)

substrate at the center of the wafer was confirmed by *C–V* and SIMS measurements (Figure 5.18). The XRD rocking curves (Figure 5.19) reveal a high screw dislocation density at the center and a low one at the edge of the wafer, whereas the edge dislocation density is high at the edge and low at the center, that is, a relatively high screw dislocation density and a relatively low edge dislocation density at the leakage region, indicating that it is the screw dislocations that promote the oxygen diffusion.

5.5.3 Elimination of the Buried Charge Layer

From the previous analysis, it is known that the charge in the buried charge layer is formed primarily by the accumulation of oxygen impurities diffused from substrate in the GaN buffer layer closely neighboring to the nucleation layer. The formation of high-resistivity GaN buffer layer requires first the elimination of the buried charge layer and suppression of the oxygen diffusion from substrate into the GaN buffer layer. Such blocking can be best realized by the nucleation layer, typically including the low-temperature GaN (LT-GaN) layer, low-temperature AlN (LT-AlN) layer, and high-temperature AlN (HT-AlN) layer, whose surface morphologies are shown in Figure 5.20. In low-temperature buffer layers, the orientation consistence of the grains in nucleation islands is very poor and a relatively consistent grain orientation requires recrystallization at elevated temperature, so the low-temperature nucleation layer should not be too thick. The low-temperature GaN nucleation layer can achieve a dense coverage of the sapphire substrate, which is comparatively difficult for the low-temperature AlN nucleation layer, and therefore high quality GaN epitaxial layers can be grown on low-temperature AlN nucleation layers but among them high-resistivity samples are difficult to obtain. Whereas the high-temperature AlN nucleation layer performs simultaneously the nucleation layer epitaxy and annealing recrystallization of the underlying AlN nucleation islands to guarantee the quality of the nucleation layer, thus a complete coverage of the sapphire substrate surface and good suppression of the oxygen out-diffusion from the substrate. Figure 5.21 shows the *C*–*V* characteristic

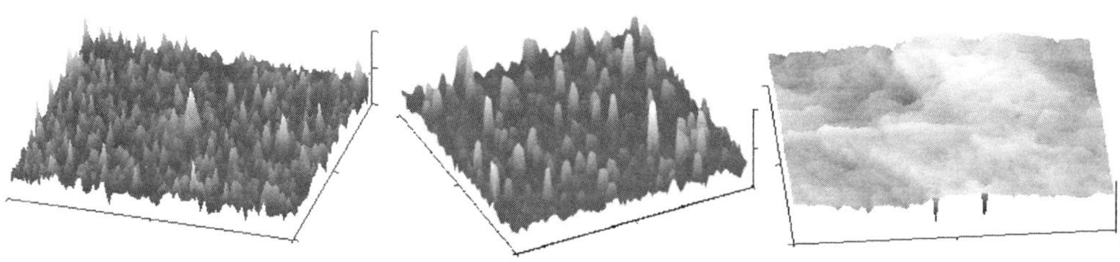

Figure 5.20 Surface morphologies of LT-AlN, LT-GaN, and HT-AlN nucleation layers. (Courtesy of Hao Wang.)

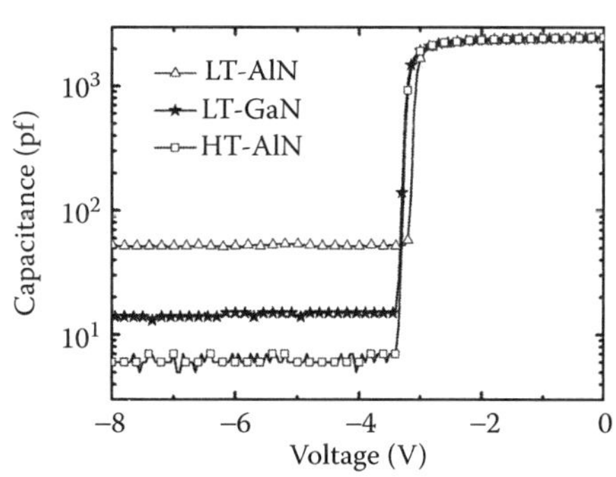

Figure 5.21 *C*–*V* characteristic curves of AlGaN/GaN heterostructures based on different nucleation layers. (Courtesy of Hao Wang.)

curves of AlGaN/GaN heterostructures with different nucleation layers under the same conditions, and a great difference is observed in the depletion capacitances of the GaN buffer layers, a proof of the impact of the nucleation layers on the leakage.

The high-temperature growth of AlN nucleation layers also involves the growth of three-dimensional islands and the inter-island coalescence to form continuous films, and a minimal thickness of the high-temperature AlN nucleation layer exists for the full coverage of the sapphire substrate surface. A series of AlGaN/GaN heterostructure samples were grown under the same conditions but for the growth time of high-temperature AlN nucleation layer (10, 15, 20, 25, 30, and 35 min, respectively). The minimum thickness of the high-temperature AlN nucleation layer for the effective suppression of the formation of buried charge layers is found to be 100 nm (growth time of 25 min) according to the measured *C–V* characteristic curves (Figure 5.22) of these samples coupled with the oxygen concentration profiles measured by SIMS (Figure 5.23).

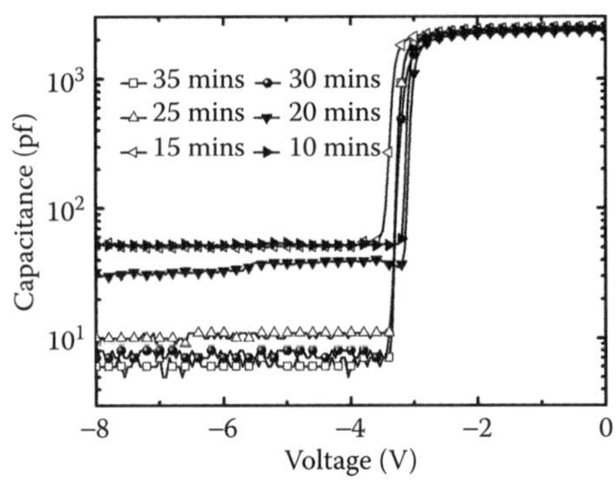

Figure 5.22 *C–V* characteristics of AlGaN/GaN samples with HT-AlN nucleation layers of different thicknesses. The nucleation layers are grown for 10, 15, 20, 25, 30, and 35 min, respectively. (Courtesy of Hao Wang.)

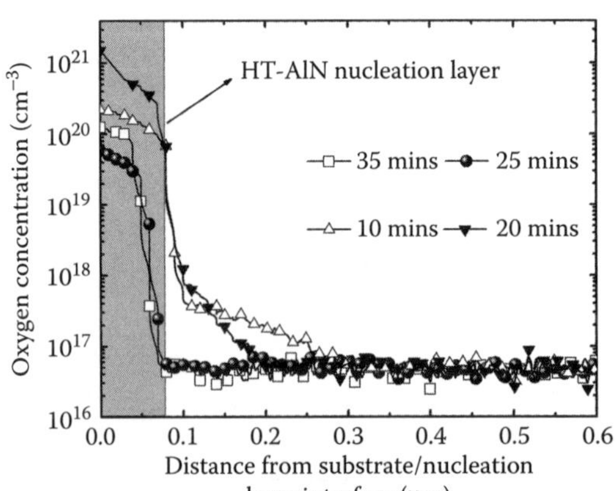

Figure 5.23 SIMS measured oxygen concentrations in samples with AlN nucleation layers grown for 10, 20, 25, and 35 min, respectively. (Courtesy of Hao Wang.)

5.5.4 Suppression of Background n-Type Doping in GaN Buffer Layers

The typical impurities in the unintentionally doped GaN films are carbon, silicon, and oxygen. Carbon comes mainly from the decomposition of MO sources such as TMGa ($Ga(CH_3)_3$) and possibly from the graphite susceptor in the reaction chamber (in the case of our self-built MOCVD system). Silicon is mostly from the growth ambience, such as adsorption from the atmosphere or decomposition of the outer SiC coating of the graphite susceptor. Oxygen may originate from the oxygen in the carrier gas and the adsorbed atmospheric oxygen including such impurities as water vapor and oxygen gas, but mainly from the upward diffusion of oxygen in the sapphire substrate (Al_2O_3). Of these impurities, oxygen and silicon are shallow donors while carbon is an amphoteric impurity. In n-type GaN, interstitial carbon is a deep acceptor which can be used for self-compensating.

Besides the elimination of the buried charge layers, the formation of high-resistivity GaN buffer layers also requires a reduced background carrier concentration. The introduction of large amounts of edge dislocations in GaN generates deep acceptor traps to capture background electrons thus forming high resistivity, but at the cost of declined crystal quality. High resistivity can also be achieved by changing the MOCVD growth conditions to obtain epitaxial GaN with a high carbon concentration. However, the low pressure and low-temperature conditions favorable for the massive admission of carbon also lead to poorer crystal quality of epitaxially grown GaN.

In the methods of neutralizing unnecessary electrons by compensating p-type doping of GaN, iron doping is rather successful. Iron in GaN exhibits the valence state of $Fe^{3+/2+}$ with acceptor levels of 2.6 and 1.7 eV, respectively, at the top of valence band. Iron doping is currently the best method for achieving high-resistivity GaN buffer layer without significant loss in material crystal quality and device reliability in spite of its memory effect, that is, in the epitaxial process the iron doping concentration does not drop immediately at the shutoff of the iron source owing to the adhesion of iron and its compounds on the reaction chamber walls or tubes, causing the doping of the materials that should not be doped.

The SIMS results of a series of iron-doped samples are shown in Figure 5.24, where the shaded part is the target doping region with the rest regions unintentionally doped. Since in the alternate iron doping and non-iron-doping experiments the iron atom content in the non-iron-doped sample is below the detection limit of SIMS measurement (3×10^{15} cm^{-3}), it is suggested that the unintentional iron doping beyond the target doping region is not necessarily attributed to the generally understood memory effect caused by the adsorption and desorption of iron impurities on the reaction chamber walls, but to a slow on/off effect induced by the iron doping. An iron-rich deposited layer forms on the epitaxial film surface when the reaction source of iron (Cp_2Fe) is passed into the reaction chamber. The low incorporation rate of iron impurities in the film before the establishment of this deposited layer results in a gradual increase of the iron impurity concentration in iron doping, namely the slow on effect; when

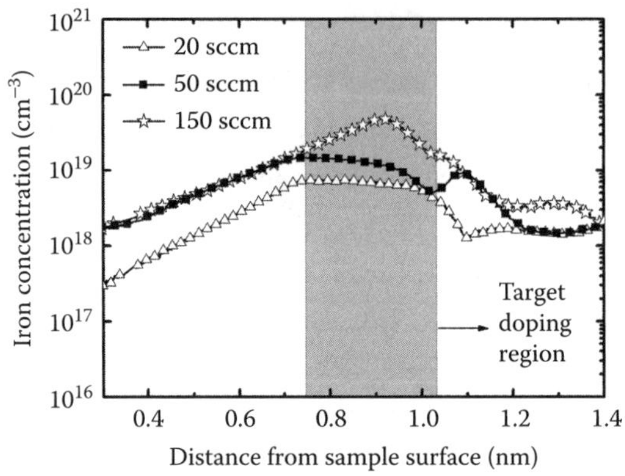

Figure 5.24 SIMS results of samples with different iron source flow rates. At Cp_2Fe flow rates of 150, 50, and 20 sccm, the iron concentrations in the doped region are 4.93×10^{19} cm^{-3}, 1.48×10^{19} cm^{-3}, and 7.29×10^{18} cm^{-3}, respectively. (Courtesy of Hao Wang.)

the Cp_2Fe flux is cut off, iron atoms deposited on the epitaxial layer surface serve as the iron source and further diffuse into the subsequently grown epitaxial layer until entirely depleted, namely the slow off effect. Another important contributing factor to the slow on/off effect of the iron impurity concentration is the symmetrical diffusion of iron into the undoped regions on both sides of the iron-doped region where the material experiences identical thermal treatment.

5.6 Optimization of AlGaN Barrier Layers

5.6.1 Influence of Al Content and Thickness of AlGaN Barrier on 2DEG Properties

The increase in the Al content and thickness of the AlGaN barrier layer causes a rise in 2DEG density, which if too large may induce strain relaxation in the barrier layer, thus a degradation of the material characteristics of the heterostructure. In the absence of strain relaxation, an optimization of the Al content or thickness of the AlGaN layer is required for better 2DEG conductivity.

The reported experimental data (Smorchkova et al. 1999, Hiroki et al. 2002, Jang et al. 2002, Heikman et al. 2003) and our research results are presented in Figure 5.25, based on which the dependence of the 2DEG density on the Al content and thickness of the AlGaN layer is given with fitting curves that agree very well with the experimental data. It can be seen that before the emergence of strain relaxation, the 2DEG density increases with the Al content in an approximately linear manner. Theoretically 2DEG can form with Al content $x > 0$, and experimentally even with as low an Al content as 5%, 2DEG does arise if the barrier layer is thick enough (>40 nm). A minimum thickness of AlGaN exists for the 2DEG to form, which decreases with rising Al content. The 2DEG density increases quickly with greater AlGaN thickness and then saturates, and for the heterostructures with an Al content of 0.27 the 2DEG density generally saturates at an AlGaN thickness in excess of 25 nm.

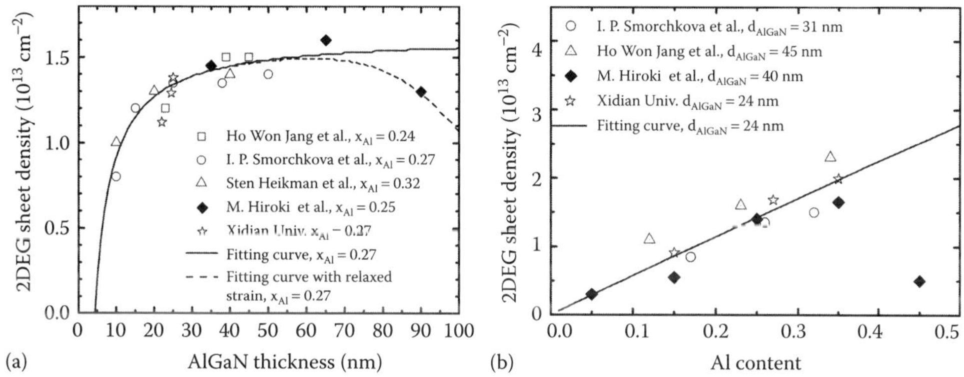

Figure 5.25 Dependence of 2DEG density on (a) Al content and (b) thickness of AlGaN layer in AlGaN/GaN heterostructures. (From Smorchkova, I.P. et al., *J. Appl. Phys.*, 86, 4520–4520, 1999; Heikman, S. et al., *J. Appl. Phy.*, 93, 10114–10118, 2003; Jang, H.W. et al., *Appl. Phys. Lett.*, 81, 1249–1251, 2002; Hiroki, M. et al., *J. Cryst. Growth*, 237–239, 956–960, 2002; Some experimental data are reused by permission of American Institute of Physics.)

The 2DEG mobility also varies with the Al content or thickness of the AlGaN layer. We studied such effect using a theoretical model combining multiple scattering mechanisms including acoustic deformation-potential scattering, acoustic wave piezoelectric scattering, polar optical phonon scattering, alloy disorder scattering, and interface roughness scattering, finding that the change in scattering strengths and the decrease in 2DEG mobility are due mainly to the higher density, narrower distribution, and closer population of the 2DEG to the AlGaN/GaN heterointerface induced by the increase in barrier Al content or thickness (Zhang et al. 2008) (see Section 4.3.2 for details).

The optimum Al content and thickness of the AlGaN barrier layer may vary in different MOCVD systems. We optimized the barrier thickness within the range of 10–30 nm for many AlGaN/GaN heterostructures with an Al content of 30% grown in the same MOCVD system under identical growth conditions but with different barrier layer growth times. The electrical characteristics of these samples are shown in Figures 5.26 and 5.27. It is found that with a 2DEG density range

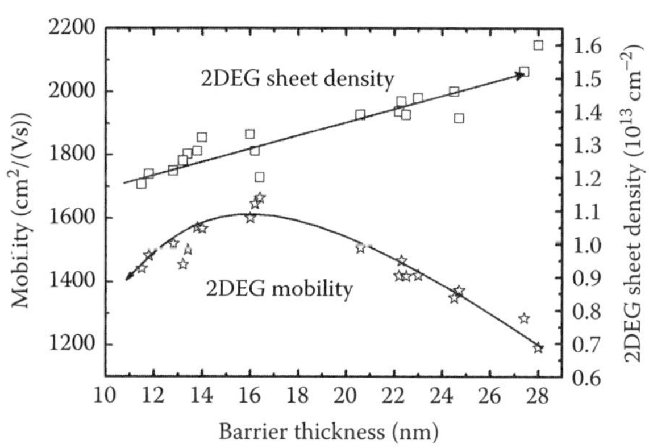

Figure 5.26 2DEG sheet density and mobility of AlGaN/GaN heterostructure as a function of barrier thickness. (Courtesy of Hao Wang.)

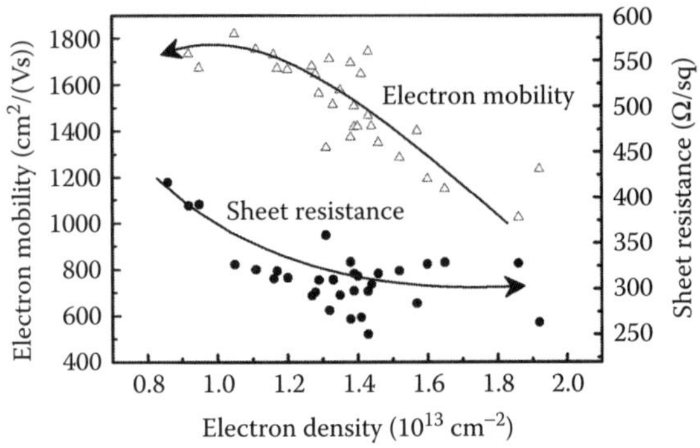

Figure 5.27 Sheet resistance and carrier mobility in the AlGaN/GaN samples as a function of 2DEG density. (Courtesy of Hao Wang.)

of 1.0–1.4 × 10^{13} cm^{-2}, the carrier mobility is 1600–1800 cm^2/Vs with no obvious decrease; while with a 2DEG density in excess of 1.4 × 10^{13} cm^{-2}, the carrier mobility drops dramatically. Such a changing pattern renders a minimum sheet resistance of 240–300 Ω/sq with a carrier mobility of 1.4–1.6 × 10^{13} cm^{-2} corresponding to a barrier thickness of ~20 nm, that is, a barrier thickness of 20 nm can yield the best conductivity of the AlGaN/GaN heterostructure samples at a high mobility.

5.6.2 Effect of AlN Interlayers

In 2001, Shen et al. at University of California, St. Barbara, suggested that the introduction of ~1 nm thick AlN interlayers into the AlGaN/GaN interfaces to form the AlGaN/AlN/GaN heterostructures is favorable for the improvement of heterostructure material characteristics (Shen et al. 2001). Under the polarization effect, this interlayer improves the effective conduction band offset between the AlGaN barrier layer and GaN channel layer (Figure 5.28), on the one hand yielding deeper and narrower quantum wells in favor of a higher channel electron density and on the other hand suppressing the alloy disorder scattering to the 2DEG penetrating into the AlGaN alloy barrier layer, thus increasing the channel electron mobility.

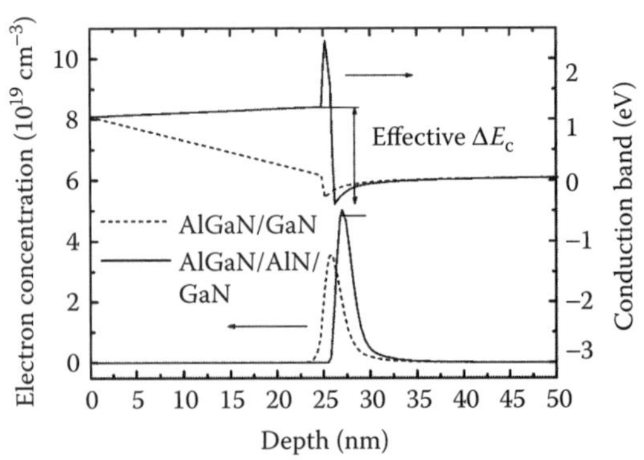

Figure 5.28 Conduction band profile and electron density distribution of AlGaN/GaN and AlGaN/AlN/GaN heterostructures. (After Zhang, J.-F. et al., *Chin. Phys.* 15 1060–1066, 2006.)

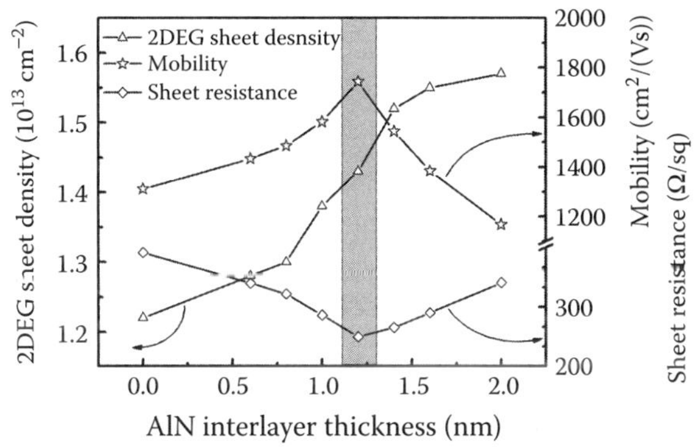

Figure 5.29 Measured electrical properties of AlGaN/GaN heterostructure as a function of AlN interlayer thickness. (Courtesy of Hao Wang.)

Too thin an AlN interlayer makes no real sense while too thick an AlN interlayer introduces great strain into the barrier (~2.4% lattice mismatch between AlN and GaN), which lowers the epitaxy quality of AlGaN layers and hence a decreased mobility. Therefore, the proper choice of the AlN interlayer thickness is of great importance for the improvement of the carrier mobility. The variation of the electrical properties of AlGaN/GaN samples grown by our group with AlN interlayer thickness is presented in Figure 5.29, showing the optimum AlN thickness to be ~1.2 nm.

Inserting an AlN interlayer between the barrier layer and the channel layer has greater effect on the improvement of material properties of InAlN/GaN heterostructures: the channel electron mobility increases from ~70 to 1170 cm^2/Vs (Gonschorek et al. 2006) with generally no or little change in the channel electron density and the temperature-dependence of the Hall electron mobility also changes from the quasi-bulk electron behavior (increasing and then decreasing with dropping temperature) into the 2DEG behavior (increasing and then saturating gradually with dropping temperature). We compared the properties of the nearly lattice-matched InAlN/GaN and InAlN/AlN/GaN heterostructures grown by pulsed MOCVD (Zhang Jin-Feng 2011). Distinct atomic steps were observed in the microscopic surface morphologies of both samples with a surface RMS roughness of ~0.3 nm over an AFM scan area of $2 \times 2\ \mu m^2$, whereas some short linear depressions spreading mainly along the atomic steps were seen on the surface of the InAlN/GaN sample (Figure 5.30a), which are assumed to be relative to the penetration of defects from GaN layer into InAlN layer. A relevant phenomenon is that in the optimization of the AlN interlayer thickness in the InAlN/AlN/GaN structures such linear depressions are reduced and eventually disappear (Figure 5.30b) as the AlN thickness grows gradually from 0 to 1.2 nm, but reappear with further increase in the AlN thickness. Song et al. investigated the effect of AlN interlayer thickness on the AlGaN barrier by TEM (Song et al. 2009), finding that at the optimum AlN interlayer thickness the AlGaN barrier layer possesses the most uniform stress distribution and the lowest threading dislocation density, which they attributed to the bending and annihilation of dislocations in the vicinity of the

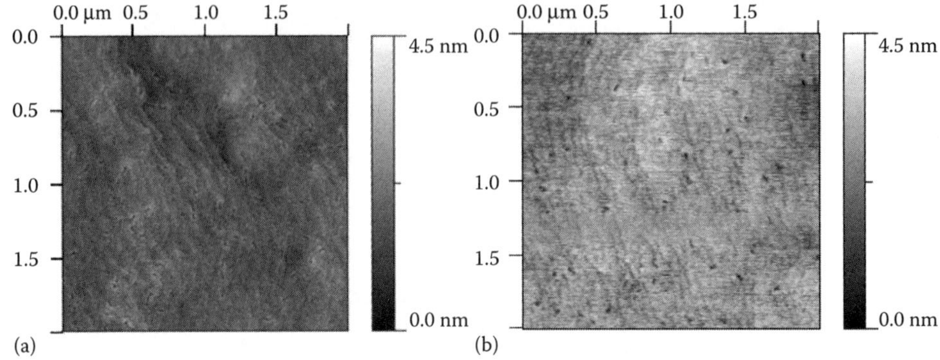

Figure 5.30 AFM surface images of (a) InAlN/GaN and (b) InAlN/AlN/GaN samples. (After Zhang, J.-F. et al., *Acta Phys. Sinica* 60, 611–616, 2011.)

heterointerface due to the relatively large mismatch strain between AlN and GaN. It is shown that the AlN interlayer not only yields a smooth heterointerface, but also reduces the epitaxial defects in the AlGaN or AlInN barrier layer, hence remarkably improving the crystal quality of the barrier layer. Both effects may contribute much to the improvement of the channel electronic transport characteristics of InAlN/GaN heterostructures by tens of times by the AlN interlayer.

The introduction of ~1 nm AlN interlayer has become a common practice in the optimization of the material properties of AlGaN/GaN and InAlN/GaN heterostructures. However, such practice is by no means without negative effects. It was reported that the interlayer may affect the reliability of HEMT devices. According to their research on the reliability of more than 300 AlGaN/GaN and AlGaN/AlN/GaN HEMT samples grown by MOCVD and MBE (Coffie et al. 2006), Coffie et al. at US Northrop Grumman Co. found that under the radio frequency power stress 78% of the AlGaN/AlN/GaN HEMTs and only 28% of the AlGaN/GaN HEMTs exhibited an increased gate leakage. It is proposed that the higher probability of an increased gate leakage in AlGaN/AlN/GaN HEMTs is due to the nonuniform thickness of the AlN interlayer under the gate, more specifically, the inability to control the fluctuation of the AlN interlayer thickness within a monolayer during the material growth. Even such a tiny thickness fluctuation may significantly influence the effective barrier height between the channel and the gate under the polarization effect (Figure 5.31), causing localized breakdown (hot spots in Figure 5.32) of the gate fingers where the AlN layer is comparatively thin, hence an increased gate leakage.

Further investigation is needed into the influence of the AlN interlayer on the HEMT materials and devices.

5.6.3 *Influence of Cap Layers on Heterostructure Material Properties*

The cap layer is widely used in the material structure optimization of GaN-based devices. Research shows that the InGaN cap layer on top of the AlGaN/GaN HEMT

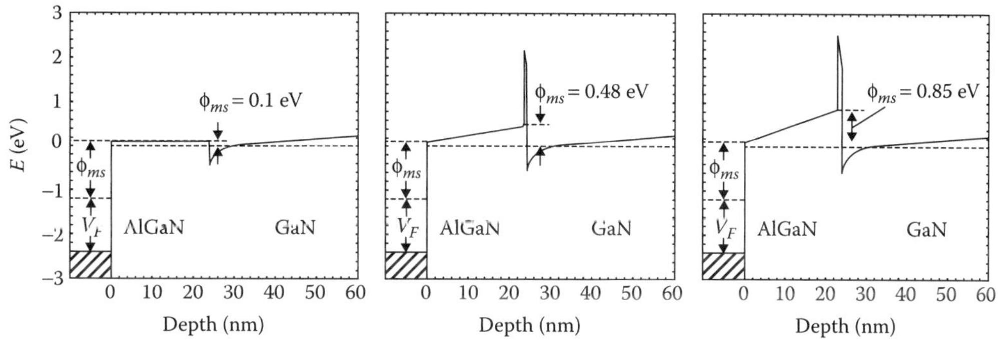

Figure 5.31 Schematic band diagrams of AlGaN/AlN/GaN heterostructure with an AlN interlayer thickness of 0, 6 Å, and 10 Å, respectively, under a forward bias $V_F = 1.2V$. ϕ_{ms} is the effective barrier height between the channel and the gate. (Reproduced from Coffie, R. et al., Impact of AlN interalayer on reliability of AlGaN/GaN HEMTs, in *44th Annual IEEE International Reliability Physics Symposium*, March 26–30, 2006, San Jose, CA.)

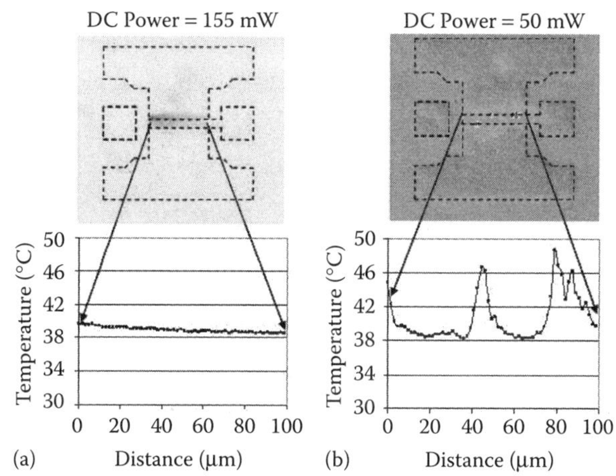

Figure 5.32 Infrared surface images (plan view) of AlGaN/GaN HEMTs under a forward gate voltage. Uniform emission is observed for (a) the device without an AlN interlayer while hot spots along the gate finger are seen for (b) the device with a 6 Å AlN interlayer. (Reproduced from Coffie, R. et al., Impact of AlN interalayer on reliability of AlGaN/GaN HEMTs, in *44th Annual IEEE International Reliability Physics Symposium*, March 26–30, 2006, San Jose, CA.)

material can effectively reduce the ohmic contact resistance while the GaN cap layer can improve the 2DEG mobility at the cost of a slight drop in the carrier concentration under the polarization effect as well as increase the effective potential barrier of the AlGaN/GaN heterostructures, thus significantly decreasing the gate leakage current (Yu et al. 1998); moreover, the low-temperature AlN cap layer is a good candidate for gate insulation and passivation in AlGaN/GaN HEMTs and GaN-based photovoltaic devices (Selvaraj et al. 2007).

We analyzed the effect of the GaN and AlN cap layers on the material quality and 2DEG transport properties of high Al content AlGaN/GaN heterostructures (Liu et al. 2011). Both cap layers were 2 nm thick and grown under the same growth temperature and pressure as the GaN buffer layer with 22 nm $Al_{0.37}Ga_{0.63}N$ as the AlGaN barrier layer and a ~1.5 μm thick GaN buffer layer.

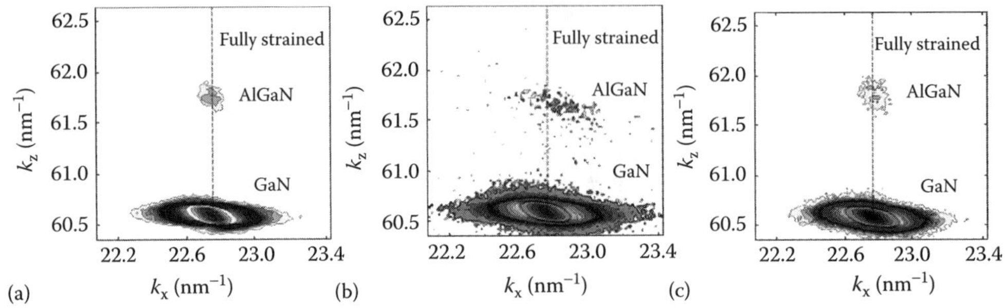

Figure 5.33 Reciprocal space maps (RSMs) around an asymmetric (105) reflection of AlGaN/GaN structures. The dashed lines show the position of completely strained AlGaN. (a) Sample A with GaN cap layer, (b) sample B with AlN cap layer, and (c) reference sample C without cap layer. (After Liu, Z.-Y. et al., *Chin. Phys. B*, 20, 097701, 5 pp., 2011.)

According to the RSMs around a (105) reflection of the AlGaN/GaN structures with and without the cap layer (Figure 5.33), the AlGaN in the capless sample C displays a slight relaxation (22 nm approaches the critical thickness for strain relaxation of $Al_{0.37}Ga_{0.63}N$ on GaN [Lee et al. 2004]). By introducing a GaN cap (sample A), the AlGaN layer is fully strained, while an AlN cap (sample B) causes a severe relaxation of the AlGaN barrier, thus a higher dislocation density and poorer crystal quality. The degrees of relaxation for samples A, B, and C were estimated to be 0%, 33%, and 6.5%, respectively. It is attributed to that GaN and AlN cap layers apply different strains on AlGaN, with the former enhancing the coherent state, whereas the latter intensifying the lattice mismatch between AlGaN and GaN during post-growth cooling, leading to different strain state in AlGaN.

The AFM surface images (Figure 5.34) of the three samples reveal that B has the roughest surface followed by C and A, which is consistent with the prediction that the strain relaxation is accompanied by stress release along with more dislocations and poorer surface morphologies. The interface roughnesses obtained by fitting the X-ray reflectivity results also show the same trend as the AFM surface roughnesses.

The room temperature Hall mobility and sheet density as a function of the degree of relaxation (R) of the samples are given in Figure 5.35, from which it can be seen that the mobility is more correlated to R in the barrier. Although the GaN cap layer in sample A increases the effective AlGaN barrier height with a lower electron density than that in sample C, the complete strain state ($R = 0$) of AlGaN causes a relatively low dislocation density and interface roughness and a relatively weak interface roughness scattering and Coulomb scattering of charged dislocations as well as scattering of the strain field in the vicinity of the dislocations, making its 2DEG mobility the highest. In contrast, the AlN cap layer induces strain relaxation in the AlGaN barrier which impairs the piezoelectric polarization, so sample B has a lower carrier sheet density than sample C, equivalent to that of sample A. However, sample B has a significantly lower mobility than sample A as a result of the increase in dislocations and

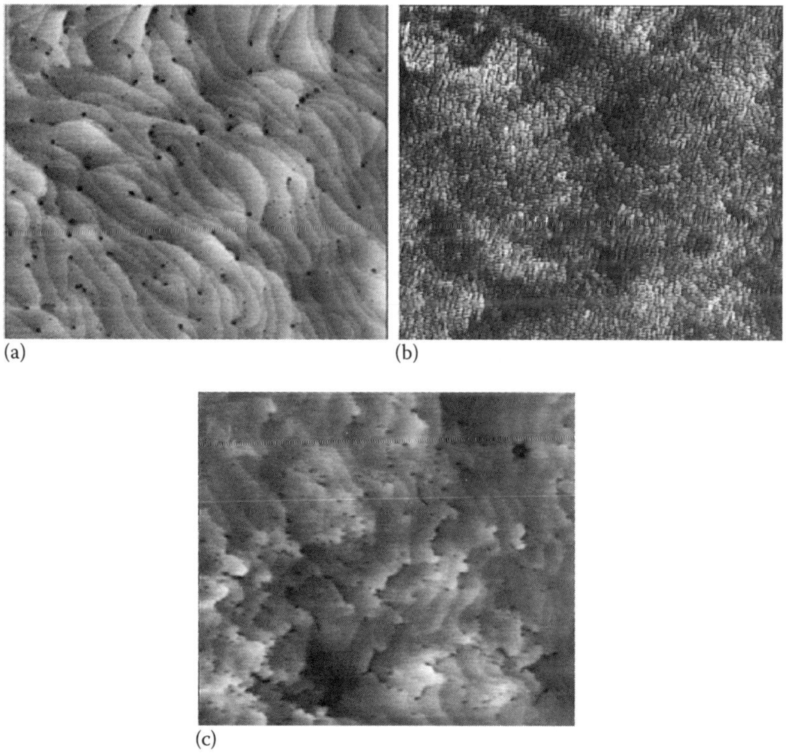

Figure 5.34 5 × 5 μm² AFM images of three samples. (a) Sample A, (b) sample B, and (c) sample C. (After Liu, Z.-Y. et al., *Chin. Phys. B,* 20, 097701, 5 pp., 2011.)

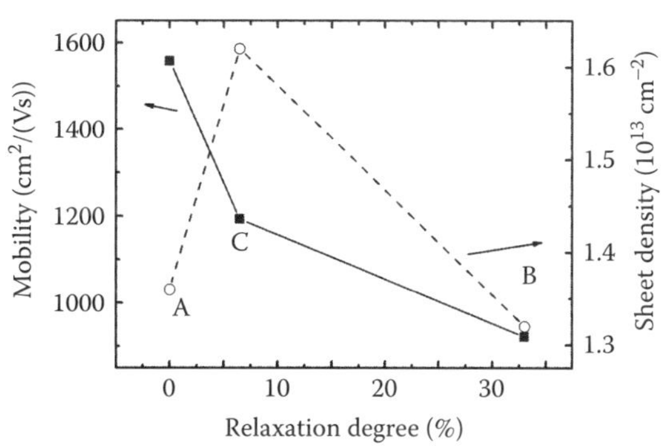

Figure 5.35 Hall mobility and sheet carrier density as a function of degree of relaxation (*R*). (After Liu, Z.-Y. et al., *Chin. Phys. B,* 20, 097701, 5 pp., 2011.)

interface roughness induced by the strain relaxation in the AlGaN barrier layer, which greatly intensify the scattering to 2DEG.

In summary, the strain state of barrier layer may vary with cap layers of different lattice constants, accompanied by changes in the dislocation density and interface morphology. These structural changes in the barrier layer have an impact on the 2DEG transport properties.

References

Ambacher, O., B. Foutz, J. Smart, J. R. Shealy, N. G. Weimann, K. Chu, M. Murphy et al. 2000. Two dimensional electron gases induced by spontaneous and piezoelectric polarization in undoped and doped AlGaN/GaN heterostructures. *Journal of Applied Physics* 87(1):334–344. doi: 10.1063/1.371866.

Chakraborty, A., K. C. Kim, F. Wu, J. S. Speck, S. P. DenBaars, and U. K. Mishra. 2006. Defect reduction in nonpolar a-plane GaN films using in situ SiN_x nanomask. *Applied Physics Letters* 89(4):041903-3.

Coffie, R., Y. C. Chen, I. Smorchkova, M. Wojtowicz, Y. C. Chou, B. Heying, and A. Oki. 2006. Impact of AlN interalayer on reliability of AlGaN/GaN HEMTs. *44th Annual IEEE International Reliability Physics Symposium,* March 26–30, San Jose, CA.

Gonschorek, M., J. F. Carlin, E. Feltin, M. A. Py, and N. Grandjean. 2006. High electron mobility lattice-matched AlInN/GaN field-effect transistor heterostructures. *Applied Physics Letters* 89(6):062106. doi: 10.1063/1.2335390.

Heikman, S., S. Keller, W. Yuan, J. S. Speck, S. P. DenBaars, and U. K. Mishra. 2003. Polarization effects in AlGaN/GaN and GaN/AlGaN/GaN heterostructures. *Journal of Applied Physics* 93(12):10114–10118. doi: 10.1063/1.1577222.

Hiroki, M., N. Maeda, and N. Kobayashi. 2002. Metalorganic vapor phase epitaxy growth of AlGaN/GaN heterostructures on sapphire substrates. *Journal of Crystal Growth* 237–239(1–4 II):956–960. doi: 10.1016/s0022-0248(01)02020-6.

Huang, J.-J., K.-C. Shen, W.-Y. Shiao, Y.-S. Chen, T.-C. Liu, T.-Y. Tang, C.-F. Huang, and C. C. Yang. 2008. Improved a-plane GaN quality grown with flow modulation epitaxy and epitaxial lateral overgrowth on r-plane sapphire substrate. *Applied Physics Letters* 92(23):231902.

Jang, H. W., C. M. Jeon, K. H. Kim, J. K. Kim, S.-B. Bae, J.-H. Lee, J. W. Choi, and J.-L. Lee. 2002. Mechanism of two-dimensional electron gas formation in $Al_xGa_{1-x}N$/GaN heterostructures. *Applied Physics Letters* 81(7):1249–1251. doi: 10.1063/1.1501162.

Kato, Y., S. Kitamura, K. Hiramatsu, and N. Sawaki. 1994. Selective growth of wurtzite GaN and $Al_xGa_{1-x}N$ on GaN/sapphire substrates by metalorganic vapor phase epitaxy. *Journal of Crystal Growth* 144(3–4):133–140. doi: 10.1016/0022-0248(94)90448-0.

Lee, S. R., D. D. Koleske, K. C. Cross, J. A. Floro, K. E. Waldrip, A. T. Wise, and S. Mahajan. 2004. In situ measurements of the critical thickness for strain relaxation in AlGaN/GaN heterostructures. *Applied Physics Letters* 85(25):6164–6166. doi: 10.1063/1.1840111.

Liu, Z.-Y., J.-C. Zhang, H.-T. Duan, J.-S. Xue, Z.-Y. Lin, J.-C. Ma, X.-Y. Xue, and Y. Hao. 2011. Effects of the strain relaxation of an AlGaN barrier layer induced by various cap layers on the transport properties in AlGaN/GaN heterostructures. *Chinese Physics B* 20(9): 097701(5pp.). doi: 10.1088/1674-1056/20/9/097701.

Moram, M. A., M. J. Kappers, Z. H. Barber, and C. J. Humphreys. 2007. Growth of low dislocation density GaN using transition metal nitride masking layers. *Journal of Crystal Growth* 298(1):268–271.

Selvaraj, S. L., T. Ito, Y. Terada, and T. Egawa. 2007. AlN/AlGaN/GaN metal-insulator-semiconductor high-electron-mobility transistor on 4 in. silicon substrate for high breakdown characteristics. *Applied Physics Letters* 90(17):173506-1. doi: 10.1063/1.2730751.

Shen, L., S. Heikman, B. Moran, R. Coffie, N. Q. Zhang, D. Buttari, I. P. Smorchkova, S. Keller, S. P. DenBaars, and U. K. Mishra. 2001. AlGaN/AlN/GaN high-power microwave HEMT. *IEEE Electron Device Letters* 22(10):457–459. doi: 10.1109/55.954910.

Xu, S. R. 2010. Growth and properties study of nonpolar and Semipolar GaN. PhD thesis, Xidian University, Xi'an. China.

Smorchkova, I. P., C. R. Elsass, J. P. Ibbetson, R. Vetury, B. Heying, P. Fini, E. Haus, S. P. DenBaars, J. S. Speck, and U. K. Mishra. 1999. Polarization-induced charge and electron mobility in AlGaN/GaN heterostructures grown by plasma-assisted molecular-beam epitaxy. *Journal of Applied Physics* 86(8):4520–4526. doi: 10.1063/1.371396.

Song, J., F. J. Xu, Z. L. Miao, Y. Wang, X. Q. Wang, and B. Shen. 2009. Influence of ultrathin AlN interlayer on the microstructure and the electrical transport properties of $Al_xGa_{1-x}N$/GaN heterostructures. *Journal of Applied Physics* 106(8):083711 (5 pp.). doi: 10.1063/1.3246866.

Xu, S. R. 2010. Growth and properties study of nonpolar and Semipolar GaN. PhD thesis, Xidian University, Xi'an. China.

Xu, S. R., J. C. Zhang, L. A. Yang, X. W. Zhou, Y. R. Cao, J. F. Zhang, J. S. Xue et al. 2011. Defect reduction in (11–20) nonpolar a-plane GaN grown on r-plane sapphire using TiN interlayers. *Journal of Crystal Growth* 327(1):94–97. doi: 10.1016/j.jcrysgro.2011.06.013.

Xu, Z.-H., J.-C. Zhang, Y. Hao, Z.-F. Zhang, Q.-W. Zhu, and H.-T. Duan. 2009. The effects of vicinal sapphire substrates on the properties of AlGaN/GaN heterostructures. *Chinese Physics B* 18(12):5457–5461. doi: 10.1088/1674-1056/18/12/054.

Yu, E. T., X. Z. Dang, L. S. Yu, D. Qiao, P. M. Asbeck, S. S. Lau, G. J. Sullivan, K. S. Boutros, and J. M. Redwing. 1998. Schottky barrier engineering in III-V nitrides via the piezoelectric effect. *Applied Physics Letters* 73(13):1880–1882. doi: 10.1063/1.122312.

Zhang, J.-F., Y. Hao, J. C. Zhang, and J. Y. Ni. 2008. The mobility of two-dimensional electron gas in AlGaN/GaN heterostructures with varied Al content. *Science in China Series F (Information Science)* 51(6):177–186. doi: 10.1007/s11432-008-0056-7.

Zhang, J.-F., C. Wang, J.-C. Zhang, and Y. Hao. 2006. Effects of donor density and temperature on electron systems in AlGaN/AlN/GaN and AlGaN/GaN structures. *Chinese Physics* 15(5):1060–1066. doi: 10.1088/1009-1963/15/5/032.

Zhang, J.-F., P.-Y. Wang, J.-S. Xue, Y.-B. Zhou, J.-C. Zhang, and Y. Hao. 2011. High electron mobility lattice-matched InAlN/GaN materials. *Acta Physica Sinica* 60(11):611–616.

6 AlGaN/GaN Multiple Heterostructure Materials and Electronic Devices

The HEMT device performance based on conventional AlGaN/GaN heterostructure can be further improved by the optimization of the heterostructure. An important approach is to use the modulation effect of polarization on the energy bands and carrier distribution to obtain high-performance AlGaN/GaN multiple heterostructure materials. This chapter discusses the layered structure optimization of multiple heterostructures in theory based on the self-consistent solution of Schrödinger–Poisson equations and analyzes the properties of the channels and potential barriers formed in the multiple heterostructures. Currently, there are two major development trends of the multiple heterostructures: one is the formation of the back barrier under the channel in the single-channel heterostructure, and the other is the direct formation of multichannel heterostructures to increase the conductivity of the channel region of the HEMT.

6.1 Al(Ga, In)N/InGaN/GaN

As Figure 6.1 shows, GaN serves as the material for both the buffer layer and the channel in conventional AlGaN/GaN heterostructures. Owing to the lower barrier height formed in GaN under the channel, the easy spillover of channel carriers into the buffer layer to become three-dimensional electrons at high temperature or a relatively high gate/drain voltage leads to a reduced 2DEG confinement and performance degradation. The short channel effect also causes soft channel pinch-off and large off-state leakage currents in deep submicrometer GaN HEMTs.

One solution to this problem is the replacement of GaN by InGaN with a narrower bandgap as the channel material to form AlGaN/InGaN/GaN(/nucleation layer/substrate) structures. The conduction band offset and polarization effect yield a relatively high GaN back barrier to InGaN channel (Figure 6.2) in favor of better 2DEG confinement, hence reducing the off-state leakage current and improving the pinch-off characteristics of the device. The stronger polarization effect in AlGaN/InGaN/GaN than that in AlGaN/GaN is also advantageous for a higher 2DEG density.

In actual material growth, due to the lower growth temperature of InGaN than that of GaN and AlGaN and the high sensitivity of indium content to temperature,

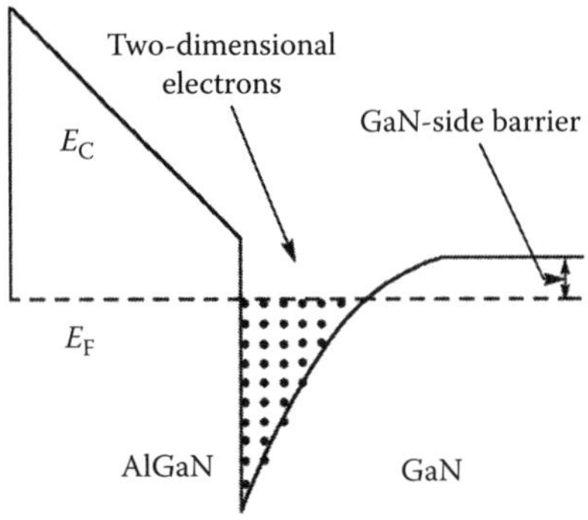

Figure 6.1 Band diagram of an AlGaN/GaN heterostructure. Low potential barrier under the 2DEG channel makes it easier for two-dimensional carriers in the channel to spill over and become bulk electrons.

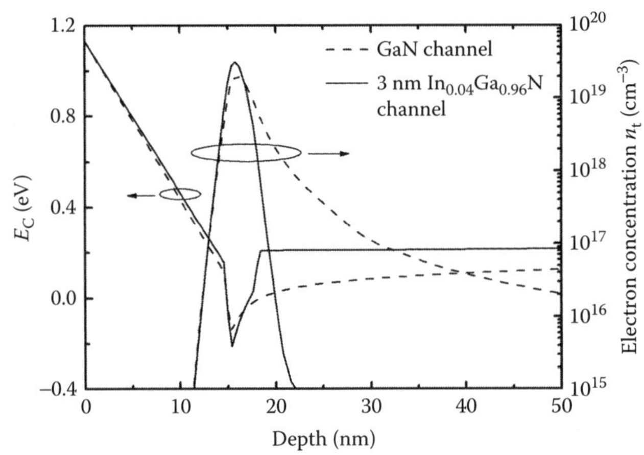

Figure 6.2 Schematic band diagram and carrier distribution of fully strained AlGaN/InGaN/GaN heterostructure showing a higher GaN back barrier height than that of AlGaN/GaN heterostructure.

the subsequent high temperature growth of AlGaN barrier may lead to degraded quality of the InGaN layer and indium segregation as well as a rougher heterointerface, thus affecting the 2DEG electrical properties. Therefore, a thin film of AlN is generally grown on top of the InGaN channel at the same growth temperature for InGaN to protect the InGaN surface, followed by the high temperature growth of the AlGaN barrier layer to form the whole heterostructure (Song et al. 2011). And the dislocation density in the GaN template under the InGaN channel layer should be reduced to lower the possibility of indium segregation at defect sites.

Experiments show that the AlGaN/InGaN/GaN double heterostructure (DH) generally has higher Hall electron concentrations but lower room temperature Hall mobility (500–800 $cm^2/[Vs]$) than the AlGaN/GaN heterostructure. Okamoto et al. found that a reduction of the InGaN thickness of the $Al_{0.3}Ga_{0.7}N/In_{0.06}Ga_{0.94}N/GaN$ structure from 15 to 3 nm resulted in a significant increase in mobility from 500 to 1110 $cm^2/(Vs)$ (Okamoto et al. 2004), which they attributed to the smaller fluctuation of indium content in thin InGaN layer.

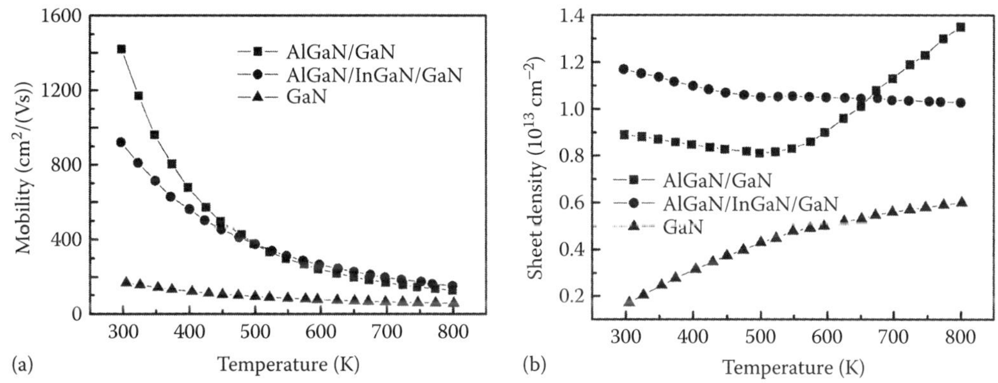

Figure 6.3 Temperature-dependent electrical properties of $Al_{0.25}Ga_{0.75}N/In_{0.03}Ga_{0.97}N/GaN$ and $Al_{0.25}Ga_{0.75}N/GaN$ heterostructures. (a) Temperature-dependent mobility showing higher mobility in InGaN channel at high temperature. (b) Temperature-dependent electron density showing no sign of thermally excited parallel conductivity in InGaN channel at high temperature. (Reproduced from Song, J. et al., *Chin. Phys. B*, 20, 057305, 5 pp., 2011.)

Xie et al. compared the effects of AlGaN, InAlN, and AlInGaN quaternary barrier layers on the electrical properties of the InGaN channel heterostructures (Xie et al. 2007) and found that the room temperature Hall mobility of the heterostructure was improved from 870 to 1230 cm^2/(Vs) by replacing the $Al_{0.25}Ga_{0.75}N$ barrier layer grown at high temperature (1030°C) with the quaternary $Al_{0.24}In_{0.01}Ga_{0.75}N$ alloy grown at a lower growth temperature (900°C), and the room temperature Hall electron concentration was improved from 1.26×10^{13} to 2.12×10^{13} cm^{-2} by replacing the AlGaN barrier with an InAlN barrier layer which is also grown at a relatively low-temperature (750°C–800°C). It is believed that even with the protection of the low temperature AlN interlayer over the InGaN surface, the high temperature in AlGaN barrier layer growth may still induce degradation of the quality of the InGaN channel. Therefore, AlInN or AlInGaN are better barrier layer material than AlGaN for InGaN channel heterostructures.

The research on the temperature-dependent electrical properties of $Al_{0.25}Ga_{0.75}N/In_{0.03}Ga_{0.97}N/GaN$ and $Al_{0.25}Ga_{0.75}N/GaN$ heterostructures by Song Jie et al. (2011) showed that at high temperatures the InGaN channel heterostructure exhibited better current confinement (Figure 6.3b) possibly because the electron depletion layer formed at the InGaN/GaN interface effectively suppress the parallel conduction due to the thermal excitation in GaN at high temperatures. It is shown that the $Al_xGa_{1-x}N/In_yGa_{1-y}N/GaN$ heterostructures are advantageous in high temperature device applications (Figure 6.3a).

6.2 Introduction of AlGaN Back Barrier under GaN Channel

For nitride HEMT structures still using GaN as the channel material, a foreign material can also be inserted under the GaN channel to form the back barrier layer, then the 2DEG will be better confined in the channel between the top barrier and the back barrier.

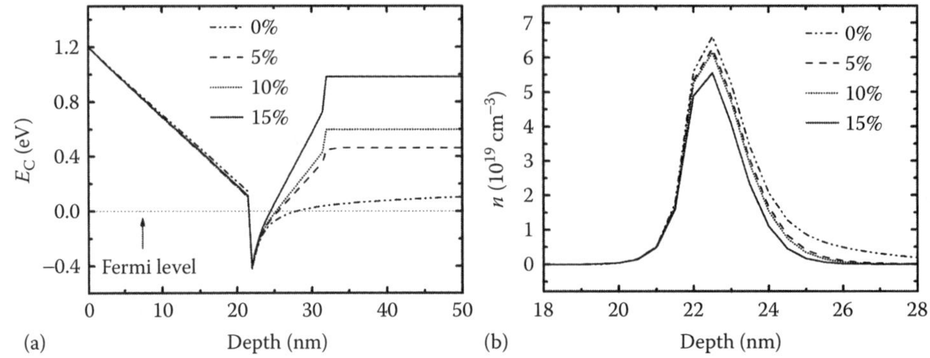

Figure 6.4 2DEG distribution of 22 nm $Al_{0.32}Ga_{0.68}N$/10 nm GaN/1.3 μm $Al_xGa_{1-x}N$ DH as a function of the Al content in the $Al_xGa_{1-x}N$ buffer layer. (a) Band diagram and (b) carrier distribution. (Courtesy of Juncai Ma.)

The introduction of a relatively thin layer of AlGaN under the GaN channel into the AlGaN/GaN heterostructure with a GaN buffer yields the AlGaN/GaN/AlGaN/GaN(/nucleation layer/substrate) structure with the lower AlGaN layer acting as the back barrier of the top channel, but meanwhile it results in the formation of a parasitic channel at the lower AlGaN/GaN interface. Consequently, a thick AlGaN buffer layer is generally introduced as the back barrier between the GaN channel and the nucleation layer to form a single-channel AlGaN/GaN/AlGaN(/nucleation layer/substrate) DH.

Self-consistent Schrödinger–Poisson calculations of this structure (Figure 6.4) reveal an increase in back barrier height and better 2DEG confinement yet a reduced sheet density with increasing Al content of the buffer layer. This is attributed to the reduced lattice mismatch and piezoelectric polarization in the subsequently grown AlGaN/GaN heterostructures owing to the relatively thin compressively strained GaN channel layer grown on the fully relaxed thick AlGaN buffer layer. Thus, the optimization of the Al composition of the AlGaN back barrier is required for a trade-off between the 2DEG confinement and density. A higher Al composition also leads to greater difficulty in the growth of thick AlGaN buffer layer material.

An example is provided to illustrate the optimization procedures of AlGaN/GaN/AlGaN DHs (Meng et al. 2012). Sample A is a 22 nm $Al_{0.32}Ga_{0.68}N$/1.3 μm GaN single heterostructure (SH) without the AlGaN back barrier, and sample B is a 22 nm $Al_{0.32}Ga_{0.68}N$/10 nm GaN/1.3 μm $Al_{0.07}Ga_{0.93}N$ DH grown on an AlGaN back barrier with an Al content of 7%. The reciprocal space maps of the (105) reflection of sample B are given in Figure 6.5a, showing a distinct separation of the diffraction pattern of the $Al_{0.07}Ga_{0.93}N$ buffer layer from that of the $Al_{0.32}Ga_{0.68}N$ barrier layer with the diffraction patterns of both located exactly on the same straight line along the k_z orientation, indicating complete strain in the $Al_{0.32}Ga_{0.68}N$ barrier layer relative to the $Al_{0.07}Ga_{0.93}N$ buffer layer. The contactless Hall measurement results (Table 6.1) demonstrate a decreased sheet electron density and an improved electron mobility of sample B compared to sample A, which agree with the calculated results from Figure 6.4.

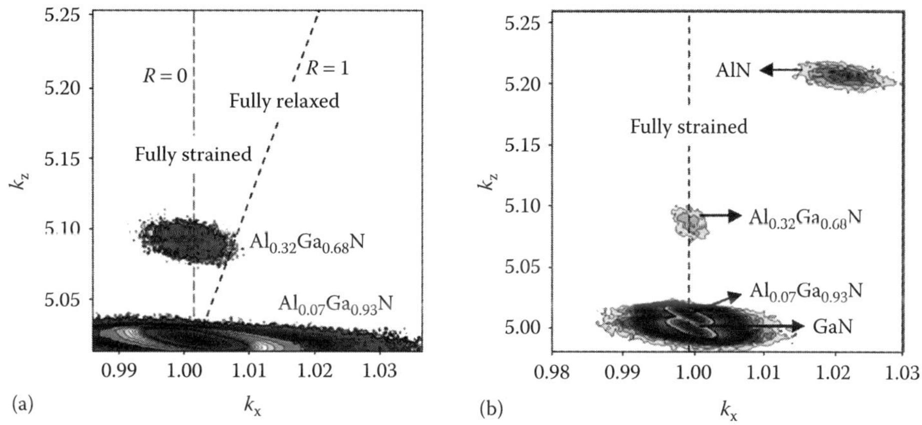

Figure 6.5 (105) reciprocal space maps of DH samples. (a) Sample B. (After Ma, J. et al., *J. Semiconduct.*, 33, 014002, 5 pp., 2012.) (b) Sample C. (Courtesy of Juncai Ma.)

Table 6.1 Electrical Properties Obtained from Contactless Hall Measurements

SAMPLE	SAMPLE STRUCTURE	HALL MOBILITY ($cm^2/[Vs]$)	HALL ELECTRON CONCENTRATION (10^{13} cm^{-2})	SHEET RESISTANCE (Ω/sq)
A	22 nm $Al_{0.32}Ga_{0.68}N$/1.3 μm GaN	1356	1.17	394
B	22 nm $Al_{0.32}Ga_{0.68}N$/10 nm GaN/1.3 μm $Al_{0.07}Ga_{0.93}N$	1508	0.85	450
C	22 nm $Al_{0.32}Ga_{0.68}N$/10 nm GaN/600 nm $Al_{0.07}Ga_{0.93}N$/graded $Al_xGa_{1-x}N$/700 nm GaN	1752	1.06	354
D	22 nm $Al_{0.32}Ga_{0.68}N$/14 nm GaN/600 nm $Al_{0.07}Ga_{0.93}N$/graded $Al_xGa_{1-x}N$/700 nm GaN	1821	1.04	340
E	22 nm $Al_{0.32}Ga_{0.68}N$/14 nm GaN/0.7 nm AlN/600 nm $Al_{0.07}Ga_{0.93}N$/graded $Al_xGa_{1-x}N$/700 nm GaN	1862	0.95	362

In order to improve the crystalline quality of the $Al_{0.07}Ga_{0.93}N$ buffer layer, the 1.3 μm $Al_{0.07}Ga_{0.93}N$ buffer layer was replaced by a composite buffer layer (sample C) consisting of a 700 nm GaN buffer layer and a 600 nm $Al_{0.07}Ga_{0.93}N$ buffer grown on top of it. A thin film of $Al_xGa_{1-x}N$ with a linearly graded Al composition x (x varying gradually from 0 to 0.07 along the growth direction) is grown between the $Al_{0.07}Ga_{0.93}N$ and GaN buffer layer to avoid the generation of a parasitic channel in between. It can be seen from the (105) reciprocal space maps of sample C in Figure 6.5b that the diffraction pattern centers of the GaN buffer layer, the $Al_{0.07}Ga_{0.93}N$ buffer layer and the $Al_{0.32}Ga_{0.68}N$ top barrier layer lie on a straight line perpendicular to the k_x-axis, suggesting both AlGaN layers maintain the complete strain state with the same lattice constant as that of GaN. $Al_{0.07}Ga_{0.93}N$ has an equivalent diffraction spot area to GaN, which indicates a significant increase in the crystalline quality of $Al_{0.07}Ga_{0.93}N$ buffer layer comparable to that of the GaN buffer layer. The measured

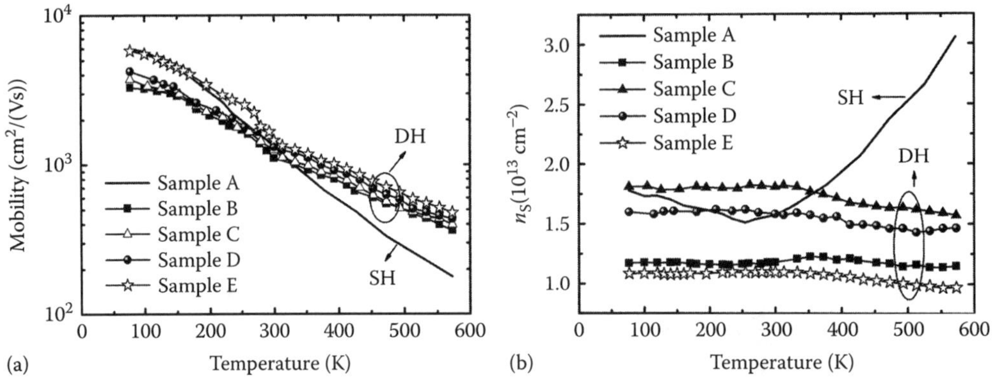

Figure 6.6 Temperature-dependent Hall characteristics of AlGaN/GaN single-heterojunction (SH) and double-heterojunction (DH) samples. (a) Mobility as a function of temperature. (b) Sheet density as a function of temperature. (After Meng, F. et al., *J. Appl. Phys.*, 112, 023707, 6 pp., 2012.)

electrical properties show a significantly higher Hall electron concentration and mobility of the sample with composite buffer layer than that with the $Al_{0.07}Ga_{0.93}N$ single buffer layer with a reduced sheet resistance ununiformity from 6.53% to 1.95%.

The mobility rises as the GaN channel layer thickness increases to 14 nm (sample D) which may be explained by reduced interface roughness scattering to the 2DEG from the back barrier side. A slight increase in mobility is observed again when a 0.7 nm AlN interlayer is introduced between the GaN channel of sample D and the AlGaN back barrier (sample E) to reduce the possible alloy disorder scattering to the 2DEG by the back barrier.

The van der Pauw measurement results of samples A~E at temperatures 77–573 K are listed in Figure 6.6, showing a slightly lower mobility of the AlGaN/GaN/AlGaN DH than that of the AlGaN/GaN SH at low temperatures due primarily to the greater impact of the interface roughness scattering and the alloy disorder scattering on the 2DEG resulting from the presence of the top barrier/channel and channel/back barrier interfaces in the DH. The interface roughness scattering and alloy disorder scattering were greatly reduced by a series of optimizations of the DHs from sample B to E, achieving in the DH sample E a Hall mobility of 5973 $cm^2/(Vs)$ at 77 K close to that of a conventional SH. At temperatures above 300 K, the AlGaN/GaN/AlGaN DH has a remarkably higher mobility than the AlGaN/GaN SH with little variation in its Hall electron concentration with temperature, while the Hall electron concentration of the AlGaN/GaN SH increases from 1.57×10^{13} cm^{-2} at room temperature to 3.06×10^{13} cm^{-2} at 573 K.

It is suggested that in the AlGaN/GaN SH, the background electron density may increase due possibly to the further ionization of the background dopants or some deep level impurities as the temperature rises; on the other hand, the two-dimensional electrons in the channel after acquiring sufficient energy spill over into the buffer layer and become bulk electrons. Thus, the effect of low mobility bulk electron conduction is significantly intensified at high temperature, leading to a dramatic decline of the

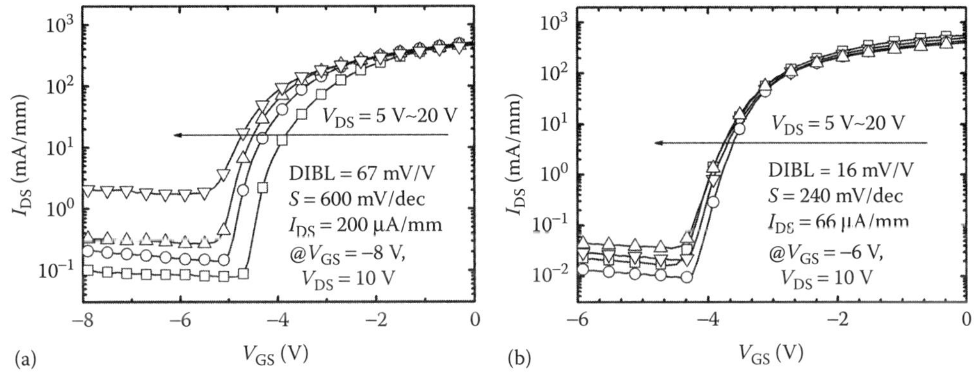

Figure 6.7 Transfer characteristics as a function of drain voltage of AlGaN/GaN HEMTs. (a) Conventional SH-HEMT and (b) DH-HEMT with AlGaN buffer layer as back barrier. (Courtesy of Linyu Shi.)

Hall mobility and an increase of the Hall electron density at high temperature. In the DH samples, the enhanced confinement of the 2DEG by the AlGaN back barrier layer effectively suppresses the spillover of electrons from channel into the buffer layer, thus better quantum conductance stability at high temperature. Meanwhile, the comparatively large energy separation between the conduction band edge of the AlGaN back barrier and the Fermi level curbs the ionization of background dopants or the possible deep level impurities. Consequently, the DH samples have greater Hall mobilities than SH ones at high temperatures and show little variation in electron density. At 573 K, the DH sample E has a mobility of 478 $cm^2/(Vs)$, more than twice that of the SH sample A (179 $cm^2/[Vs]$).

In terms of the properties of the fabricated DH and SH HEMTs, the AlGaN/GaN/AlGaN DH-HEMT enjoys a significantly lower off-state leakage current both at room temperature and at high temperatures and a smaller temperature rise caused degradation of on-state characteristics than the AlGaN/GaN SH-HEMT. The AlGaN/GaN/AlGaN DH-HEMT also has an improved breakdown voltage, a markedly suppressed current collapse, and a reduced drain-induced barrier lowering (DIBL) and subthreshold slope S (the increase of gate voltage required for the drain current in the subthreshold region to increase by one order of magnitude), all benefiting from the improved channel carrier confinement by the AlGaN back barrier (Figure 6.7).

6.3 InGaN Back Barrier

Besides AlGaN, InGaN can also be used to form back barrier structure. Though InGaN has narrower bandgap than GaN, in the AlGaN/GaN/InGaN/GaN (/nucleation layer/substrate) multiple heterostructure formed by inserting a thin InGaN layer underneath the GaN channel of the AlGaN/GaN heterostructure, compressively strained InGaN has an opposite piezoelectric polarization field to that in the AlGaN layer, significantly elevating the energy bands of the InGaN interlayer and

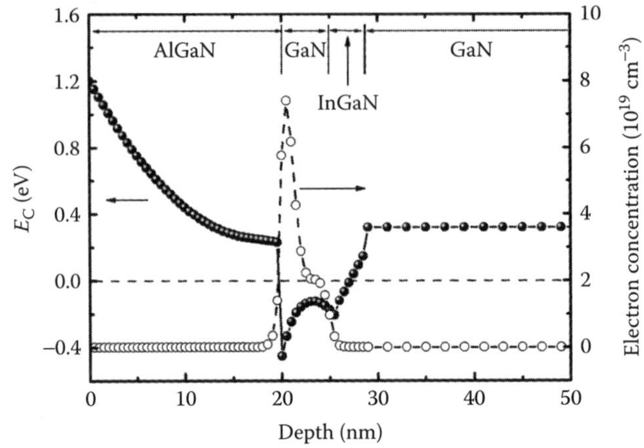

Figure 6.8 Band profiles and carrier distributions of AlGaN/GaN/InGaN/GaN multiple heterostructures. (Courtesy of Juncai Ma.)

the underlying GaN buffer layer via polarization-induced band bending, thus forming a InGaN back barrier structure, which improves the 2DEG confinement and prevents the spillover of electrons to the buffer layer (Figure 6.8). The polarization effect and the narrower bandgap of InGaN also leads to the formation of a very shallow subchannel at the interface between the GaN channel and the InGaN back barrier, making it easy for the electrons in the subchannel to get into the main channel. The AlGaN/GaN/InGaN/GaN heterostructure with InGaN back barrier has the advantages over the AlGaN/InGaN/GaN heterostructure with the InGaN channel that the interconnecting main channel and subchannel expands the whole channel in favor of the device linearity and that there is generally little degradation of 2DEG mobility due to the weak alloy disorder scattering in the InGaN subchannel since the GaN main channel still dominates.

In actual material growth, the InGaN interlayer is required to be relatively thin (generally not more than 10 nm) to avoid indium segregation. A relatively thin InGaN layer suffices to achieve a high back barrier thanks to its strong polarization strength. To avoid indium segregation during the GaN growth at a rising temperature after the InGaN back barrier growth, a thin GaN film can be grown first at the same growth temperature as InGaN to protect the InGaN surface followed by the high temperature growth of the GaN channel layer.

A comparison of the material and HEMT device properties is made between the AlGaN/GaN SH and the AlGaN/GaN/InGaN/GaN multiple heterostructure. According to the mercury probe *C*–*V* profiles shown in Figure 6.9, the heterostructure with InGaN back barrier shows no sign of the impact of the parasitic channel but a larger absolute value of threshold voltage than that of the SH, and a lower background electron density under the effect of the InGaN back barrier. The measured HEMT properties (Shi et al. 2010) indicate the InGaN back barrier has similar effects to

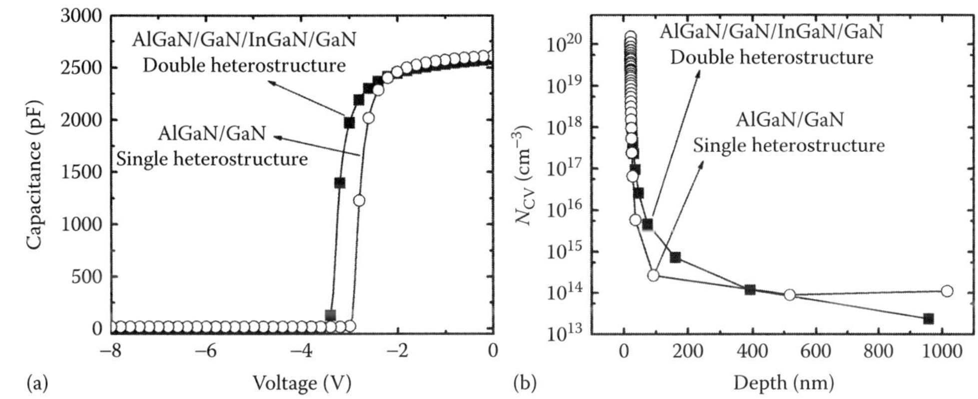

Figure 6.9 *C-V* profiles and carrier distributions of AlGaN/GaN SH and AlGaN/GaN/InGaN/GaN multiple heterostructure. (a) *C-V* profiles and (b) Carrier distributions. (Courtesy of Linyu Shi.)

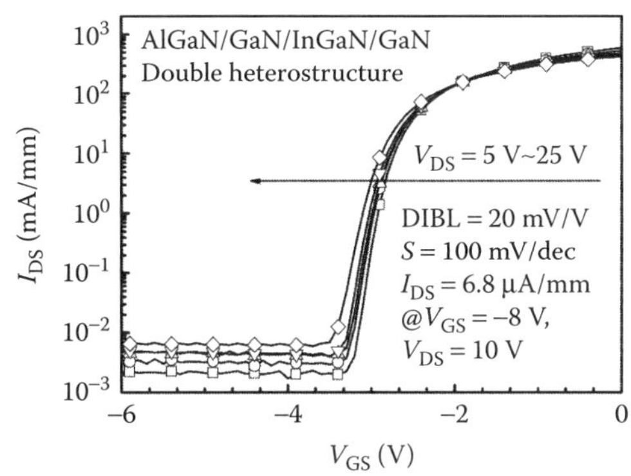

Figure 6.10 Transfer characteristics for AlGaN/GaN/InGaN/GaN HEMT at source-drain voltages of 5, 10, 15, 20, and 25, respectively. (Courtesy of Linyu Shi.)

the AlGaN back barrier, that is, a reduced off-state leakage current (Figure 6.10), a decreased DIBL and subthreshold slope, and an improved breakdown voltage and high temperature characteristics of the device.

6.4 Double/Multiple Channel AlGaN/GaN Heterostructures

What have been discussed above involves only the single-channel heterostructures. If the single channel is replaced by the double/multiple channels while maintaining an equivalent 2DEG density and mobility in each channel to that of the single channel, the material conductivity will increase linearly with the number of channels, significantly improving the current driving capability of the device. In the ungated regions of the double/multiple channel HEMTs, the reduced sheet resistance is also favorable for the decrease in source-gate and gate-drain access resistance and the enhancement of transconductance and cut-off frequency linearity (i.e., a relatively high transconductance and cut-off frequency under a relatively large drain current), thereby a better power performance and linearity at high frequency.

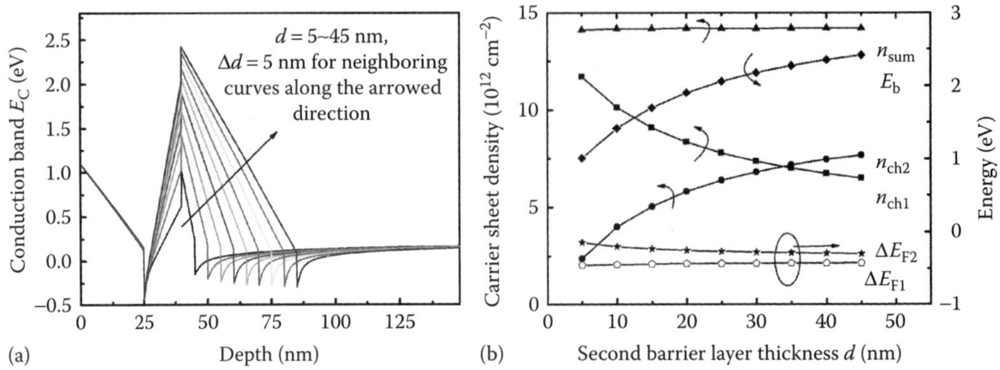

Figure 6.11 Effect of second barrier thickness d on double-channel AlGaN/GaN heterostructures. (a) **(See color insert.)** Conduction band edge E_C as a function of thickness d of second barrier layer. (b) Quantum well depths ΔE_{F1} and ΔE_{F2}, the channel carrier sheet densities n_{ch1} and n_{ch2}, the sum of the sheet densities n_{sum}, and the barrier height E_b between the two channels as a function of second barrier thickness d. (Courtesy of Xiaofan Fu.)

The theoretical simulation (Figure 6.11) of the unintentionally doped double-channel AlGaN/GaN heterostructure, that is, AlGaN/GaN/AlGaN/GaN(/nucleation layer/substrate), shows changes in the 2DEG density and band structure in the two channels under the polarization effect with varied structural parameters such as the Al compositions and thicknesses of the two AlGaN barrier layers and the thicknesses of the two GaN channel layers, respectively. However, except for an increase in the sum of the 2DEG densities in the two channels resulting from the increase in the Al composition or thickness of the top AlGaN barrier layer, the total 2DEG density of the heterostructure displays little variation (as shown in Figure 6.11b), which equals that of the AlGaN/GaN SH with the same top AlGaN barrier layer. This indicates that although the second 2DEG can be generated under the polarization effect by inserting a relatively thin and unintentionally doped second AlGaN barrier layer into the AlGaN/GaN heterostructure (in contrast, a relatively thick and heavily doped second AlGaAs barrier layer is required for the generation of the second 2DEG in the AlGaAs/GaAs heterostructure), no electrons are produced by the polarization effect which changes only the distributions of carriers and energy bands, so the electrons in undoped AlGaN/GaN heterostructures should originate from the surface states. Consequently, the variation of the second AlGaN layer in either composition, thickness, or position cannot significantly increase the total electron density (a slight rise in total electron density may be seen in a multichannel heterostructure under the effect of background doping ionization in the barrier layers under the first channel), and doping the AlGaN barrier layer is required for the linear increase of the 2DEG density with the number of channels in the AlGaN/GaN heterostructures. Additionally, AlGaN with gradually increasing Al composition from top to bottom should be employed for the second barrier layer to achieve a relatively low energy barrier in the electron transfer between channels (since the source and drain electrodes are fabricated on top surface of the heterojunction, the interchannel transfer of electrons is necessary for the

Table 6.2 Electrical Properties of Double-Channel Heterostructures with Varied Barrier Layer Doping and Single-Channel Heterostructure Samples

SAMPLE	SILANE FLOW FOR BARRIER LAYER DOPING (sccm)	HALL EFFECT MEASUREMENTS		
		SHEET RESISTANCE (Ω/sq)	SHEET DENSITY (10^{13} cm^{-2})	MOBILITY (cm^2/ [Vs])
R	Unintentionally doped single channel	354	1.22	1398
A	Unintentionally doped double-channel	328	1.23	1552
B	Double-channel, first barrier, 50	263	1.92	1234
C	Double-channel, second barrier, 50	143	3.91	1112
D	Double-channel, second barrier, 5	145	3.22	1331

Source: Courtesy of Xiaofan Fu.

current on the electrodes to flow in and out of the channels) and better connectivity between two channels.

On the basis of theoretical analysis, we prepared unintentionally doped single- and double-channel AlGaN/GaN heterostructures, whose Hall electrical properties are listed in Table 6.2. The double-channel samples have virtually the same electron density as, but a significantly higher mobility than the single-channel samples. The growth and properties of the double-channel structures with varied barrier layer doping show that with the same doping concentration and total doping amount, doping the second barrier layer has greater effect on the improvement of the electron density than doping the first barrier; but an excessively heavy doping of the second barrier layer poses a negative influence on the crystalline quality and the surface/interface, leading to an obvious decrease in the mobility of the heterostructure. Therefore, a relatively low doping concentration is desirable for good quality of DHs and carrier transport properties.

Since the variation of the electrical properties of the double-channel heterostructure with rising temperature may differ from that of the single-channel heterostructure, double-channel AlGaN/GaN circular Schottky diodes were prepared with the *C-V* measurement results (300–673 K) shown in Figure 6.12a. The two platforms in the *C-V* curves correspond to the two sheets of 2DEG. A visible increase in the background carrier concentration in GaN buffer layer is seen in the *C-V* carrier profile (Figure 6.12b) at temperatures above 570 K, similar to that of the single-channel AlGaN/GaN heterostructure. The carrier concentration distributions of the first and second channel displayed in Figure 6.12c and d, respectively, show a very small shift of the carrier concentration peak toward the surface in the first channel but a large shift in the second one, indicating a far greater effect of rising temperature on the second channel than on the first one. This is because the relatively shallow quantum well in the second channel in this double-channel structure becomes even shallower at high temperature, making it easy for the carriers in the channel to spill over into the buffer layer. Therefore, deepening the quantum well of the second channel or adding a back barrier under the second channel by structural optimization may suppress this effect.

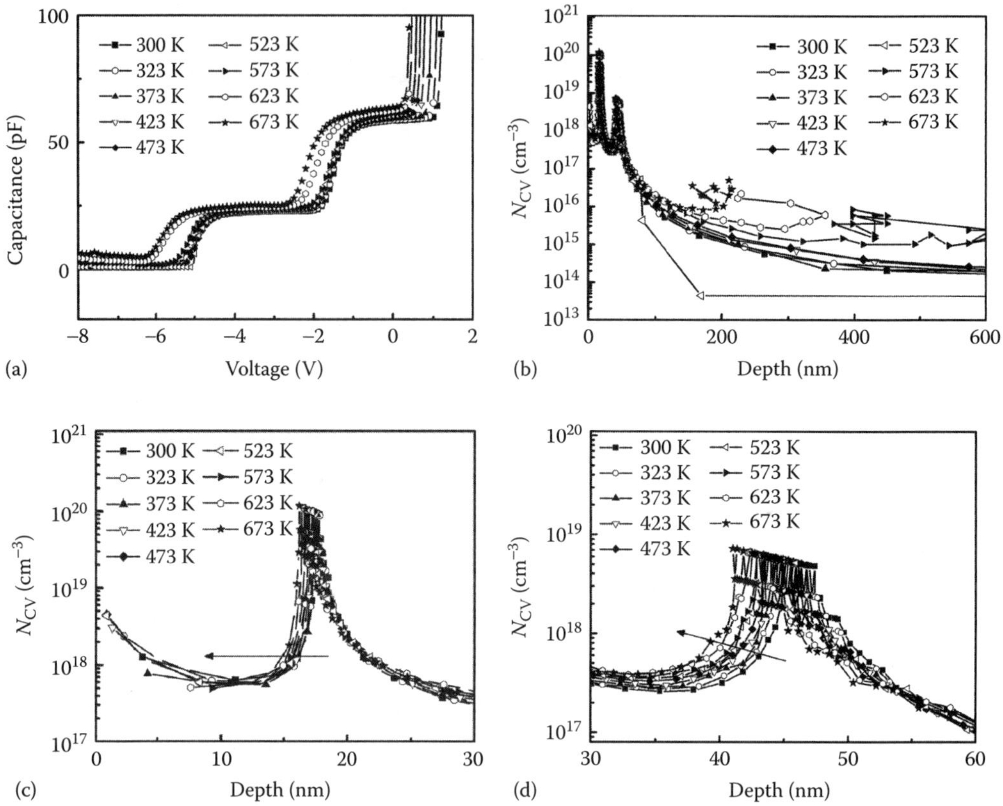

Figure 6.12 *C-V* measurement results of unintentionally doped double-channel AlGaN/GaN heterostructure at varied temperatures. (a) *C-V* curves, (b) *C-V* carrier concentration distributions, (c) carrier concentrations in first channel, and (d) carrier concentrations in second channel. (Courtesy of Xiaofan Fu.)

Single/double-channel AlGaN/GaN HEMTs were prepared for a comparison of their properties. The regions under the Schottky gates of the unintentionally doped double-channel HEMTs were etched for different lengths of time to obtain devices with the planar double-channel (etched for 0s), recessed-gate double-channel (etched for 26s) and recessed-gate single-channel (etched for 55s with the first channel etched off) structures. The output and transfer characteristics of these devices are shown in Figure 6.13. The planar double-channel device achieves a maximal saturation drain current of 1400 mA/mm, suggesting good current driving capacity of the double-channel HEMTs; its two peaks of the transconductance (g_m) are 203 mS/mm (−4.7 V) and 289 mS/mm (−1.6 V), respectively, and the broadened high transconductance region is much favorable for better device linearity. However, as shown in Figure 6.14 the planar double-channel device has a low off-state breakdown voltage of only 58 V. The recessed-gate double-channel device has a current output and transconductance comparable to that of the planar double-channel device, but with a breakdown voltage of 114 V, nearly twice that of the planar one. It suggests that the recessed gate etching doesn't bring obvious etching damage, and meanwhile the improved breakdown voltage may be attributed to the removing of the leaky film surface under the gate, or the

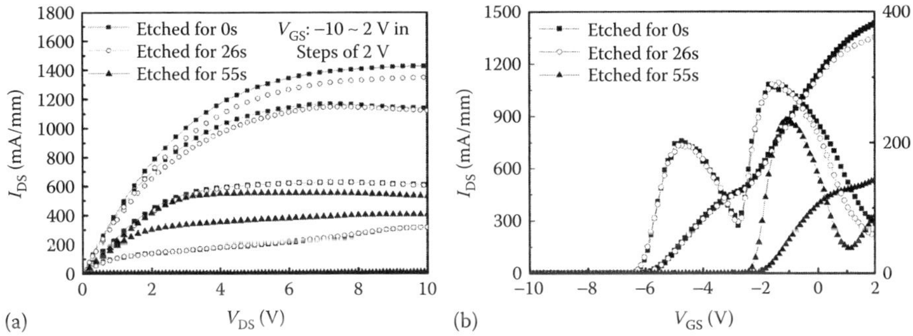

Figure 6.13 Direct-current characteristic curves of double-channel HEMTs with varied recess etching times. (a) Output characteristic curves and (b) transfer characteristic curves. (Courtesy of Xiaofan Fu.)

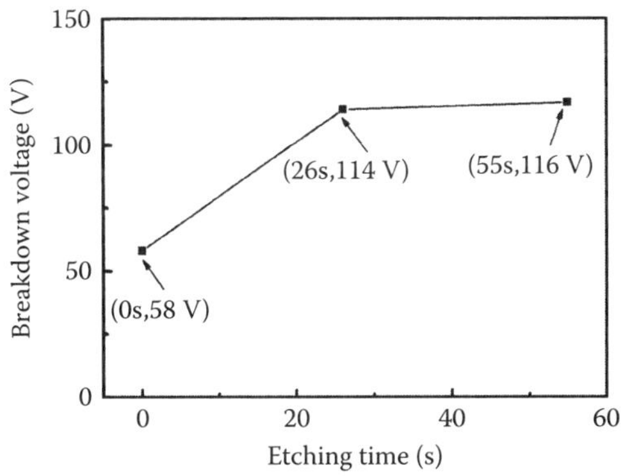

Figure 6.14 Breakdown voltage versus gate etching time. (Courtesy of Xiaofan Fu.)

change of electric field distribution in the device induced by the closer gate to the two channels. The recessed-gate single-channel device has a breakdown voltage comparable to the recessed-gate double-channel device but a significantly lower current output and transconductance due to the quite poor conductivity of only the second channel, which is also another proof of the large on-state current, high transconductance, good linearity, and high off-state breakdown voltage of the recessed-gate double-channel device.

References

Ma, J., J. Zhang, J. Xue, Z. Lin, Z. Liu, X. Xue, X. Ma, and Y. Hao. 2012. Characteristics of AlGaN/GaN/AlGaN double heterojunction HEMTs with an improved breakdown voltage. *Journal of Semiconductors* 33(1):014002 (5 pp.). doi: 10.1088/1674-4926/33/1/014002.

Meng, F., J. Zhang, H. Zhou, J. Ma, J. Xue, L. Dang, L. Zhang, M. Lu, S. Ai, X. Li, and Y. Hao. 2012. Transport characteristics of AlGaN/GaN/AlGaN double heterostructures with high electron mobility. *Journal of Applied Physics* 112(2):023707 (6 pp.). doi: 10.1063/1.4739408.

Okamoto, N., K. Hoshino, N. Hara, M. Takikawa, and Y. Arakawa. 2004. MOCVD-grown InGaN-channel HEMT structures with electron mobility of over 1000 cm2/V s. *Journal of Crystal Growth* 272(1–4):278 (7 pp.). doi: 10.1016/j.jcrysgro.2004.08.071.

Shi, L., J. Zhang, Y. Hao, H. Wang, J. Xue, X. Ou, X. Fu, and K. Chen. 2010. Growth of InGaN and double heterojunction structure with InGaN back barrier. *Journal of Semiconductors* 31(12):123001 (4 pp.). doi: 10.1088/1674-4926/31/12/123001.

Song, J., F.-J. Xu, C.-C. Huang, F. Lin, X.-Q. Wang, Z.-J. Yang, and B. Shen. 2011. Different temperature dependence of carrier transport properties between AlxGa1-xN/InyGa1-yN/GaN and AlxGa1-xN/GaN heterostructures. *Chinese Physics B* 20(5):057305 (5 pp.). doi: 10.1088/1674-1056/20/5/057305.

Xie, J., J. H. Leach, X. Ni, M. Wu, R. Shimada, U. Ozgur, and H. Morkoc. 2007. Electron mobility in InGaN channel heterostructure field effect transistor structures with different barriers. *Applied Physics Letters* 91(26):262102-1. doi: 10.1063/1.2824461.

7

Growth of InAlN/GaN Heterostructures by Pulsed MOCVD

The ultrahigh frequency operation is an important goal for the nitride semiconductor devices. For the microwave power GaN HEMTs to work at higher operation frequencies, especially in millimeter wave or even entering the THz band, the gate length is often required to be less than 150 nm, and in some cases 50 nm or even 10 nm, where the short channel effects of the devices become a major problem. The suppression of the short channel effect on the millimeter wave devices requires a thinning of the AlGaN barrier (reducing the distance between the gate and the 2DEG channel). The back barrier, high-resistance buffer layer, and so on are also employed to obtain better pinch-off characteristics and high output resistance. However, in order to maintain a high channel conductance of the heterostructure at a reduced barrier thickness, the Al content in the barrier needs to be increased to maintain a high 2DEG density. In terms of material properties, the strain induced by the increased lattice mismatch between high Al content AlGaN and GaN causes an increase in the AlGaN/GaN interface roughness as well as a degraded AlGaN barrier crystalline quality and even strain relaxation, hence decreasing the 2DEG sheet density and mobility and reducing device performance and reliability under long time, high bias, and high temperature. And in terms of device fabrication, the thin and highly defected AlGaN barrier makes the 2DEG highly sensitive to the barrier surface condition and strain, and thus the 2DEG depletion may arise from surface degradation caused by oxidization and etching damage in the fabrication; on the other hand, the greater insulativity of high Al content AlGaN inevitably leads to a higher ohmic contact resistance, which also imposes difficulty in the development of millimeter wave devices.

The AlGaN/GaN HEMT has strong spontaneous and piezoelectric polarization. The inverse piezoelectric effect may arise from too strong an electric field in the device (especially aggressively scaling-down device) at high voltage, resulting in material degradation and in worse cases microcracks (see Chapter 11); therefore it is desired to find an HEMT structure that can eliminate the piezoelectric polarization without seriously affecting the 2DEG density so as to avoid the negative influence of strong inverse piezoelectric effect on the device reliability.

The solution to the above issues requires the development of better group III nitride HEMT heterostructures and devices for high frequency, high temperature, and high reliability applications. The substitution of InAlN for AlGaN barrier layers to form the InAlN/GaN HEMT heterostructures proves a successful practice.

7.1 The Advantages of Nearly Lattice-Matched InAlN/GaN Heterostructures and the HEMT Properties

The nearly lattice-matched InAlN/GaN heterostructure has become a highly favored structure for GaN HEMTs lately. Shown in Figure 7.1 is the InAlN bandgap as a function of the lattice constant and the structure of the InAlN/GaN heterojunction. $In_xAl_{1-x}N$ has the greatest range of bandgap (0.7–6.2 eV), lattice constant (3.112–3.522 Å), and spontaneous polarization intensity (−0.090 to −0.042 C/m^2) of all III nitride alloys. InAlN with an indium content of ~17% can be lattice-matched to GaN, and the generated $In_{0.17}Al_{0.83}N$/GaN heterostructure has the spontaneous polarization and very little piezoelectric polarization, thus significantly diminishing the inverse piezoelectric effect in strong electric field.

Although the strictly lattice-matched InAlN/GaN eliminates the lattice mismatch strain in the barrier and has only spontaneous polarization and no piezoelectric polarization as compared to AlGaN/GaN, the former has a stronger total polarization effect than the latter because $In_{0.17}Al_{0.83}N$ has very strong spontaneous polarization. Its greater conduction band offset of 0.65 eV at the heterointerface between the InAlN barrier and the GaN channel (the conduction band offset at the $Al_{0.25}Ga_{0.75}N$/GaN interface is 0.34 eV) also yields a higher 2DEG density and better conductivity.

As a result, with good barrier crystalline quality, AlGaN/GaN generally needs a barrier thickness of ~25 nm to obtain a 2DEG density of approximately 1.0–1.8 × 10^{13} cm^{-2}, while the lattice-matched InAlN/GaN with only a 5–15-nm-thick unintentionally doped InAlN barrier can achieve high conductivity characteristics with a 2DEG density higher than 2.5 × 10^{13} cm^{-2} and a sheet resistance lower than 220 Ω/sq (Katz et al. 2004, Gonschorek et al. 2006, Jeganathan et al. 2007). As for the HEMT characteristics, the InAlN/GaN HEMT with only 13-nm-thick

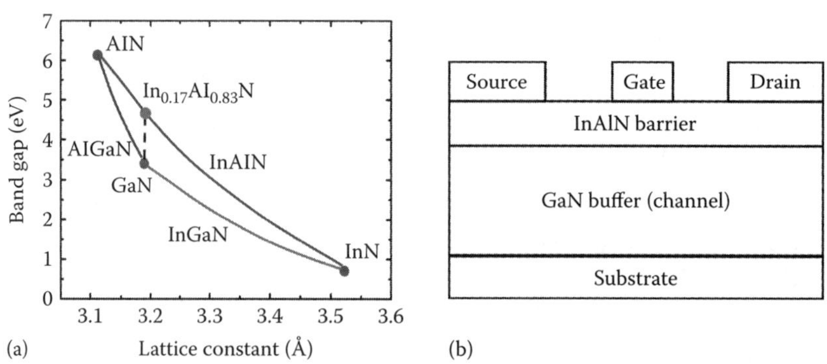

Figure 7.1 Lattice-matched InAlN/GaN heterostructure. (a) **(See color insert.)** Physical mechanism of lattice match. (b) Device structure.

InAlN can yield a maximum output current up to 2.0 A/mm while the AlGaN/GaN HEMTs on sapphire, SiC and Si substrates generally achieve only about 0.8–1.6 A/mm (Gaquiere et al. 2007). InAlN/GaN HEMTs also exhibit better current driving capability and apparent superiority in device scaling, very desirable characteristics for high-frequency power devices (Gaquiere et al. 2007, Lee et al. 2011). InAlN/GaN HEMTs with InGaN back barrier were reported to achieve f_T of 290–300 GHz at a gate length of 30 nm (Lee et al. 2011), and even up to 370 GHz at a gate length of 20 nm (Yue et al. 2012).

The strainless InAlN barriers offer the InAlN/GaN HEMTs much higher reliability than the AlGaN/GaN HEMTs owing to the absence of the inverse piezoelectric effect induced by the strong electric field and to the better high temperature performance of the InAlN/GaN HEMT with an output current in excess of 600 mA at 1000°C in vacuum (the growth temperature of InAlN is merely ~840°C) and a full recovery from degradation as the temperature drops to room temperature (Medjdoub et al. 2006). Such high temperature stability stems directly from the lattice match between InAlN and GaN (Gadanecz et al. 2007).

7.2 Growth, Defects, and Electrical Properties of Nearly Lattice-Matched InAlN/GaN Heterostructures

7.2.1 Growth and Defects of Nearly Lattice-Matched InAlN/GaN Heterostructures

It is very difficult to grow high quality monocrystal InAlN thin films because of the conflicting growth conditions for AlN and InN: AlN requires a high temperature and a low V/III ratio while InN demands a low temperature and a high V/III ratio. The significant difference between AlN and InN covalent bonds also leads to a very low cosolvency in the growth of monocrystalline $In_{0.17}Al_{0.83}N$, thus a poor crystal quality due to the phase separation and nonuniform distribution of alloy compositions. As a result, the epitaxy of high quality $In_{0.17}Al_{0.83}N$ monocrystalline thin film involves the trade-off and optimization of growth conditions based on the different growth behaviors of InN and AlN. The dominant epitaxy techniques for InAlN alloys are the conventional MOCVD and the plasma-enhanced MBE. In this chapter, we propose the pulsed MOCVD (PMOCVD) for InAlN/GaN, which is much superior in fabricating high mobility InAlN/GaN heterostructures.

When $In_xAl_{1-x}N$ is in tensile or compressive strain after it is epitaxially grown on a GaN template, the state of strain has a significant effect on the surface morphology and microstructure of the $In_xAl_{1-x}N$ film (Figure 7.2). Miao et al. (2010) reported that good crystal quality and smooth surface of the $In_xAl_{1-x}N$ thin film (200 nm thick) was achieved under nearly lattice-matched conditions with an indium content of 0.166–0.208, and cracks were seen in the InAlN under greater tensile strain but with no change in the surface roughness, while an increase in lattice relaxation and surface roughness was observed in InAlN under greater compressive strain, as shown in Figure 7.2f and i.

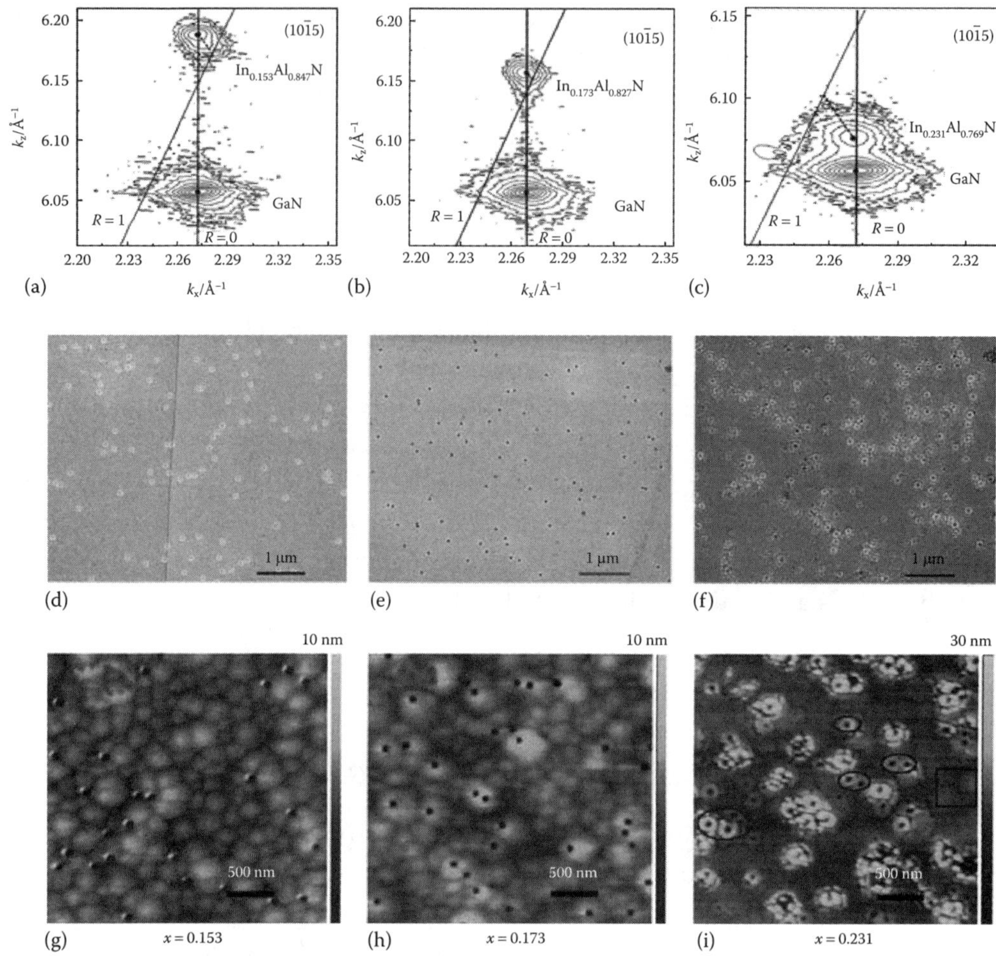

Figure 7.2 Effect of indium content x on the strain state, surface morphology and microstructure of 200 nm $In_xAl_{1-x}N$ thin film on GaN template. (a–c) are the reciprocal space maps, (d–f) the SEM images, and (g and h) the 3 × 3 μm² AFM images of the three samples, respectively. Two types of pits are shown in (i): the larger ones with bulges around (highlighted by circles), and the smaller ones with smooth surface around (highlighted by rectangles). (Reproduced from Miao, Z.L. et al., *J. Appl. Phys.*, 107, 043515, 5 pp., 2010.)

V-shape pits of uniform dimension are often observed on the surface of nearly lattice-matched InAlN with the defect size increasing in a linear manner with InAlN thickness (Figure 7.3) (Miao et al. 2009). It is suggested that the V-pits formed at the initial growth of $In_xAl_{1-x}N$ owing to the threading dislocations having screw components (Figure 7.4). The segregation of the indium atoms at these dislocation cores during the InAlN growth prevents the growth of $In_xAl_{1-x}N$ here, and the V-pit morphology is caused by the lower growth rate of the {10$\bar{1}$1} sidewalls (Figure 7.5). Therefore, the growth of nearly lattice-matched InAlN/GaN requires both optimized growth conditions for $In_{0.17}Al_{0.83}N$ and a high crystal quality of the GaN templates.

InAlN/GaN heterostructures generally have several orders of magnitude greater Ni/Au Schottky reverse-bias leakage currents than AlGaN/GaN heterostructures, even if no V-pits arise on the microscopic surface of nearly lattice-matched InAlN

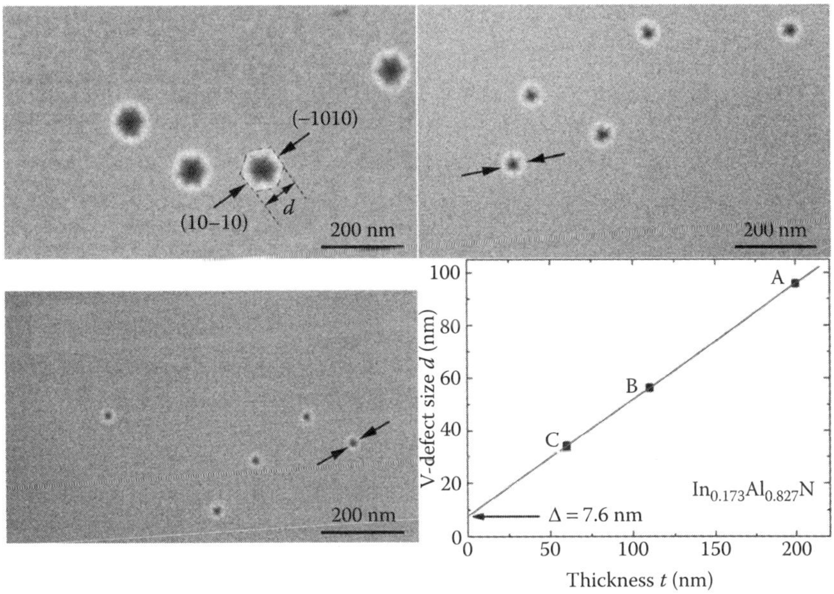

Figure 7.3 SEM images of InAlN surface and V-pit diameter as a function of InAlN film thickness. (Reproduced from Miao, Z.L. et al., *Appl. Phys. Lett.,* 95, 231909, 3 pp., 2009.)

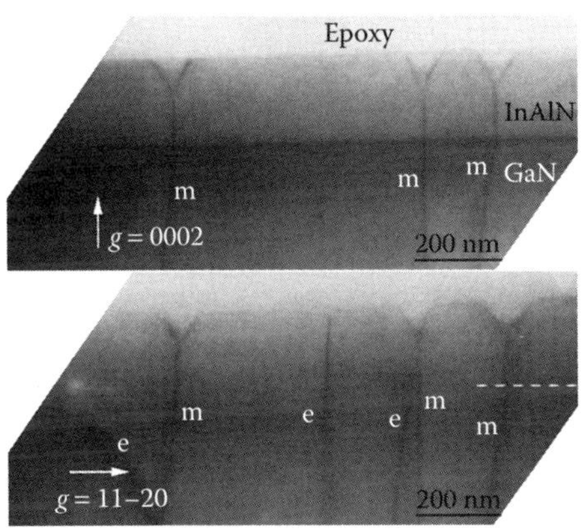

Figure 7.4 Cross-sectional TEM images of V-pits. (Reproduced from Miao, Z.L. et al., *Appl. Phys. Lett.,* 95, 231909, 3 pp., 2009.)

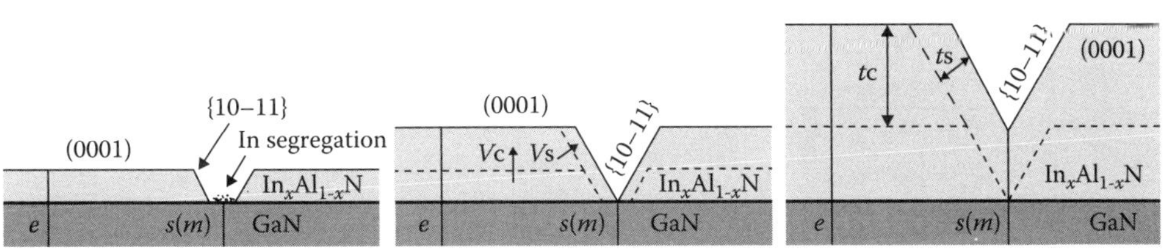

Figure 7.5 Schematic evolution of V-pit morphology, *e* refers to edge dislocations, and *s*(*m*) to screw (mixed) dislocations. (Reproduced from Miao, Z.L. et al., *Appl. Phys. Lett.,* 95, 231909, 3 pp., 2009.)

on GaN templates. The conductive AFM and TEM observations suggest the large reverse-bias Schottky leakage currents in the InAlN/GaN heterostructure is due to the high conductivity leakage path formed by screw dislocations owing to the indium segregation and accumulation nearby.

7.2.2 Electrical Properties of Nearly Lattice-Matched InAlN/GaN Heterostructures

The nearly lattice-matched InAlN/GaN heterostructure obtained by growing InAlN directly on the GaN template typically has a high electron density (>2.5 × 10^{13} cm^{-2}) and relatively low electron mobility (70–500 cm^2/[Vs]) as shown in Figure 7.6a, and its Hall mobility displays mostly the characteristics of the bulk electrons (increasing and then decreasing with dropping temperature) (Xie et al. 2007) with few reports showing the 2DEG characteristics (increasing and then gradually saturating with dropping temperature). AlGaN, AlN/GaN/AlN, and AlN interlayers are already employed in InAlN/GaN structures, just as the ultrathin AlN interface interlayer introduced in the conventional AlGaN/GaN heterostructure to improve 2DEG mobility. By introducing ~1 nm AlN layer at the InAlN/GaN interface, the resulting InAlN/AlN/GaN obtains an improved channel electron mobility from ~70 to 1170 cm^2/(Vs) with only a slight increase or no change in the channel electron density, and the Hall electron mobility exhibits the 2DEG characteristics (Figure 7.6b) (Gonschorek et al. 2006). The introduction of AlN interlayer has become a common practice in the growth of high-performance InAlN/GaN heterostructures in light of the simplicity and effective control of the epitaxial growth.

Miao et al. were the first to observe the SdH oscillation and obvious double subband occupancy of the 2DEG in the lattice-matched InAlN/AlN/GaN heterostructure (Miao et al. 2011). The energy separation between the two subbands is up to 191 meV, far greater than that in the regular AlGaN/GaN heterostructure (~70 meV),

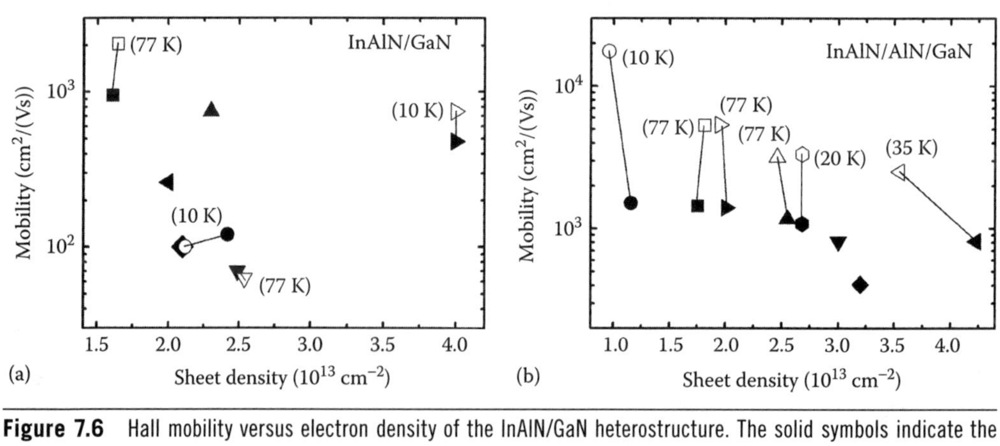

(a) (b)

Figure 7.6 Hall mobility versus electron density of the InAlN/GaN heterostructure. The solid symbols indicate the experimental data at room temperature, and the open symbols indicate low-temperature experimental data with corresponding measurement temperatures. (a) InAlN/GaN and (b) InAlN/AlN/GaN. (After Zhang, J.-F. et al., *Acta Phys. Sinica,* 60, 611–616, 2011.)

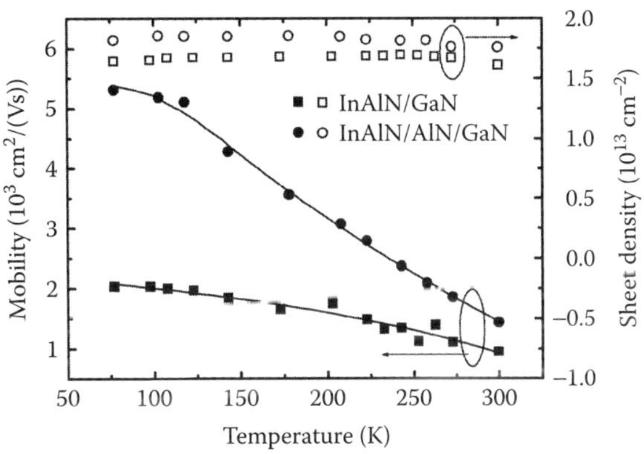

Figure 7.7 Temperature-dependent Hall measurement results of InAlN/GaN and InAlN/AlN/GaN grown by pulsed MOCVD. (After Zhang, J.-F. et al., *Acta Phys. Sinica,* 60, 611–616, 2011.)

and the mobility in the first subband is far smaller than that in the second. They suggested that the 2DEG occupancy of the lattice-matched InAlN/AlN/GaN heterostructure originates from the strong spontaneous polarization of InAlN, and that the very strong polarization-induced electric field at the heterointerface makes the interface roughness scattering the dominant mechanism that determines the 2DEG mobility.

We prepared nearly lattice-matched InAlN/GaN heterostructures by pulsed MOCVD with Hall mobilities of 949 cm^2/(Vs) (room temperature) and 2032 cm^2/(Vs) (77 K), respectively. The formation of 2DEG mobility behavior indicates significantly higher InAlN crystalline quality by pulsed MOCVD. The further introduction of 1.2 nm AlN interlayers yielded InAlN/AlN/GaN structures with the Hall mobilities improved to 1437 cm^2/(Vs) (room temperature) and 5308 cm^2/(Vs) (77 K), respectively (Figure 7.7) (Wang et al. 2011). Based on the Hall measurement results, crystalline quality evaluation and measured surface/interface roughnesses of this series of samples, an analysis is made of the mechanisms of electron mobility improvement by AlN interlayers for InAlN/GaN heterostructures. It is found that the improvement is attributed to both the reduced alloy disorder scattering and the suppressed interface roughness scattering as a result of significantly improved heterointerface by the introduction of the AlN interlayer.

7.3 PMOCVD with Enhanced Surface Reaction

The advantage of PMOCVD for growing high-quality InAlN/GaN heterostructures with good interface properties has much to do with its unique growth mode and mechanism. This section presents an introduction to this technique, and the experimental results of the InAlN/GaN heterostructures grown by PMOCVD are given in Section 7.4.

In the growth of nitrides, a pre-reaction of the group V and group III atoms exists before they arrive at the major reaction zone for film deposition on top of the substrate,

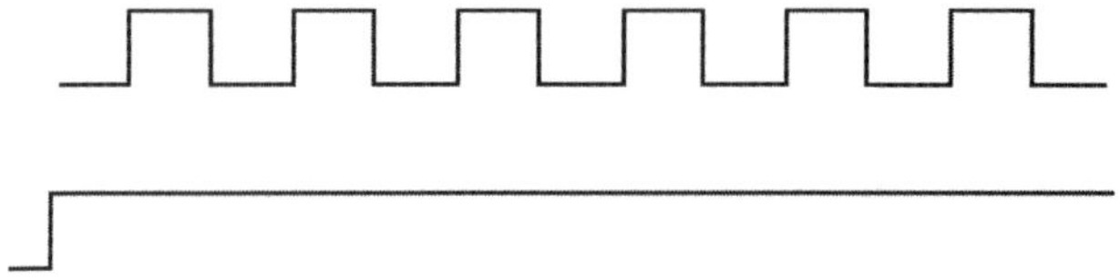

Figure 7.8 Pulse sequences for growth of AlN by PMOCVD.

where both the chemical reaction between these atoms to form III-N bonds and the physical processes of their lateral migration, adsorption, and desorption at the surface are going on. The intensities of these co-existing processes are subject to ambient conditions (such as the growth temperature and pressure) and the V/III ratio.

In the conventional epitaxial growth by MOCVD, the group V and group III sources (i.e., NH_3 and the MO source) are injected simultaneously and continuously into the reaction chamber. The pre-reaction of the reactants (the aluminum source in particular) causes serious problems as the solid particles of the parasitic products may fall on the surface of epitaxial film on the substrate to form defects, thus a degraded crystalline quality. The pre-reaction is generally suppressed by gas path separation through special design of the reaction chamber, that is, the reactant gases are spatially separated and transported on top of the substrate, where they are mixed for reaction.

The PMOCVD employs the time-sharing pulsed transport of V- and III-reactant sources, which flow at different moments into the reaction chamber, as shown in Figure 7.8. Such a transport method not only suppresses the defect-inducing pre-reaction product deposition by reducing the chance for the reactants to react before arriving at the substrate, but also provides adequate time for the metal atoms from the MO source to migrate on the epifilm surface before bonding with nitrogen atoms, then they are more likely to take up the most favorable nucleation sites such as the kinks of the steps or the defect sites. Therefore, a regular arrangement of atoms can be realized in their integration into the crystal in favor of better two-dimensional surface coverage to achieve an atomically smooth surface and a higher crystalline quality. The enhanced surface reaction leads to a significant decrease in the required growth temperature, which further suppresses the pre-reaction.

In terms of process control, PMOCVD divides the growth process of a single epilayer into multiple repetitive cycles, in each of which different reactant sources are given respective pulse settings (primarily the pulsewidth and number of pulses). This growth technique increases in itself the flexibility of process control, and the precision of process control is also improved by adjusting the parameters (such as the pulsewidth, number of pulses, pulse interval, and pulse overlap) to regulate the reactant source flow and the V/III flow ratio, realizing ~0.1 nm growth thickness in one cycle and accurate thickness control on the atomic monolayer level.

This makes PMOCVD much favorable for the growth of high Al-content nitride alloys and AlN materials. AlN grown by the conventional MOCVD typically have few screw dislocations ($10^7\ cm^{-2}$) and many edge dislocations ($10^{10}\ cm^{-2}$) with the film

Table 7.1 Crystalline Quality of AlN Grown by Conventional MOCVD and PMOCVD

	XRC FWHM (arcsec)		
SUBSTRATE AND GROWTH TECHNIQUE	(002)	(102)	SOURCE OF DATA
C-plane 6H-SiC, pulsed laser deposition (PLD)	210	252	Kim et al. (2008)
Sapphire, solid source HVPE	103	828	Eriguchi et al. (2008)
Sapphire, pulsed NH_3 gas flow multilayer growth	200	370	Hirayama et al. (2009)
Sapphire, modified migration-enhanced epitaxy (MMEE)	43	250	Banal et al. (2009)
Sapphire, PMOCVD	65	236	Zhou (2010)

epitaxied by preferential island vertical growth and high-density island coalescence. We obtained high-quality epitaxial AlN on sapphire by PMOCVD via the optimization of the V/III ratio and flow rates of the reactant sources, and the FWHMs of (002) and (102) XRD rocking curve (XRC) are 65 and 236 arcsec, respectively (Zhou 2010), one of the best reported of the time as shown in Table 7.1.

7.4 InAlN/GaN Heterostructures Grown by PMOCVD

In group III nitrides, InAlN and AlInGaN can form with GaN nearly lattice-matched heterojunctions with high-density 2DEGs, a feature finding important applications in GaN HEMT techniques. A high-quality InAlN barrier in the InAlN/GaN heterostructure requires a smooth InAlN surface, an indium content of about 17% without indium phase separation and a good interface with the GaN template so as to obtain good transport properties, which lay a solid foundation for the HEMT fabrication.

A balance can be achieved between the conflicting epitaxial parameters for AlN and InN by PMOCVD since the time-sharing transport and alternate entry of the group V and group III sources into the reaction chamber can largely prevent the Al-related pre-reaction and enhance the surface migration of aluminum atoms, which effectively decreases the InAlN epitaxy temperature and avoids indium phase separation. Very accurate control of the InAlN composition and thickness is also available in PMOCVD owing to the flexible adjustment and design of the number of cycles and the length of the pulses.

We reported for the first time high-quality nearly lattice-matched InAlN/GaN heterostructures obtained by epitaxial growth using PMOCVD based on a series of optimizations of the epitaxial growth temperature and pressure, V/III and Al/In ratios, and the pulse duration/duty cycle (Xue et al. 2011). The optimized pulse sequences for the epitaxial growth of InAlN by PMOCVD, or from a microscopic view the deposition of ultrathin AlN/InN superlattices, is shown in Figure 7.9. The indium content was estimated to be ~17% from the results in Figure 7.10, and a smooth surface with visible atomic step flows free of large hillocks and bulges was observed by AFM as shown in Figure 7.11. Further optimization of the thickness and growth conditions for the AlN interlayer at the InAlN/GaN interface yielded InAlN/GaN heterostructures with a room temperature Hall mobility of 1402 $cm^2/(Vs)$, a sheet

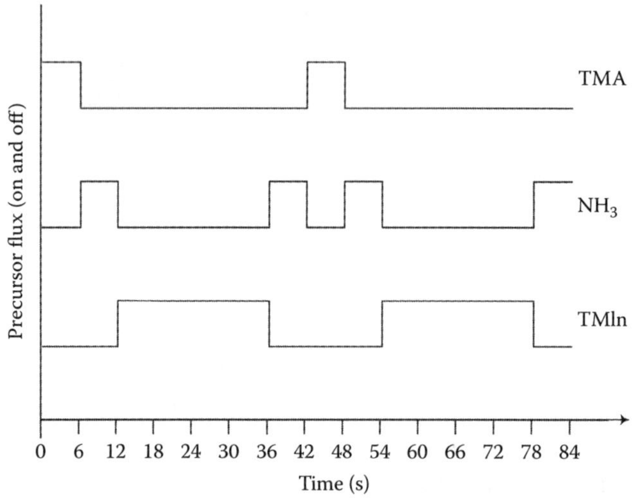

Figure 7.9 Growth sequence of PMOCVD pulses for InAlN barrier layer. (After Xue, J.S. et al., *J. Cryst. Growth,* 314, 359–364, 2011.)

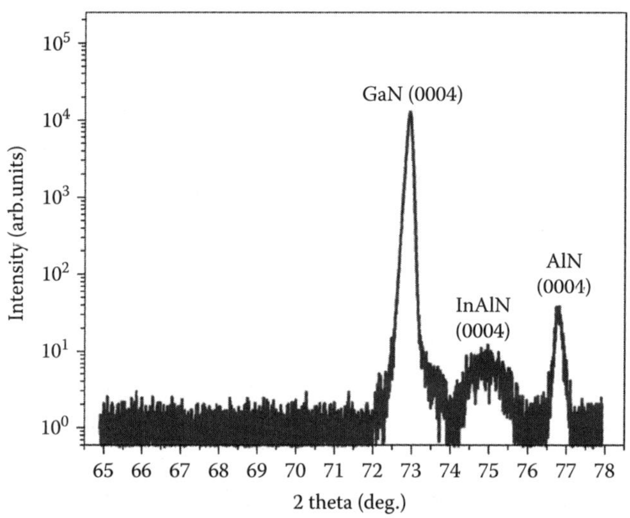

Figure 7.10 High-resolution XRD (0004) scan of InAlN/GaN heterostructures (In content ≈17%). (After Xue, J.S. et al., *J. Cryst. Growth,* 314, 359–364, 2011.)

resistance of 231 Ω/sq, and a sheet resistance nonuniformity of 1.22% over a wafer of 2 in. diameter (Figures 7.12 and 7.13).

Our experiments on the epitaxial growth of high-quality InAlN/GaN heterostructures by PMOCVD revealed significant effects of the InAlN growth temperature and pressure as well as the combination and design of the pulse sequence on the InAlN surface morphology and the indium content. Such influences transfer directly onto the transport properties of the InAlN/GaN heterostructures. In the optimization of the growth parameters, we prepared InAlN/GaN heterostructures with structural parameters that meet the requirements of HEMTs to optimize the InAlN material and heterostructure at one time.

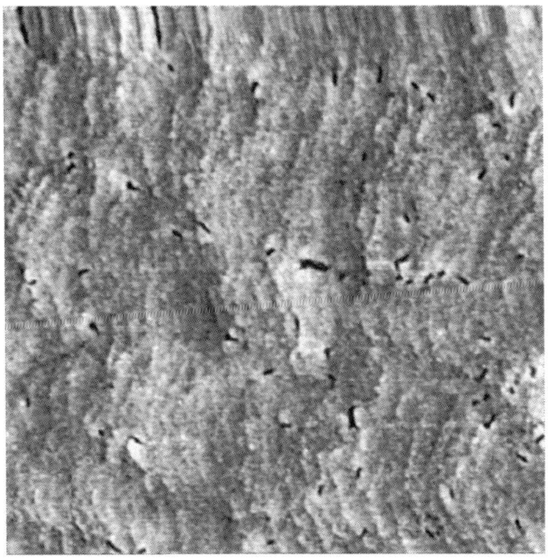

Figure 7.11 AFM images of surface morphology of InAlN/GaN heterostructures with a scan area of 2 × 2 μm² and an RMS roughness of 0.257 nm. (After Xue, J.S. et al., *J. Cryst. Growth*, 314, 359–364, 2011.)

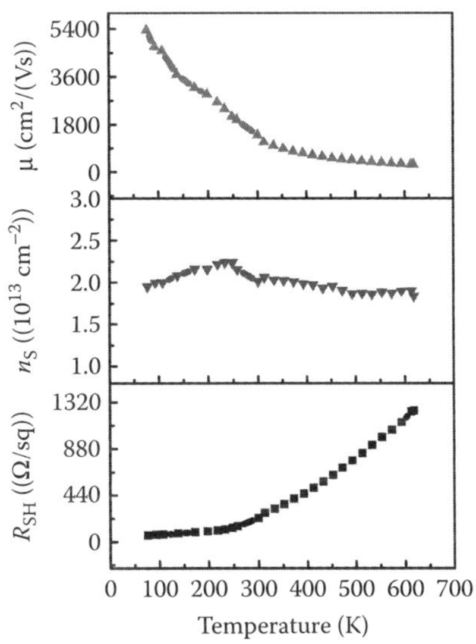

Figure 7.12 Hall mobility (μ), sheet carrier density (n_s), and sheet resistance (R_{SH}) of PMOCVD-grown InAlN/GaN as a function of temperature. (After Xue, J.S. et al., *J. Cryst. Growth*, 314, 359–364, 2011.)

7.4.1 Effect of Epitaxial Growth Pressure on InAlN/GaN Properties

The most important parameter for PMOCVD-grown InAlN and a trade-off between the low pressure required by AlN growth and the high pressure demanded by InN growth, the growth pressure has a major impact on the indium incorporation efficiency. The growth was performed in the pulse cycle sequence shown in Figure 7.9 at a growth temperature of 720°C with the pressure increasing from 100 to 250 Torr in steps of 50 Torr. It can be seen from Figure 7.14 that the increase in pressure is highly favorable for indium incorporation. The surface morphology of the InAlN film grown

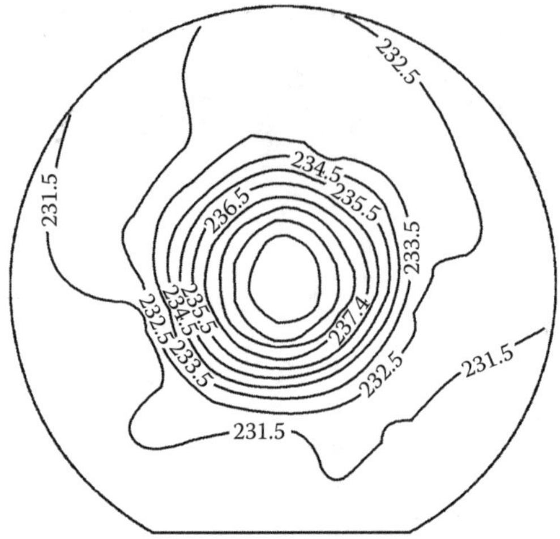

Figure 7.13 Sheet resistance mapping of the InAlN/GaN heterostructure wafer of 2 in. diameter. (After Xue, J.S. et al., *J. Cryst. Growth*, 314, 359–364, 2011.)

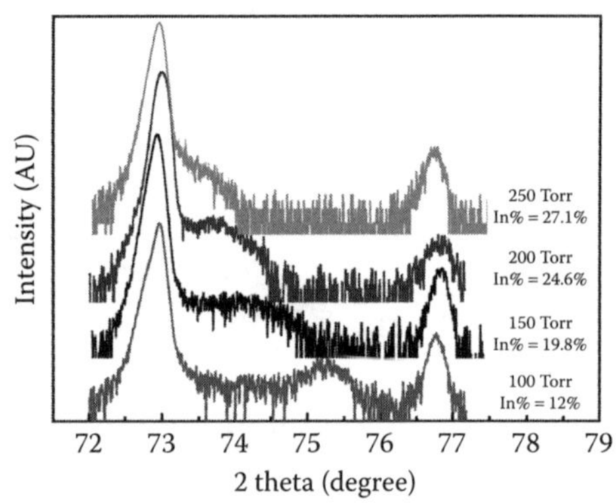

Figure 7.14 Effect of varied growth pressures on indium content of InAlN barrier. (Courtesy of Dr. Junshuai Xue.)

at high pressure is also greatly improved. As Figure 7.15 shows, the atomic steps are more obvious and the V-pit density is considerably reduced. The *C-V* characteristics of the InAlN/GaN shown in Figure 7.16 reveals a very steep falling edge of the capacitance corresponding to the depletion region in GaN under the 2DEG channel for the high growth pressure sample while an sloped capacitance for the low growth pressure sample, indicating a remarkable leakage in the Schottky diode induced by substantial V-defects in the low growth pressure sample.

7.4.2 Effect of Indium Source Pulse Duration on InAlN/GaN Properties

In the epitaxial growth of InAlN by PMOCVD, various combinations of pulses of the MO source and NH_3 have influence on the film surface morphology, crystalline quality, and composition. For simplicity of the experimental procedures, pulse overlap is avoided by design and the major parameter in variation is the pulsewidth

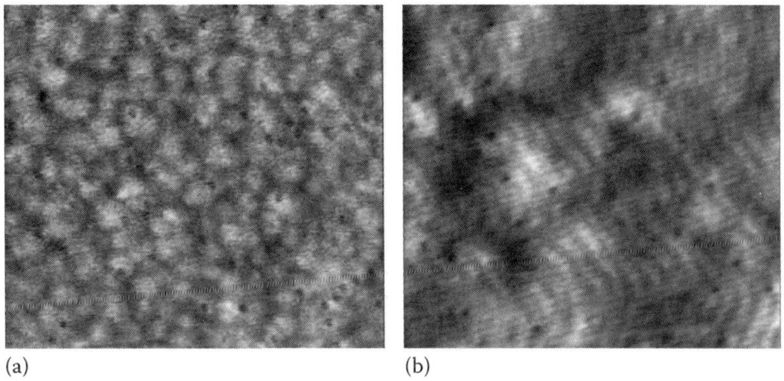

Figure 7.15 2×2 μm^2 AFM surface images of epitaxial InAlN/GaN heterostructures. (a) 100 Torr and (b) 200 Torr. (Courtesy of Dr. Junshuai Xue.)

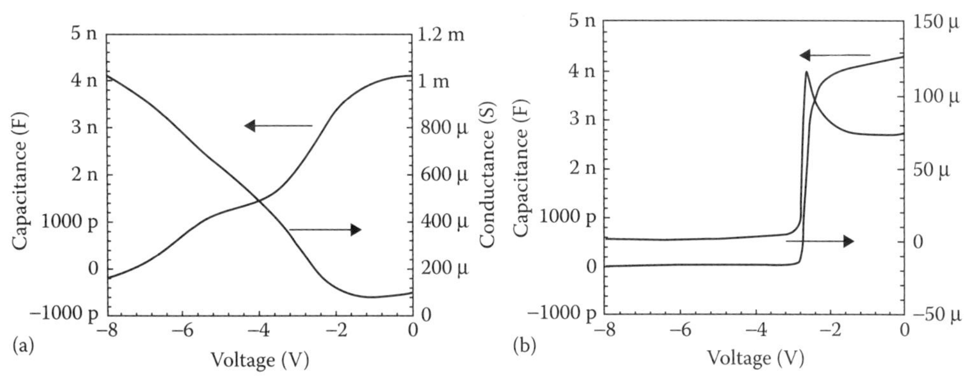

Figure 7.16 *C-V* and *G-V* characteristics of epitaxial InAlN/GaN heterostructures. (a) 100 Torr and (b) 200 Torr. (Courtesy of Dr. Junshuai Xue.)

of the indium source (TMIn) while other pulsewidths remain constant so as to determine the relation between the indium content of InAlN film and the TMIn pulsewidth. InAlN/GaN heterostructures were grown with TMIn pulsewidths of 0.1, 0.2, 0.3, and 0.4 min, respectively, and the characterization results are shown in Figures 7.17 and 7.18.

The XRD 2θ-ω curves in Figure 7.17 show that the InAlN peak position is obviously closer to the GaN peak and of greater intensity as the TMIn pulsewidth increases, suggesting effective modulation of the indium content by the TMIn pulsewidth with the indium content increasing from 7.4% to 18.8% as the TMIn pulsewidth increases from 0.1 to 0.4 min. The surface morphology, thickness, and crystalline quality of the InAlN barrier layer vary accordingly, the immediate influence of which is the change in the 2DEG electrical properties of the InAlN/GaN heterostructure (Figure 7.18). At a TMIn pulsewidth of 0.3 min, the 2DEG mobility reaches 1506 cm^2/(Vs) with an AFM surface RMS roughness of 0.24 nm over an area of 2×2 μm^2.

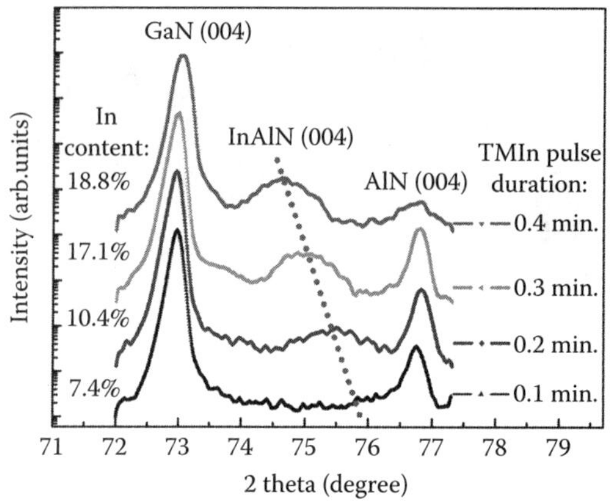

Figure 7.17 XRD 2θ-ω scan curves for epitaxial InAlN/GaN with different TMIn pulsewidths. (After Xue, J. S. et al., *J. Cryst. Growth*, 401, 661–664, 2014.)

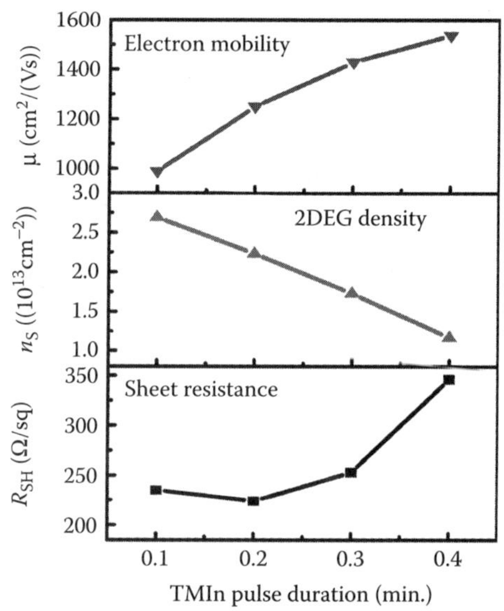

Figure 7.18 Room temperature Hall characteristics of InAlN/GaN heterostructure versus TMIn pulsewidth. (After Xue, J. S. et al., *J. Cryst. Growth*, 401, 661–664, 2014.)

7.4.3 Effect of Epitaxial Growth Temperature on InAlN/GaN Properties

In the InAlN epitaxy by PMOCVD, five InAlN/GaN samples were fabricated with the growth temperature elevated from 650°C to 730°C in steps of 20°C, and the characterization results are shown in Figures 7.19 through 7.23.

The (004) XRD 2θ-ω scan curves of the InAlN/GaN heterostructures obtained at different temperatures are shown in Figure 7.19. The InAlN peak gets gradually closer to the AlN peak with rising temperature, indicating a decrease in the indium content in the InAlN barrier layer. The indium content drops in an approximately linear manner from 20.7% to 15.3% (Figure 7.20) as the temperature increases from

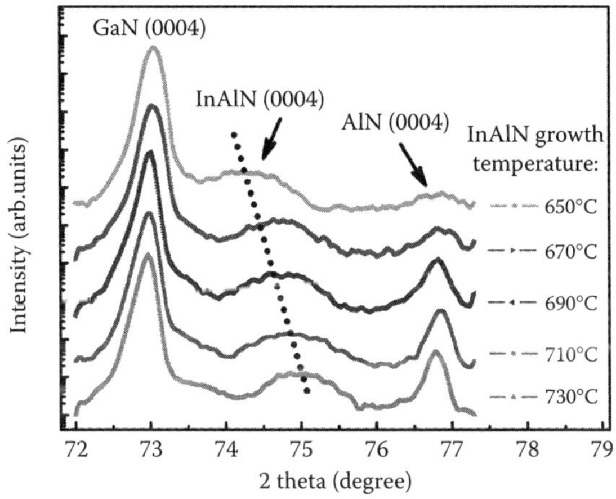

Figure 7.19 XRD 2θ-ω curves of InAlN/GaN heterostructure grown at different temperatures. (After Xue, J.S. et al., *Jpn. J. App. Phys.* 52, 08JB04, 2013.)

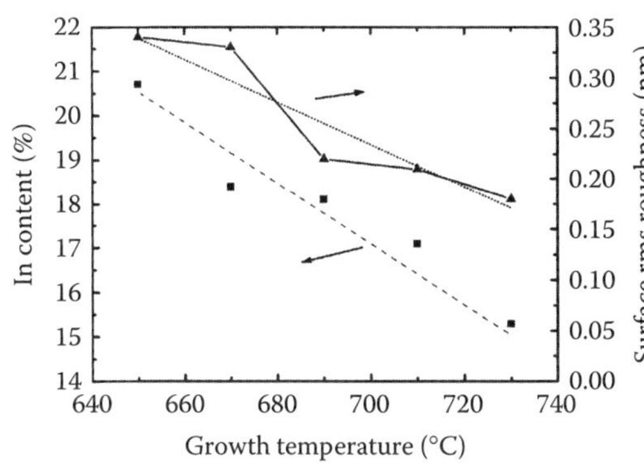

Figure 7.20 Indium content and surface roughness of InAlN barrier layer as a function of temperature (the dashed line is the fitting results). (After Xue, J.S. et al., *Jpn. J. App. Phys.* 52, 08JB04, 2013.)

650°C to 730°C with the lattice-matched content of 17% obtained at 710°C. Such a temperature dependence of the indium content agrees with that by the conventional MOCVD.

The growth temperature also has an impact on the surface morphology as well as on the indium content of InAlN barrier layer. Under lattice match conditions, a smooth surface with atomic steps is observed in the 2 × 2 μm^2 AFM images shown in Figure 7.21. Pits are observed at low temperatures and are also seen at very high temperatures but of lower density. Such morphologics are related to the lattice mismatch between InAlN and GaN, and more surface defects are seen under greater lattice mismatch. As shown in Figure 7.20, the InAlN surface RMS roughness decreases with rising temperature owing to the fact that surface adatoms can find better incorporation sites due to their greater surface mobility at higher temperature. No evident decrease in surface roughness is observed at temperatures higher than 690°C.

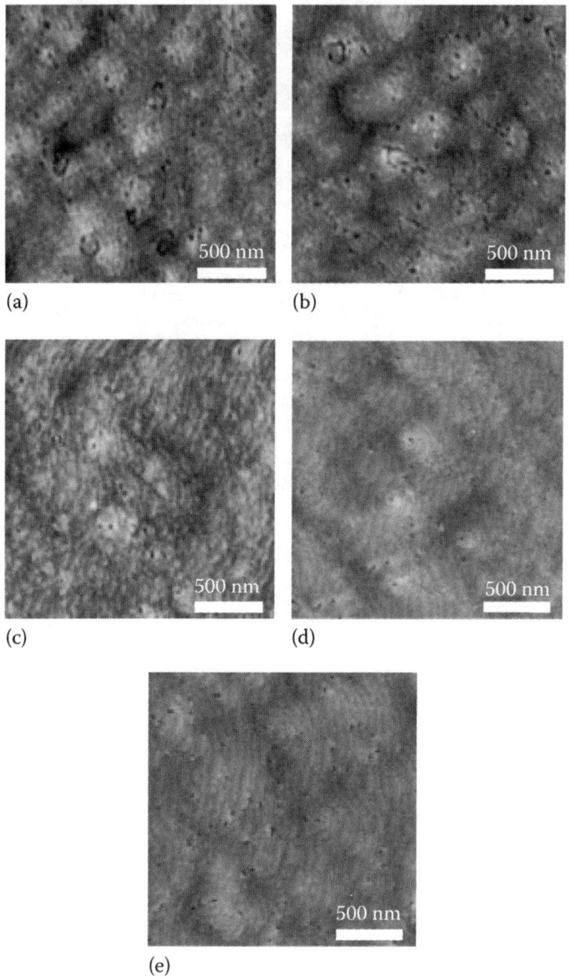

Figure 7.21 $2 \times 2\ \mu m^2$ AFM surface images of InAlN barrier layer at different temperatures. (a) $T = 650°C$, (b) $T = 670°C$, (c) $T = 690°C$, (d) $T = 710°C$, and (e) $T = 730°C$. (After Xue, J.S. et al., *Jpn. J. App. Phys.* 52, 08JB04, 2013.)

The room temperature Hall and *C-V* measurement results of the InAlN/GaN heterostructures grown at different temperatures are shown in Figures 7.22 and 7.23, demonstrating greatly varied 2DEG mobility increasing from 385 to 1587 $cm^2/(Vs)$ but little change in the sheet density with a slight drop at high temperatures due primarily to the strain relaxation induced by too high an aluminum content. The sheet resistance deceases with sheet resistances <300 Ω/sq obtainable at growth temperatures 690°C–730°C. The *C-V* curves suggest the presence of leakage in the Schottky diodes in the low temperature InAlN/GaN heterostructures, which is attributed to the large number of surface pits.

In summary, a tradeoff between the indium content and surface morphology and proper growth conditions are requisite for fabricating high quality lattice-matched InAlN/GaN heterostructures. The electrical properties such as the mobility are also important indices.

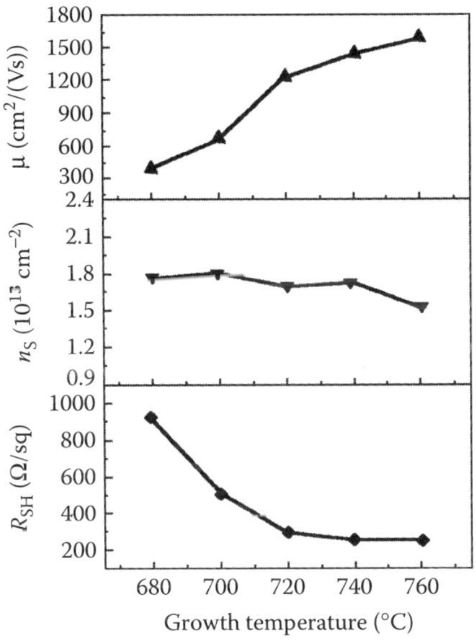

Figure 7.22 Room temperature Hall characteristics of InAlN/GaN heterostructures as a function of growth temperature. (After Xue, J.S. et al., *Jpn. J. App. Phys.* 52, 08JB04, 2013.)

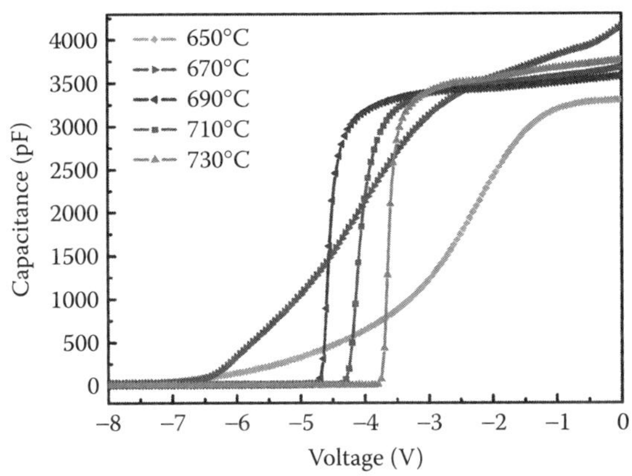

Figure 7.23 *C-V* characteristics of InAlN/GaN heterostructures grown at different temperatures. (After Xue, J.S. et al., *Jpn. J. App. Phys.* 52, 08JB04, 2013.)

7.5 Growth of Double-Channel InAlN/GaN Heterostructure by PMOCVD

We propose the use of double-channel InAlN/GaN heterostructures to further reduce the sheet resistance, thereby decreasing the on resistance R_{on} of the HEMT and improving device linearity and maximum operation frequency. InAlN/GaN-based double-channel heterostructures can be grown by PMOCVD on the basis of the growth of high-quality InAlGaN/GaN-based single-channel heterostructures using PMOCVD. Shown in Figure 7.24 is the first high electrical property GaN-based double-channel heterostructure using only InAlN barrier layers (including an AlN interlayer) (Xue et al. 2012a).

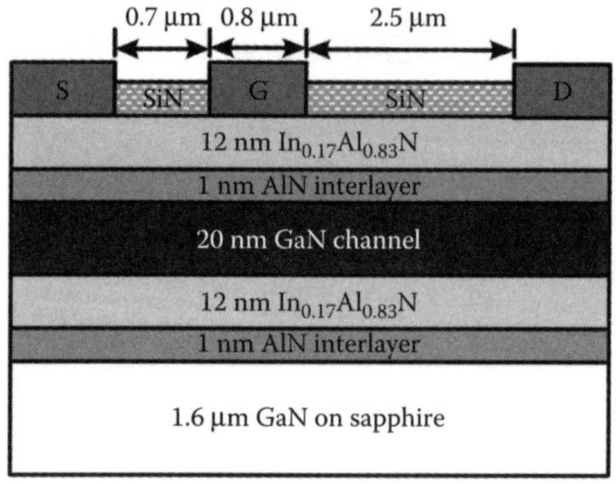

Figure 7.24 Schematic structure of InAlN/GaN double-channel HEMT. (After Xue, J. S. et al., *J. App. Phys.,* 111, 14513, 2012a.)

As the epitaxial growth of InAlN/GaN single-channel heterostructures is already susceptible to indium phase separation, the growth of InAlN/GaN double-channel heterostructures is even more challenging. The lower InAlN barrier layer need to be grown at a relatively low temperature but the subsequent growth of the GaN channel layer requires a high temperature, which may degrade the quality of the lower InAlN barrier layer leading to indium segregation and surface roughening as well as poorer 2DEG channel quality and degraded electrical properties of the heterostructure. As a result, it is necessary to employ innovative techniques for growing double-channel materials based on the PMOCVD growth of InAlN.

As shown in Figure 7.25, different from the high temperature hydrogen gas (H_2) used in the conventional epitaxial growth of GaN, a combination of low temperature nitrogen gas (N_2) and high temperature H_2 is adopted for growing the 20 nm upper GaN channel in the InAlN/GaN/InAlN/GaN heterostructure. First, after the epitaxy of the 12 nm lower InAlN barrier layer a ~2 nm protective GaN layer is grown in the same N_2 and low temperature ambient instead of shifting immediately into the H_2 ambient and raising the temperature to that for conventional GaN. This thin layer of low-temperature GaN can effectively protect the lower InAlN barrier from an annealing effect of the subsequent high temperature GaN channel growth, and the resulting surface morphology degradation and indium phase redistribution. Then the temperature is elevated in H_2 ambient for the growth of an 18 nm GaN channel layer in high temperature H_2 ambient. The epitaxy of this GaN channel layer adopts a lower growth rate than that of the conventional GaN in order to obtain a high crystalline quality and smooth surface thereby effectively improving the 2DEG mobility. Both InAlN barrier layers are grown with optimized growth parameters by PMOCVD at an indium content of ~17% for a close lattice match.

As-grown InAlN/GaN/InAlN/GaN exhibits a room temperature Hall electron concentration up to 2.55×10^{13} cm^{-2} (Figure 7.26), nearly twice the 2DEG sheet density obtainable in conventional AlGaN/GaN heterostructures, thus an effective reduction in the sheet resistance owing to the higher 2DEG sheet density. Its Hall

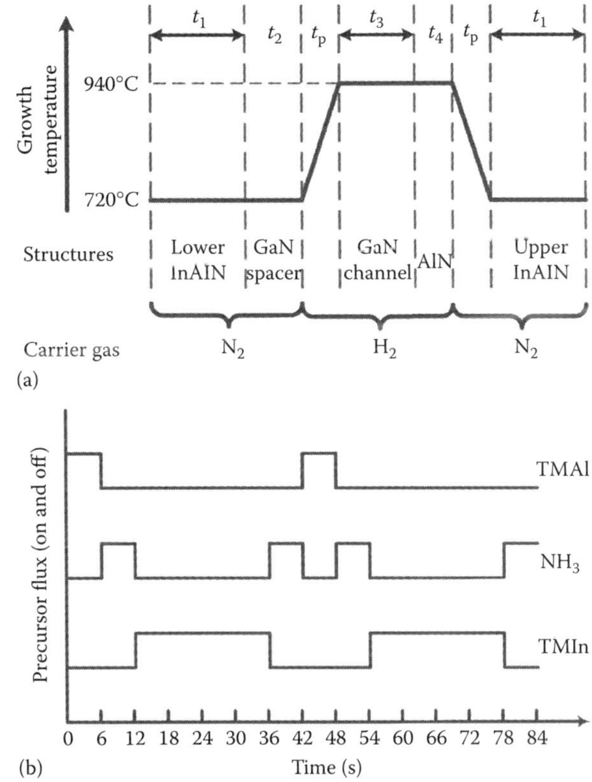

Figure 7.25 Schematic diagram of epitaxial growth of InAlN/GaN double-channel heterostructures by PMOCVD. (a) Growth temperature, layer structures, and carrier gas for the growth of double-channel heterostructures and (b) growth sequence of InAlN barrier layer by PMOCVD. (After Xue, J. S. et al., *App. Phys. Lett.*, 100, 013507, 2012b.)

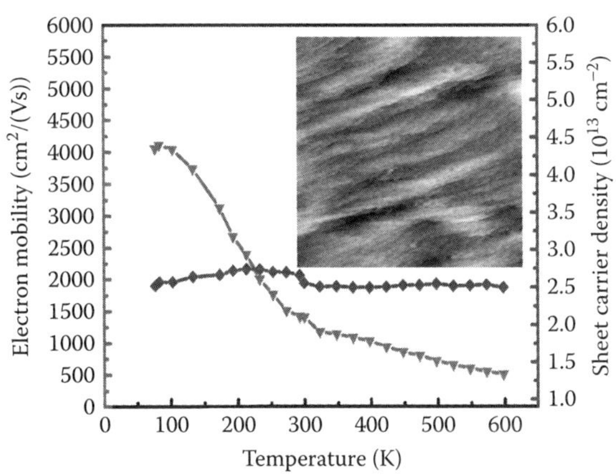

Figure 7.26 Temperature-dependent Hall characteristics of InAlN/GaN double-channel heterostructures. The inset is the AFM surface image. (After Xue, J. S. et al., *App. Phys. Lett.*, 100, 013507, 2012b.)

mobility is also as high as 1414 $cm^2/(Vs)$ with a sheet resistance as low as 172 Ω/sq, which is close to that reported for ultrathin pure binary AlN/GaN heterostructures (170 Ω/sq). The surface morphology of the double-channel heterostructure (the inset in Figure 7.26) is also similar to that of the PMOCVD-grown single-channel

InAlN/GaN without surface roughening or degradation induced by the stacking growth of multiple layers of different materials in different epitaxial ambients. The surface roughness is merely 0.2 nm over a 2 × 2 μm^2 AFM scan area.

In the HRXRD reciprocal space maps (Figure 7.27), the maximum intensities of diffraction of the InAlN barrier layers, the GaN channel layer, and the GaN template are on exactly the same vertical line, indicating fully strained InAlN barrier layers without relaxation or phase separation. The presence of the two channels is verified by the *C-V* measurement results (Figure 7.28), and a concentrated population of 2DEG exists at each InAlN/GaN interface.

In summary, high-quality InAlN/GaN/InAlN/GaN heterostructures can be obtained by an innovative epitaxial growth of GaN channels in a combination of low

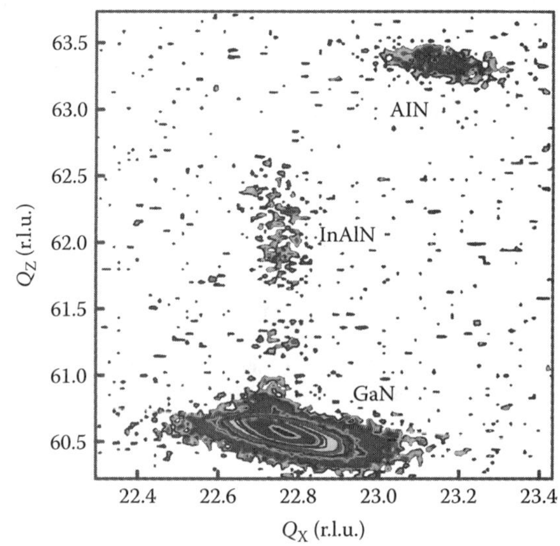

Figure 7.27 **(See color insert.)** Reciprocal space maps of (105) reflection of InAlN/GaN/InAlN/GaN double-channel heterostructure. (After Xue, J. S. et al., *App. Phys. Lett.*, 100, 013507, 2012b.)

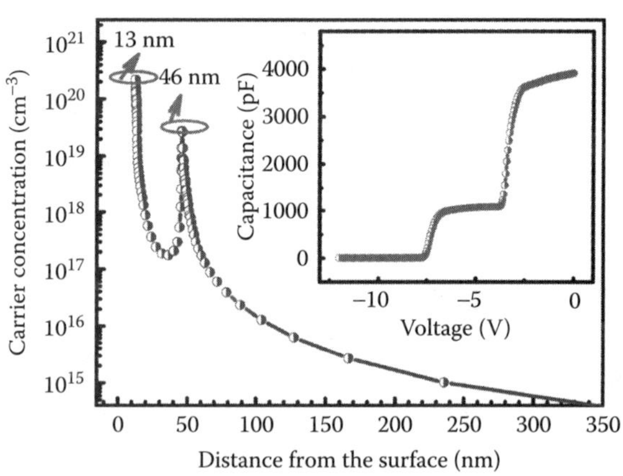

Figure 7.28 *C-V* carrier profile of InAlN/GaN double-channel heterostructure. The inset is the *C-V* characteristics. (After Xue, J. S. et al., *App. Phys. Lett.*, 100, 013507, 2012b.)

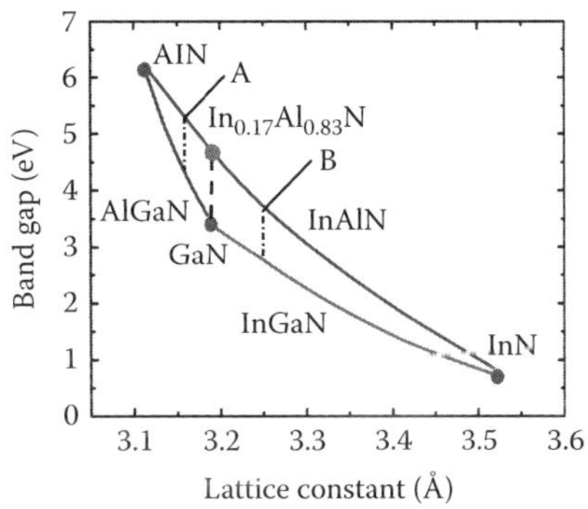

Figure 7.29 **(See color insert.)** Schematic lattice-matched structures of nitrides.

temperature N_2 and high temperature H_2 and the growth of InAlN barrier layers by PMOCVD. This technique can be applied to the epitaxy of optoelectronic devices based on multilayer nitride heterostructures such as the distributed Bragg reflector. It is worth noting that further research can be made according to the fundamental principles of nearly lattice-matched nitride heterostructures (Figure 7.1) in other two directions shown in Figure 7.29, that is, the lattice-matched $Al_xIn_{1-x}N/Al_xGa_{1-x}N/GaN$ structure (point A in Figure 7.29) to the left of the match point of $In_{0.17}Al_{0.83}N/GaN$ featuring a wider bandgap of $Al_xIn_{1-x}N$ suitable for high temperature high breakdown voltage HEMTs, and the lattice-matched $In_xAl_{1-x}N/In_xGa_{1-x}N/GaN$ structure (point B in Figure 7.29) to the right characterized by a high electron mobility of the $In_xGa_{1-x}N$ channel layer (InN has an electron mobility up to 5000 $cm^2/[Vs]$) favorable for higher speed higher frequency HEMTs. For whatever structures mentioned here PMOCVD is a good method for obtaining high-quality materials.

References

Banal, R. G., M. Funato, and Y. Kawakami. 2009. Growth characteristics of AlN on sapphire substrates by modified migration-enhanced epitaxy. *Journal of Crystal Growth* 311(10):2834–2836. doi: 10.1016/j.jcrysgro.2009.01.023.

Eriguchi, K.-I., T. Hiratsuka, H. Murakami, Y. Kumagai, and A. Koukitu. 2008. High-temperature growth of thick AlN layers on sapphire (0 0 0 1) substrates by solid source halide vapor-phase epitaxy. *Journal of Crystal Growth* 310(17):4016–4019. doi: 10.1016/j.jcrysgro.2008.06.033.

Gadanecz, A., J. Blasing, A. Dadgar, C. Hums, and A. Krost. 2007. Thermal stability of metal organic vapor phase epitaxy grown AlInN. *Applied Physics Letters* 90(22):221906-1. doi: 10.1063/1.2743744.

Gaquiere, C., F. Medjdoub, J. F. Carlin, S. Vandenbrouck, E. Delos, E. Feltin, N. Grandjean, and E. Kohn. 2007. AlInN/GaN a suitable HEMT device for extremely high power high frequency applications. *2007 International Microwave Symposium*, June 3–8, 2007, Piscataway, NJ.

Gonschorek, M., J. F. Carlin, E. Feltin, M. A. Py, and N. Grandjean. 2006. High electron mobility lattice-matched AlInN/GaN field-effect transistor heterostructures. *Applied Physics Letters* 89(6):062106. doi: 10.1063/1.2335390.

Hirayama, H., S. Fujikawa, N. Noguchi, J. Norimatsu, T. Takano, K. Tsubaki, and N. Kamata. 2009. 222–282 nm AlGaN and InAlGaN-based deep-UV LEDs fabricated on high-quality AlN on sapphire. *Physica Status Solidi (A) Applications and Materials Science* 206(6):1176–1182. doi: 10.1002/pssa.200880961.

Jeganathan, K., M. Shimizu, H. Okumura, Y. Yano, and N. Akutsu. 2007. Lattice-matched InAlN/GaN two-dimensional electron gas with high mobility and sheet carrier density by plasma-assisted molecular beam epitaxy. *Journal of Crystal Growth* 304(2):342–345. doi: 10.1016/j.jcrysgro.2007.03.035.

Katz, O., D. Mistele, B. Meyler, G. Bahir, and J. Salzman. 2004. InAlN/GaN heterostructure field-effect transistor DC and small-signal characteristics. *Electronics Letters* 40(20): 1304–1305. doi: 10.1049/el:20045980.

Kim, M., J. Ohta, A. Kobayashi, H. Fujioka, and M. Oshima. 2008. Low-temperature growth of high quality AlN films on carbon face 6H-SiC. *Physica Status Solidi (RRL)—Rapid Research Letters* 2(1):13–15. doi: 10.1002/pssr.200701246.

Lee, D. S., L. Bin, M. Azize, G. Xiang, G. Shiping, D. Kopp, P. Fay, and T. Palacios. 2011. Impact of GaN channel scaling in InAlN/GaN HEMTs. *2011 IEEE International Electron Devices Meeting*, December 5–7, 2011, Piscataway, NJ.

Medjdoub, F., J. F. Carlin, M. Gonschorek, E. Feltin, M. A. Py, D. Ducatteau, C. Gaquiere, N. Grandjean, and E. Kohn. 2006. Can InAlN/GaN be an alternative to high power/high temperature AlGaN/GaN devices? *2006 International Electron Devices Meeting*, December 10–13, 2006, San Francisco, CA.

Miao, Z. L., N. Tang, F. J. Xu, L. B. Cen, K. Han, J. Song, C. C. Huang et al. 2011. Magnetotransport properties of lattice-matched In0.18Al0.82N/AlN/GaN heterostructures. *Journal of Applied Physics* 109(1):016102 (3 pp.). doi: 10.1063/1.3525989.

Miao, Z. L., T. J. Yu, F. J. Xu, J. Song, C. C. Huang, X. Q. Wang, Z. J. Yang et al. 2009. The origin and evolution of V-defects in InxAl1-xN epilayers grown by metalorganic chemical vapor deposition. *Applied Physics Letters* 95(23):231909 (3 pp.). doi: 10.1063/1.3272017.

Miao, Z. L., T. J. Yu, F. J. Xu, J. Song, L. Lu, C. C. Huang, Z. J. Yang et al. 2010. Strain effects on InxAl1-xN crystalline quality grown on GaN templates by metalorganic chemical vapor deposition. *Journal of Applied Physics* 107(4):043515 (5 pp.). doi: 10.1063/1.3305397.

Wang, P.-Y., J.-F. Zhang, J.-S. Xue, Y.-B. Zhou, J.-C. Zhang, and Y. Hao. 2011. Transport properties of two-dimensional electron gas in lattice-matched InAIN/GaN and InAIN/AIN/GaN materials. *Acta Physica Sinica* 60(11):117304 (6 pp.).

Xie, J., X. Ni, M. Wu, J. H. Leach, U. Ozgur, and H. Morkoc. 2007. High electron mobility in nearly lattice-matched AlInN/AlN/GaN heterostructure field effect transistors. *Applied Physics Letters* 91(13):1–3. doi: 10.1063/1.2794419.

Xue, J. S., Y. Hao, X. W. Zhou, J. C. Zhang, C. K. Yang, X. X. Ou, L. Y. Shi et al. 2011. High quality InAlN/GaN heterostructures grown on sapphire by pulsed metal organic chemical vapor deposition. *Journal of Crystal Growth* 314(1):359–364. doi: 10.1016/j.jcrysgro.2010.11.157.

Xue, J. S., J. C. Zhang, and Y. Hao. 2013. Effects of growth temperature on structural and electrical properties of InAlN/GaN heterostructures grown by pulsed metal organic chemical vapor deposition on c-plane sapphire. *Japanese Journal of Applied Physics* 52(8 PART 2):08JB04. doi: 10.7567/jjap.52.08jb04.

Xue, J. S., J. C. Zhang, and Y. Hao. 2014. Investigation of TMIn pulse duration effect on the properties of InAlN/GaN heterostructures grown on sapphire by pulsed metal organic chemical vapor deposition. *Journal of Crystal Growth* 401:661–664. doi: 10.1016/j.jcrysgro.2014.01.025.

Xue, J. S., J. C. Zhang, Y. Hao, K. Zhang, Z. Yi, L. X. Zhang, X. H. Ma, X. G. Li, and F. N. Meng. 2012a. Fabrication and characterization of InAlN/GaN-based double-channel high electron mobility transistors for electronic applications. *Journal of Applied Physics* 111(11):114513 (5 pp.). doi: 10.1063/1.4729030.

Xue, J. S., J. C. Zhang, Y. Hou, H. Zhou, J. Zhang, and Y. Hao. 2012b. Pulsed metal organic chemical vapor deposition of nearly latticed-matched InAlN/GaN/InAlN/GaN double-channel high electron mobility transistors. *Applied Physics Letters* 100 (1):013507. doi: 10.1063/1.3675453.

Yue, Y., Z. Hu, J. Guo, B. Sensale-Rodriguez, G. Li, R. Wang, F. Faria et al. 2012. InAlN/AlN/GaN HEMTs with regrown ohmic contacts and fT of 370 GHz. *IEEE Electron Device Letters* 33(7):988–990. doi: 10.1109/led.2012.2196751.

Zhang, J.-F., P.-Y. Wang, J.-S. Xue, Y.-B. Zhou, J.-C. Zhang, and Y. Hao. 2011. High electron mobility lattice-matched InAlN/GaN materials. *Acta Physica Sinica* 60(11):611–616.

Zhou, X. 2010. Growth of high Al fraction AlGaN and GaN semiconductor materials. PhD Thesis, Xidian University, Xi'an, China.

8

Defects and Physical Properties of Group III Nitrides

Group III nitride materials are typically in the form of epitaxial films, mostly heteroepitaxial ones, which may contain various types of lattice defects. Material physical properties are closely related to the defects, polarity, strain, and so on. This chapter deals with the dislocations, point defects, and the relevant physical properties of Group III nitrides.

8.1 Evaluation of Types and Densities of Dislocations in GaN by Etching

Selective wet etching has a strong effect on the strained or nonstoichiometric regions of defective surfaces, such as dislocations and impurity stripes. Defects can be revealed by etching due to the different etching rates in the defective and defect-free regions. Wet etching is widely employed in the defect characterization of semiconductor single crystals for its low cost and simple processes. Currently, all the industrial evaluations of the quality of silicon and GaAs single crystal wafers adopt wet etching, the key technologies of which are:

1. The judgment of the types of defects based on the correlation between the etch pits and the defects.
2. The determination of the optimum conditions of etching for all types of defects to be revealed as much as possible.
3. The accurate microscopic observation.

The polarity of GaN should be taken into consideration in the etching of GaNs. The N-polar GaN has higher chemical activity than Ga-polar GaN, and the former is readily etched in alkaline solutions for surface polishing or even peeling off. Whereas Ga-polar GaN is hardly etched in common acid or alkaline solutions even when heated, and only at the surface terminations of dislocations selective etching is possible in molten KOH or hot strong phosphoric acid while other regions remain intact.

The correspondence of etch pits to defects was not thoroughly settled in early studies of GaN characterization by wet etching due to the lack of in-depth investigation into the mechanisms of etch pit formation. Our systematic experiments on the type and density of GaN dislocations characterized by etching (Gao et al. 2007a,b, Lu et al. 2008) show that the difference in chemical properties of different polar surfaces plays a pivotal role in the generation of etch pits.

8.1.1 Correspondence between Etch Pit Shapes and Dislocation Types

Three forms of etch pits, referred to as α-, β-, and γ- type, respectively, can be observed in the surface morphology of Ga-polar GaN by SEM after being etched in molten KOH at a proper temperature and for an appropriate period of time. According to the topographic contrast principle of the secondary electron imaging, the brightest region represents a steep incline, the gray area should be a plane or a gentle incline, and the black core indicates the presence of a hole or a sharp groove in which the secondary electrons cannot be collected. Therefore, the α-type etch pit may be an inversed truncated hexagonal pit, and the β-type an inversed hexagonal pyramid, while the γ-type pit has a line profile of a combination of triangular and trapezoidal shapes as the result of a combination of α and β types according to further AFM observation (Figure 8.1).

The etching mechanism of the three types of pits is closely correlated with polarity. α-type etch pits can correspond to pure screw dislocations. A screw dislocation creates a spiral step at the surface termination as shown in Figure 8.2, and the step will associate with the natural surface terrace structure. These spiral steps are so vulnerable as to be attacked by KOH thus increasing the size of the dislocation pit, while a small facet is left at the bottom of the step. This small facet must be Ga-face, otherwise it would be further etched due to the chemical activity of N-face. Once this small smooth facet is formed,

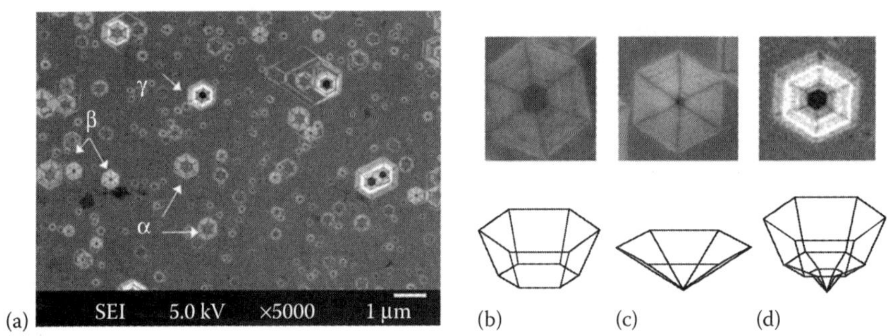

Figure 8.1 SEM image of KOH etched GaN film surface, α, β, and γ type etch pits and their corresponding schematic views. (a) SEM image of KOH etched GaN film surface, (b) α-type, (c) β-type, and (d) γ-type etch pits. (After Gao, Z. et al., *Chin. J. Semiconduct.*, 28, 473–479, 2007a.)

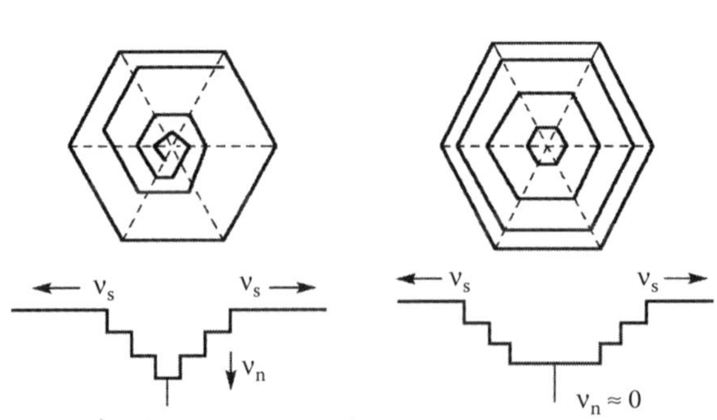

Figure 8.2 Etching of a pure screw dislocation. (After Gao, Z. et al., *Chin. J. Semiconduct.*, 28, 473–479, 2007a.)

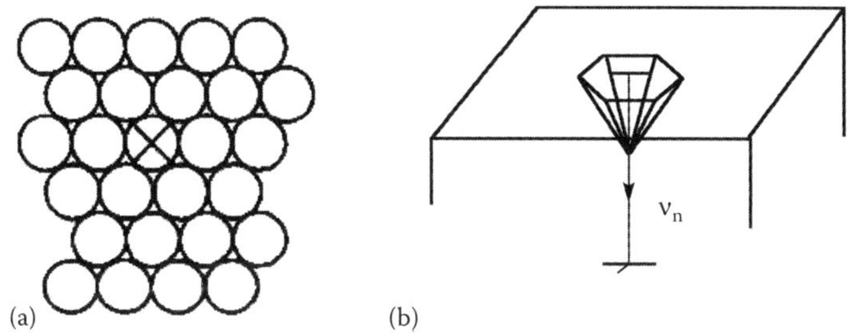

Figure 8.3 Etching of a pure edge dislocation. (a) Plan view of a pure edge dislocation. (b) Etching is easier along the vertical dislocation line. (After Gao, Z. et al., *Chin. J. Semiconduct.*, 28, 473–479, 2007a.)

the Ga-face chemical stability prevents further vertical etching while the transverse one remains, and finally this small facet will turn into a big one which can be observed by microscopy, thus the formation of the α-type etch pit. The β-type etch pits can correspond to pure edge dislocations. In the plan view in Figure 8.3a, "×" denotes the edge dislocation line, which is vertical to the surface. Since each atom in this line has a dangling bond, they are the most vulnerable to attack. The etching carries on along the vertical dislocation line to form an inversed pyramid-shaped β-type etch pit. As the γ-type is a combination of α and β types, it can be associated with a mixed screw/edge dislocation.

The above conclusions are well supported by TEM images. Based on the visibility criteria of TEM diffraction contrast for dislocations, the screw and mixed dislocations are observable with $g = [0002]$ while edge and mixed dislocations observable with $g = [11\bar{2}0]$, and with other diffraction vectors such as $g = [10\bar{1}1]$ all the types of dislocations are visible. The TEM images in Figure 8.4 show the agreement of the above conclusions with the correspondence of the cross-sectional shape of each etch pit at dislocation termination to its dislocation type, indicating that chemical stability of Ga face plays an important role in the formation of different types of dislocation pits.

8.1.2 Accurate Evaluation of Dislocation Densities by Wet-Etching Technique

As the correspondence between the etch pit and the dislocation type is known, it is possible to measure the GaN dislocation density by optimizing etching parameters. The surface morphologies corresponding to different levels of etching shown in Figure 8.5

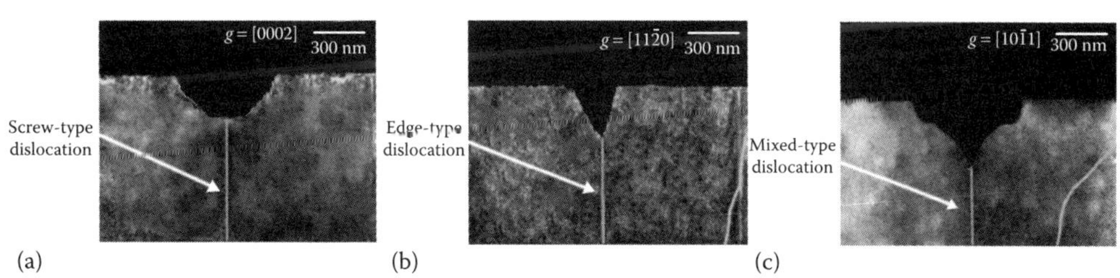

Figure 8.4 Cross-sectional TEM images of three types of etch pits. The corresponding diffraction vectors are given. (a) α-type, (b) β-type, and (c) γ-type. (After Lu, L. et al., *J. Appl. Phys.*, 104, 123525, 4 pp., 2008.)

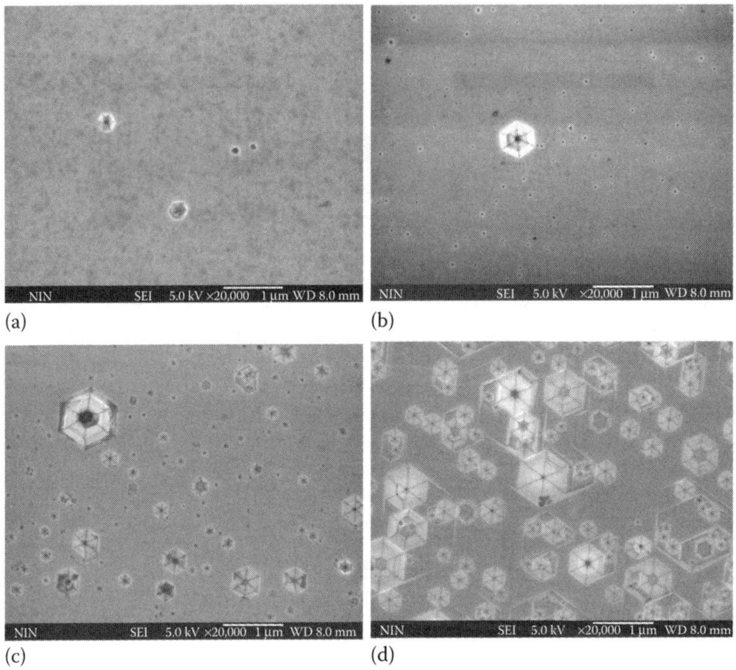

Figure 8.5 SEM images of the morphologies of etched GaN films on sapphire substrates. (a) Unetched surface, (b) etched at 210°C in molten KOH for 2.5 min, (c) etched at 210°C in molten KOH for 5 min, and (d) etched at 235°C in molten KOH for 2.5 min. (After Gao, Z.-Y. et al., *J. Funct. Mater. Device*, 14, 742–750, 2008.)

suggest that insufficient etching may lead to an underestimation of the dislocation density, and so may overetching due to the coalescence of etch pits. Therefore, it is necessary to judge constantly and repeatedly whether optimal etching has been achieved throughout the whole etching process. In fact, a mild overetching is the most desirable, that is, the etch pits have a large diameter with few coalesced and with no or few tiny holes in the background. Such condition (as shown in Figure 8.5d) allows the distinct recognition of the type of each etch pit by SEM at low magnification and avoids missing counts induced by insufficient etching, and the coalescence of very few pits does not threaten the accurate evaluation of the number of pits. If optimal etching is not achieved, further etching can be carried out for an extra period of time.

The etching time is as great an influencing factor on the etched surface as the etching temperature, and an increase in either of the two leads to greater dimension and density of the etch pits until all the etch pits are exposed. An observed relation between the etching time and etching temperature for the maximum etch pit density to be achieved is shown in Figure 8.6. High quality samples require higher temperature or longer time for all the dislocation to be revealed.

The obtained dislocation densities of the GaN samples in Figure 8.6 by measuring the (002) and (302) XRD rocking curves and fitting with pseudo-Voigt function (Figure 8.7) are listed in Table 8.1, which agree with the estimated dislocation densities according to etch pit densities, verifying the accuracy of the evaluation of the densities of different types of dislocations by wet etching.

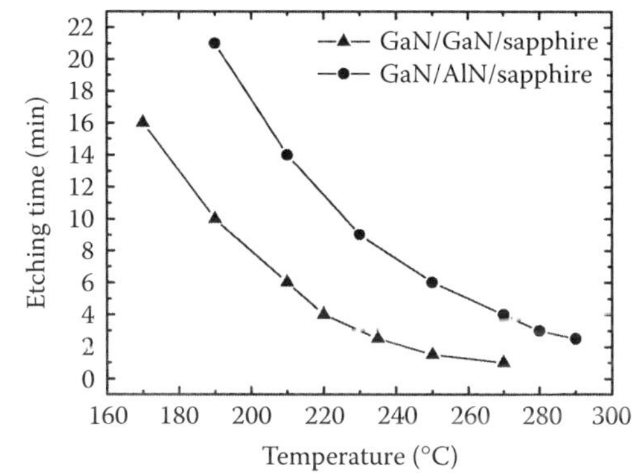

Figure 8.6 Temperature versus time required for maximum etch pit density of GaN epiwafers. (After Gao, Z.-Y. et al., *J. Funct. Mater. Device*, 14, 742–750, 2008.)

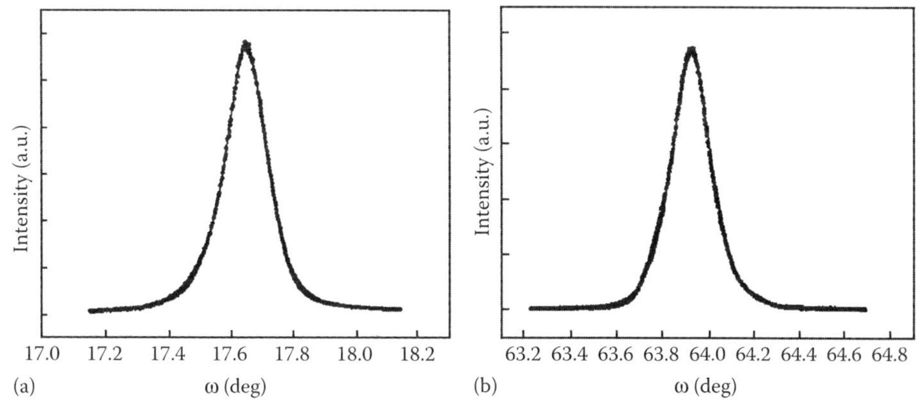

Figure 8.7 (002) and (302) rocking curves (scattered spots) and fitting curves (solid line) of the samples in Figure 8.6. (a) (002) reflection and (b) (302) reflection. (After Gao, Z. et al., *Chin. J. Semiconduct.*, 28, 473–479, 2007a.)

Table 8.1 Densities of α-, β-, and γ-Type Etch Pits of Two Typical Samples and Their Dislocation Densities Obtained from the XRD Measurements

	ETCH PIT DENSITY (CM^{-2})			XRD-MEASURED DISLOCATION DENSITY (CM^{-2})	
SAMPLE	α-TYPE	β-TYPE	γ-TYPE	SCREW DISLOCATION	EDGE DISLOCATION
GaN/GaN/sapphire	4×10^7	5×10^8	5×10^6	4.36×10^7	4.98×10^8
GaN/AlN/sapphire	3×10^6	6×10^7	1×10^6	3.89×10^6	5.86×10^7

Source: Gao, Z.-Y. et al., *J. Funct. Mater. Device*, 14(4), 742–750, 2008.

8.1.3 Analysis of Other Types of Defects in GaN by Etching: IDBs and LAGBs

For polar GaN films, the difference in chemical properties between the two polarities can characterize the inversion domain boundary (IDB). The N-face region is much more vulnerable to oxidization than the Ga-face one. The region enclosed by the IDBs within the Ga-face GaN is N-polar, where the surface is soon etched off to generate deep holes with continual etching into the depth; while the Ga-polar region remains intact with selective etching only at the dislocation sites. The IDB revealed by wet

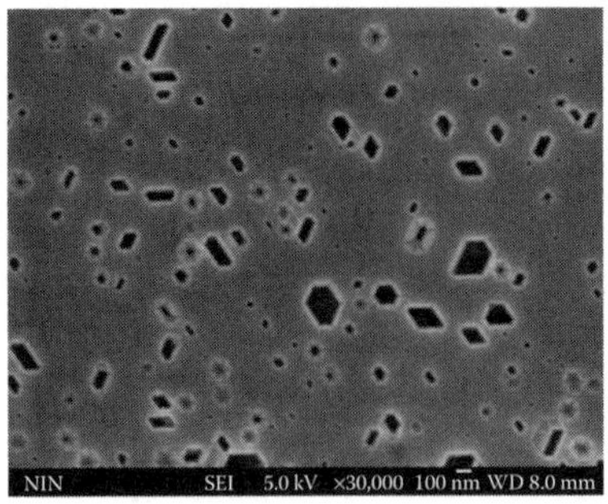

Figure 8.8 SEM image of the surface morphology of GaN etched in molten KOH for 1 min. The irregular deep etch pits show the IDBs. (After Gao, Z. et al., *Mater. Rev.*, 23, 1–5, 2009.)

etching is shown in Figure 8.8. The large area region is the selectively etched Ga-face, and the inversed hexagonal pyramid pits are the etch pits of the above-mentioned dislocations. Besides the dislocation-generated etch pits in regular shape, substantial etch pits in the irregular shape of arbitrary polygons with black centers are observed, which are penetrating deep holes where the emitted electrons cannot be collected by the detector according to the SEM contrast principle, and thus such type of etch pits is formed by the IDB.

The low-angle grain boundary (LAGB) formed by the coalescence of the slightly misoriented columnar subgrains consists of substantial threading dislocations, so the presence of the LAGB can be indicated by the distribution of the surface etch pits of the dislocations. Shown in Figure 8.9 is the morphology image of GaN after etching. The dislocations are situated along the LAGBs on the verge of the crystal grains within which there are no or few dislocations, and thus a series of dislocation etch pits outline how the grains join with each other. The low angle dislocations characterized by plan-view TEM are shown in Figure 8.10, where the dislocation lines are demonstrated by short segments, and they are interconnected delineating the LAGBs. The results have the same effect as those revealed by etching.

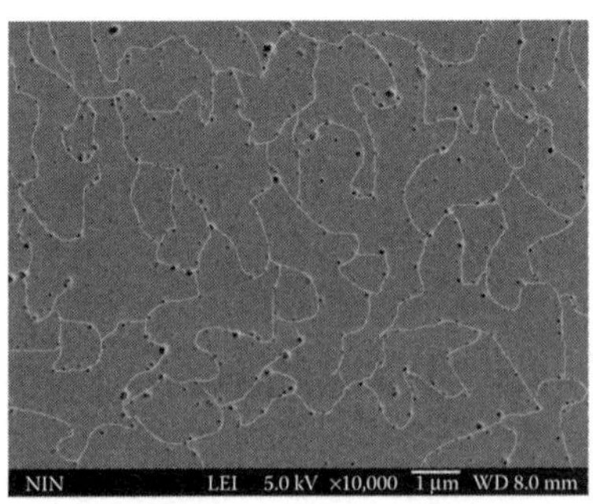

Figure 8.9 SEM backscattered electron image of the surface morphology of the Ga-face GaN sample after etching. The dark spots are the etch pits formed at the surface terminations of dislocations in the experiment, and these etch pits are linked up by the white line to show how the grains are connected. (After Gao, Z. et al., *Mater. Rev.*, 23, 1–5, 2009.)

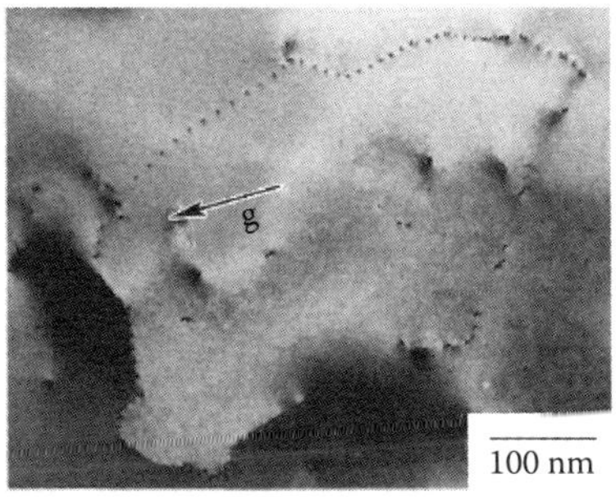

Figure 8.10 Plan-view TEM image ($g = [2\bar{1}\bar{1}0]$) of the GaN film. The dislocation lines are shown as short segments, which interconnect delineating the LAGB. (Reproduced from Qian, W. et al., *Appl. Phys. Lett.*, 66, 1252–1524, 1995.)

8.2 Etched Morphologies of Nitrides of Different Polarities and Their Origin

It can be seen from the above section that the chemical stability of Ga-face polarity plays a vital role in the formation of different types of dislocation etch pits. When it comes to the N-face and nonpolar materials, new features of etching are found (Xu et al. 2010b).

8.2.1 Etching Characteristics of N-Face Materials

N-face materials enjoy easy fabrication of low resistance contacts on their surfaces and have much weaker yellow luminescence (YL) in the PL spectrum than Ga-face ones, thus their great potentials in both optoelectronics and microwave power devices. MOCVD can be employed to grow N-face materials under similar growth procedure to those for Ga-face materials, and on sapphire substrate a greater flow of ammonia gas and a longer time of substrate nitrification in the substrate treatment are the most important to obtain N-face materials.

The typical surface morphology of N-face GaN before etching is shown in Figure 8.11a. The surface consists of many hexagonal grains. The N-face morphology is drastically altered after the same KOH etching as that for Ga-face materials (for Ga-face materials,

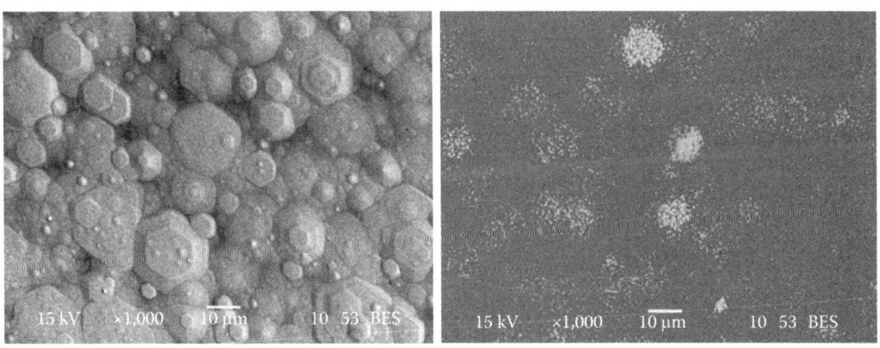

(a) (b)

Figure 8.11 Morphologies of N-face GaN before and after etching. (a) Before etching and (b) after etching. (After Xu, S.-R. et al., *Chin. Phys. B*, 19, 107204, 5 pp., 2010b.)

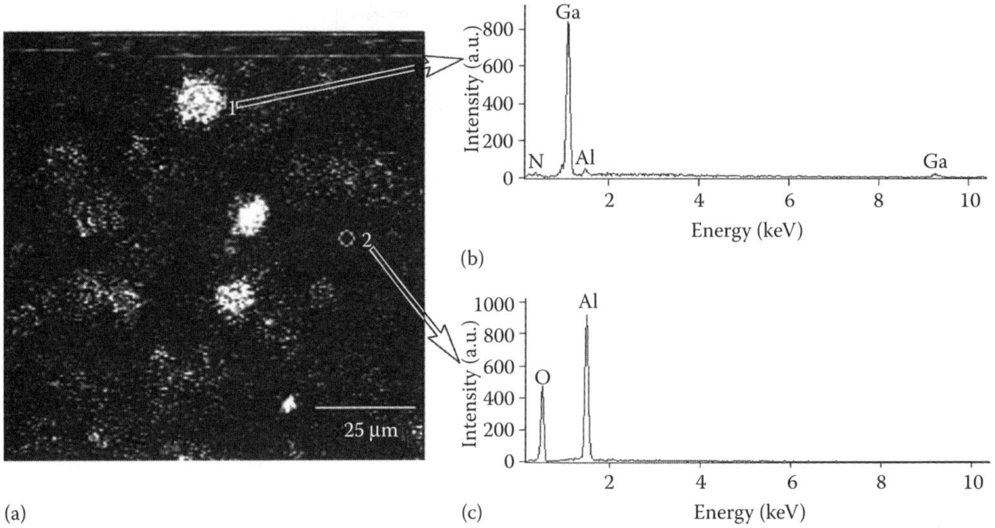

Figure 8.12 Element analysis results. (a) Selected spots for element analysis, (b) element analysis of Spot 1, and (c) element analysis of Spot 2. (Courtesy of Dr. Shengrui Xu.)

no overetching occurs with only etch pits formed on the surface under this condition, thus no change in the material thickness), as shown in Figure 8.11b. The etch seems to have reached the substrate surface owing to the active chemical property of N-face with only some minor nucleation grains left. In order to verify this assumption, we made an element analysis of two representative spots from the surface: Spot 1 chosen from the surface of a granular material whilc Spot 2 from the surface without granular morphology. The results of the analysis are shown in Figure 8.12. It is seen that the elements at Spot 1 are mainly nitrogen and Ga as well as some Al, suggesting the crystal grains here are part of the nucleation layer; while those at Spot 2 consist only of Al and oxygen without peaks of any other elements, which indicates that Spot 2 is at the surface of the sapphire substrate. Accordingly N-face GaN does have a very high etching rate.

This disparity in the etching rates of Ga-face and N-face nitrides can be explained by their atomic arrangements. As shown in Figure 8.13, each layer of the Ga-face

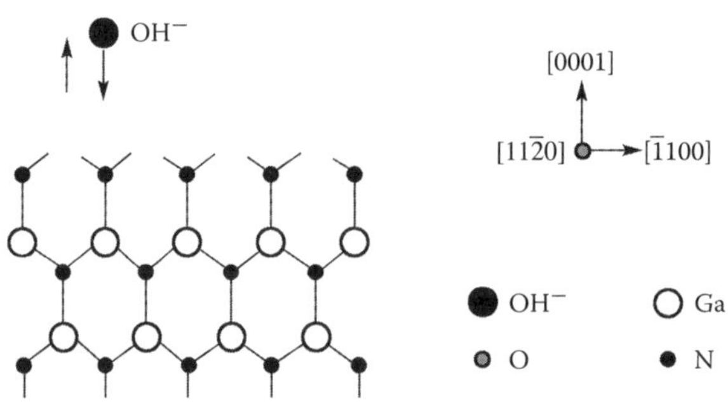

Figure 8.13 Schematic atomic arrangements and etching of Ga-face GaN. (After Xu, S.-R. et al., *Chin. Phys. B*, 19, 107204, 5 pp., 2010b.)

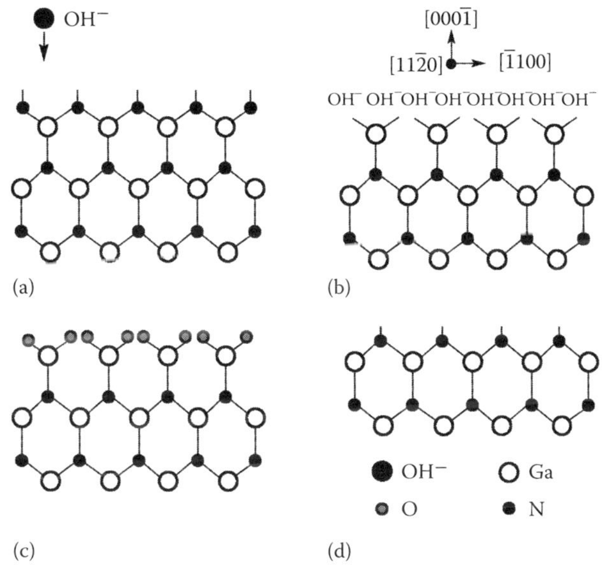

Figure 8.14 (a–d) Schematic atomic arrangements and etching of N-face GaN. (After Xu, S.-R. et al., *Chin. Phys. B*, 19, 107204, 5 pp., 2010b.)

material consists of atoms of the same kind (gallium or nitrogen) and the outermost layer of nitrogen has three dangling bonds, which impose so great a repulsive effect on the OH^- ions that it is very difficult for the OH^- ions to break the N-Ga bonds under the nitrogen atoms. Gallium atoms are in effective protection of the nitrogen atoms in such atomic arrangements. Only in the vicinity of the dislocation line where the atomic arrangement is disordered can the OH^- ions attack along the dislocation line, as a result of which the dislocation pits are exposed by etching while other regions are difficult to attack thus remaining largely intact.

The atomic arrangements of N-face GaN is shown in Figure 8.14. The outmost layer of nitrogen atoms has only one dangling bond, thus a weak repulsive effect on the OH^- ions, which can attack the surface Ga-N bonds and be adsorbed on the GaN surface where it reacts with GaN forming Ga_2O_3 and NH_3. Ga_2O_3 can be dissolved in alkali solution, and the reaction equation is

$$2GaN + 3H_2O \xrightarrow{KOH} Ga_2O_3 + 2NH_3 \tag{8.1}$$

After the adsorption and solution processes shown schematically in Figure 8.14b–d, the OH^- ions attack the Ga-N bonds of the next layer. This repeated etching process is continued down through the layers until the N-face material is rapidly depleted.

8.2.2 Selective Etching of Nonpolar A-Plane GaN Films

Different from the polar GaN, nonpolar a-plane GaN has hybrid arrangements of gallium and nitrogen atoms in each layer of atomic planes parallel with the surface.

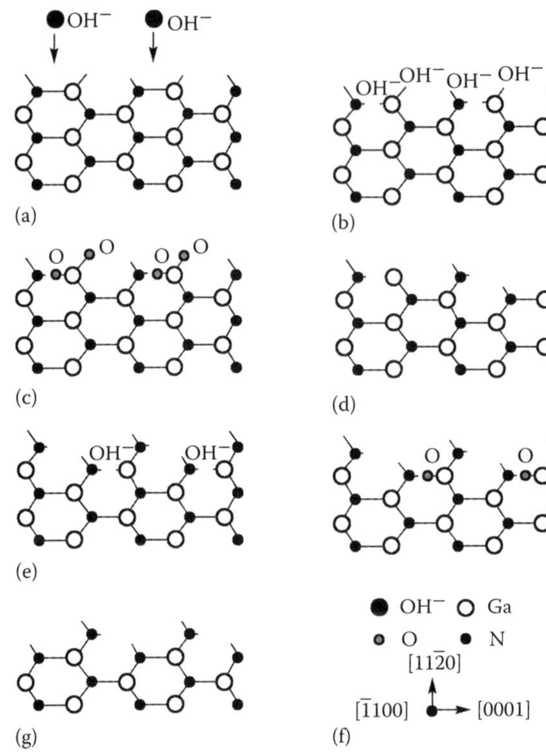

Figure 8.15 (a–f) Ideal etching model of nonpolar a-plane GaN. (After Xu, S.-R. et al., *Chin. Phys. B*, 19, 107204, 5 pp., 2010b.)

The atomic arrangements shown in Figure 8.15 may lead to the etching of nonpolar a-plane GaN as similar as that in Figure 8.14: attack of the Ga-N bonds by the OH^- ions and adsorption → reaction of the OH^- ions with GaN forming Ga_2O_3 and NH_3 and dissolution of Ga_2O_3 → continued attack by the OH^- ions down through the layers.

However, under the same etching conditions as those for the Ga-face and N-face, many micro- and nano-meter lying columns were observed on the high quality V-defect-free nonpolar a-plane GaN surface as shown in Figure 8.16b in comparison with its surface morphology before etching in Figure 8.16a. Although of different diameters, these columns are highly regular and all oriented along the *c*-axis. This morphology was universal over a large area, suggesting the anisotropy of the etching of nonpolar materials.

The etching model shown in Figure 8.15 is an ideal one without the lattice imperfections in consideration, which apparently fails to explain the origin of the etched morphology in Figure 8.16. For a more accurate interpretation of the etching mechanism of nonpolar a-plane GaN, it is necessary to consider an important type of defects in nonpolar materials: stacking faults.

The morphology of nucleation islands at various times in the initial growth of the nucleation layer of nonpolar a-plane GaN is shown in Figure 8.17. These nucleation

Figure 8.16 SEM surface morphologies of the nonpolar a-plane GaN before etching and after etching over a small and a large area. (a) Before etching. (After Xu, S.-R. et al., *Chin. Phys. B*, 19, 107204, 5 pp., 2010b.) (b) After etching over a small area. (After Xu, S.-R. et al., *Chin. Phys. B*, 19, 107204, 5 pp., 2010b.) (c) After etching over a large area. (Courtesy of Dr. Shengrui Xu.)

Figure 8.17 Morphologies of nonpolar a-plane GaN in the initial nucleation stage. (a) Nucleation layer grown for 120 s and (b) nucleation layer grown for 600 s. (Reproduced from Johnston, C.F. et al., *J. Appl. Phys.*, 105, 073102, 5 pp., 2009.)

islands consist mostly of (000$\bar{1}$), (10$\bar{1}$1), and (01$\bar{1}$1) planes. The (000$\bar{1}$) plane maintains its structures in growth and grows very slowly, which explains the existence of stacking faults in nonpolar a-plane nitrides. Nonpolar nitrides generally have a very high stacking fault density which could be up to 4×10^5 cm^{-1}, making it difficult to eliminate the faults even by ELOG. It is the presence of (000$\bar{1}$) planes in the stacking faults that has a decisive influence on the etching characteristics of nonpolar nitrides.

The side walls of the V pits on the surface of nonpolar a-plane GaN correspond very well to some crystalline orientations. A piece of GaN film with V pits was etched to show the influence of the (000$\bar{1}$) planes of the faults in the etching of nonpolar a-plane GaN. The surface morphologies before and after the etching are shown in Figures 8.18 and 8.19. With the dark spot inside the V-pit as a reference, etching carries on primarily along the [000$\bar{1}$] orientation, while no sign of etching is observed along the [0001] orientation. It can be seen that the (000$\bar{1}$) plane of the V-pit has disappeared, in other words, has been etched a long distance along the [000$\bar{1}$] orientation. Thus for nonpolar GaN, the (000$\bar{1}$) plane still has a high etching rate, and the anisotropic etching morphology of nonpolar a-plane GaN results from the different chemical properties of Ga-face and N-face nitrides.

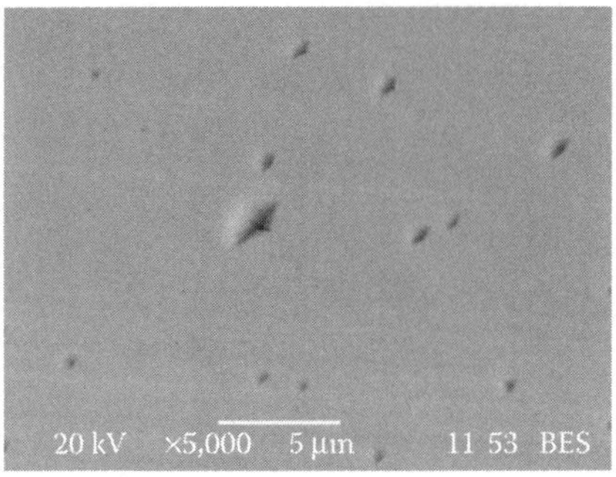

Figure 8.18 Morphology of nonpolar a-plane GaN with V-pits before etching. (Courtesy of Dr. Shengrui Xu.)

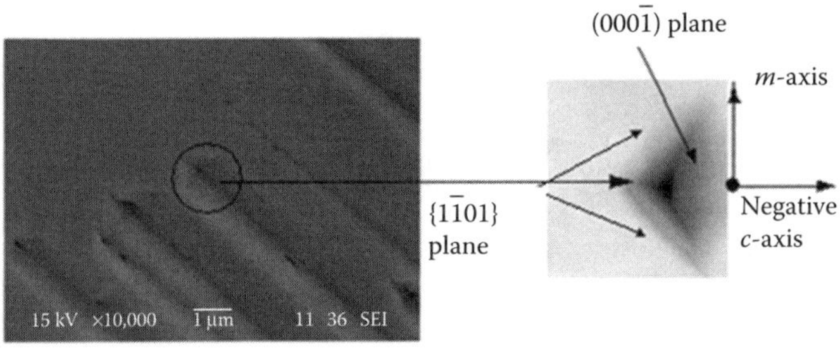

Figure 8.19 Correspondence of the morphology of nonpolar a-plane GaN with V-pits after etching to the original V-pit. (Courtesy of Dr. Shengrui Xu.)

Generally speaking, N-face GaN has the highest etching rate followed by nonpolar a-plane GaN with Ga-face GaN having the lowest rate of the three. The presence of the N-face in the stacking faults of nonpolar a-plane GaN makes its etching much different from that of c-plane GaN, and results in the lying columns on the etched surface.

8.3 Mechanism of Dislocation Reduction by Vicinal Substrates

The superiority of vicinal substrates to conventional on-axis substrates for the growth of low defect GaN films has been discussed in Chapter 5. In this section, we analyze the mechanism of reducing the dislocation density of the epitaxial GaN films by the vicinal substrate with the help of TEM observation.

8.3.1 Dislocation Types and Dislocation Clusters in GaN on Vicinal Substrates

The sample for analysis is a 1.4 μm unintentionally doped GaN film grown on a c-plane sapphire substrate with 0.5° misorientation toward m-plane. Figure 8.20a–c show the cross-sectional TEM images of GaN near the $[1\bar{1}00]$ orientation with $g = [0002]$, $g = [11\bar{2}0]$, and $g = [11\bar{2}2]$, respectively. According to the visibility criteria of TEM diffraction contrast for dislocations, the dislocations only visible in the TEM image of the hexagonal GaN with $g = [0002]$ are the screw dislocations, those visible

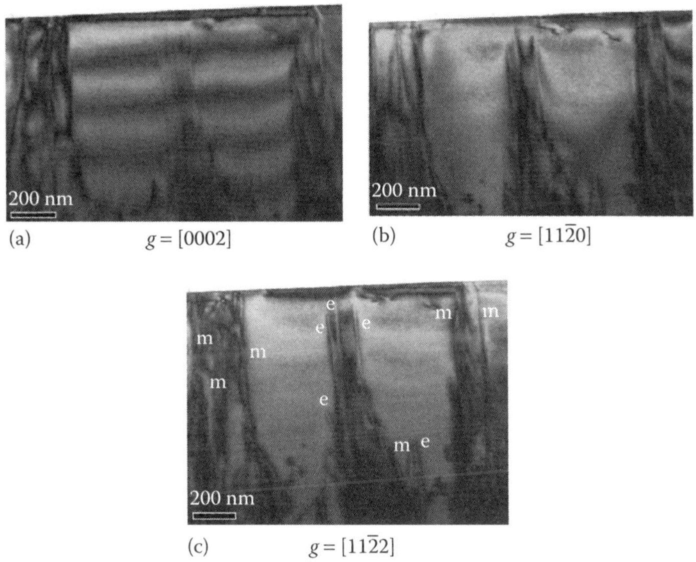

Figure 8.20 Cross-sectional TEM images of the GaN film with (a) $g = [0002]$, (b) $g = [11\bar{2}0]$, and (c) $g = [11\bar{2}2]$, respectively. The e's stand for edge dislocations and m's for mixed dislocations. (Courtesy of Xinxiu Ou.)

with $g = [11\bar{2}0]$ are the edge dislocations, and those visible under both conditions are the mixed dislocations. All the dislocations are typically characterized by the cross-sectional TEM image with $g = [11\bar{2}2]$. The type of dislocations can be determined by the variation in the dislocation contrast of the same location on the specimen with different diffraction vectors.

It is observed in Figure 8.20 that the majority of the dislocations are mixed dislocations with edge components and there are few pure screw dislocations in the epitaxial GaN on the vicinal substrate. Dislocation clusters are also seen in some regions on the epitaxial GaN film on the vicinal sapphire substrate and the dislocation bundles gradually narrowing down. It is visible in Figure 8.21 that there are few dislocations in some local regions of the epitaxial GaN film, and among these regions dislocation

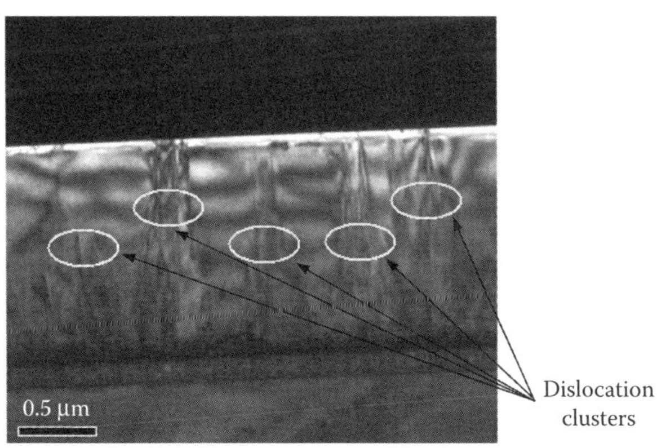

Figure 8.21 TEM image of GaN grown on vicinal sapphire substrate nearby $[1\bar{1}00]$ orientation with $g = [0002]$. (Courtesy of Xinxiu Ou.)

cluster arise in an alternative manner. Since dislocations are typically generated where two crystal grains coalesce, the presence of dislocation clusters and relatively large dislocation-free regions indicates the relatively large grain size of GaN grown on the vicinal sapphire substrate.

8.3.2 Collective Annihilation of Dislocations in GaN Epifilms on Vicinal Substrates

The region near the surface of the considered GaN film grown on the vicinal sapphire substrate has a small dislocation density, as most of the dislocations are annihilated by the formation of dislocation loops in the growth of the epitaxial layers and thus fail to penetrate to the top layer. Two regions where massive annihilation of dislocations occurs can be found by a closer look at the TEM image in Figure 8.22. One is within 100 nm from the AlN nucleation layer where substantial dislocations form dislocation loops without extending to the GaN epitaxial layer. This is a universal phenomenon in GaN growth as the dislocation density is significantly reduced by the massive annihilation of dislocations during the transition from the 3D growth mode of the AlN nucleation layer on the sapphire substrate into the 2D mode of GaN growth due to the faster lateral growth rate. Another region of massive annihilation of dislocations is ~0.8 μm from the nucleation layer, with only a few dislocations penetrating to the surface of the subsequently grown GaN film. It is suggested that the misorientation leads to the second massive dislocation annihilation in the epitaxial films at a specific thickness, a phenomenon also found in nitride epifilms on sapphire substrates with other miscut angles, and the position of the second dislocation annihilation region is related to the miscut angle. The TEM image of a GaN epitaxial film grown on an on-axis sapphire substrate is given for comparison in Figure 8.23, in which the second region of the collective annihilation of dislocations is not observed.

According to the XRD analysis (Figure 5.13) of GaN grown on vicinal substrates with different miscut angles in Section 5.2 and the TEM analysis in this section, the epitaxial GaN film grown on vicinal sapphire substrates has a significantly lower dislocation density than that on the on-axis sapphire substrate. The vicinal substrate enhances

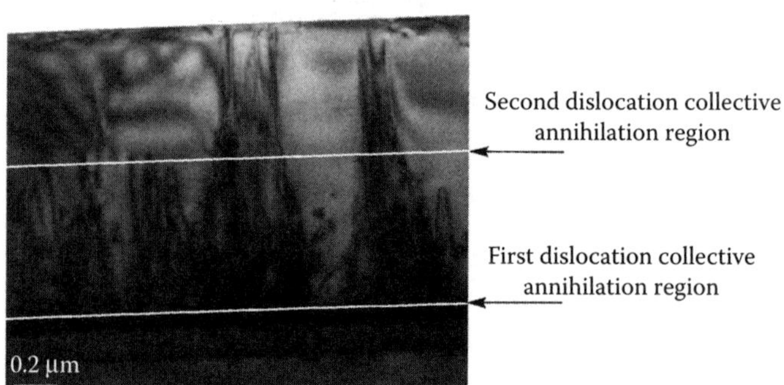

Figure 8.22 TEM image of collective annihilation of dislocations in GaN on vicinal sapphire substrate near $[1\bar{1}00]$ orientation with $g = [11\bar{2}0]$. (Courtesy of Xinxiu Ou.)

Figure 8.23 TEM image of GaN grown on on-axis sapphire substrate near $[1\bar{1}00]$ orientation with $g = [11\bar{2}0]$. (Courtesy of Xinxiu Ou.)

the step flow growth of the epitaxial GaN film, and the coalescence of steps induces an inclination of the dislocations to form a closed loop in a certain depth preventing their penetration to the surface, thus an effective reduction in dislocation density.

In summary, the dislocations in epitaxial GaN films on the vicinal substrate have the distinctive features of clustering and collective annihilation, and multiple collective annihilation regions may be observed in thicker epitaxial films.

8.4 Influence of Polarity on Impurity Incorporation and YL

An important part of the GaN photoluminescence (PL) spectrum, the yellow band is in the approximate range of 2.2–2.3 eV. Numerous reports demonstrate YL is a defect-related deep level luminescence, therefore the intensity and distribution of the YL serve as a direct indicator of the point defect density in GaN materials. High quality GaN materials used in electronic devices are generally required to be free of YL. Although extensive YL studies have yielded a consensus that YL originates from the recombination of donors and acceptors, the physical nature of the deep acceptors that gives rise to YL has been inconclusive.

YL can be observed in the PL spectra of polar, nonpolar, and semi-polar GaNs, all of which have their growth and properties subject notably to the polarity or more essentially to the atomic arrangement. Thus, YL and the related incorporation of impurities also need to be viewed from the perspective of polarity.

8.4.1 Impurity Incorporation Model Related to Polarity

The impurity incorporation of polar, semi-polar, and nonpolar GaN films grown on c-plane, m-plane, and r-plane sapphire substrates, respectively, was investigated. The samples were grown by the identical technique with the same growth pressure, temperature, and nucleation layer to eliminate the influence of growth technique. The element analysis was performed on silicon, carbon, and oxygen impurities by secondary

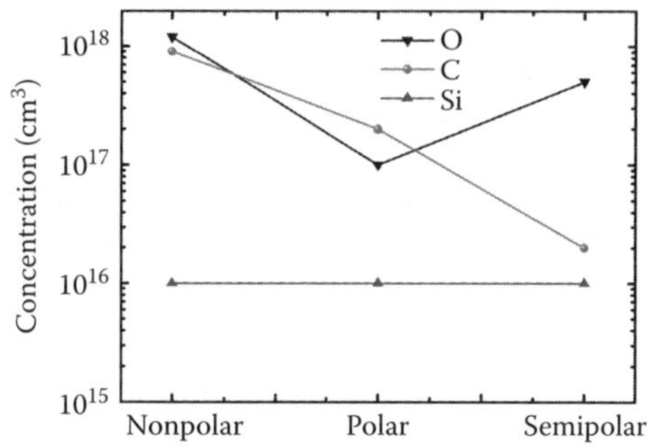

Figure 8.24 Oxygen, carbon, and silicon impurities incorporation with respect to the different polarities of GaN epifilms. (After Xu, S.R. et al., *J. Cryst. Growth*, 312, 3521–3524, 2010a.)

ion mass spectrometry (SIMS) with the detection limits of [C] 2×10^{16} cm^{-3}, [O] 1×10^{16} cm^{-3}, and [Si] 1×10^{16} cm^{-3}.

The concentrations of oxygen, carbon, and silicon impurities in polar (0001), semipolar ($11\bar{2}2$), and nonpolar ($11\bar{2}0$) GaNs are shown in Figure 8.24. The concentrations of silicon in all of the three GaN layers are very low, close to detection limits, indicating the universal presence of small silicon incorporation in GaNs grown by typical techniques. Therefore, its impact on material characteristics is negligible compared to the carbon and oxygen incorporation.

It can be seen in Figure 8.24 that the concentration of oxygen in the semipolar sample is approximately 6 times that of oxygen in the polar sample, and the concentration of carbon in the polar sample is 15 times higher than that of carbon in the semipolar sample with the nonpolar sample having the highest concentration of both carbon and oxygen. Since the three samples are grown under the same conditions, the clear distinction in the impurity incorporation may be attributed to polarity. It should be noted that as the semipolar GaN is N-polar, the N-face atomic arrangement is employed to analyze its impurity incorporation. Fichtenbaum et al. predicted that for the Ga-face, oxygen atoms impinging on a group V site will form only a single bond to the gallium atoms, while on the N-face, an atom impinging on a group V site will form three bonds to the gallium atoms, leading to a stronger bonding of oxygen atoms on the N-polar surface (Fichtenbaum et al. 2008). However, it failed to explain the difference in oxygen and carbon incorporation between the three different polarities.

Xu et al. proposed a model to interpret such carbon and oxygen incorporation by the origin of the impurities and the atomic arrangements of GaN films with different polarities (Xu et al. 2010a). Oxygen comes mostly from decomposition of the "downside" sapphire, while the carbon incorporation most likely stems from decomposition of the precursors, namely it is from the "upside." For oxygen and carbon in the n-type GaN film, the most stable site is C_N and O_N. For the polar Ga-face GaN as shown in Figure 8.25a, the oxygen atom from the bottom attempts to replace nitrogen to form O_N; however, every time there is a bigger gallium atom offering an effective protection to the upper nitrogen. For the N-face GaN as shown in Figure 8.25b, no protection

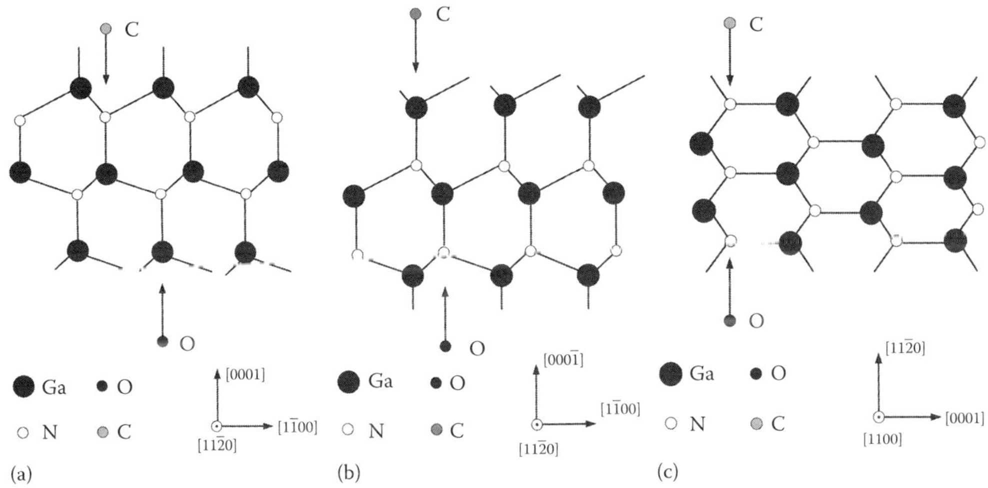

Figure 8.25 Schematic model for considering the origin of impurities to explain their incorporation in GaN films with different polarities. (a) Ga-face, (b) semipolar, (c) nonpolar a-plane. (After Xu, S.R. et al., *J. Cryst. Growth*, 312, 3521–3524, 2010a.)

for nitrogen is available from gallium atoms due to the opposite atomic arrangement to the Ga-face GaN, thus a significantly higher concentration of oxygen. The carbon incorporation can also be explained in the same way, which happens only in a different top-down direction since the carbon atom comes from the decomposition of the precursors upside. There is no protection from gallium for nitrogen atoms when the carbon atom from the upside attempts to replace nitrogen to form C_N, thus a high concentration of carbon in Ga-face GaN owing to the replacement of nitrogen by carbon. For the N-face GaN as shown in Figure 8.25b, there is a gallium atom to provide effective protection for the lowered nitrogen every time the carbon atom from the top attempts to replace nitrogen to form C_N, thus a low carbon concentration.

For the nonpolar a-plane GaN as shown in Figure 8.25c, no protection is available from gallium for nitrogen every time the carbon or oxygen atom attempts to replace nitrogen to form a substitution because each longitudinal array of atoms are of the same type, so the carbon and oxygen concentrations are the highest in the nonpolar sample.

8.4.2 Influence of Impurity Incorporation on YL

The normalized PL measurements of polar (0001), semipolar (11$\bar{2}$2), and nonpolar (11$\bar{2}$0) GaN samples are given in Figure 8.26 with the band edge peak as the reference. Much different luminescence behaviors are observed on different polar GaN samples grown under the same conditions. The semipolar sample displays a very weak YL while both nonpolar and polar samples exhibit very strong YL.

As is discussed earlier, YL stems from the recombination of the defect-induced donors and acceptors, but the deep acceptors leading to YL remains inconclusive. There are three major theories: that the deep acceptor introduced by Ga vacancies

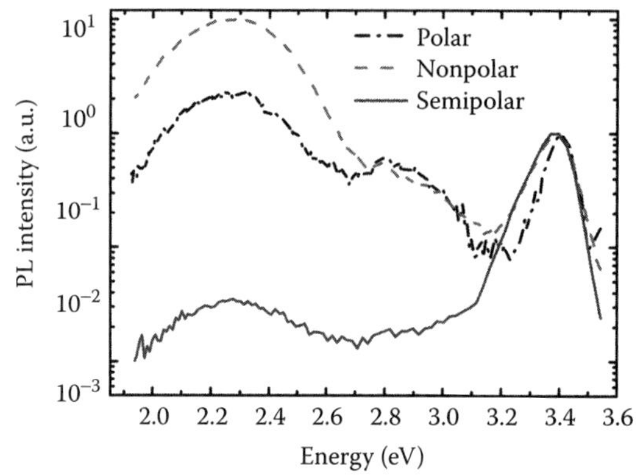

Figure 8.26 PL spectra of samples of different polarities. (After Xu, S.R. et al., *J. Cryst. Growth*, 312, 3521–3524, 2010a.)

(V_{Ga})–O_N complexes is the origin of YL (V_{Ga} is Ga vacancy), that YL originates from the carbon-related impurities, and that the edge dislocations have a major influence on YL in GaN.

To investigate the impact of various mechanisms on YL, we performed HRXRD measurements on several samples with the FWHM values of the (102) reflection measured to be 0.18° for c-plane, 0.4° for a-plane, and 0.58° for semipolar GaN, respectively, as shown in Figure 8.27. It has been reported that the densities of edge dislocation are indirectly represented by the FWHMs of XRD in the (102) reflection (Heinke et al. 2000). The results show that c-plane GaN with very strong YL has a much smaller FWHM of the (102) reflection, that is, a much lower dislocation density, than those of other samples; while the semipolar GaN with the weakest YL has the greatest FWHM of the (102) reflection, that is, the highest dislocation density, which cannot be explained by the third theory that edge dislocations are the major influencing factor of YL in GaN. So there must be a more decisive factor than the edge dislocations in the origin of YL in GaN.

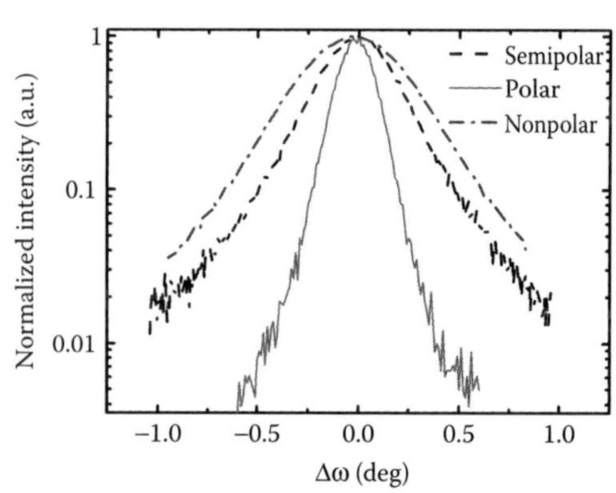

Figure 8.27 (102) HRXRD rocking curves of samples of different polarities. (After Xu, S.R. et al., *J. Cryst. Growth*, 312, 3521–3524, 2010a.)

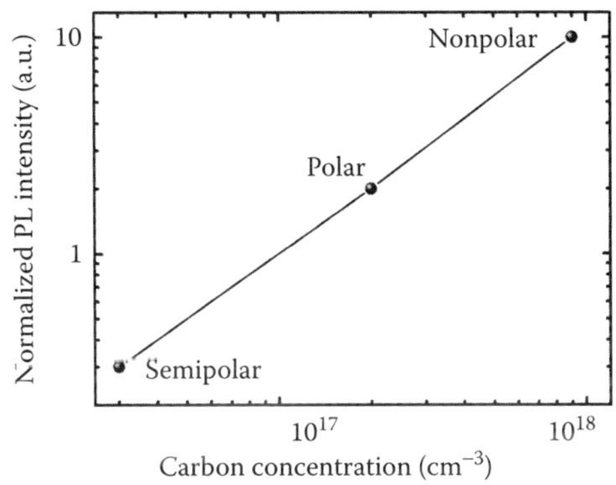

Figure 8.28 The relation between YL intensity and the carbon concentration of samples of different polarities. (After Xu, S.R. et al., *J. Cryst. Growth*, 312, 3521–3524, 2010a.)

It can be found in Figure 8.24 that the semipolar GaN has a high oxygen concentration, indicating the presence of more V_{Ga}–O_N complexes. According to the theory that YL originates from the V_{Ga}–O_N complexes, strong YL should be observed, but Figure 8.26 shows YL in the semipolar $(11\bar{2}2)$ GaN is the weakest. This indicates that the V_{Ga}–O_N complexes are not responsible for YL in GaNs.

As shown in Figure 8.28, the YL intensity agrees well with carbon concentration in GaN films with different polarities, suggesting the carbon-introduced deep acceptors being the origin of YL. Carbon in the MO source such as $Ga(CH_3)_3$ is inevitably introduced in MOCVD. Additionally, numerous reports on the HVPE grown nonpolar a-plane GaN samples showed very weak YL even for poor quality a-plane GaN. This further verifies the theory that the origin of YL is carbon, as carbon is difficult to be introduced in HVPE. Wright also proved in his research the good correspondence of the energy levels of the carbon-related defects to that of the YL band (Wright 2002).

The above discussions lead to the conclusion that carbon is the decisive factor of the origin of YL in GaN.

8.5 Origin of Deep Acceptors Inducing YL in GaN

SIMS can be used to make a quantitative analysis of the distribution of various types of point defects. SIMS characterization of point defect impurities and PL spectra of substantial samples were performed to analyze the origin of the deep acceptors inducing YL from the perspective of point defects (Yang 2011).

8.5.1 Relevance of YL to Carbon Impurities in GaN

The comparison of YL between different samples should focus on the ratio of the YL peak to the band edge peak (YL/BL) in view of the difference in measurement conditions and instrument status. Seven typical samples were chosen from numerous samples with YL, involving different polarities, growth techniques, and relative YL

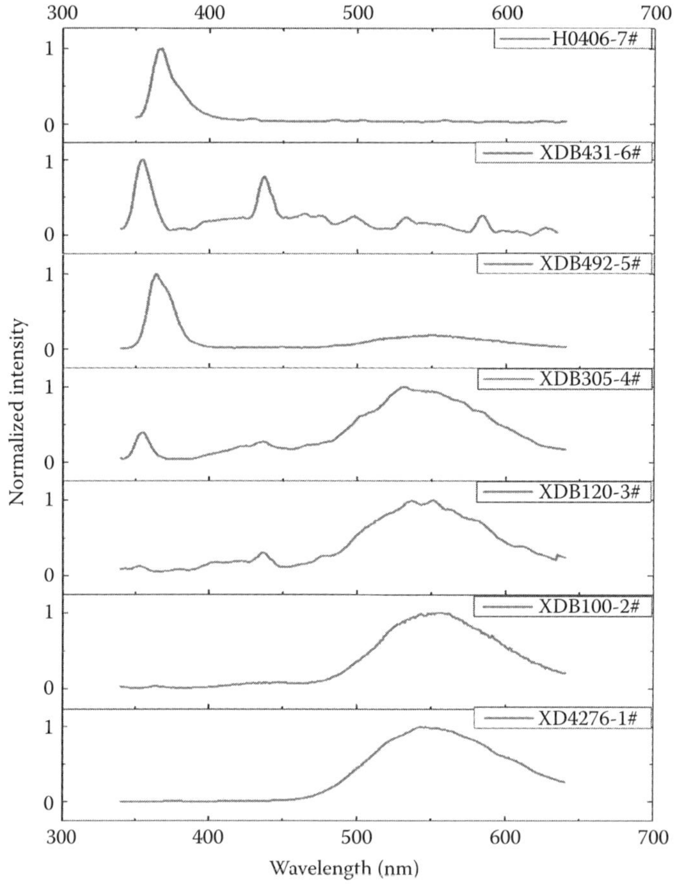

Figure 8.29 PL spectra of the seven samples. (Courtesy of Chuankai Yang.)

intensities, that is, YL/BLs. For convenience in comparison, the PL spectra of the seven samples were normalized based on the respective highest luminescence peak as shown in Figure 8.29, and arranged in an ascending order with 1# having the greatest YL/BL at the bottom and 7# having the least at the top. The cross-sectional layer structures of these samples are shown in Figure 8.30. 1# is the nonpolar a-plane GaN sample, and the rest are polar c-plane ones; 7# was the epitaxial GaN sample grown by HVPE, and all the other samples by MOCVD.

As is discussed in the above section, point defects responsible for YL may be the V_{Ga} and C. Since no means of direct characterization of V_{Ga} concentration is currently available, analysis of V_{Ga} concentration is generally performed by a combination of theoretical analysis based on first principles and the indirect experimental method. The carbon concentration can be characterized by SIMS analysis. Here, we investigate first the correlation between carbon impurities and YL.

Since the seven samples vary greatly in thickness and the SIMS measurements also show nonuniform distribution of carbon in the samples, a reasonable scheme for comparing their carbon concentrations is to normalize the epifilm thicknesses and take the carbon concentration at the 20%–50% of depth away from the surface (enclosed by the

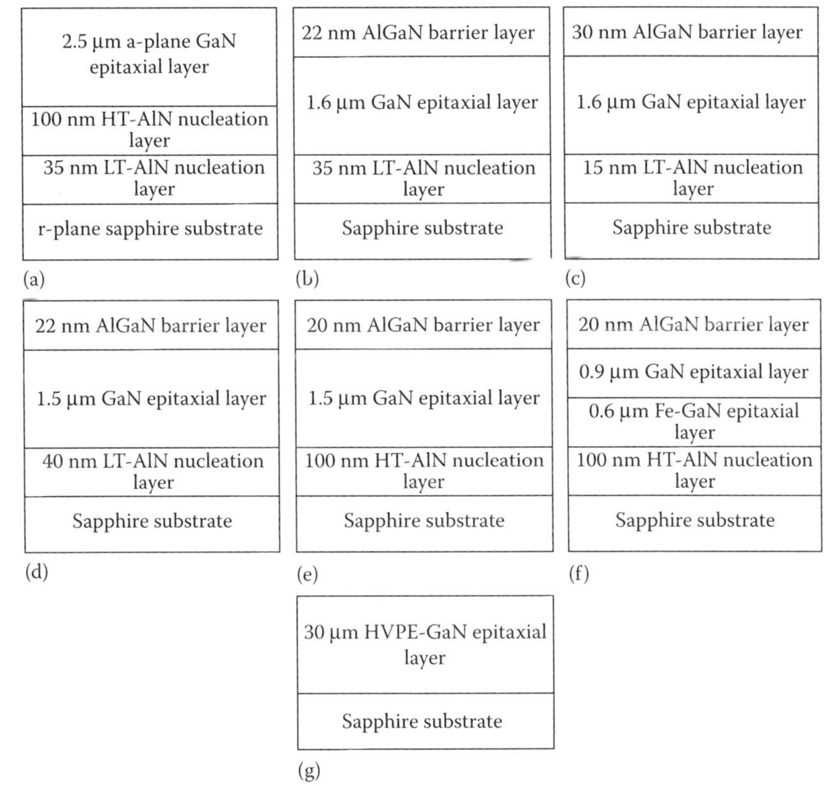

Figure 8.30 Schematic layer structures of the seven samples. (a) XD4276-1#, (b) XDB100-2#, (c) XDB120-3#, (d) XDB305-4#, (e) XDB492-5#, (f) XDB431-6#, and (g) H0406-7#. (Courtesy of Chuankai Yang.)

broken line rectangle in Figure 8.31) as the source data. According to the layer structure in Figure 8.30, this range of depth falls mostly within the GaN epilayer, where the carbon concentration is relatively stable. These fluctuating source data were filtered by fast Fourier transform (FFT) and then the average carbon concentration for each sample is calculated. The SIMS result of 3# is shown in Figure 8.31. The SIMS results of the carbon concentrations of the seven samples after data processing are shown in Figure 8.32.

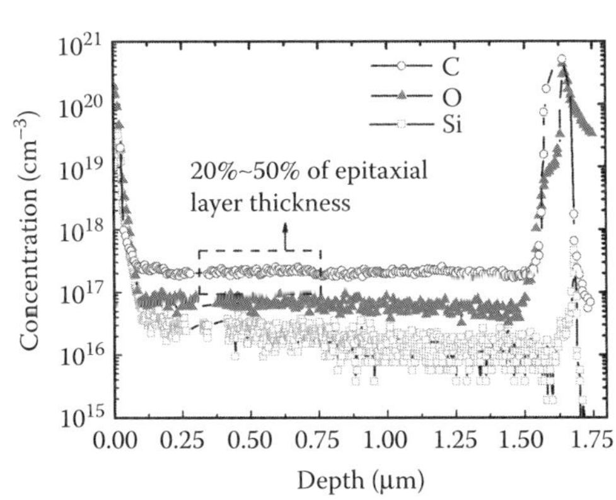

Figure 8.31 Typical SIMS result of 3#. (Courtesy of Chuankai Yang.)

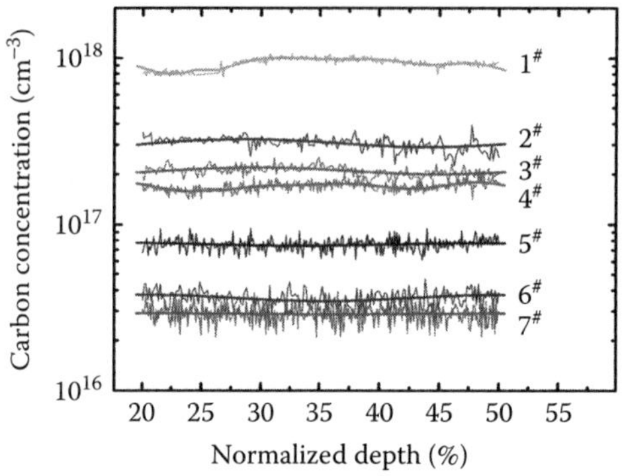

Figure 8.32 SIMS data and the FFT smoothed data of the carbon concentrations of seven samples. (Courtesy of Chuankai Yang.)

Table 8.2 YL/BL's, Average Carbon Concentration, and Characteristic Structures of the Samples

SAMPLE	YL/BL	CARBON CONCENTRATION (10^{16} CM^{-3})	CHARACTERISTIC STRUCTURE
1# XD4276	64.30699	~90	Nonpolar a-plane GaN
2# XDB100	32.855	~31	LT-GaN nucleation layer
3# XDB120	8.8018	~20	LT-AlGaN nucleation layer
4# XDB305	2.4697	~17	LT-AlN nucleation layer
5# XDB492	0.6412	~8.0	HT-AlN nucleation layer
6# XDB431	0.1819	~4.0	HT-AlN nucleation layer, Fe doped
7# H0406	0.05479	~3.0	HVPE-grown GaN

Source: Courtesy of Chuankai Yang.

A great difference in carbon concentration between the seven samples is visible in Figure 8.32. Their YL/BLs, average carbon concentrations, and characteristic structures are given in Table 8.2.

It is observed in Figure 8.33 that the carbon concentrations agree very well with the YL intensities of these samples with significantly different structures and growth conditions, a strong support of the theory that carbon-related impurities are primarily responsible for YL. Thus, we conclude that the dominant mechanism inducing YL in GaN is the carbon-related impurities.

8.5.2 Disproof of YL Induction by V_{Ga}

Gallium vacancy concentration can be characterized by three indirect methods:

1. The growth condition of the GaN epilayer. The probability of V_{Ga} formation varies with the V/III ratio when using triethyl gallium (TEGa) as the gallium source and NH_3 as the nitrogen source. The greater the V/III ratio (i.e., under nitrogen-rich condition), the easier for V_{Ga} to occur.
2. The formation energy of the eigen point defects in GaN, which varies with the Fermi level according to first principles. The stronger the n-type characteristics of GaN, the higher the Fermi level is and thus the lower formation

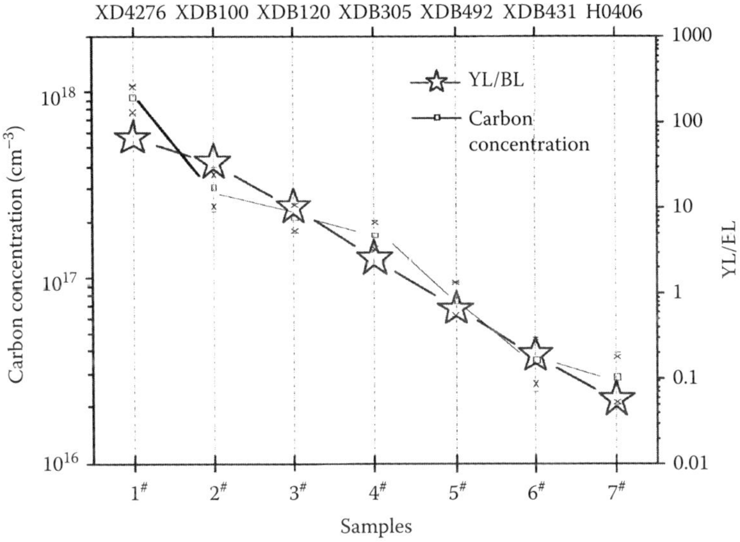

Figure 8.33 Carbon concentration versus relative YL intensity (YL/BL). The average, maximum, and minimum carbon concentrations are given. (Courtesy of Chuankai Yang.)

energy of Gallium vacancies and greater probability of V_{Ga} formation. So for unintentionally doped GaN, SIMS characterized concentrations of oxygen and silicon as shallow donors can be used to estimate the background electron concentration, and the sample with a high background electron concentration is supposed to have a high V_{Ga} concentration.

3. The atom ratio of gallium to nitrogen calculated based on the N1s and Ga3d peaks obtained by measuring the X-ray photoelectron spectrum (XPS) of GaN. Ideally, the atom ratio of gallium to nitrogen should be 1:1. In the presence of V_{Ga}, the nitrogen concentration is higher and the Ga/N should be less than 1, in other words, the Ga/N ratio value is indicative of the V_{Ga} concentration. The smaller the ratio, the higher the V_{Ga} concentration is. However, the XPS quantitative analysis has a poor accuracy with an error of 10%–20% and the results may vary with fitting algorithms, so similar measurement conditions and the same fitting algorithm should be adopted if possible in using this approach.

Two representative samples 2# and 5# with different V/III ratios in the GaN epilayer growth were chosen to determine their V_{Ga} concentrations by the three methods mentioned above. It can be concluded from the GaN epilayer growth conditions of samples 2# and 5# (Table 8.3) and the oxygen concentration (background electron concentration) from the SIMS measurements (Figure 8.34) that the V_{Ga} concentration of 2# should be lower than that of 5#.

Based on the measured XPS of 2# and 5# (not given in this book), a fitting to their Ga3d and N1s peaks was performed and the results are listed in Table 8.4. Both samples have a Ga/N ratio less than 1, indicating the possibility of V_{Ga} formation in

Table 8.3 Growth Conditions of $2^{\#}$ and $5^{\#}$

SAMPLE	NH_3 FLOW (SCCM)	TEG FLOW (SCCM)	V/III RATIO
$2^{\#}$	4897	299	1547
$5^{\#}$	4896	219	2116

Source: Courtesy of Chuankai Yang.

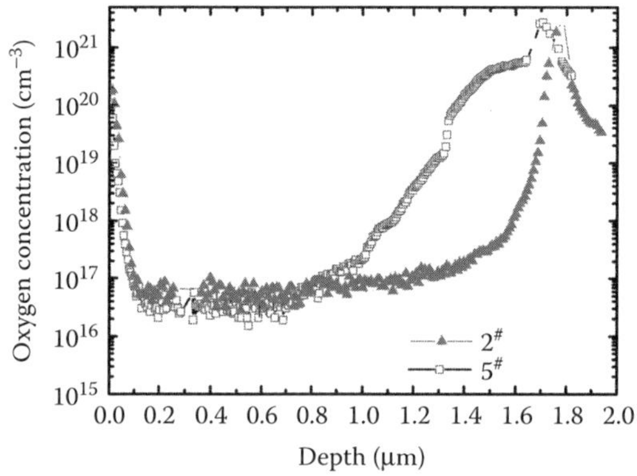

Figure 8.34 Oxygen concentrations of $2^{\#}$ and $5^{\#}$. (Courtesy of Chuankai Yang.)

Table 8.4 Calculated Ga/N Ratios of $2^{\#}$ and $5^{\#}$

SAMPLE	PEAK	BINDING ENERGY (EV)	HEIGHT (CPS)	FWHM (EV)	PERCENTAGE OF ATOM	GA/N
$2^{\#}$	N1s	394.38	43186.36	1.92	50.49	0.98
	Ga3d	18.08	45416.13	2.58	49.51	
$5^{\#}$	N1s	394.38	43797.93	2.62	51.16	0.95
	Ga3d	18.08	41494.29	2.04	48.84	

Source: Courtesy of Chuankai Yang.

both. The lower Ga/N of $5^{\#}$ suggests its higher V_{Ga} concentration, which agrees with the results by the other two methods.

Results achieved by all three methods show a lower V_{Ga} concentration of $2^{\#}$ than that of $5^{\#}$ while the PL spectra results in Figure 8.29 indicate $2^{\#}$ has a far greater YL intensity than $5^{\#}$, suggesting that YL has little to do with V_{Ga} concentration, which disproves the assumption that the V_{Ga}-related deep acceptors are the origin of YL.

References

Fichtenbaum, N. A., T. E. Mates, S. Keller, S. P. DenBaars, and U. K. Mishra. 2008. Impurity incorporation in heteroepitaxial N-face and Ga-face GaN films grown by metalorganic chemical vapor deposition. *Journal of Crystal Growth* 310(6):1124–1131. doi: 10.1016/j.jcrysgro.2007.12.051.

Gao, Z., Y. Hao, and J. Zhang. 2009. Effect of structural defects in GaN epitaxial layer on its surface morphology. *Materials Review* 23:1–5.

Gao, Z., Y. Hao, J. Zhang, J. Zhang, H. Chen, and J. Ni. 2007a. Observation of dislocation etch pits in GaN epilayers by atomic force microscopy and scanning electron microscopy. *Chinese Journal of Semiconductors* 28(4):473–479.

Gao, Z., Y. Hao, J. Zhang, J. Zhang, H. Chen, and J. Ni. 2007b. Polarity results in different etch pit shapes of screw and edge dislocations in GaN epilayers. *2007 International Workshop on Electron Devices and Semiconductor Technology*, June 3–4, 2007, Beijing, China.

Gao, Z.-Y., Y. Hao, J.-C. Zhang, J.-F. Zhang, and J.-Y. Ni. 2008. Reliable evaluation of dislocation densities in GaN epilayers by molten KOH etching. *Journal of Functional Materials and Devices* 14(4):742–750.

Heinke, H., V. Kirchner, S. Einfeldt, and D. Hommel. 2000. X-ray diffraction analysis of the defect structure in epitaxial GaN. *Applied Physics Letters* 77(14):2145–2147. doi: 10.1063/1.1314877.

Johnston, C. F., M. J. Kappers, and C. J. Humphreys. 2009. Microstructural evolution of non-polar (11–20) GaN grown on (1–102) sapphire using a 3D-2D method. *Journal of Applied Physics* 105(7):073102 (5 pp.). doi: 10.1063/1.3103305.

Lu, L., Z. Y. Gao, B. Shen, F. J. Xu, S. Huang, Z. L. Miao, Y. Hao et al. 2008. Microstructure and origin of dislocation etch pits in GaN epilayers grown by metal organic chemical vapor deposition. *Journal of Applied Physics* 104(12):123525 (4 pp.). doi: 10.1063/1.3042230.

Qian, W., M. Skowronski, M. De Graef, K. Doverspike, L. B. Rowland, and D. K. Gaskill. 1995. Microstructural characterization of α-GaN films grown on sapphire by organometallic vapor phase epitaxy. *Applied Physics Letters* 66(10):1252–1524.

Wright, A. F. 2002. Substitutional and interstitial carbon in wurtzite GaN. *Journal of Applied Physics* 92(5):2575–2585. doi: 10.1063/1.1498879.

Xu, S. R., Y. Hao, J. C. Zhang, Y. R. Cao, X. W. Zhou, L. A. Yang, X. X. Ou, K. Chen, and W. Mao. 2010a. Polar dependence of impurity incorporation and yellow luminescence in GaN films grown by metal-organic chemical vapor deposition. *Journal of Crystal Growth* 312(23):3521–3524. doi: 10.1016/j.jcrysgro.2010.09.026.

Xu, S.-R., Y. Hao, J.-C. Zhang, X.-W. Zhou, Y.-R. Cao, X.-X. Ou, W. Mao, D.-C. Du, and H. Wang. 2010b. The etching of a-plane GaN epilayers grown by metal-organic chemical vapour deposition. *Chinese Physics B* 19(10):107204 (5 pp.). doi: 10.1088/1674-1056/19/10/107204.

Yang, C. K. 2011. Study on dependence of electrical, optical properties on point defects in GaN epitaxial films. Master's thesis, Xidian University, Xi'an, China.

9

Principles and Optimization of GaN HEMTs

Two-dimensional electron gases (2DEGs) with high density and high electron mobility can form in AlGaN/GaN heterostructures used in HEMTs without intentional doping owing to strong piezoelectric and spontaneous polarization, and it is the high conductivity of the 2DEG and the high breakdown voltage of AlGaN/GaN heterostructures that make the GaN HEMT microwave power devices possible. Systematic theories of HEMT principles and structure optimization have been available for the conventional GaAs material system. This chapter introduces the principles of the GaN HEMT and gives some examples of optimization of the field plate (FP) structure of GaN HEMTs.

9.1 Principles of GaN HEMTs

The cross-sectional structure of the GaN HEMT is illustrated in Figure 9.1. The source and drain electrodes form ohmic contacts on the nitride heterostructure and actually low-resistance access to 2DEG, and the transverse electric field generated by the source-drain voltage V_{DS} leads to the 2DEG transport along the heterojunction interface to form the output current I_{DS}; the Schottky barrier gate controls the switching on/off of the 2DEG channel by using the depletion effect of the gate voltage V_{GS}.

The physical model for the channel modulation by gate voltage (Delagebeaudeuf and Linh 1982) is described as follows. As shown in Figure 9.2, suppose an AlGaN barrier layer of thickness d is n-type modulation doped; N_D and d_d are the doping concentration and thickness of the barrier; $e\varphi_b$ is the Schottky barrier height, ΔE_C is the band offset between the conduction band edges of GaN and AlGaN at the interface, σ_{pol} is the total polar charge density induced by the changes in piezoelectric polarization strength P_{PZ} and spontaneous polarization strength P_{SP} at the AlGaN/GaN interface (assuming a complete strain relaxation of GaN, that is, $P_{PZ}(\text{GaN}) = 0$, we have $\sigma_{pol} = P_{PZ}(\text{AlGaN}) + P_{SP}(\text{AlGaN}) - P_{SP}(\text{GaN})$), and n_{s2D} is the 2DEG sheet density.

Suppose that the dielectric constant and electric field at AlGaN side of the AlGaN/GaN heterointerface are ε_1 and F_1, respectively, and that those at GaN side are ε_2 and F_2, respectively. Defining the unit area capacitance between gate and channel as

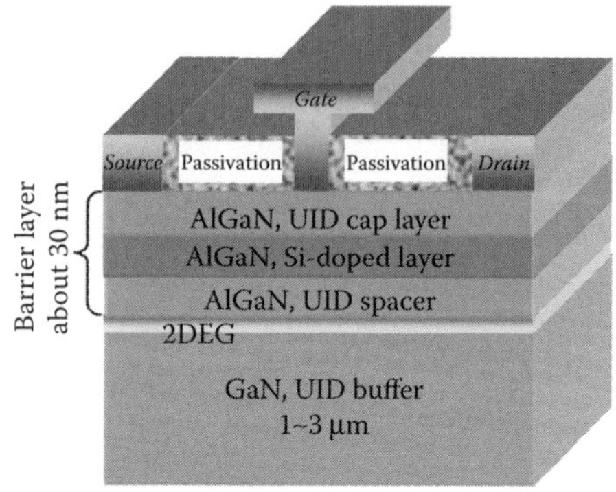

Figure 9.1 Schematic structure of GaN HEMT. (After Zhang, J. F. et al., *Sci. Chin. Ser. F* [*Inform. Sci.*], 51, 177–186, 2008.)

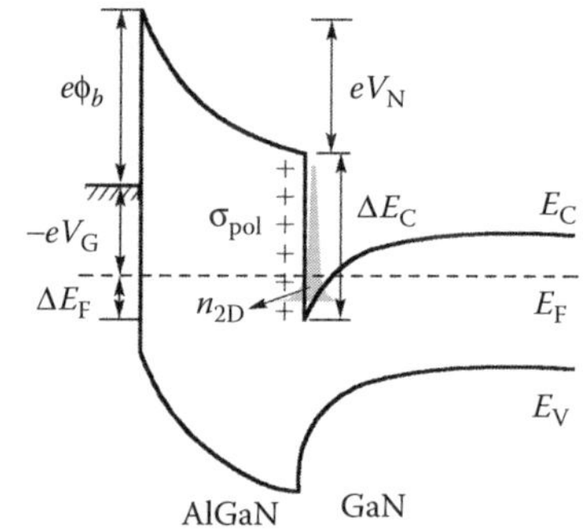

Figure 9.2 Relation between gate voltage and AlGaN/GaN energy bands.

C_1 and assuming that the distance of the 2DEG from the AlGaN/GaN interface Δd is negligible as compared to the barrier thickness d, we have

$$C_1 = \frac{\varepsilon_1}{(d + \Delta d)} \approx \frac{\varepsilon_1}{d} \tag{9.1}$$

From the Poisson equation it follows

$$\varepsilon_2 F_2 = e n_{s2D} \tag{9.2}$$

$$\varepsilon_1 F_1 = C_1 \left[V_N - \frac{(e\phi_b - eV_G + \Delta E_F - \Delta E_C)}{e} \right] \tag{9.3}$$

$$V_N = e \int_0^d \frac{N_D(x)}{\varepsilon(x)} x dx = \frac{e N_D}{2\varepsilon_1} d_d^{\,2} \tag{9.4}$$

where e is the elementary charge.

An integration of the Gauss equation along the direction perpendicular to the AlGaN/GaN heterointerface yields (Ridley 2004)

$$\varepsilon_2 F_2 = \varepsilon_1 F_1 + \sigma_{pol} \tag{9.5}$$

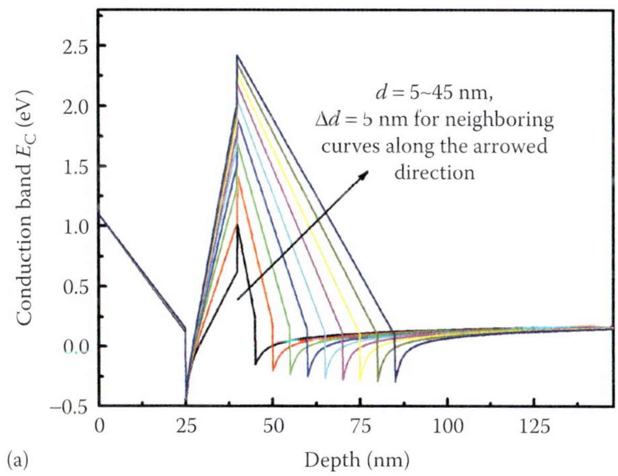

Figure 6.11 Effect of second barrier thickness d on double-channel AlGaN/GaN heterostructures. (a) Conduction band edge E_C as a function of thickness d of second barrier layer.

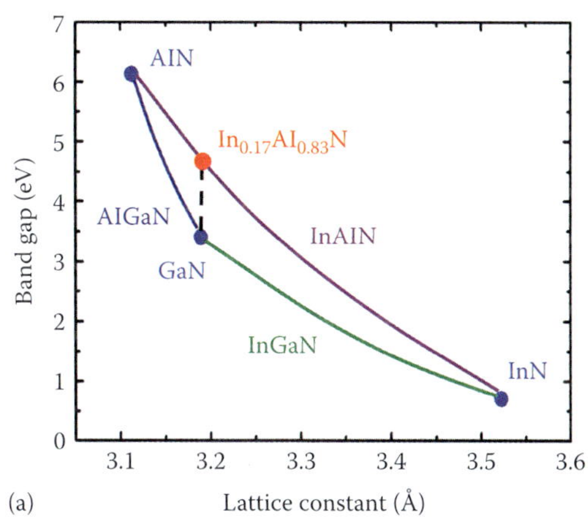

Figure 7.1 Lattice-matched InAlN/GaN heterostructure. (a) Physical mechanism of lattice match.

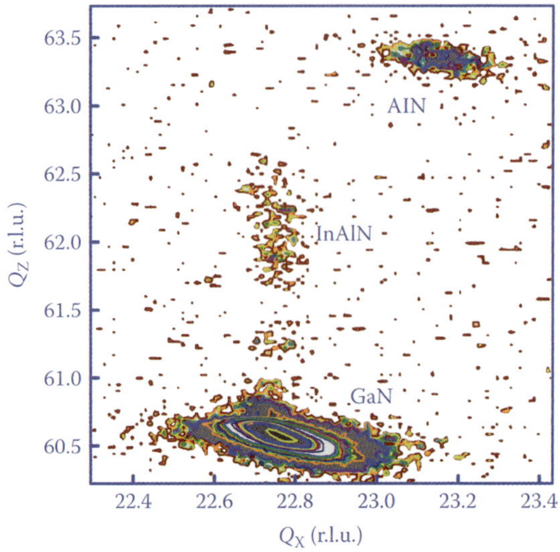

Figure 7.27 Reciprocal space maps of (105) reflection of InAlN/GaN/InAlN/GaN double-channel heterostructure. (After Xue, J. S. et al., *App. Phys. Lett.*, 100, 013507, 2012b.)

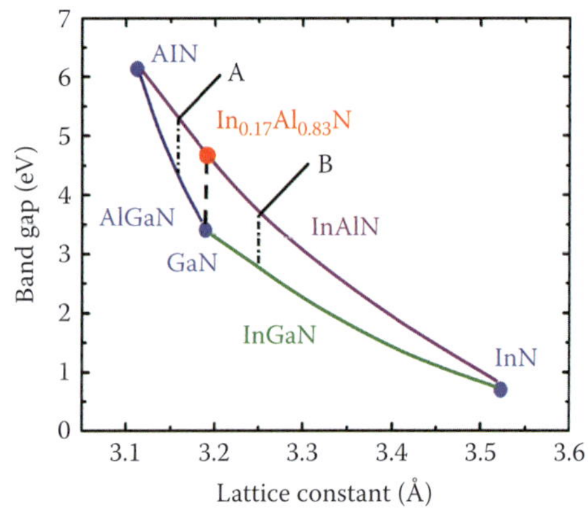

Figure 7.29 Schematic lattice-matched structures of nitrides.

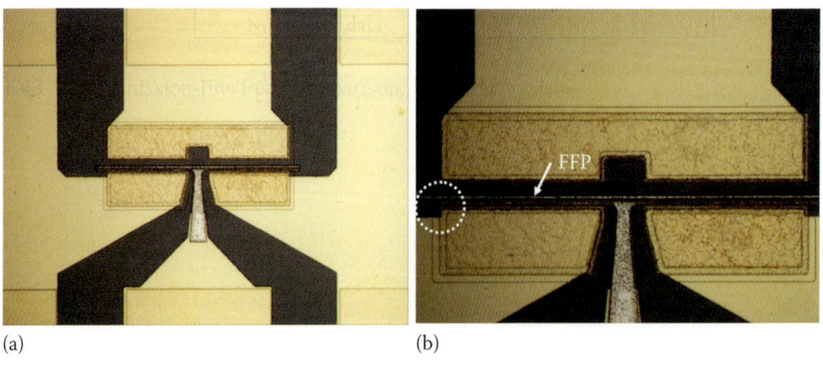

Figure 9.19 Photograph of a FFP-HEMT with one floating FP. (a) Overall view and (b) detail view. (Courtesy of Dr. Wei Mao.)

Figure 10.1 Window for gate fingers etched on a multi-finger HEMT by optimized photolithography showing well-developed straight lines. (Courtesy of Dr. Ling Yang.)

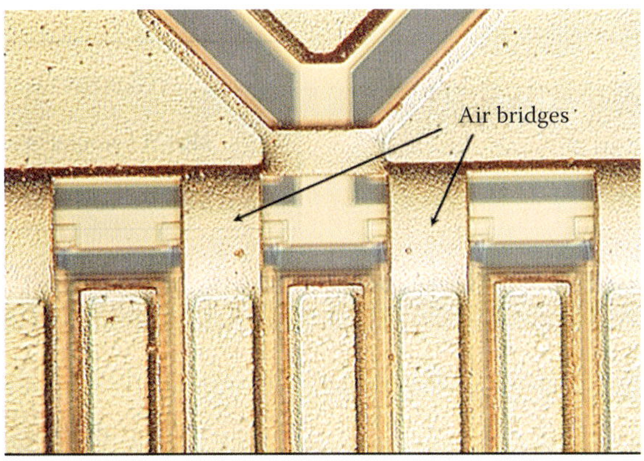

Figure 10.17 Plan view of the air bridge in a HEMT sample.

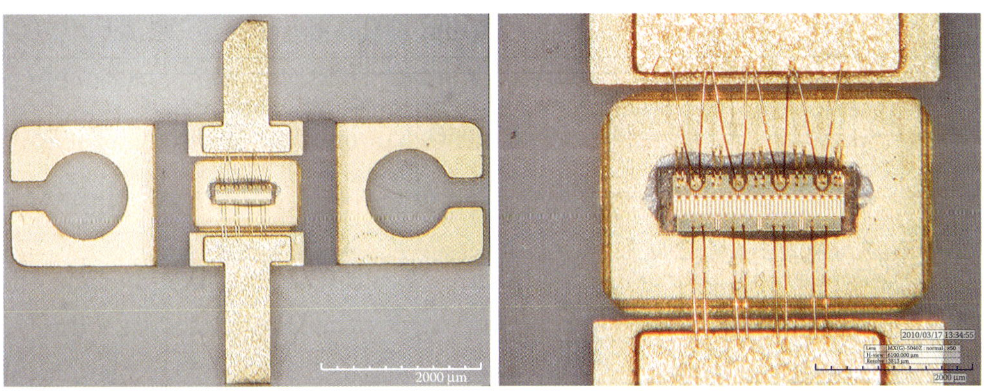

(a) (b)

Figure 10.20 (a–b) Photograph of multifinger gate HEMT with a gate width of 1.25 mm after bonding and packaging.

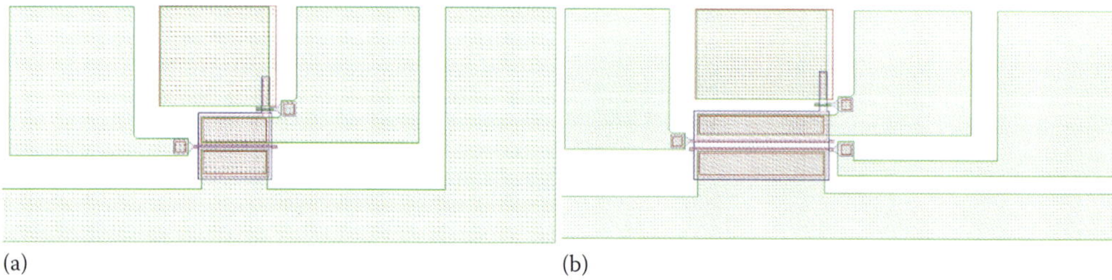

Figure 12.27 Digital circuit layout of inverter and NAND gate. (a) Inverter and (b) NAND gate. (After Xie, Y. et al., *J. Semiconduct.*, 32(6), 065001 (4 pp.), 2011.)

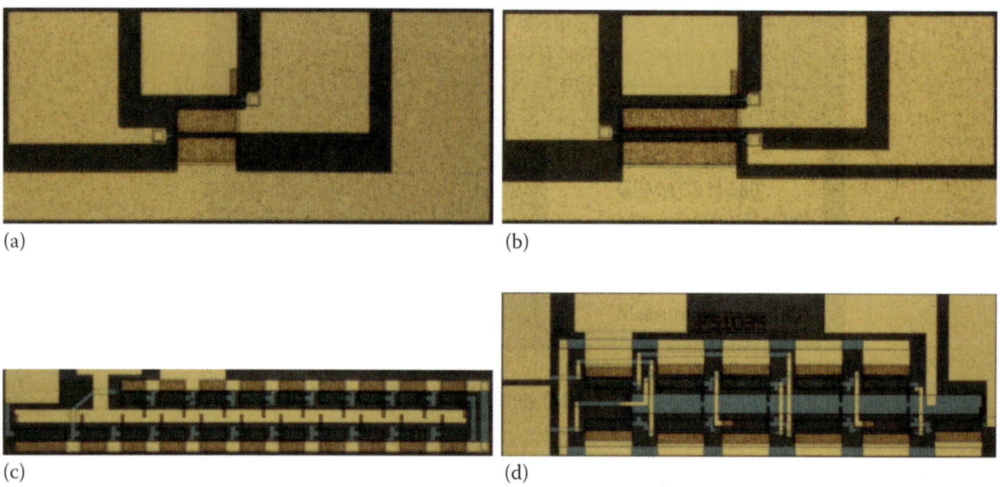

Figure 12.29 Photographs of prepared circuit modules. (a) Inverter, (b) NAND gate, (c) ring oscillator, and (d) flipflop. (After Xie, Y. et al., *J. Semiconduct.*, 32(6), 065001 (4 pp.), 2011.)

which is rearranged into

$$en_{s2D} = \sigma_{pol} + C_1\left(V_N - \phi_b + V_G - \frac{\Delta E_F}{e} + \frac{\Delta E_C}{e}\right) \tag{9.6}$$

When the gate voltage reaches the threshold voltage V_T leading to 2DEG depletion, it is assumed $n_{s2D} \approx 0$ and $\Delta E_F \approx 0$, which are substituted into the above equation to yield

$$V_T = \frac{-\sigma_{pol}}{C_1} - V_N + \phi_b - \frac{\Delta E_C}{e} \tag{9.7}$$

Since ΔE_F in Equation 9.6 as a function of gate voltage V_{GS} is typically rather small and thus negligible compared to other terms, Equation 9.6 can be rearranged into

$$en_{s2D} = C_1(V_G - V_T) = C_1 V_{GT} \tag{9.8}$$

Upon the application of the source-drain voltage, it is necessary to take into account the variation of the channel potential V_C along the channel. Suppose that the coordinates along the channel orientation is x, and $x = 0$ at the source terminal of the gate and $x = L$ at the drain terminal of the gate. Then Equation 9.8 should be rectified as

$$en_{s2D}(x) = C_1(V_{GT} - V_C(x)) \tag{9.9}$$

According to the general definition of current density, the source drain current I_{DS} in the 2DEG channel is expressed as

$$I_{DS} = eWv(x)n_{s2D}(x) \tag{9.10}$$

where:

W is the gate width
$v(x)$ is the electron velocity

The simplest approximation of electron velocity in form is given by a two stage model as

$$v(F) = \begin{cases} \mu F, & F < F_S \\ v_S, & F \geq F_S \end{cases} \tag{9.11}$$

When the electric field $F(x) = dV_C(x)/dx$ in the channel is lower than the critical electric field F_S, there is

$$I_{DS} = W\mu C_1(V_{GT} - V_C(x))\frac{dV_C(x)}{dx} \tag{9.12}$$

Considering the series resistances R_S between source and gate and R_D between gate and drain, we have at the boundary of the channel under the gate

$$V_C(0) = R_S I_{DS}, \quad V_C(L) = V_{DS} - R_D I_{DS} \tag{9.13}$$

If the variation of potential along the channel (i.e., the term $V_C(x)$ in Equation 9.12) is neglected in case of a very small V_{DS}, we get

$$I_{DS} = \beta V_{GT}[V_{DS} - (R_S + R_D)I_{DS}] \tag{9.14}$$

$$\beta = \frac{W\mu\varepsilon_1}{dL} \tag{9.15}$$

where β is referred to as the transconductance coefficient. A linear relationship between I_{DS} and V_{DS} can be seen in Equation 9.14, which is the linear region of the current–voltage relation. The variation of the potential along the channel cannot be neglected with a large V_{DS}, in which case Equation 9.13 is substituted into Equation 9.12 for an integration along the channel, and the resulting channel potential satisfies

$$V_C(x) = V_{GT} - \sqrt{(V_{GT} - V_C(0))^2 - 2I_{DS}x/(\beta L)} \tag{9.16}$$

In this way, the electric field in the channel $F(x)$ can be derived. For $V_L = F_S L$, if $F(L) = F_S$ at $x = L$, the electron drift velocity saturates and the channel current reaches its saturation value I_{Dsat}, which is given by

$$I_{Dsat} = \beta V_L\left[\sqrt{V_L{}^2 + (V_{GT} - R_S I_{Dsat})^2} - V_L\right] \tag{9.17}$$

The magnitude of the channel electric field is related to the gate length L as well as the voltage V_{DS}. At the long channel limit $V_L \gg V_{GT} - R_S I_{Dsat}$, nowhere under the gate the electric field can reach F_S and thus Equation 9.17 is reduced to

$$I_{Dsat} = \frac{\beta}{2}(V_{GT} - R_S I_{Dsat})^2 \tag{9.18}$$

At the short-channel limit, $V_L \ll V_{GT} - R_S I_S$, the electrons travel over the channel with the saturation velocity, in which case Equation 9.17 is simplified as

$$I_{Dsat} = \beta V_L(V_{GT} - R_S I_{Dsat}) \tag{9.19}$$

Without the source/drain series resistances R_S and R_D in consideration, the saturation voltage V_{Dsat} are V_{GT} and V_L for the long-channel and short-channel devices, respectively, and V_{Dsat} may increase with the series resistances taken into account.

The above simplified model presents the current–voltage characteristics of the HEMT in operation from the perspective of its principles with definite physical significance, but is by no means accurate. For the 2DEG sheet density n_{s2D}, the linear charge modulation model (Equations 9.8 and 9.9) is employed to describe its modulation by the gate voltage. A more accurate model needs to take into consideration the nonlinear modulation by gate voltage (Roblin et al. 1990) in the discussion of subthreshold characteristics, where the expressions for the threshold voltage may also vary.

An improvement can be made of the two stage velocity model in Equation 9.11, which is much too simple for the description of the velocity–field relationship with negative differential characteristics as shown in Figure 2.6. The Schottky gate leakage also needs to be considered in some cases (Ruden 1990). The impact of heat dissipation should also be included in the analysis of the behaviors of the GaN HEMT grown on sapphire due to its notable self-heating effects. There were some reports on the even more accurate model that takes the channel outside the gate and the diffusion current into account (Albrecht et al. 2000).

9.2 Performance Parameters of GaN HEMTs

9.2.1 Direct Current Performance Parameters

The major direct current (DC) performance parameters of the HEMT are the maximum output saturation current density (I_{DS}/W) and the threshold voltage.

It follows from the threshold voltage expression (Equation 9.7) as well as Equations 9.1 and 9.4 that the threshold voltage is closely related to the Schottky barrier height $e\varphi_b$, the polar charge density σ_{pol}, and the conduction band discontinuity ΔE_C at the AlGaN/GaN interface, and the AlGaN layer thickness (influencing C_1) and doping (influencing V_N). The magnitudes of $e\varphi$, σ_{pol}, and ΔE_C are directly affected by the Al content in the AlGaN layer, therefore the structure and doping of the barrier layer plays a decisive role in the modulation of the threshold voltage. In addition, from Equation 9.8 and the expression of C_1 in follows $en_{s2D} = -(\varepsilon_1/d)V_T$ at $V_G = 0$, thus as for the relationship between the material properties and device performance parameters, the value of the 2DEG density (also decided by the barrier layer structure and doping) has a positive correlation with the absolute value of the threshold voltage.

With the source/drain series resistances R_S and R_D in consideration, the determination of the influencing factors for the saturation current requires first the determination of the origin and contributing factors of R_S and R_D. Essentially, R_S and R_D originate not only from the access resistance of gateless channel regions between the gate-source and gate-drain regions but also from the source-drain ohmic contact resistance R_C, that is,

$$R_S = R_C + R_{SH}\frac{L_{GS}}{W} \tag{9.20}$$

$$R_D = R_C + R_{SH}\frac{L_{GD}}{W} \tag{9.21}$$

where:

$R_{SH} = 1/\mu n_{s2D} q$ is the sheet resistance
W is the gate width
L_{GS} is the gate-source spacing
L_{GD} is the gate-drain spacing

Accordingly, the magnitudes of R_S and R_D per unit gate width are decided by R_C and R_{SH} as well as the source-drain spacing.

Knowing the mechanisms of R_S and R_D, according to Equations 9.18 and 9.19 for the saturation current at the long-channel limit and short-channel limit, we may achieve a large saturation current either by improving the AlGaN/GaN mobility (long-channel device), saturation velocity (short-channel device) and 2DEG density in terms of heterojunction properties, or by reducing the gate length (enhancing the electric field in the channel) and source-drain spacing (decreasing R_S and R_D) in terms of the device size.

9.2.2 Alternate Current Small Signal Transconductance

The alternate current (AC) small signal equivalent circuit of an actual HEMT is shown in Figure 9.3, containing the intrinsic FET and the parasitic components such as the series resistances R_S and R_D. The transconductance g_m reflects the ability of the gate to modulate the channel current, and the transconductance obtained with R_S and R_D neglected is the intrinsic transconductance g_m^*. According to Equations 9.18 and 9.19, the saturation region transconductance of the HEMT is

$$g_m^* = \frac{\partial I_{\text{DS}}}{\partial V_{\text{GS}}} = \begin{cases} \beta V_{\text{GT}} & \text{long-channel devices} \\ \beta V_L & \text{short-channel devices} \end{cases} \tag{9.22}$$

According to this equation and the definition of transconductance coefficient by Equation 9.15, the improvement of the intrinsic transconductance per unit gate width requires better transport properties and smaller barrier thickness and gate length.

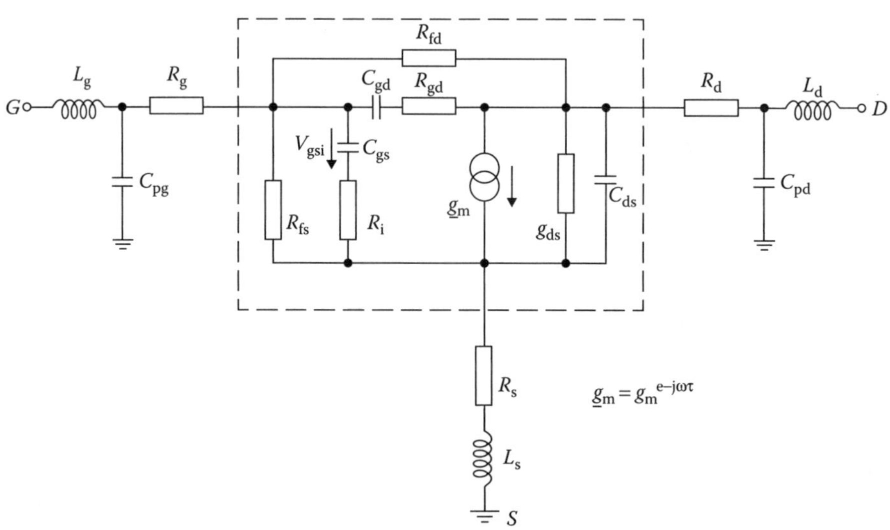

Figure 9.3 AC small signal equivalent circuit of an actual HEMT. The intrinsic FET is highlighted by the dashed rectangle. R_i is the input resistance. (Reproduced from Berroth, M. and Bosch, R., *IEEE Trans. Microw. Theor. Tech.*, 39, 224–229, 1991.)

In the actual device operation, voltage drops across R_S and R_D are generated by I_{DS}. R_S causes a drop in the active gate voltage applied across the gate and source, exerting an influence on the saturation region transconductance g_m; while R_D increases the source-drain voltage for the current to start to saturate V_{Dsat}, but for $V_{DS} > V_{Dsat}$, V_{DS} has no influence on the output current, that is, R_D has no effect on g_m. The measured transconductance satisfies

$$g_m = \frac{g_m^*}{\left(1 + R_S g_m^*\right)} \tag{9.23}$$

where g_m^* is the intrinsic transconductance which can be derived by the measured transconductance (Pingying and Zhou 1985).

9.2.3 *Cut-Off Frequency f_T and Maximum Oscillating Frequency f_{max}*

The cut-off frequency f_T is defined as the frequency at which the current through the input capacitance equals that of the current source $g_m V_{gs}$ in the common source equivalent circuit (Pingying and Zhou 1985), that is, at which the magnitude of the current gain $|h_{21}|$ drops to unity. Without considering the parasitic components, we have

$$f_T = \frac{g_m^*}{2\pi C_G} \tag{9.24}$$

where g_m^* is the intrinsic transconductance and $C_G = C_1 WL$ the gate capacitance.

For short-channel devices, the frequency limit is subject mainly to the velocity saturation. When the channel length reduces to the extent where the drift velocity of the channel carriers saturates, the electron transit time under the gate is $\tau = L/v_s$ and the transconductance is $g_m^* = C_1 v_s W$; whence the cut-off frequency is

$$f_T = \frac{1}{2\pi\tau} = \frac{v_s}{2\pi L} \tag{9.25}$$

When the parasitic effect is taken into account, the f_T expression can be changed according to its definition into

$$f_T = \frac{g_m^* / 2\pi}{[C_{GS} + C_{GD}][1 + (R_S + R_D)/R_{DS}] + C_{GD} g_m^* (R_S + R_D)} \tag{9.26}$$

The above relation shows that the improvement of f_T needs to increase the transconductance and reduce the gate capacitance as well as R_S and R_D, in other words, to improve the carrier mobility and to reduce the gate length along with the source-drain spacing and ohmic contact resistance.

The maximum oscillating frequency f_{max} is defined as the frequency where the unilateral power gain (UPG) equals unity, and expressed by

$$f_{\max} \cong \frac{f_T}{2\sqrt{(R_G + R_S + R_I)/R_{DS} + 2\pi f_T R_G C_{GD}}} \tag{9.27}$$

where $R_I = \partial V_G / \partial I_G$ is the input resistance of the intrinsic FET, and the other physical quantities bear the same notations as in Figure 9.3. In actual devices, the improvement of f_{max} requires an increased f_T and reduced gate series resistance R_G and source resistance R_S.

In the measurement of the frequency characteristics of an actual device, h_{21} and UPG can be obtained by the S-parameter measurements, hence f_T and f_{max} of the device (Cheng'en 1994). The HEMT is viewed as a two-terminal network with the input signal applied on the gate and source and the output measured between the source and drain. Defining a_1 and a_2 as the incident waves and b_1 and b_2 as the reflected waves, which are independent from each other as shown in Figure 9.4, we have the S-parameter linear network equation

$$\begin{cases} b_1 = S_{11}a_1 + S_{12}a_2 \\ b_2 = S_{21}a_1 + S_{22}a_2 \end{cases} \tag{9.28}$$

The S-parameters here are defined as

$$S_{ij} = \left.\frac{b_i}{a_j}\right|_{a_m=0} \quad S_{11} = \left.\frac{b_1}{a_1}\right|_{a_2=0} \quad S_{12} = \left.\frac{b_1}{a_2}\right|_{a_1=0} \quad S_{21} = \left.\frac{b_2}{a_1}\right|_{a_2=0} \quad S_{22} = \left.\frac{b_2}{a_2}\right|_{a_1=0} \tag{9.29}$$

Physically speaking, S_{11} is the input terminal voltage reflection coefficient with matched load at the output, S_{12} is the reverse transmission coefficient with matched load at the input, S_{21} is the positive transmission coefficient with matched load at the output, and S_{22} is the output voltage reflection coefficient with matched load at the input.

S parameters can be measured with the network analyzer and their magnitudes depend on the voltage and current of the operation point as well as the frequencies. The correlation between S parameters and h_{21} and *UPG* is given by

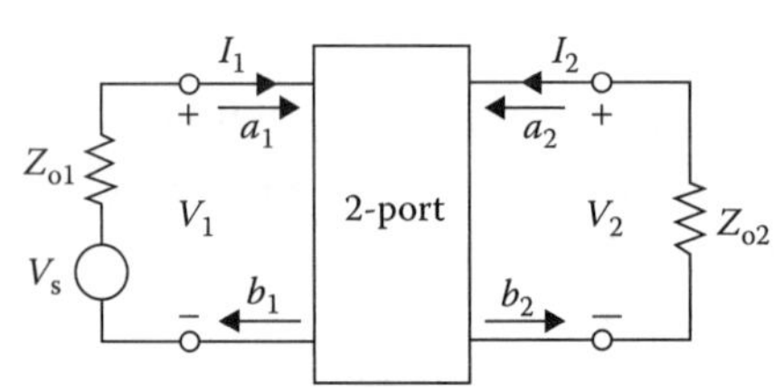

Figure 9.4 Schematic diagram of S-parameters. (Reproduced from Sadiku, M.N.O., *IEEE Trans. Educ.*, 46, 399–404, 2003.)

$$h_{21}(dB) = 20\log\left(\frac{-2S_{21}}{(1-S_{11})(1+S_{22})+S_{12}S_{21}}\right) \tag{9.30}$$

$$UPG(dB) = 10\log\left[|S_{21}|^2\left(\frac{1}{1-|S_{11}|^2}\right)\left(\frac{1}{1-|S_{22}|^2}\right)\right] \tag{9.31}$$

f_T and f_{max} can be determined by the frequency when h_{21} and UPG go to unity obtained by S-parameter measurements at different frequencies.

9.2.4 Power Performance Parameters

For the sinusoidal waveform, the maximum power output can be obtained from the maximum swings of the dynamic output voltage and current as

$$P_{om} = \frac{1}{8} I_{D\max}(BV_{DS} - V_{Dsat}) \tag{9.32}$$

The power obtained in this way is measured in units of mW. There is also another common expression of power: $10\log P_{om}$ (in units of dBm obtained by the substitution of values in mW). A large output power requires not only a large saturation output current but also as great as possible an increase of the breakdown voltage and a reduction of the saturation voltage.

The other two commonly used power performance parameters are the gain G and power added efficiency PAE. G is defined as the ratio of the output power P_{out} to the input power P_{in} of AC signal

$$G = \frac{P_{out}}{P_{in}} \tag{9.33}$$

which is often expressed in $10\log G$ (in units of dB obtained by the substitution of power values in mW). Generally, the gain decreases with larger P_{in} signals. A device is said to have a good linearity if its gain varies little over a broad range of P_{in} signal, and thus a small harmonic component in the output signal.

The power added efficiency is defined as

$$PAE = \frac{(P_{out} - P_{in})}{P_{DC}} \tag{9.34}$$

It can be seen that PAE is the ratio of the difference between the output power and the input power to the DC power consumption P_{DC}, as a measure of the transformation efficiency from DC power into AC output power. With increasing input signal P_{in}, PAE and P_{out} generally rise gradually, with PAE saturating first and then beginning to drop followed by the saturation of P_{out}.

9.3 Optimizations of the Performance of GaN HEMTs

The performance optimization approach depends on where the device is applied. The relationship between the properties of GaN and GaN-based heterostructures and the target characteristics of GaN-based HEMTs in microwave power applications is shown in Figure 9.5.

The optimization of AlGaN/GaN heterostructure for applications in HEMTs has been discussed in detail in Chapter 5. The parameters for AlGaN barrier layer have been optimized within a range (the barrier layer Al content is about 15%–30% and the thickness about 10–30 nm) to obtain a high 2DEG density while maintaining high mobility and velocity without inducing strain relaxation in the barrier. Generally, the AlN interlayer and GaN cap layer are also introduced for further optimization of the barrier. Thus, the output current density, frequency characteristics, and power characteristics of HEMTs are basically guaranteed. High HEMT output impedance demands a small buffer layer leakage. High quality high-resistance buffer layers can be obtained by employing high crystalline quality intrinsic materials or by compensation of unintentional n-type doping with deep acceptors. Good pinch-off characteristics require an off-state current at least three orders of magnitude less than the on-state current, which can be realized either by using low leakage buffer layers or by adopting the back barrier structure under the heterojunction channel as described in Chapter 6.

Preparation techniques also have direct influence on the performance of GaN HEMTs. The source/drain ohmic contact resistance, whose value should be no greater than the channel resistance, has an effect on the on-resistance, transconductance, and frequency characteristics. The gate modulation ability, breakdown voltage, and reliability are all influenced by the reverse leakage of Schottky contact, which should be reduced as much as possible. A reversible degradation called current collapse exists in the GaN HEMT mainly under the influence of surface traps and requires a stabilization of the surface states in the gateless surface region by depositing dielectric films on the surface, a technique referred to as surface passivation. Although the maximum

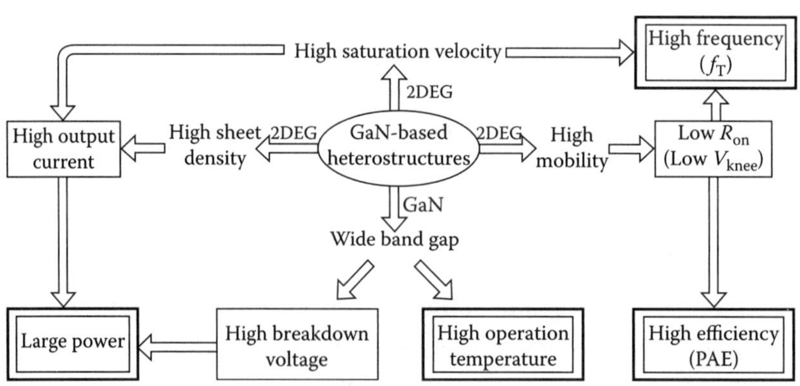

Figure 9.5 Relationship between the properties of GaN and GaN-based heterostructure and the target characteristics of HEMTs.

output current and breakdown voltage of the passivated devices have been reported to increase, remain constant, or decrease in various experiments, passivation has proved by many experimental observations to be an important method for eliminating current collapse and improving the output power density and *PAE*. It has become a standard procedure in GaN HEMT preparation.

Besides the optimization of the material structure and device preparation process, the optimization of the HEMT structure, mainly the gate structure, can also improve the performance of GaN-based HEMTs.

The frequency characteristics can be enhanced by reducing the gate length (gate-to-channel separation should be reduced simultaneously to avoid the short-channel effect) and source-drain spacing. The different shapes of the gate also have an impact on the cut-off frequency f_T and maximum oscillating frequency f_{max}. A very high f_T can be obtained by using the I-gate (with a uniform longitudinal section width) at a small enough gate length, but the extremely narrow gate leads to a very large series resistance, and thus f_{max} is not high (significantly lower than f_T). In order to maintain or even to reduce the gate series resistance as the gate gets narrower, the mushroom gate, T-gate, or Y-gate are generally adopted for a very small gate foot (small gate length) along with a large gate head. This leads to a prominent increase of f_{max}, but the introduced parasitic capacitance and so on between the gate head and the semiconductor poses some negative influence on f_T (f_T is generally smaller than f_{max}). In summary, the improvement of frequency characteristics requires a trade-off between increasing f_T and reducing the ratio of f_T to f_{max}. The shape of the gate should be designed with such factors in mind as the frequency characteristics requirements and the mechanical stability of the gate, and the more complicated shape the gate has, the more complex fabrication technique is required.

A recessed gate can bring higher transconductance and better frequency characteristics by reducing the gate-channel spacing. The recessed gate of the GaN HEMT is often prepared by dry etching techniques such as reactive ion etching (RIE), and the etching leads to surface damages which can be recovered to some extent by annealing at proper temperatures. It is observed in experiments that the recessed gate helps diminish current collapse, which may be attributed to the separation of the gate from the surface states. The major design parameters of a recessed gate are the recess width and depth. As a rule, the recess width should be designed to allow the edge of the gate foot to be as close to the recess sidewall as possible in order to minimize the surface area subject to etching. If the recess width is less than the gate length, a so-called buried gate is formed, which better suppresses current collapse and achieves a higher output current and breakdown voltage than the recessed gate. The recess depth should be set according to the target threshold voltage, which shifts positively with the increasing recess depth. Deep recess is one of the design methods of enhancement-mode HEMTs.

The maximum electric field of an operating GaN HEMT occurs at the gate edge on the drain side (Figure 9.6). Lowering this electric field peak is in favor of increasing

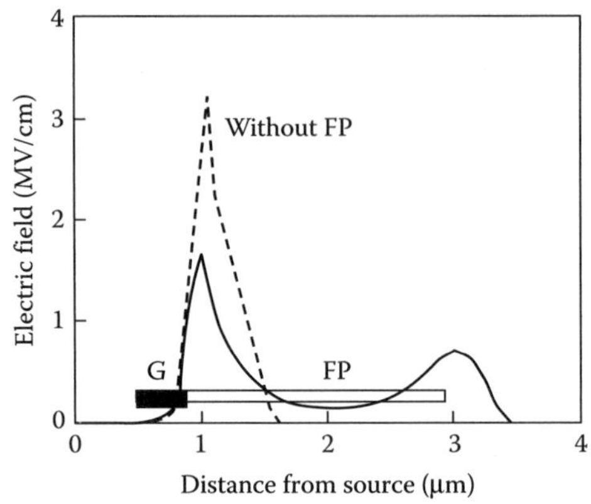

Figure 9.6 Electric field reduction by gate FP. (Reproduced from Karmalkar, S. and Mishra, U.K., *IEEE Trans. Elect. Device*, 48, 1515–1521, 2001.)

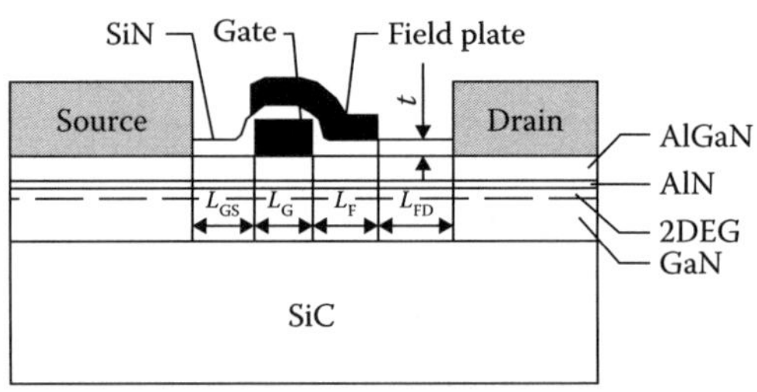

Figure 9.7 Schematic structure of the gate FP. (Reproduced from Wu, Y.F. et al., *IEEE Elect. Device Lett.*, 25, 117–119, 2004.)

the breakdown voltage to weaken the influence of high field on electrons, thereby suppressing current collapse and improving output power density and PAE. To this end, the FP structure can be adopted (Figure 9.7). Generally, the FP refers to the metal plate connected to the electrode of the device and can be prepared together with the preparation of the electrodes or metal interconnects (see Chapter 10 for detail). The FP is often connected with the gate or source. The gate FP that lies between gate and drain can decrease the strong electric field at the drain-side gate edge but increase the gate-drain feedback capacitance, which has an adverse effect on power gain. The source FP extending to the region between gate and drain on a dielectric layer of a thickness greater than the gate height can also diminish the strong electric field at the drain side of the gate, and it increases the source drain capacitance but the negative influence is lesser than the gate FP, thus a slightly higher large-signal gain of the source FP-HEMT than that of the gate FP-HEMT. The design parameters for the FP are the FP length (i.e., the length of the part of FP extending out of the drain side of the gate) and the thickness of the dielectric layer between the FP and the device surface. A rational design of the FP structure is necessary to ensure its effective adjustment of the distribution of the channel electric field with minimum influence on the device capacitance to obtain the best performance of HEMTs.

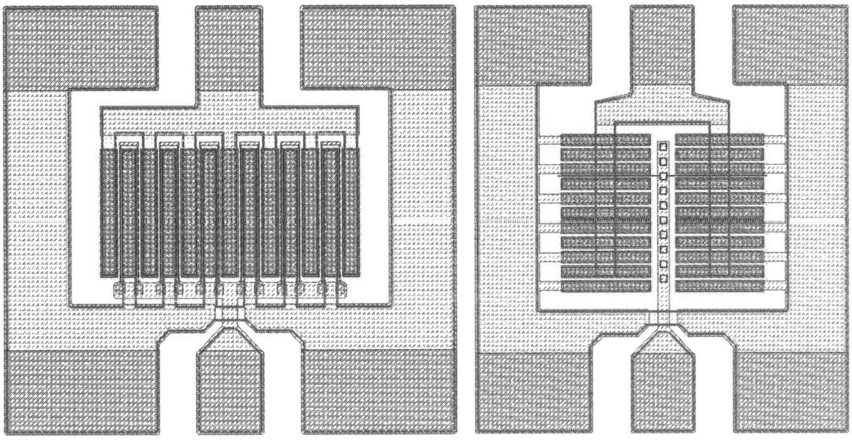

(a) (b)

Figure 9.8 Two structures of large gate width HEMTs. (a) Parallel gate structure and (b) fishbone gate structure.

The above optimizations of performance parameters are for the HEMTs with rather small gate width. To increase the total output power a larger gate width, say a total gate width greater than 1 mm, is needed. Multifinger gate structures, that is, the combination of multiple gate fingers of a small width into one with a relatively larger gate width, are adopted for a smaller chip size, higher yield, and reduced electric signal phase shift. The two common structures of the parallel gates and the fishbone gates are shown in Figure 9.8. In the parallel structure, signals from the gate electrode are assigned to each gate finger and collected at the drain fingers after amplification. The distinct signal phase difference between the gate fingers in the middle and those on both sides leads to the phase shift of amplified signals. The problem aggravates with increasing number of gate fingers, thus a greater drop in power gain efficiency. In the fishbone structure, the total distance signals traveling along each gate finger are equal (for the gate finger nearest to the input signal, the distance for the signal to be passed from the drain finger to the bonding pad of the drain electrode is the greatest, and for farther gate fingers the signal transmission distance increases on the gate but decreases on the drain, hence an invariant total distance), as a result of which the phase shift between signals from different gate fingers is greatly reduced, thus improving the power gain efficiency. This structure is also more favorable for heat dissipation as the gate fingers spread on both sides. The disadvantage of the fishbone structure lies in the fact that the each air bridge (the metallic connection bridge across electrodes with air as the dielectric under the bridge) standing on two drain fingers has to go across a gate finger and a source finger, leading to a large parasitic capacitance and thus a decreased signal gain. Overall the fishbone gate structure is superior in power gain efficiency to the parallel structure. In the multifinger gate devices, electricity delay increases with the number of gate fingers posing a deleterious effect on gain and efficiency, while the increase in the width of individual gate finger may cause a greater parasitic resistance as well as additional phase shift. Therefore, both the gate finger width and the number of fingers along with heat dissipation should be taken into account in the design of multifinger gate HEMTs.

9.4 Simulation and Implementation of FP Structure for Higher Breakdown Voltage

As is stated in the above section, the gate FP can lower the electric field peak at the gate edge on the drain side in favor of a higher breakdown voltage. Theoretical simulations and examples of device fabrication are given in this section to demonstrate the effect of gate FP optimization.

9.4.1 Optimization of FP-HEMTs by Simulation

The optimization procedures of the FP are described based on HEMT devices. A baseline HEMT and a FP-HEMT based on the same heterostructure are employed in simulation to compare the device properties before and after optimization. The heterostructure consists from bottom to top of a 1.4 μm GaN buffer layer, a 1.5 nm AlN interlayer, a 23 nm AlGaN barrier, and a 2 nm GaN cap layer, all unintentionally doped (the basic structure is the same as shown in Figure 9.1 with the addition of an AlN interlayer at the AlGaN/GaN interface and a GaN cap layer on the top). The AlGaN layer has an Al content of 30%. A numerical simulation model for FP-HEMTs is built in the two-dimensional numerical simulation software ATLAS@Silvaco. The Shockley–Read–Hall compound model, the Caughey and Thomas mobility model, and the Van Overstraeten–de Man impact ionization model are employed in the simulation with a background donor concentration of 1×10^{15} cm^{-3} for each layer, and the gate current is primarily contributed by the two mechanisms of Schottky barrier tunneling and thermionic emission. The polarization effect is simulated by placing a sheet of positive charges at the AlGaN/GaN heterointerface. Since this charge density is closely relate to the threshold voltage and transconductance, a comparison of the simulated transfer characteristic curve with the measured results (Figure 9.9) shows that the best agreement of the two curves is achieved with a sheet charge density

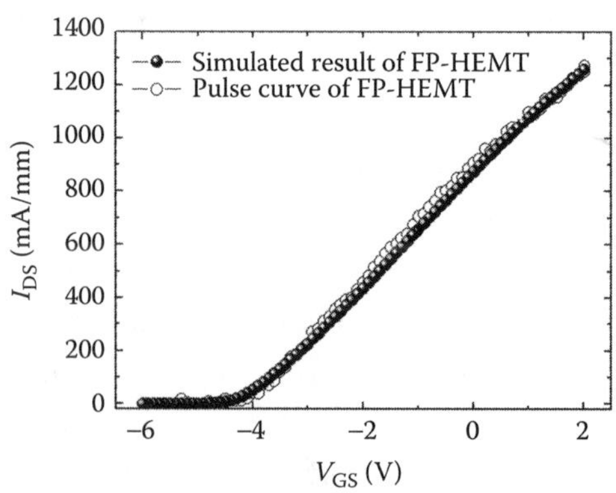

Figure 9.9 Simulated transfer characteristic curve and experimental results for the FP-HEMT. (After Mao, W. et al., *Chin. Phys. B*, 20, 097203, 7 pp., 2011.)

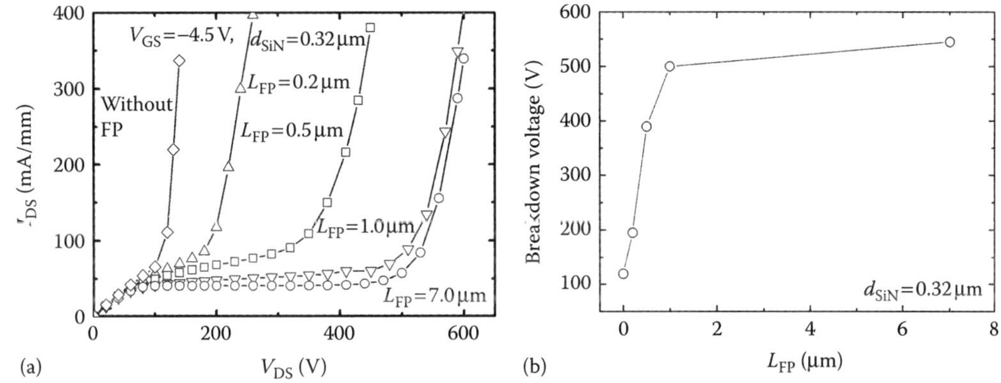

Figure 9.10 On-state breakdown characteristics of HEMTs with different lengths of gate FP. (a) Comparison of I-V curves. The breakdown voltage is defined as the intersection point of the extrapolation line of the saturation drain current with that of the sharp rising current in the impact ionization. (b) Breakdown voltage versus length of field plate. (Courtesy of Dr. Wei Mao.)

of 1.22×10^{13} cm^{-2}. We take a source-drain spacing of 16 μm for the convenience of discussion in the simulation of FP device.

In the biased device, the FP has an electric potential as same as that of the connected electrode and can modulate the electric field in the region under the FP, thus the ability of the gate FP to decrease the peak of the electric field at the gate edge on the drain side. It can be seen from the simulated on-state breakdown characteristics of HEMTs with different gate FP structures in Figure 9.10 that FPs can increase the breakdown voltage. For a FP length $L_{FP} < 1$ μm, the breakdown voltage rises sharply with L_{FP} and arrives at 500 V at $L_{FP} = 1$ μm; for $L_{FP} > 1$ μm, the improvement of breakdown voltage by increasing the FP length weakens and the breakdown voltage tends to saturate gradually.

1. Comparison of *I-V* curves. The breakdown voltage is defined as the intersection point of the extrapolation line of the saturation drain current with that of the sharp rising current in the impact ionization.
2. Breakdown voltage versus length of FP.

The electric field distributions in the channel of the off-state HEMTs of Figure 9.10a versus distance from source under the voltage $V_{DS} = 500$ V are shown in Figure 9.11. For the HEMT without FP only one peak in excess of the breakdown field of GaN is observed at the gate edge on the drain side as breakdown occurs. While for the FP-HEMTs, two peaks of the electric field are seen at the gate edge on the drain side and at the FP edge on the drain side, respectively. The two peaks are gradually separated with the peak at the gate edge dropping and the one at the FP edge rising as L_{FP} increases, and they are approximately equally high at $L_{FP} = 1$ μm with both less than the breakdown field of GaN. Thus, theoretically the FP with a length of 1 μm can be regarded as the optimum gate FP structure.

The thickness of the passivation layer under the FP also has significant influence on the breakdown characteristics. The breakdown voltage as a function of passivation

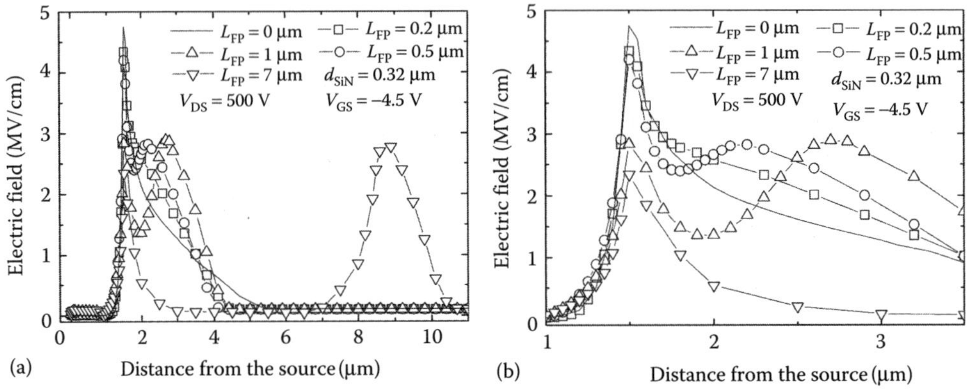

Figure 9.11 Distributions of the electric field in the channel of the HEMTs versus distance from the source. (a) Overall view and (b) detail view. (Courtesy of Dr. Wei Mao.)

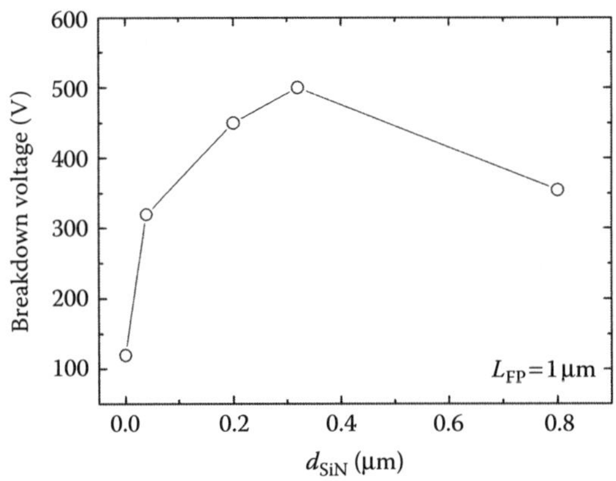

Figure 9.12 Breakdown voltage versus thickness of passivation layer under FP. (Courtesy of Dr. Wei Mao.)

layer thickness (with 1 μm FPs for the simulated devices) is shown in Figure 9.12. With increasing thickness of the passivation layer, the breakdown voltage of the FP-HEMTs gradually increases and then decreases. The maximum breakdown voltage is obtained at a passivation layer thickness of 0.32 μm. The effect of the passivation layer thickness on the channel electric field is presented in Figure 9.13, where a very similar distribution of channel electric field is observed of the 0.8 μm SiN FP-HEMT to that of the HEMT without the FP, indicating that device breakdown has happened as the FP can hardly modulate the channel electric field. The electric field peak at the gate edge drops while the peak at the FP edge increases with decrease in the passivation layer thickness, showing an increasing modulation ability of the FP. The heights of the peaks at the gate edge and at FP edge are approximately equal when the passivation layer thickness increases to 0.32 μm, where the optimal FP structure is obtained. A smaller passivation layer thickness leads to a lower peak of the electric field at the gate edge but a higher peak at the FP edge, also decreasing the breakdown voltage.

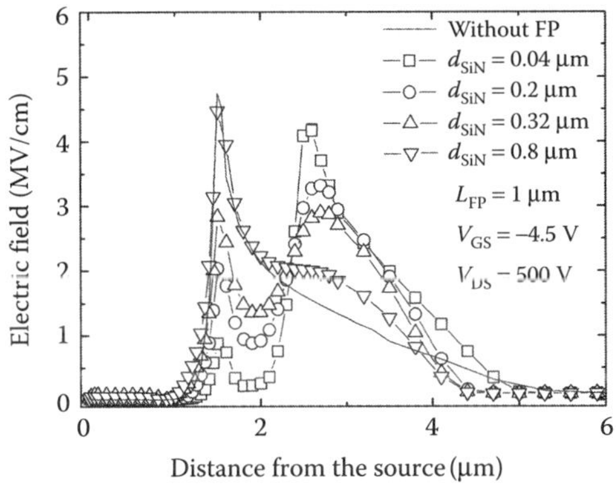

Figure 9.13 Channel electric field as a function of thickness of passivation layer under FP. (Courtesy of Dr. Wei Mao.)

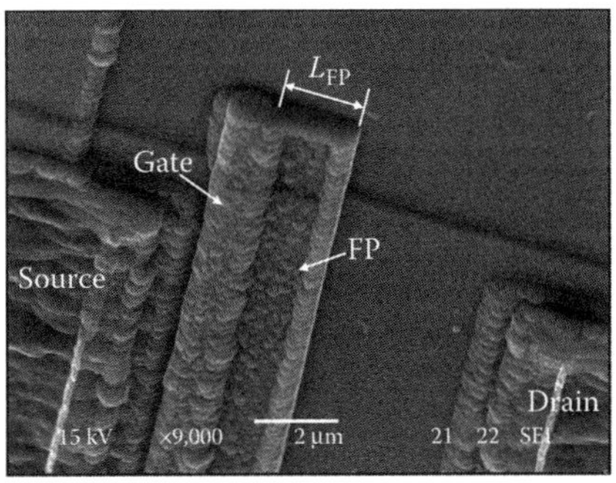

Figure 9.14 Detail view obtained by SEM of a FP-HEMT. (Courtesy of Dr. Wei Mao.)

9.4.2 Implementation of FP-HEMTs

The effect of FP length and source-drain spacing on FP-HEMTs should be considered. A plan-view SEM image with an effective FP length L_{FP} is shown in Figure 9.14.

The off-state breakdown characteristics of the FP-HEMTs with FP integrated with gate are given in Figure 9.15, displaying an average breakdown voltage (measured at I_{DS} = 1 mA/mm) more than 120 V. The increase in gate-drain spacing along with the increase in L_{FP} can improve the breakdown voltage and thereby the output power, but the capacitance between the FP and the channel also increases and causes degraded frequency characteristics (not shown here).

9.4.3 Introduction, Optimization, and Implementation of FFPs

By introducing the concept of the field limiting ring of the silicon planar device in the 1960s (Kao and Wolley 1967), we propose a FP structure more advantageous in improving the breakdown voltage for HEMTs compared with the conventional FP-HEMT: the floating compound FP (Wei Mao et al. 2010a, b). As shown in Figure 9.16, a floating compound FP consists of a conventional FP (CFP) connected

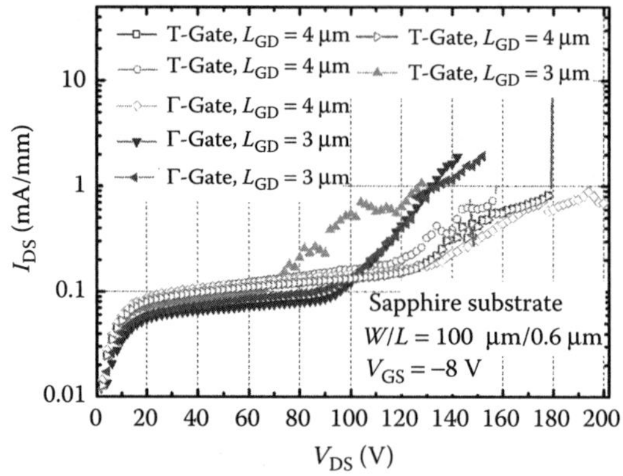

Figure 9.15 Off-state breakdown characteristics for FP-HEMTs.

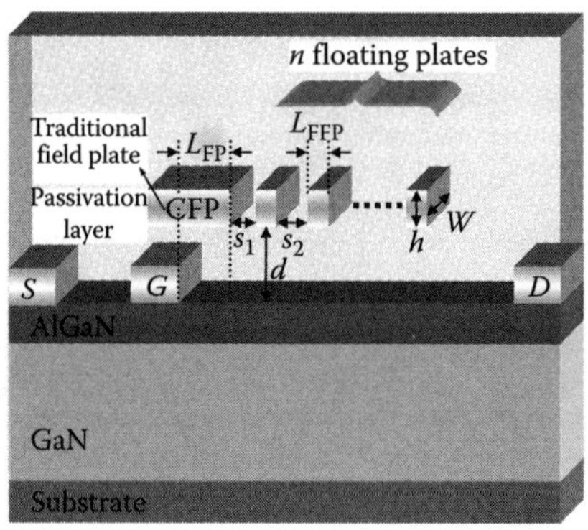

Figure 9.16 HEMT based on the floating compound FP structure. (Courtesy of Dr. Wei Mao.)

to gate or source combined with a number of floating FPs (FFPs) which are not connected to any electrode. The FFP-HEMTs has the advantages over conventional FP-HEMTs with one or more FPs of sustainable improvement of breakdown voltage by increasing the number of FFPs and simple fabrication processes fully compatible with those for the conventional FP-HEMTs with a single FP. In Figure 9.16, *W* denotes the FP width, L_{FP} the length of the CFP, L_{FFP} the length of the FFP, s_n ($n = 1, 2, 3,\ldots$) the spacing between two adjacent FPs, *d* the passivation layer thickness, *h* the FP thickness, and *n* the number of the FFPs.

The parameters of the FFP structure to be optimized are the FP spacing, the FP thickness, and the passivation layer thickness. The optimization principles are exemplified by the FP spacing optimization of a conventional FP-HEMT and a FFP-HEMT. Assume the CFP and the FFP have the same length of 1 μm. The breakdown characteristics of the HEMTs as a function of FP spacing *s* are given in Figure 9.17. The breakdown voltage increases first and then decreases with increasing FP spacing due primarily to the change in electrostatic induction between the CFPs and FFPs, which can be analyzed with the help of the electric field distributions in the channel of HEMTs in Figure 9.18. For a very small FP spacing, the electrostatic induction

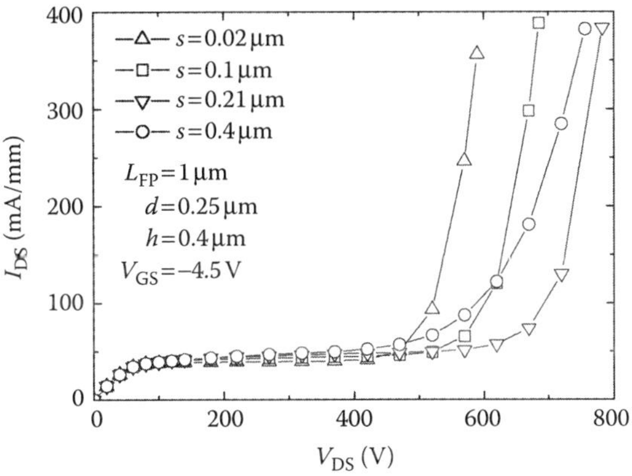

Figure 9.17 HEMT breakdown characteristics versus FP spacing *s*. (Courtesy of Dr. Wei Mao.)

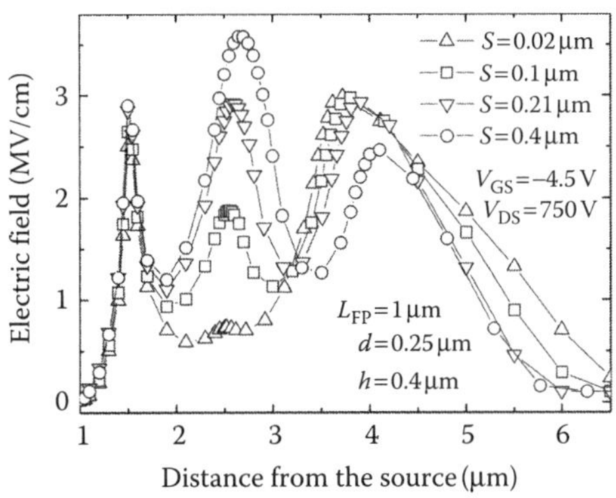

Figure 9.18 Electric field distribution in HEMT channel versus FP spacing *s*. (Courtesy of Dr. Wei Mao.)

between CFPs and FFPs are very strong and two major electric field peaks occur, one near the gate edge and the other near the FFP edge on the drain side, respectively, along with a weak peak near the CFP edge on the drain side. At this time the electric field distribution very much resembles that of the CFP-HEMT with a long FP, indicating that the effect of the FFP to modulate the channel electric field to be more even or disperse the very high electric field peak is very weak. As the FP spacing increases, the electrostatic induction between the CFPs and FFPs is weakened. More electric fluxlines spread under the FPs, and the electric field distribution along the channel is changed resulting in the increase of the peak near the CFP edge on the drain side and the decrease of the other two peaks. So the channel electric field distribution becomes more even as modulated by the FFP. At $s = 0.21$ μm, the heights of the three peaks of electric fields approximately equal, suggesting an optimal structure of the FFP with a maximum breakdown voltage of 750 V, better than the maximum breakdown voltage obtainable in CFP-HEMTs (<600 V) with an equivalent FP length (2.21 μm). This proves the floating compound FP to be a more effective FP structure. With $s >$ 0.21 μm, the peak of the electric field in the middle continues to increase, whereas the peak at the edge of the FFP keeps dropping. It can be predicted that the electrostatic induction between the CFPs and FFPs will fade away and finally the electric fields under and at the edge of the FFP reduce to zero as the FP spacing increases infinitely, in which case the device is equivalent to a conventional FP-HEMT.

Theoretical simulation of the FFP-HEMT with four FFPs yields two valuable laws applicable for HEMTs with n ($n \geq 1$) FFPs:

1. The FFP-HEMT has a slightly smaller optimum thickness of passivation layer under the FP than that of the CFP-HEMT, and this optimum thickness is not influenced by the number of FFPs.
2. The optimum FP spacings, which increase gradually from the gate to the drain and are independent to a certain extent from each other, can be obtained by successively adding the FFPs.

The FFP-HEMT devices are designed and fabricated based on the theoretical analysis. We adopt a GaN-based heterostructure (similar in structure to that in Section 9.4.1) wafer of a sheet resistance of 331 Ω/sq, an electron mobility of 1391 $cm^2/(V{\cdot}s)$, and an electron sheet density of 1.35×10^{13} cm^{-2}. The devices have a uniform gate length of 0.6 μm, a source-drain spacing of 7 μm, a gate-source spacing of 0.8 μm, and a gate width of 100 μm. The picture of a developed FFP-HEMT with one FFP is shown in Figure 9.19, and the SEM image of the dotted circular area in Figure 9.19b is given in Figure 9.20.

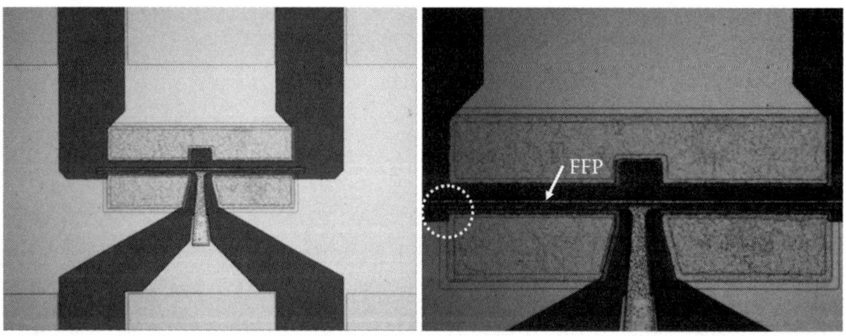

Figure 9.19 (See color insert.) Photograph of a FFP-HEMT with one floating FP. (a) Overall view and (b) detail view. (Courtesy of Dr. Wei Mao.)

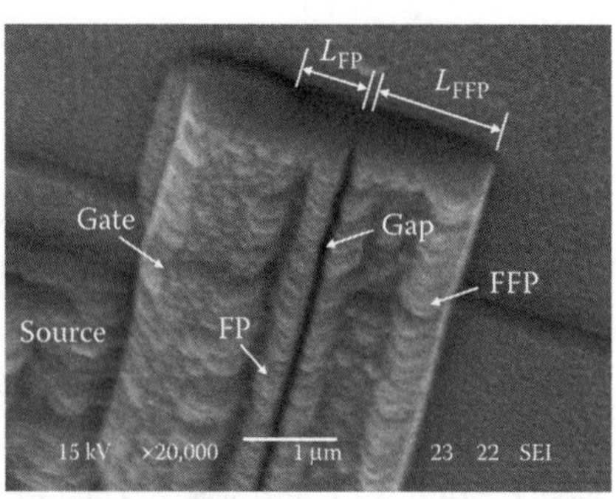

Figure 9.20 SEM image of the FFP-HEMT. (Courtesy of Dr. Wei Mao.)

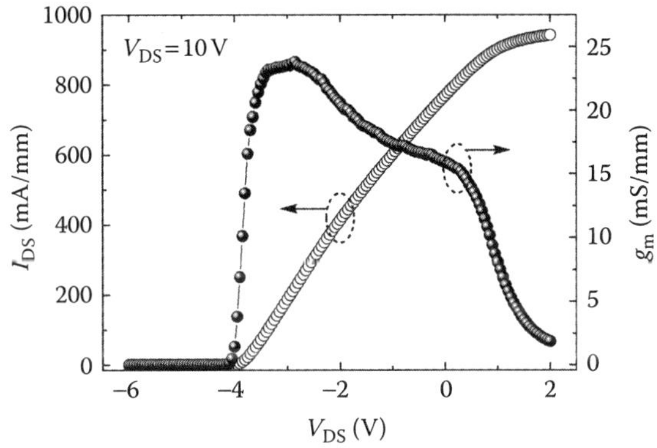

Figure 9.21 Transfer characteristics and transconductance characteristics of FFP-HEMT. (Courtesy of Dr. Wei Mao.)

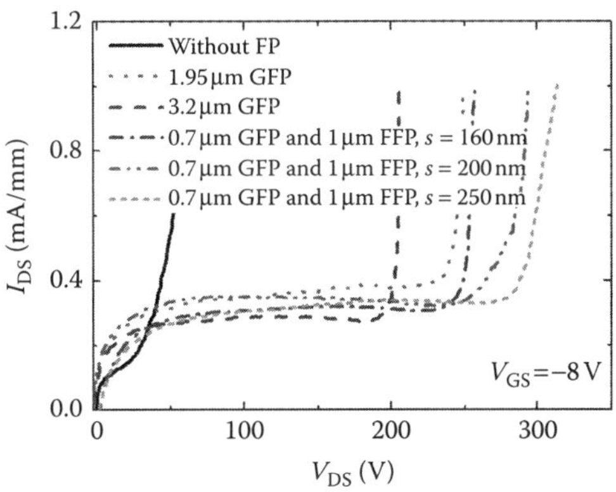

Figure 9.22 Breakdown characteristics of FFP-HEMT. (Courtesy of Dr. Wei Mao.)

Shown in Figure 9.21 are the transfer characteristics and transconductance characteristics of the FFP-HEMT with a saturation output current of 942 mA/mm and a peak transconductance of 238 mS/mm at $V_{GS} = 2$ V. The breakdown characteristics of the HEMT without FP, the CFP-HEMT, and the FFP-HEMT are presented in Figure 9.22. The HEMT without FP has a mere breakdown voltage of 65 V, and the CFP-HEMT has a higher breakdown voltage of 249 V. The FFP-HEMT achieves a breakdown voltage of 294 V when the FP spacing increases to 200 nm, a proof of the channel field modulation by the FFP. A maximum breakdown voltage up to 313 V is achieved in the FFP-HEMT with a spacing of 250 nm.

The microwave small signal measurements were performed on a HEMT without FP, a 1.95 μm gate FP-HEMT (GFP-HEMT) and a FFP-HEMT with a FP spacing of 250 nm, and the results are shown in Figure 9.23. All the HEMTs were measured at a drain voltage of 10 V with the gate-source bias taking the voltage at the peak transconductance. It can be seen that the HEMT without field gate has the best frequency characteristics with $f_T = 17.4$ GHz and $f_{max} = 57$ GHz and the 1.95 μm GFP-HEMT has a lower $f_T = 10.3$ GHz and $f_{max} = 21.8$ GHz. The FFP-HEMT with FPs of an identical equivalent length (1.95 μm) has its f_T and f_{max} significantly improved to 13.8 and 34 GHz, respectively, which is attributed primarily to the

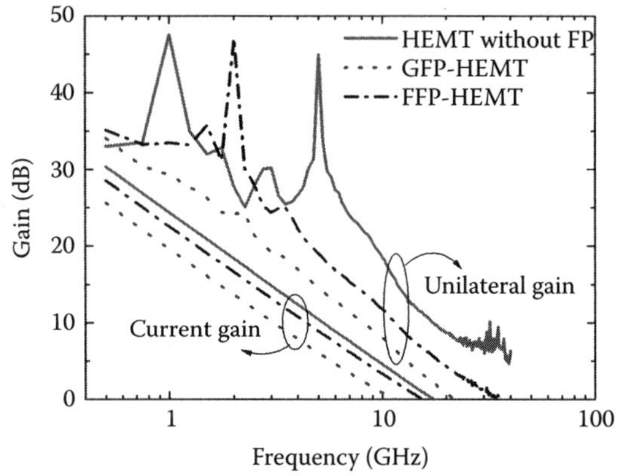

Figure 9.23 Gain versus frequency at $V_{DS} = 10$ V for the HEMT without FP and the FP-HEMTs. (Courtesy of Dr. Wei Mao.)

smaller additional capacitance introduced by the floating compound FP structure than that by the CFP structure.

The above results show that the FFP-HEMT is favorable for high power microwave device applications with a high breakdown voltage, good suppression of current collapse, remarkable frequency characteristics, and simple fabrication processes.

References

Albrecht, J. D., P. P. Ruden, S. C. Binari, and M. G. Ancona. 2000. AlGaN/GaN heterostructure field-effect transistor model including thermal effects. *IEEE Transactions on Electron Devices* 47(11):2031–2036. doi: 10.1109/16.877163.

Berroth, M. and R. Bosch. 1991. High-frequency equivalent circuit of GaAs FETs for large-signal applications. *IEEE Transactions on Microwave Theory and Techniques* 39(2):224–229. doi: 10.1109/22.102964.

Delagebeaudeuf, D. and N. T. Linh. 1982. METAL-(n) AlGaAs-GaAs two-dimensional electron gas fet. *IEEE Transactions on Electron Devices* ED-29(6):955–960.

Kao, Y. C. and E. D. Wolley. 1967. High-voltage planar p-n junctions. *Proceedings of the IEEE* 55(8):1409–1414. doi: 10.1109/proc.1967.5842.

Karmalkar, S. and U. K. Mishra. 2001. Enhancement of breakdown voltage in AlGaN/GaN high electron mobility transistors using a field plate. *IEEE Transactions on Electron Devices* 48(8):1515–1521. doi: 10.1109/16.936500.

Liao, C. E. 1994. *Fundamentals of Microwave Technology*. Xi'an, China: Xidian University Press.

Mao, W., C. Yang, Y. Hao, and R. Guo. 2010a. Recessed-gate metal-insulator-semiconductor heterojunction field effect transistor with composite source field plate. Edited by State Intellectual Property Office of the People's Republic of China.

Mao, W., C. Yang, Y. Hao, and R. Guo. 2010b. Recessed-gate metal-insulator-semiconductor high electron mobility transistor with composite source & drain field plates. Edited by State Intellectual Property Office of the People's Republic of China.

Mao, W., C. Yang, Y. Hao, X.-H. Ma, C. Wang, J.-C. Zhang, H.-X. Liu et al. 2011. The effect of a HfO2 insulator on the improvement of breakdown voltage in field-plated GaN-based HEMT. *Chinese Physics B* 20(9):097203 (7 pp.). doi: 10.1088/1674-1056/20/9/097203.

Ridley, B. K. 2004. Analytical models for polarization-induced carriers. *Semiconductor Science and Technology* 19(3):446–450. doi: 10.1088/0268-1242/19/3/027.

Roblin, P., S. C. Kang, and H. Morkoc. 1990. Analytic solution of the velocity-saturated MOSFET/MODFET wave equation and its application to the prediction of the microwave characteristics of MODFET's. *IEEE Transactions on Electron Devices* 37(7):1608–1622. doi: 10.1109/16.55746.

Ruden, P. P. 1990. Heterostructure FET model including gate leakage. *IEEE Transactions on Electron Devices* 37(10):2267–2270. doi: 10.1109/16.59919.

Sadiku, M. N. O. 2003. Deficiencies in the way scattering parameters are taught. *IEEE Transactions on Education* 46(3):399–404.

Wu, Y. F., A. Saxler, M. Moore, R. P. Smith, S. Sheppard, P. M. Chavarkar, T. Wisleder, U. K. Mishra, and P. Parikh. 2004. 30-W/mm GaN HEMTs by field plate optimization. *IEEE Electron Device Letters* 25(3):117–119. doi: 10.1109/led.2003.822667.

Zhang, P. Y. and Y. Zhou. 1985. *Principles of Transistors*. Shanghai, China: Shanghai Scientific & Technical Publishers.

Zhang, J. F., Y. Hao, J. C. Zhang, and J. Y. Ni. 2008. The mobility of two-dimensional electron gas in AlGaN/GaN heterostructures with varied Al content. *Science in China Series F (Information Science)* 51(6):177–186. doi: 10.1007/s11432-008-0056-7.

10

Preparation and Performance of GaN HEMTs

The preparation of GaN HEMTs are largely similar to that of GaAs HEMTs, including such major processes as active region isolation, preparation of ohmic contact for source/drain electrodes, preparation of Schottky contact for gate electrode, surface passivation, and metal interconnecting. The performance of the HEMT is influenced by each process step as well as by cleaning and photolithograph. Each single step must be fully optimized to form a process flow both reasonable and stable for the prepared HEMT die to operate normally and demonstrate high performance.

10.1 Surface Cleaning, Photolithograph, and Metal Lift-Off

10.1.1 Surface Cleaning

Surface cleaning prior to HEMT processing is a very important link in the whole process that has a decisive effect on the adhesiveness of the metal or dielectric deposited on the wafer surface and the characteristics of the final devices. Cleaning with organic solvents and acid/alkaline solutions may not guarantee an atomic-level cleanliness, but can effectively remove the oxide layer and contamination from the GaN surface.

Uncleaned GaN surfaces contain mostly organic and inorganic pollutants and the oxide layer. Surface organics are often removed by acetic acid, acetone, and ethanol while the oxide layer and inorganics by NH_4OH, $(NH_4)_2S$, and NaOH solutions. Additionally, the HCl solution can effectively reduce the residual oxides and the HF solution can remove the contamination by carbon and hydrocarbons.

Prior to the AlGaN/GaN-HEMT processing, the heterojunction sample is generally cleaned in acetone (MOS grade) by ultrasonic followed by a treatment in heated isopropanol and finally washed with deionized water and dried under nitrogen gas flow. Residuals from mesa etching should also be completely cleaned. Surface treatment of the sample wafer prior to gate metal evaporation can improve the ideality factor of the Schottky barrier.

10.1.2 Photolithograph and Metal Lift-Off

Photolithograph is performed before both etching and metal deposition to form patterns of the etching zone or metallic contact on the wafer surface. The area to be etched or connected to metal is exposed by optical lithography while the rest

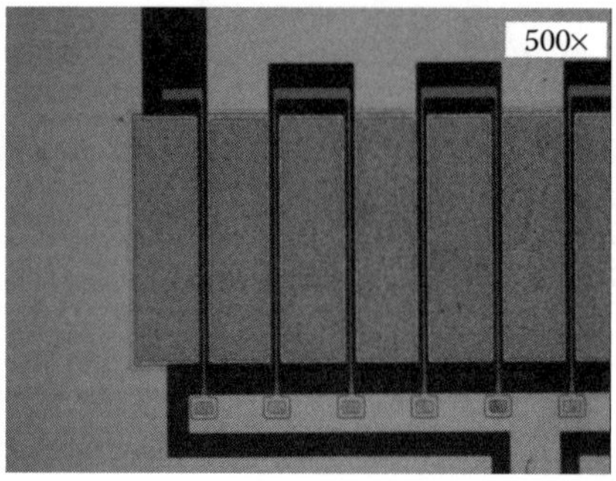

Figure 10.1 (See color insert.) Window for gate fingers etched on a multi-finger HEMT by optimized photolithography showing well-developed straight lines. (Courtesy of Dr. Ling Yang.)

of the wafer is covered and protected by photoresist. The exposure and developing time need to be optimized in photolithograph for both overexposure and insufficient exposure have deleterious effects on photolithographic patterns. An insufficient exposure leads to incomplete exposure of the semiconducting material in developing, which may cause a total failure of the lift-off process. Overexposure results in irregular borderlines of the photolithographic pattern especially in case of metal deposition for narrow gate fingers, thus severely affecting the device characteristics. High level of photolithography is required for HEMTs with multiple gate fingers. The window for gate fingers patterned on a multifinger HEMT by optimized photolithographic process is shown in Figure 10.1, where both the straight lines and the rectangular window corners are well developed.

The metal lift-off process refers to the removal of residual photoresists on the wafer surface with unwanted metal on the photoresist after photolithograph and metal deposition in the preparation of metal contacts. Lift-off is conventionally performed in acetone by ultrasonic process. Ideally, only the metal in the contact region remains after lift-off, forming the desired pattern of the metal. However, a thin film of photoresist hardly visible is often left on the semiconductor surface after a not fully optimized operation of exposure and developing, and the metal to be directly deposited on the semiconductor is actually deposited on the film and fails to form good contact to the semiconductor. Therefore, some of the metal needed for forming contact may also be lifted-off in the ultrasonic process, thus decreasing the yield of the lift-off process. For HEMTs with very thin gate fingers, a combined optimization of the photolithograph and metal lift-off process contributes much to the improved yield of gate preparation.

10.2 Device Isolation

10.2.1 Device Isolation Methods

Devices on a same wafer usually need to be separated from each other to avoid the flow of carriers between individual devices by confining each device into an "island" (referred to as isolated island) enclosed by the isolated area. Conventional device isolation is realized either by ion implantation to form a high-resistance region or by mesa etching to block the conduction path between devices, and both methods require that the barrier and channel layers be completely blocked to create perfectly isolated islands.

Ion implantation can generate perfect planar isolated islands advantageous for improving the yield and uniformity of the prepared HEMTs and their monolithic microwave integrated circuits. Compensated high-resistance regions can be produced by a combination of ion implantation with high temperature annealing for isolation based on two major mechanisms: damage compensation and chemical compensation (Pearton et al. 1999). For damage compensation, the resistance at the implantation region typically goes through a maximum with increased post-implantation annealing temperature due to the repaired damage, and at higher temperatures the defect density is further reduced below that needed to compensate the material and the resistivity decreases. For chemical compensation, the post-implantation resistance also increases with annealing temperature but it then stabilizes at higher temperatures as a thermally stable compensating deep level is formed. At present, many types of ions are available for nitride implantation isolation, such as H^+, He^+, B^+, N^+, O^+, F^+, P^+/He^+, Ar^+, and Zn^+. N^+ implantation, for example, at doses of 10^{12}–10^{13} cm^{-3} can effectively compensate both p- and n-type GaN. For both doping types the resistance first increases with annealing temperature then reaches a maximum before demonstrating a significant reduction in resistance after a 850°C anneal for n-type and a 950°C anneal for p-type GaN, a typical behavior of implant-damage compensation. The defect levels induced by implantation are 0.83 eV for n-type and 0.90 eV for p-type GaN, respectively, sufficient to realize a high sheet resistance $>10^9$ Ω/sq. Lighter ions such as H^+ are favorable for deep isolation schemes but less thermally stable, while heavier ion implantation induces high resistance of good thermostability but may cause implant surface damage and more severely amorphization or decomposition of the implanted region. It is necessary to optimize the implantation conditions (such as the dose, ion density, and energy) and annealing conditions.

In mesa etching, inter-device trenches significantly lower than channels are etched to form the isolated area. This process finds wide applications in GaN HEMTs owing to its low cost processing equipment and simple implementation. Dry etching is generally used for mesa etching of GaN HEMTs.

10.2.2 Common Dry Etching Techniques for GaN

GaN with bond energy up to 8.92 eV has very good chemical stability; insoluble in water, acid, or alkali at room temperature; and dissolving at a very low rate in hot alkali solution. As a result, it is difficult to achieve a satisfactory etch rate and accurate control by wet etching. Dry etching has the advantages of anisotropic profiles, relatively large difference in selectivity for dissimilar materials, good uniformity, and repeatability and ease in automated continuous production. Dry etching

Table 10.1 Common GaN Dry Etching Techniques

TECHNIQUE	PLASMA DENSITY	OPERATING PRESSURE (mTorr)	ADVANTAGES AND DISADVANTAGES
RIE	$<10^9\,cm^{-3}$	10–200	Good control of etch rate, low ionization rate, and high ion contamination
ECR	$>10^{11}\,cm^{-3}$	1–5	High ionization rate, good anisotropic property, and low ionic energy
ICP	$>10^{11}\,cm^{-3}$	1–50	Low cost, large active area, and good selectivity and directionality

Source: After Wang, C. et al., *Semiconduct. Technol.*, 31(6), 409–413, 2006.

techniques (Table 10.1) such as the reactive ion etching (RIE), electron cyclotron resonance (ECR), and inductively coupled plasma (ICP) have found use in GaN etching.

1. *RIE*: Adesida et al. (1993) were the first to report the etching of GaN by RIE, and an etch rate greater than 500 Å/min was achieved. RIE proceeds by using high-frequency plasma generated by radio-frequency discharge between two electrodes to etch the wafer on the radio-frequency electrode. Magnetically enhanced RIE (MERIE) can further decrease ion energy thus reducing etch damage as well as improving plasma density without compromising the etch rate. Mouffak et al. (2003) reported the photo-assisted RIE (PA-RIE) with significantly reduced damage level compared to conventional RIE.
2. *ECR*: Pearton et al. (1993) reported for the first time the etching of GaN by ECR, and an etch rate of 700 Å/min is achieved at a bias of 150 V in Cl_2/H_2 mixed gas. ECR has a relatively high comprehensive performance in the etch rate, selectivity, directionality, and damage.
3. *ICP*: Shul et al. (1996) were the first to report the etching of GaN by ICP, and an etch rate of 6870 Å/min is achieved in $Cl_2/H_2/Ar$ gas. The ICP system possesses two independent 13.56 MHz radio-frequency power sources, one at the top of the reaction chamber to generate plasma and the other connected to the inductive coil outside the reaction chamber, providing energy for the plasma by a bias (self-bias) for them to act perpendicular to the wafer.
4. *IBE, LE^4*: The etching technique using ion beam of certain energy to bombard the wafer surface is known as ion beam etching (IBE). IBE offers good anisotropic profiles but with poor etch selectivity and rate due to its pure physical operation. Reactive IBE (RIBE) replaces inert gases with other reactant gases to improve etch rate and selectivity. Low-energy electron-enhanced etching (LE^4) utilizes electrons of a lower energy (<15 eV) to react with the material surface thereby causing less etching damage.

10.2.3 Mechanism and Evaluation of Plasma Etching

Plasma etching consists of the ion physical bombardment and chemical reaction. Ion bombardment can remove the part of the material surface under attack by accelerated high-energy ion pairs in the plasma, and its intensity depends on the self-bias. Physical bombardment is in favor of anisotropic etching, but brings damage to the surface leading to rough surfaces, dangling bonds, and defects along with decreased etch selectivity. Chemical reaction also exists between plasma and the material surface and produces volatile etch products for removing the surface. At relatively low ion energies, chemical reaction occur both downward and on both sides. Such isotropic etching, though deleterious for device preparation, lessens the effect of ion bombardment and thus reduces etch damages. An ideal etch process involves a perfect combination of these two etch mechanisms for optimized etch rate and morphology.

Plasma etching of GaN typically employs Cl_2, BCl_3, $SiCl_4$, I_2, Br_2, CH_4, or SF_6 as the gas sources and mixes one or more of them with Ar, H_2, and N_2 to form the etching gases. Various combinations of the constituents and compositions of etching gases give different etch rates and selectivities. The etch rate and morphology are closely related to the volatility of the etch products. The boiling points of common reaction products in GaN etching are given in Table 10.2.

For Ga- and Al-containing materials, Cl-based plasma can achieve very good etching effect. CH_3/H_2 can also be used to etch Ga-containing materials. The etch product is $Ga(CH_3)_3$, which has good volatility due to a lower melting point than $CaCl_3$, but the etch rate is rather low by using CH_3/H_2 plasma. This indicates that the etch process is very complicated where the etch rate is subject to the reaction product formation, deposition as well as vapor phase dynamics. For In-containing nitrides, satisfactory etching can be obtained by Cl-based plasma etching only at temperatures above 130°C as a result of the low volatility of $InCl_3$, while the CH_3/H_2-based plasma has good etching effect since the etch product $In(CH_3)_3$ are highly volatile.

Typical etching specifications are the etch rate, etch anisotropy, selectivity, and etch damage. The etch anisotropy is defined as the ratio of the etch rate in the vertical direction to that in the lateral direction. By selectivity is meant the ratio of the etch rates for two different materials, which can be derived from an analysis of the relationship between the etch rates of the two materials by using a microscope or surface morphology analyzer. Selectivity becomes very important when it is required for etch to stop at a specific layer of the multilayer device. Etch damage consists mainly of damage of electrical properties and physical damage. Damage of electrical properties

Table 10.2 Boiling Points of Reaction Products of GaN Etched by Different Gases

ETCH PRODUCT	$GACL_3$	GAI_3	$GABR_3$	$GA(CH_3)_3$	GAF_3	NH_3
Boiling point (°C)	201	345	279	55.7	1000	−33

Source: After Wang, C. et al., *Semiconduct. Technol.*, 31(6), 409–413, 2006.

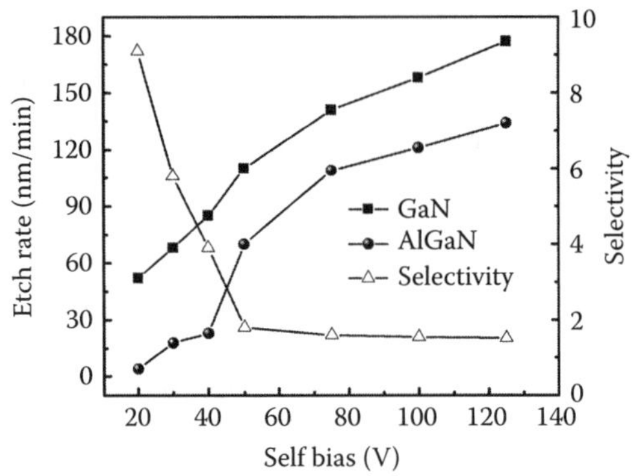

Figure 10.2 GaN and AlGaN etch rate as a function of ICP self-bias. (After Wang, C. et al., *Semiconduct. Technol.*, 31, 409–413, 2006.)

Figure 10.3 Microscopic image of the mesa edge after etching. (Courtesy of Dr. Ling Yang.)

can be measured by the conductivities before and after etching, and the recessed-gate etching damage by the reverse leakage current of the Schottky contact. Physical damage can be evaluated by the variation in surface roughness (Wang et al. 2006). Shown in Figure 10.2 is the ICP etch rate and selectivity as a function of ICP self-bias for GaN and $Al_{0.27}Ga_{0.73}N$ grown on sapphire by MOCVD at 20 sccm chlorine, 10 sccm argon, and 600 W ICP source power. GaN/AlGaN selectivity increases and AlGaN etch rate decreases with rising aluminum composition in AlGaN. Since the optimization of etch conditions depends heavily on the etching technique and equipment type, no further details are given here. The plan view of the ICP-etched GaN mesa in Figure 10.3 shows steep and straight edges and a smooth surface, meeting the device processing requirements.

10.3 Schottky Contacts

Metal electrodes are generally adopted for the source, gate, and drain of the GaN HEMT, with the Schottky contact formed at the gate and the ohmic contacts at the source and the drain. Metal electrodes are mainly deposited by vacuum evaporation and magnetron sputtering. The quality of the Schottky gate is a decisive factor of

AlGaN/GaN HEMT performance, as the gate leakage is the major source of low frequency noise and reverse gate breakdown voltage determines the operating voltage and power capacity of the device.

Experiments show that the Schottky barrier height of the metal–semiconductor contact for wide bandgap semiconductors such as AlGaN and GaN depends mainly on the difference between the work function of the metal and the affinity of the semiconductor, not on the Fermi level pinning at metal–semiconductor interface as for GaAs and InP. High work function metals such as platinum (5.65 eV), iridium (5.46 eV), nickel (5.15 eV), palladium (5.12 eV), and gold (5.1 eV) are all commonly used for Schottky gates, among which iridium and nickel have the best adhesion to AlGaN and GaN. Being comparatively cheaper, nickel is generally adopted as the underlying metal of the gate in device preparation to make sure that the gate does not fall off in the lift-off process of the gate metal. Gold is often deposited on nickel as a second layer metal to enhance the conductivity of the Schottky gate for a smaller gate series resistance, thus improving the frequency characteristics and decreasing the noise figure. The Ni/Au metal system is currently the most popular Schottky gate metals for AlGaN/GaN HEMTs.

An understanding of the characteristic parameter extraction of Schottky junction is required for the correct evaluation of the performance and current transport mechanism of Schottky junctions.

10.3.1 Extraction of the Characteristic Parameters of Schottky Junctions

The Schottky barrier height on a single epilayer of GaN or AlGaN can be extracted by *I-V* and *C-V* methods. The evaluation of Schottky junction characteristics by *I-V* measurements is usually analyzed using the thermionic emission model:

$$I = I_S\left[\exp\left\{\frac{eV - IR_S}{nk_BT}\right\} - 1\right], \quad I_S = AA^{**}T^2\exp\left(-\frac{e\phi_b}{k_BT}\right) \tag{10.1}$$

where:

e is the elementary charge
R_S is the diode series resistance
n is the ideality factor
A is the Schottky junction area
A^{**} is the Richardson constant
$e\phi_b$ is the effective barrier height
k_B is the Boltzmann constant
T is the absolute temperature
V is the applied voltage

For $V > 3k_BT$ and without considering the effect of R_S on current characteristics at high voltage, we may reduce this equation into

$$I = I_S\left[\exp\left\{\frac{eV}{nk_BT}\right\}\right] \tag{10.2}$$

Taking common logarithm on both sides we get

$$\log I = \log I_S + \frac{1}{\ln 10}\cdot\frac{eV}{nk_BT} \tag{10.3}$$

According to Equation 10.3, a linear fitting of the plotted log$I$-V curve based on the measured I-V characteristics gives the ideality factor n and saturation current I_S. According to the expression for I_S in Equation 10.1, the $\log(I_S/T^2)$~(C'/T) curve (also called Richardson curve; here C' is a constant and $\log[I_S/T^2]$ can also be $\ln[I_S/T^2]$) should be theoretically a straight line and the Schottky barrier height can be determined by its slope. Generally for $n \le 1.2$, the thermionic emission model is the dominant mechanism of current transport and a significant increase of n is seen if other mechanisms, for example tunneling, are rather visible.

The Schottky barrier height and the doping concentration N_D can be extracted by a C-V evaluation of the Schottky junction characteristics as

$$\frac{1}{C^2} = \frac{2}{A^2 e\varepsilon_s N_D}\left(\phi_b - V_n - V - \frac{k_BT}{e}\right) \tag{10.4}$$

where:

ε_s is the dielectric constant of the material

V_n is the voltage corresponding to the energy separation between the conduction band edge and the Fermi level under flat band condition

$$V_n = \frac{k_BT}{e}\ln\frac{N_C}{N_D} \tag{10.5}$$

Hence, a linear relation exists between $1/C^2$ and V. Let the slope of the straight line be K_C and the intercept of its extension line on the voltage axis be V_{int}, and we have

$$N_D = \frac{2}{A^2 e\varepsilon_s}\times\frac{1}{K_C},\quad \phi_b = \left|V_{\text{int}}\right| + \frac{k_BT}{e}\ln\frac{N_C}{N_D} + \frac{k_BT}{e} \tag{10.6}$$

10.3.2 Evaluation of the Characteristics of Schottky Junctions on GaN and AlGaN/GaN Heterostructures

Multiple mechanisms of current transport may be present for GaN Schottky contacts due to varied sample quality and growth conditions, surface conditions, and Schottky contact processes. Zhang et al. (2006) suggested that the trap-assisted electron emission near the metal–semiconductor interface is the dominant mechanism of leakage at room temperature. Sawada et al. (2003) interpreted the large leakage current by

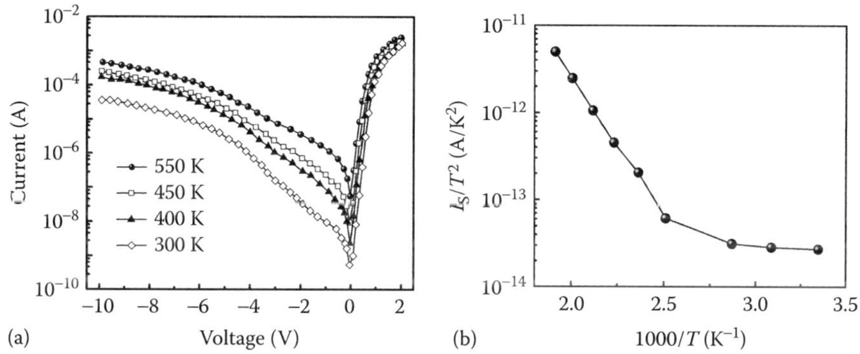

Figure 10.4 *I-V-T* characteristics of the Ni/Au Schottky diode. (a) *I-V* characteristics versus temperature. (b) Richardson curve obtained from the *I-V* curves. *T* is the temperature. (After Liu, J. et al., *Acta Phys. Sinica*, 56, 3483–3487, 2007.)

the surface patch model and attributed the significant increase of leakage current to a drop of the Schottky barrier height by 0.4 eV at the surface patch. Some analysis reports also adopted the thin surface barrier (TSB) model to explain the current transport. The TSB model assume the presence on GaN surface a high density of defect states which reduce the Schottky barrier width, and among the three current transport mechanisms of thermionic emission, thermionic field emission, and field emission of the Schottky junction on GaN, one or two may be dominant depending on the working temperature ranges and metals of the prepared Schottky diode (Benamara et al. 2006).

We obtain the characteristics of the Schottky diode prepared on n-type GaN from temperature-dependent *I-V* (*I-V-T*) and *C-V* (*C-V-T*) measurements at 300–550 K and give an analysis of Schottky characteristics based on the TBS model (Liu et al. 2007). The *I-V-T* curves and its Richardson curve are shown in Figure 10.4. The absolute value of the slope of the Richardson curve increases with rising temperature, indicating increasing electron activation energy. However, the activation energy from the fitted curve is merely 0.26 eV even at the high temperature range (400–550 K), indicating a dominant field emission (tunneling current) in the reverse current transport at room temperature. At high temperature, the surface electron trap level of 0.26 eV below the conduction band can assist the tunneling of electrons, thus the dominance of thermionic field emission (Hasegawa et al. 2003, Menozzi 2004).

Shown in Figure 10.5 are the *C-V-T* characteristics of the Schottky diode, and Figure 10.6 demonstrates the variation of extracted barrier height by *I-V-T* and *C-V-T* measurements with temperature. It can be seen that the barrier height extracted by *C-V* method is higher at low temperature and lower instead at high temperature than that by *I-V* method, which can be explained by the reduction of the field emission component in the forward current at high temperatures, leading to an increased barrier height extracted from *I-V* curves and a decreased barrier height from *C-V* curves due to the additional capacitance formed in the capture/release of electrons by the trap level at high temperature. The *I-V* extracted barrier height at high temperature

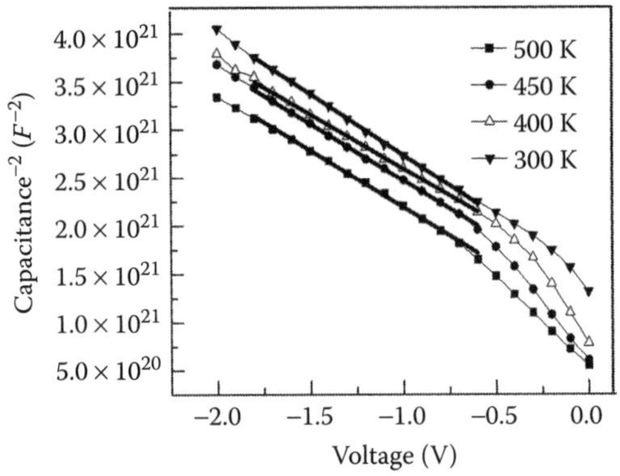

Figure 10.5 *C-V-T* characteristics of the Schottky diode. (After Liu, J. et al., *Acta Phys. Sinica*, 56, 3483–3487, 2007.)

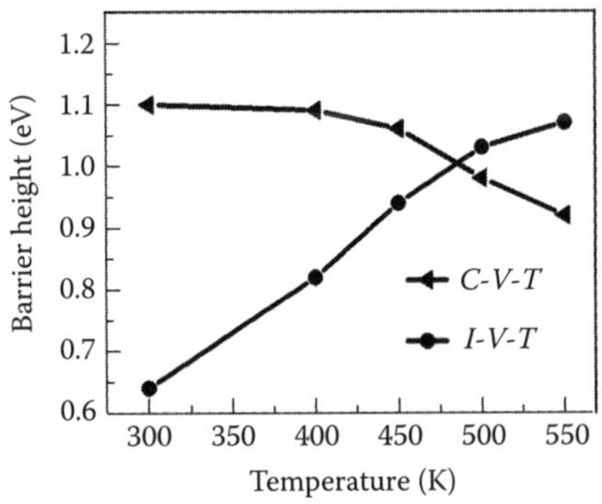

Figure 10.6 *I-V-T* and *C-V-T* extracted barrier height as a function of temperature. (After Liu, J. et al., *Acta Phys. Sinica*, 56, 3483–3487, 2007.)

is very close to the *C-V* extracted one at low temperature, both in the vicinity of 1.10 eV, which agree with the calculated barrier height using the nickel work function (5.15 eV). This indicates the barrier heights extracted in such two cases are closer to the actual values, coinciding with the results reported in Sawada et al. (2003).

In conclusion, the mechanism of current transport in the Schottky contact on GaN conforms mainly to the thermionic emission model with contribution also from the tunnel current and the surface defect-assisted tunneling effect. The actual Schottky barrier height should be determined by high temperature *I-V* or room temperature *C-V* measurements.

As for the Schottky diode on AlGaN/GaN heterostructure, a second barrier at the AlGaN/GaN heterointerface resulting from conduction band offset appears with the 2DEG except for the Schottky barrier at AlGaN surface. This makes the equivalent diode structures complicated that it is difficult to perform a quantitative analysis from the *C-V* or *I-V* measurement results to find the barrier height and ideality factor for Schottky contacts prepared on AlGaN/GaN heterostructures. However, the forward *I-V* characteristics of AlGaN/GaN heterostructures can still be used for a rough estimation of the barrier height and current transport mechanism, because the intercept of

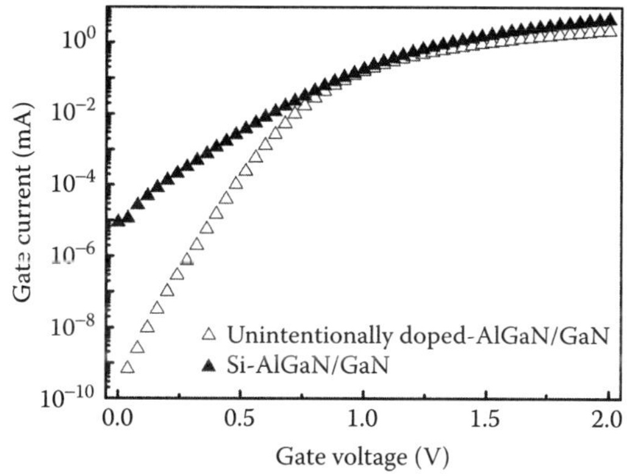

Figure 10.7 Forward characteristics of Schottky contacts on AlGaN/GaN heterojunctions. (Courtesy of Dr. Chong Wang.)

the log$I$~V curve on the logarithmic coordinate Y-axis can reflect the Schottky barrier condition: the smaller the intercept is, the smaller is the leakage current other than thermionic emission. From the AlGaN/GaN Schottky forward log$I$~V characteristics shown in Figure 10.7 we can see a relatively small forward intercept on the current axis of the forward characteristics for the Schottky contact prepared on undoped AlGaN/GaN heterojunction as well as a sharp rise of the forward current. A significantly increased intercept on the current axis for Si-AlGaN/GaN heterojunction is observed due to the existence of the tunneling current or the trap-assisted tunneling current.

10.3.3 Effect of Varied Solution Pretreatments on Schottky Junction Characteristics

Surface preparation with some solutions can be performed prior to metal deposition to guarantee the formation of tight metal–semiconductor contacts on the contamination-free semiconductor surface by the Schottky metal deposition. The I-V characteristics of the Schottky contact after surface treatment of the AlGaN/GaN heterojunction with different solutions in Figure 10.8 show better Schottky properties using HCl and HF. Element analysis of the AlGaN/GaN surface by SEM/EDS before and after HCl solution rinsing finds the oxygen element drops from 8.14% before rinsing to zero

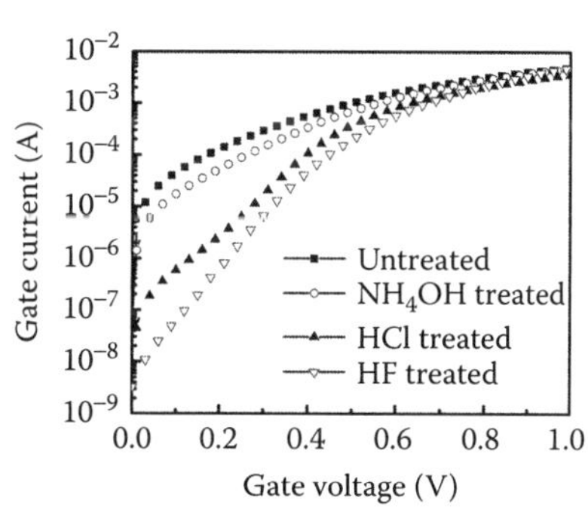

Figure 10.8 *I-V* characteristics of the Schottky contact prepared after surface treatment with different solutions. (Courtesy of Dr. Chong Wang.)

after rinsing. It is impossible for the surface to be totally free of oxides after treatment, so a reasonable explanation is that the residual oxides are under the lower detection limit. Therefore HCl can remove such oxides as Al_2O_3 and Ga_2O_3 from the AlGaN/GaN surface. The HF solution has similar effect. But surface treatment with KOH and $(NH_4)_2SO_4$ solutions is not satisfactory probably because the etching of GaN in alkaline solution magnifies the adverse effect of defects on the Schottky junction.

10.4 Ohmic Contacts

The source and drain electrodes of HEMTs are mostly formed by ohmic contacts, the performance of which has a direct influence on the output drain current and knee voltage. The majority of *n*-GaN-based ohmic contacts contain four layers of metal, typically Ti/Al/Ni/Au (from bottom to top). After the deposition of the metal layers, a rapid thermal annealing (RTA) process is needed to obtain ohmic characteristics. Each metal layer has its specific function in such a compound structure of metal contact, and the evaluation and optimization of the overall ohmic characteristics of the metal contact is of significance.

10.4.1 Design Principles of Ohmic Contacts on GaN and AlGaN/GaN Heterostructures

The electron affinity of *n*-GaN is 4.11 eV. Of the metals listed in Table 10.3, titanium and aluminum havc comparativcly low work functions and arc suitable to form ohmic contacts to GaN. But ohmic contacts on *n*-GaN formed with a single layer of titanium or aluminum or with the Ti/Al double layers are not reliable on high-power, high-temperature, and high-voltage operations since both are easily oxidized at high temperature. Resultant Al_2O_3 from aluminum oxidation forms a cap layer on top of the aluminum layer to yield high resistance, and balling-up of aluminum also tends to occur in the annealing process. The low melting point of aluminum (660°C) leads to a poor thermal stability of the aluminum single-layer or Ti/Al double-layer contacts. Therefore, ohmic contacts

Table 10.3 Work Functions, Melting Points, and Resistivities of Different Metals

METAL	WORK FUNCTION (eV)	MELTING POINT (°C)	RESISTIVITY (Ω·cm)
Gallium (Ga)	3.96	29.76	2.70×10^{-5}
Aluminum (Al)	4.25	660	2.65×10^{-6}
Titanium (Ti)	3.95	1668	4.20×10^{-6}
Nickel (Ni)	4.50	1453	6.84×10^{-6}
Gold (Au)	4.30	1063	2.35×10^{-6}
Tantalum (Ta)	4.25	3017	1.31×10^{-5}
Palladium (Pd)	5.12	1552	1.08×10^{-6}
TiN	3.74	–	1.00×10^{-6}
TaN, ZrN, VN, NbN	>4.00	–	2.25×10^{-6}

Source: Courtesy of Jue Lu.

on GaN and AlGaN/GaN require multilayer of metals to meet the performance requirements for low contact resistance, smooth surface, and good thermal stability.

For a good ohmic contact with multilayer metals, a small work function of the bottom layer metal closest to *n*-GaN is necessary for the formation of low-resistance metalized or semimetalized compound by the solid phase reaction between the metal and GaN. The compound should be chemically and thermodynamically stable and offer a protection layer making it difficult for the upper layer metals with greater work functions to diffuse to the *n*-GaN surface. Thus, this protection layer is referred to as the barrier layer. Titanium is usually chosen as the barrier layer metal for *n*-GaN ohmic contacts owing to its high activity and melting point. Semimetal TiN is formed with much ease by the solid phase chemical reaction between titanium and the nitrogen atoms diffused from *n*-GaN in annealing, and the high-density nitrogen vacancies (V_N) left in GaN can also serve as shallow donors in favor of the ohmic contact formation.

The metal layer immediately on top of the barrier layer is called the cover layer, which acts as the catalyst for the solid phase chemical reaction between nitrogen and the barrier layer metal, with which the cover layer should also be able to form a compact alloy of a low work function. Aluminum is a very good cover layer metal because it does not lead to alloys of a high work function. Both barrier layer and cover layer metals are prone to oxidation. Typically, one or more layers of cap layer metal are added on top of the cover layer to avoid the formation of nonconductive oxide/hydroxide by the barrier or cover layers. The cap layer can employ chemically stable gold, but interdiffusion occurs readily between gold and aluminum for gold to reach the GaN surface, which is disadvantageous for the formation of good ohmic contacts. So a layer of nickel is often introduced between aluminum and gold as an isolation layer to prevent diffusion of gold to GaN surface. Other alternative isolation layer metals are titanium, chromium, platinum, palladium, and molybdenum, with Ti/Al/Ni/Au as the currently most popular metal system for GaN-based ohmic contacts.

The contact system formed by multilayer metals on *n*-GaN does not gain ohmic contact characteristics automatically. A RTA process is required to transform the barrier layer metal into low-resistance nitrides on one hand, and to promote interdiffusion between the metals for solid phase interface reaction on the other hand so as to form a series of intermetallic alloys with low resistance, low work function, and good thermal stability. The actual layered structure of the contact system after the RTA process is determined by the thicknesses of the metal layers as well as the RTA time and temperature. These parameters are also to be optimized for good characteristics of ohmic contacts on GaN and AlGaN/GaN heterostructures.

10.4.2 Evaluation of Ohmic Contact Performance by Transmission Line Model

The ohmic contact performance is usually measured by the specific contact resistance ρ_C based mainly on the transmission line model (TLM). This measurement requires the preparation of specific test patterns and can also obtain the sheet resistance.

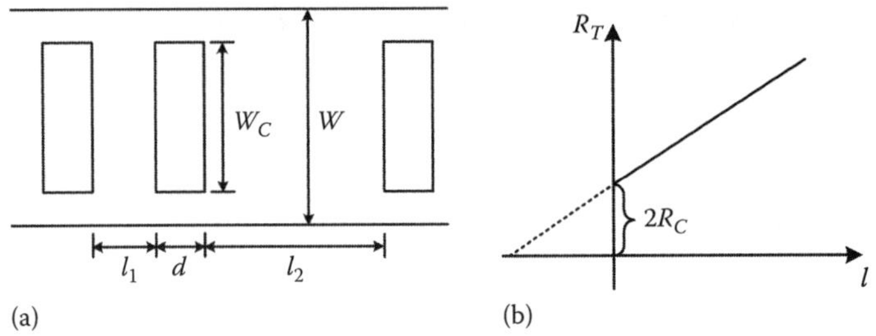

Figure 10.9 TLM measurement of the ohmic contact. (a) TLM patterns. (b) Resistance between contacts versus contact spacing for the derivation of specific contact resistivity.

A linear array of rectangle metal electrodes of length W_C and of width d was prepared on the etched mesa on the GaN surface with varied spacings between two adjacent electrodes (Figure 10.9). It is assumed that the sheet resistance of the metal electrodes is negligible as compared to that of the semiconductor in TLM measurements. The total resistance R_T between every two electrodes contains two parts, expressed as

$$R_T = 2R_C + l\frac{R_{SH}}{W_C} \tag{10.7}$$

where:

R_C is the electrode contact resistance and the second term on the right is the bulk resistance between electrodes

R_{SH} is the sheet resistance

l is the spacing of two adjacent electrodes

It is often assumed that the sheet resistance R_{SHC} for the material under the contacts is equal to that between the contacts R_{SH}, that is, $R_{SHC} = R_{SH}$. Define the transmission length as $L_T = \sqrt{\rho_C/R_{SH}}$, according to TLM theory we have

$$R_C = \frac{R_{SH} \cdot L_T}{W_C} \cdot \coth\left(\frac{d}{L_T}\right) \tag{10.8}$$

For $d \gg L_T$, $\coth(d/L_T) \to 1$, and then follows

$$R_T = \frac{2R_{SH} \cdot L_T}{W_C} + \frac{R_{SH}}{W_C} \cdot l \tag{10.9}$$

As shown in Figure 10.9b, the total resistances R_T between two adjacent electrodes with different spacing were measured respectively to plot a $R_T \sim l$ graph, where the slope of the straight line is R_{SH}/W, the intercept of the straight line on the R axis is $2R_C$. With L_x representing the absolute value of the intercept of the straight line on the l axis, we have $L_x = 2L_T$. Form the slope of the straight line follows the sheet resistance R_{SH}, and by the definition of the transmission length the specific contact resistivity ρ_C is determined by

$$\rho_C = R_{SH} L_T^2 \tag{10.10}$$

The specific contact resistivity determined by TLM is usually in units of Ω·cm². If the specific contact resistivity is close to or less than 10^{-7} Ω·cm², TLM is not accurate.

10.4.3 Optimization of Ohmic Contact Performance

For ohmic contacts formed on *n*-GaN with at least four metal layers, the thickness of each layer and the RTA time and temperature need to be optimized for low resistance characteristics. The Ti/Al/Ni/Au contact, for example, can be optimized by the following steps. First, set the initial thicknesses of the metal layers and RTA process conditions and determine the optimum Ti/Al thickness ratio according to the analysis of the effect of various Ti/Al thickness ratios on the ohmic contact by changing the thickness of the aluminum cover layer (Figure 10.10). Then determine the optimum Ti/Al thickness by a further analysis of the effect of different Ti/Al thicknesses on the ohmic contact while maintaining the Ti/Al thickness ratio constant (Figure 10.11). At last, optimize the thickness of the nickel isolation layer and so on after the Ti/Al layer optimization until optimum thicknesses of all the four layers are obtained. This is followed by the optimization of the annealing temperature (Figure 10.12), annealing time, and the heating/cooling rate as well as the annealing ambient to obtain the lowest contact resistance.

Similar to that in the Schottky contact processing, surface treatment of *n*-GaN prior to the deposition of metal layers is also of great importance to the preparation of low resistivity contacts owing to both the removal of oxides that may induce high resistance and roughening the surface that increases the contact area for better adhesion of the metal to the GaN surface. Surface preparation can be performed in HCl prior to metal deposition and RTA. The surface morphologies of the patterned ohmic

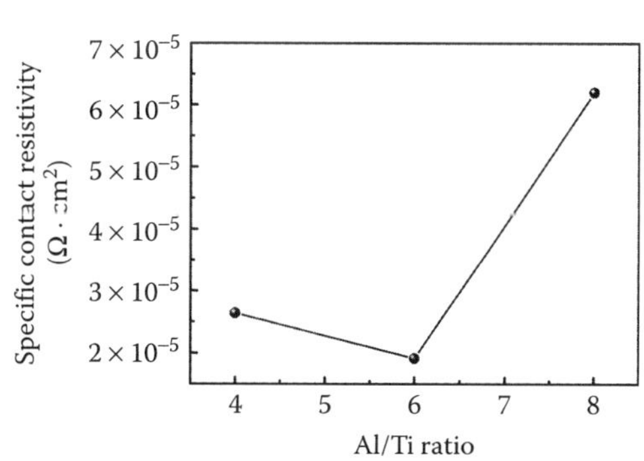

Figure 10.10 Specific contact resistivity ρ_c versus Al/Ti thickness ratio. (After Yang, Y. et al., *Chin. J. Semiconduct.*, 10, 1823–1827, 2006.)

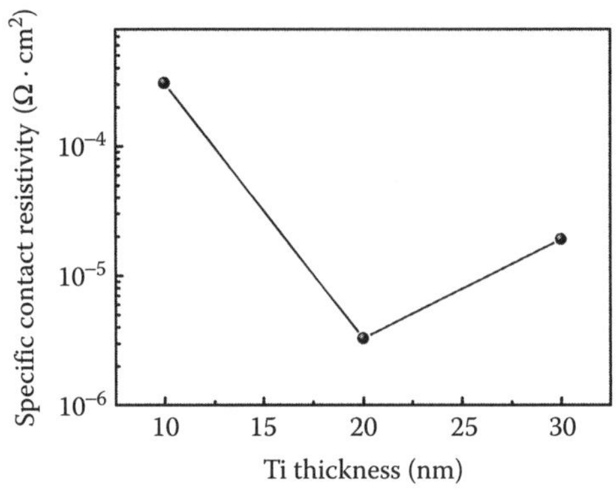

Figure 10.11 Specific contact resistivity versus thickness of titanium layer with a constant Ti/Al thickness ratio of 1/6. (After Yang, Y. et al., *Chin. J. Semiconduct.*, 10, 1823–1827, 2006.)

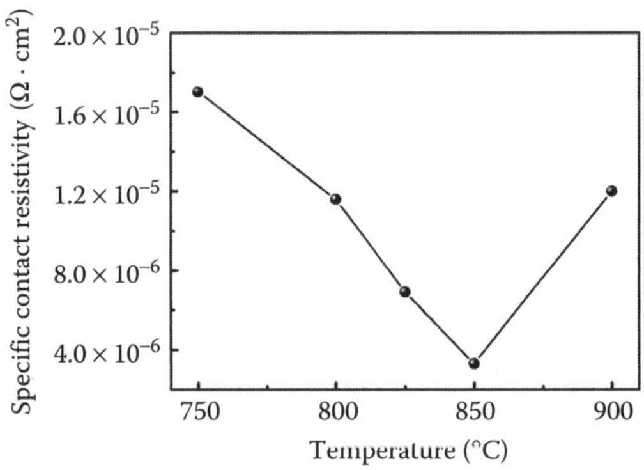

Figure 10.12 Specific contact resistivity versus RTA annealing temperature. (Courtesy of Dr. Yan Yang.)

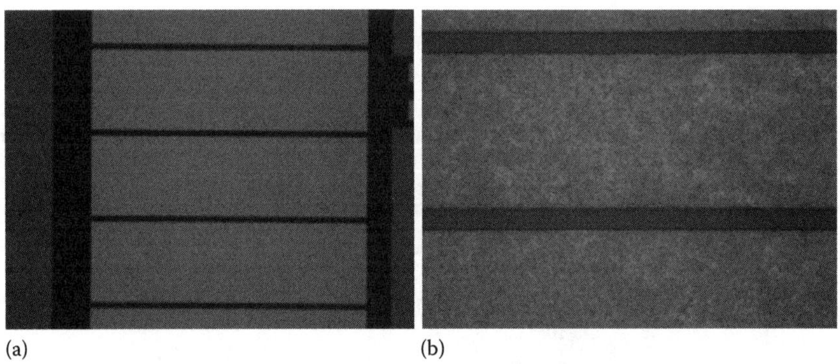

Figure 10.13 Surface morphologies of the ohmic contact (a) before and (b) after annealing. (Courtesy of Dr. Ling Yang.)

contact metal observed with high power microscope before and after annealing are shown in Figure 10.13. No metal adhesion is seen on the ohmic contact after lift-off and the electrode edges are intact before annealing. After annealing the electrode edges are smooth and burr-free without metal outflow.

Finally, the thermal stability of the ohmic contact can be estimated by the variation in its *I-V* characteristics and surface morphology after another RTA. After the above optimization procedures, a preparation process for low resistivity ohmic contacts with good surface morphology and thermal stability can be achieved.

10.5 Surface Passivation

The existence of defect-induced surface states causes current collapse in the high-frequency high-power applications of AlGaN/GaN HEMTs, decimating the output power. Surface passivation can effectively suppress current collapse and the passivation layer also lessens the effect of ambience on the electrical properties of HEMTs.

The criteria for dielectric films used as passivation layer are good adhesion to the semiconductor, a thermal expansion coefficient close to that of the semiconductor, good insulation, a small dielectric constant and high-frequency loss, high dielectric breakdown strength, a good passivation effect, and ease in photolithograph and etching.

Surface passivation for AlGaN/GaN HEMTs are typically performed with SiN_x and SiO_2 deposited by plasma-enhanced chemical vapor deposition (PECVD), which is an advanced film deposition method integrating both physical vapor deposition and chemical vapor deposition characteristics. In the operating PECVD facility, the radio-frequency electric field applied on the plate electrode causes glow discharge of the reaction gas under vacuum pressure, and many electrons are generated in the discharge region. Under the effect of the electric field these electrons acquire sufficient energies to activate the gas molecules by collision to generate plasma consisting of numerous positive and negative ions. Dielectric films are formed by the adsorption and the chemical reaction of the ions on the substrate, and the by-products are desorbed from substrate and carried out with the main flow by the vacuum pump. Such working principles enable many high-temperature-dependent reactions to be performed at lower temperatures. Compared to thermal reaction, dielectric films of uniform composition and characteristics can be prepared at a higher deposition rate. The passivation quality is evaluated by the leakage characteristics, etch rate, surface morphology, and uniformity of the passivation film as well as the breakdown voltages of the HEMT before and after passivation.

The 2DEGs of AlGaN/GaN heterostructure can be greatly influenced by the charge damage both at the AlGaN surface and at the AlGaN/GaN interface. Too high a PECVD power in the passivation may result in damage to the HEMT surface thereby a decreased current, while too low a PECVD power may affect the density of as-grown Si_3N_4 film leading to a decrease in breakdown voltage. For this reason, a thin passivation film of Si_3N_4 can be grown at low power before the growth of Si_3N_4 of adequate thickness at higher power.

The stress of the Si_3N_4 passivation layer also has profound influence on the 2DEG density in the heterostructures, so the reactant gas proportion needs to be optimized in the passivation to avoid major stress of the deposited Si_3N_4 on the surface. The stress behavior of Si_3N_4 can be adjusted in the growth of Si_3N_4 by PECVD by varied He/N_2 ratios.

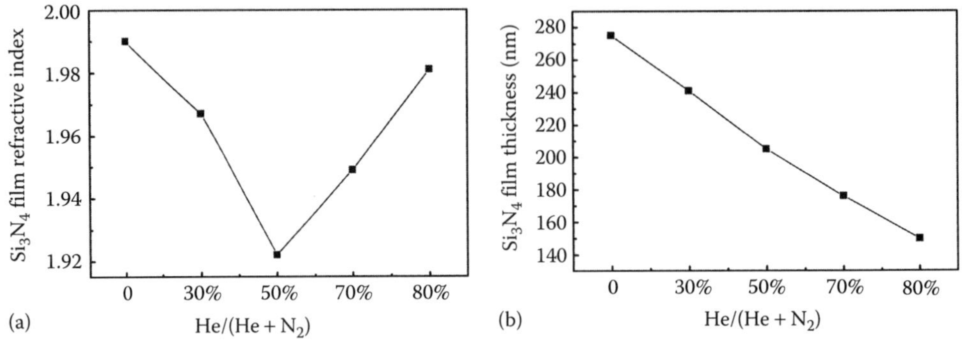

Figure 10.14 (a) Refractive index and (b) thickness of Si_3N_4 film as a function of helium proportion. (Courtesy of Dr. Ling Yang.)

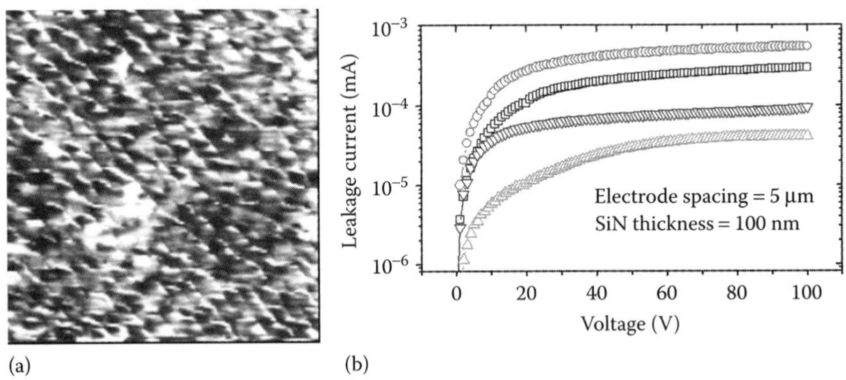

Figure 10.15 (a) AFM surface morphology (8 × 8 μm²) and (b) leakage current characteristics of the Si_3N_4 dielectric film. (Courtesy of Dr. Ling Yang.)

The Si_3N_4 growth rate drops with increasing proportion of helium with little change in the refractive index, that is, as-grown passivation film material is still Si_3N_4 but with variation in density. Generally, the lower the growth rate is, the denser the grown material and the smaller the leakage current. But a denser material also has greater stress, which requires a compromise between stress and leakage current (Figure 10.14).

The AFM image in Figure 10.15 shows smooth and uniform surface after passivation. For a 100-nm-thick SiN passivation film tested at 100 V voltage, 5 μm spacing, and 100 μm electrode width, the measured leakage current is less than 1 μA, meeting the requirements of HEMT applications.

10.6 Air Bridges Prepared by Electroplating

10.6.1 Electroplating

Since HEMTs are mostly applied in microwave high-power operations, gold is often employed as the top layer metal for HEMT electrodes for its good conductivity and chemical stability, small electromigration, and easy wire bonding. A minimum thickness of 1 μm of the gold electrode is generally required for a small sheet resistance and good microwave characteristics. Vacuum evaporation is not suitable to deposit very thick metal owing to the low vacuum evaporation rate and great waste of metal in the preparation of thick gold electrodes. The common practice is to vacuum evaporate a thin gold film as the top layer of the electrode and then to thicken the electrode by electroplating.

Electroplating effect can be evaluated from the following perspectives:

1. *Bonding*: The binding force between the electroplating layer and the substrate. Poor bonding may cause break-off of the coating.
2. *Density*: The binding force of the coated metal itself. Good density is achieved by fine crystal grains and absence of impurities.
3. *Continuity*: This indicates whether pores exist on the coating.
4. *Uniformity*: The ability of the electroplating bath to deposit on the plating piece a coating of uniform thickness.
5. *Esthetics*: The plating piece should be aesthetically pleasant with a smooth surface free of stains or bulb defects.
6. *Stress*: Too great a residual stress from the electroplating process may induce cracking or lift-off of the coating.
7. *Physical, chemical, and mechanical properties*: Such as stiffness, ductility, strength, conductivity, thermal conductivity, reflectivity, corrosion resistance and color.

Metal plating quality, which is related largely to the plating rate, has a great influence on HEMTs, especially in terms of long-term reliability and stability. The 10,000× SEM surface images of the deposited metal at varied plating rates are shown in Figure 10.16. It can be seen that the surface of sample A has rough spherical gold grains of significantly varied diameters. Sample B have gold grains in the shape of croissant. Sample C also has spherical gold grains but the grains are fine and of uniform size. The plating surface morphologies of all the three samples observed with a 400× microscope (not shown here) exhibit the color of golden yellow with increasing brightness (from A to C), suggesting that sample C has the best surface quality of the plated metal and with the optimum plating rate.

10.6.2 Air Bridges

Air bridges are employed for multifinger gate GaN HEMTs to connect the single-finger gate device units to form large dimension devices. Since air has the least dielectric constant, air bridges can considerably reduce the formation of parasitic capacitance but with greater difficulty in preparation compared to dielectric bridges. A plan view of the air bridge structure is shown in Figure 10.17.

Air bridges are prepared by two times of photolithograph and electroplating process as shown in Figure 10.18. The initial photolithograph is performed on the photoresist that supports the bridge to expose the metal electrode to be connected

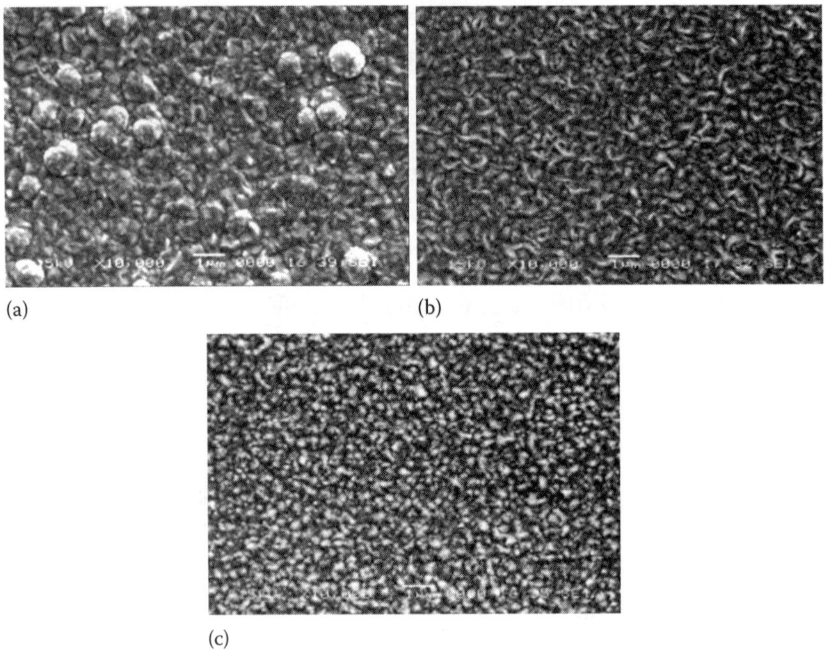

Figure 10.16 (a–c) SEM images of electroplated metal surface at different plating rates. (Courtesy of Dr. Ling Yang.)

Figure 10.17 **(See color insert.)** Plan view of the air bridge in a HEMT sample.

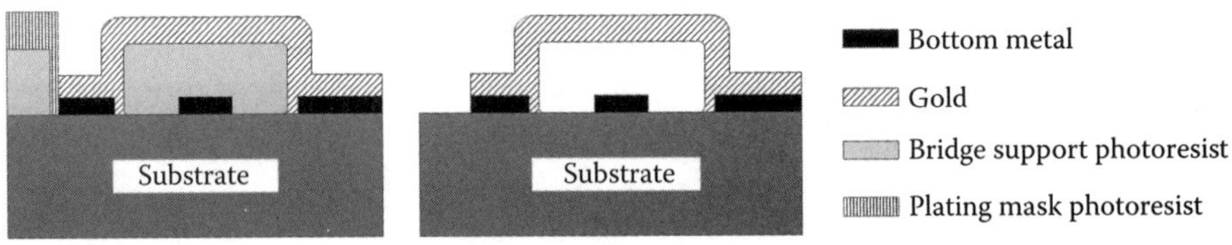

Figure 10.18 Schematic cross section of air bridge in preparation. (Courtesy of Dr. Chong Wang.)

(i.e., bridge support), followed by the vaporization of thin Ti/Au onto the whole wafer as the conducting layer for electroplating. A second photolithograph is then carried out to define the regions to be thickened by plating. A layer of ~2 μm gold is deposited to form the bridge floor and then the plating mask photoresist is removed for etching off the unused conducting layer of the plated metal, after which the support photoresist is removed. A certainty thickness of the support photoresist is needed to guarantee the height of the bridge for smaller parasitic capacitance. However, a high bridge means greater possibility for the metal to break in "climbing" on the plating mask photoresist to form the bridge floor. So an arc-like morphology of the support photoresist under the bridge is necessary to guarantee the height and strength of the air bridge. The preparation of high strength air bridges should employ two photoresists that differ greatly in thermal contraction coefficient as the sacrificial layer with the arc-like morphology optimized by adjusting the proportion, total thickness, and baking temperature of the two resists and with the air bridge-induced capacitance reduced.

A new widely used process for interconnection of single-finger gate HEMT units for a multifinger gate GaN HEMT in recent years is the source-via technology, that is, the preparation of through-holes at the source of each device unit along the direction vertical to the wafer surface, thus forming monolithic metal connections by depositing metals (using sputtering and plating, etc.) on the backside of the thinned substrate and in the via. The substitution of source-via for the source air bridge as interconnection can considerably decrease the source interconnection resistance and the parasitic inductance induced by the leads and bonding pads thus improving the gain and power added efficiency. In terms of circuit layout, such a structure can effectively reduce the size of the device and improve packaging density. Due to the great difficulty in the thinning and back hole etching of SiC, the typical substrate on which to prepare large size GaN power HEMTs, this technology has not become a standard process in the processing line of GaN HEMT monolithic microwave integrated circuit foundries until major breakthroughs have been made in recent years.

10.7 Process Flow of GaN HEMTs

Optimization of the above single step processes ultimately yields an integral process flow for GaN HEMT preparation. A feasible process flow is given in Figure 10.19 consisting of the following major steps.

1. Wafer cleaning includes removal of the grease and oxides from the sample surface.
2. Ohmic contact preparation includes metal deposition and RTA. The requirement are smooth pattern edges and no adhesion after lift-off of the ohmic contact metal, evenly distributed grains, and no side outflow of metal after annealing. TLM-measured ohmic contact resistance should be kept below 1 Ω·mm.
3. Mesa isolation is realized by mesa etching by ICP or RIE after photolithograph with etch depth greater than 100 nm to make sure the channel

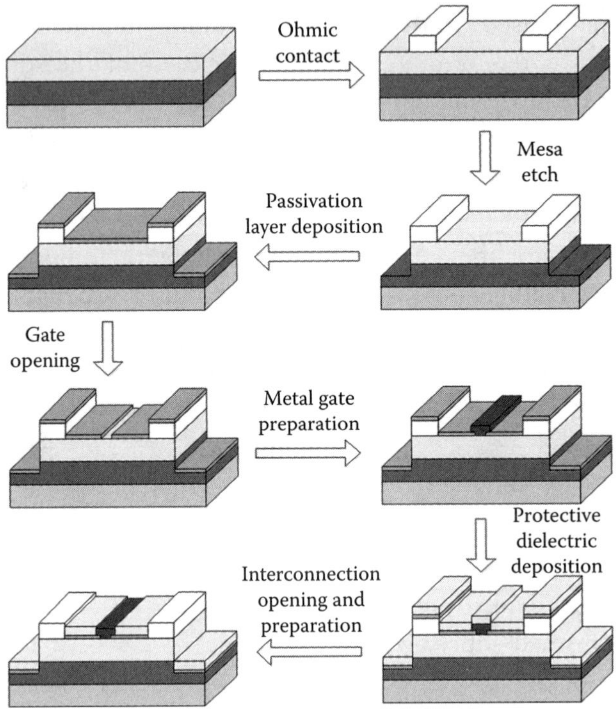

Figure 10.19 Schematic of HEMT preparation process flow. (Courtesy of Dr. Chong Wang.)

is completely etched off. The requirements are smooth mesa boundaries and steep sidewalls, and the measured mesa leakage current should be no greater than 100 nA/mm at 50 V bias.

4. Si_3N_4 passivation layer deposition is carried out by PECVD growth of Si_3N_4 passivation layer. A sacrifice wafer undergoing the same Si_3N_4 growth for process monitoring will be tested by the ellipsometer to make sure the thickness and the refractive index of silicon nitride are within the set range. The leakage current silicon nitride should also be examined to meet the device requirement.
5. Gate is prepared by surface treatment of the gate region after photolithograph followed by gate metal deposition. It is required that the gate edge should be smooth and free of adhesion or falling-off. The Schottky characteristics should be evaluated by measuring the forward and reverse current of the Schottky *C-V* test patterns on the wafer.
6. Protective Si_3N_4 layer preparation with the dielectric deposition and evaluation as same as those in step 4.
7. Interconnection is prepared by etching holes on the protective layer until the top layer metal of electrodes is exposed and then depositing or plating the interconnecting metal. For large gate width HEMTs air bridges are prepared in this step. Then the whole process flow is completed.

10.8 GaN HEMT Performance Evaluation

Methods for HEMT performance evaluation are presented with an actual HEMT as example. We employed stepper (lithography) to process a whole 2-in wafer of AlGaN/GaN HEMT dies on the SiC substrate. A 1.25 mm wide multifinger gate HEMT with 0.6 μm gate length and 4 μm source-drain spacing is bonded and packaged with its photographs shown in Figure 10.20. Tests show the gate fingers of the prepared device are intact with a small leakage current, and the device is used for microwave power measurement.

10.8.1 DC Performance

The direct current characteristics of a single-finger gate HEMT with a small gate width of 50 μm was measured using Agilent B1500A semiconductor parameter analyzer. The measured output and transfer characteristics are shown in Figure 10.21.

The maximum saturation current of the device is shown to be ~1160 mA/mm corresponding to a knee voltage of ~6 V. The self-heating effect of the device is rather

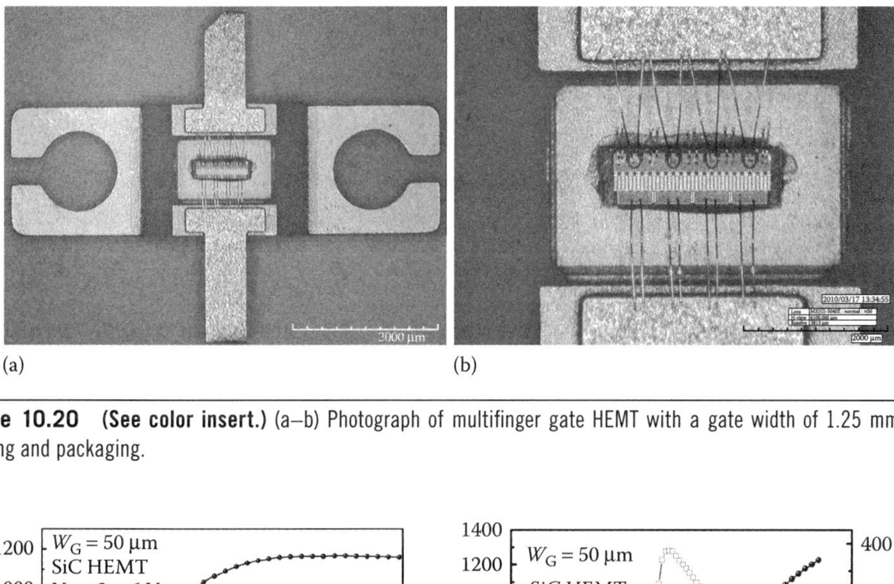

(a) (b)

Figure 10.20 **(See color insert.)** (a–b) Photograph of multifinger gate HEMT with a gate width of 1.25 mm after bonding and packaging.

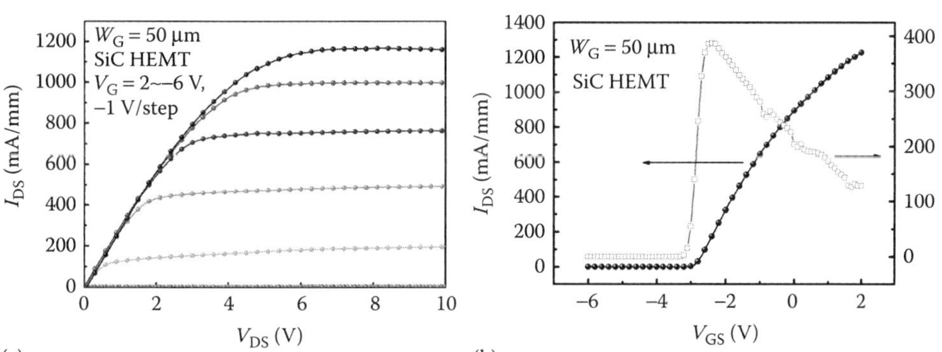

(a) (b)

Figure 10.21 (a) Output and (b) transfer characteristic curves of the single-finger gate HEMT.

small owing to the employment of the SiC substrate. The threshold voltage is approximately −3 V, the maximum transconductance is 388 mS/mm and the off-state leakage current is 1×10^{-4} mA/mm.

10.8.2 Small Signal Characteristics

The small signal measurements of a single-finger gate GaN HEMT with a gate length of 0.6 μm is performed using an Agilent 8363B vector network analyzer. The measured results in Figure 10.22 show a cut-off frequency of 19.6 GHz and a maximum oscillation frequency of 40 GHz.

10.8.3 Microwave Power Characteristics

Power measurements of the multifinger gate HEMT with 1.25 mm gate width shown in Figure 10.20 are performed at 4 GHz, and the results are given in Figures 10.23 and 10.24. The maximum output power density is 5.20 W/mm, the gain at 3 dB compression point is 14.10 dB and PAE = 58.60% with V_{DS} = 28 V and

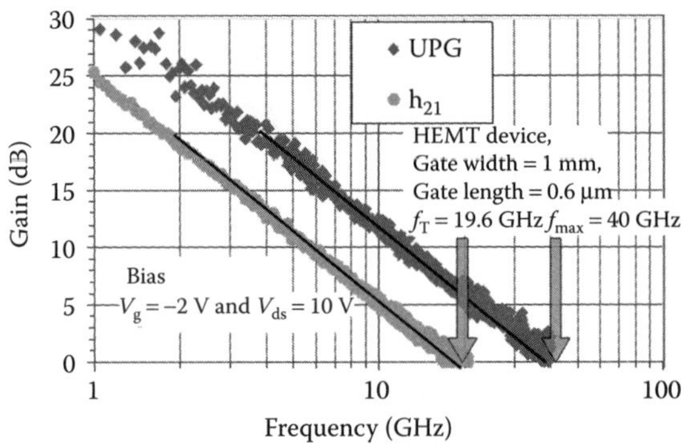

Figure 10.22 Small signal measurements of the GaN HEMT with a gate length of 0.6 μm.

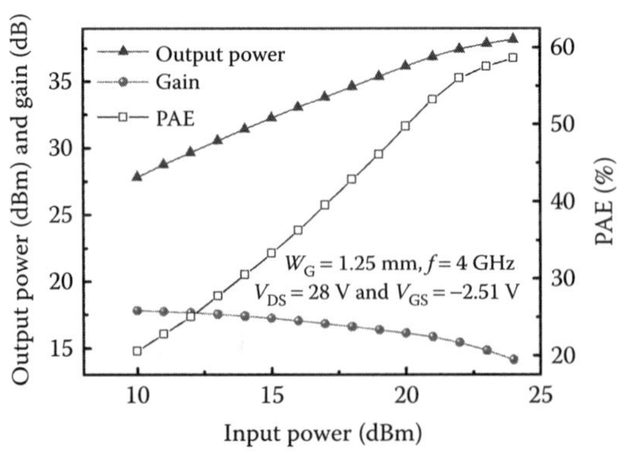

Figure 10.23 Power characteristics of the multifinger gate HEMT with 1.25 mm gate width at 4 GHz under bias conditions V_{DS} = 28 V and V_{GS} = −2.51 V.

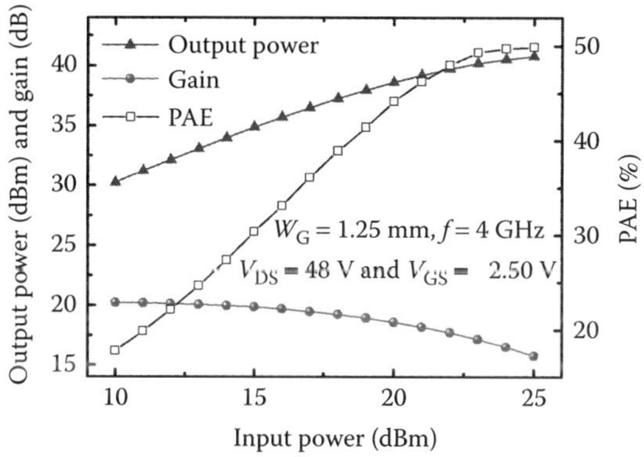

Figure 10.24 Power characteristics of the multifinger gate HEMT with 1.25 mm gate width at 4 GHz under bias conditions $V_{DS} = 48$ V and $V_{GS} = -2.50$ V.

$V_{GS} = -2.51$ V. When $V_{DS} = 48$ V, $V_{GS} = -2.50$ V is applied, the maximum output power density is 9.57 W/mm, the gain at 3 dB compression point is 15.78 dB and PAE = 49.88%.

With 1.25 mm gate width, the prepared GaN HEMT achieves a 4 GHz power density approaching 10 W/mm and a PAE of ~50%, demonstrating the feasibility of this process flow.

References

Adesida, I., A. Mahajan, E. Andideh, M. Asif Khan, D. T. Olsen, and J. N. Kuznia. 1993. Reactive ion etching of gallium nitride in silicon tetrachloride plasmas. *Applied Physics Letters* 63(20):2777–2779. doi: 10.1063/1.110331.

Benamara, Z., B. Akkal, A. Talbi, and B. Gruzza. 2006. Electrical transport characteristics of Au/n-GaN Schottky diodes. *Materials Science and Engineering C* 26(2–3):519–522. doi: 10.1016/j.msec.2005.10.016.

Hasegawa, H., T. Inagaki, S. Ootomo, and T. Hashizume. 2003. Mechanisms of current collapse and gate leakage currents in AiGaN/GaN heterostructure field effect transistors. *Journal of Vacuum Science & Technology B* 21(4):1844.

Liu, J., Y. Hao, Q. Feng, C. Wang, J.-C. Zhang, and L.-L. Guo. 2007. Characterization of Ni/Au GaN Schottky contact base on I-V-T and C-V-T measurements. *Wuli Xuebao/Acta Physica Sinica* 56(6):3483–3487.

Menozzi, R. 2004. Off-state breakdown of GaAs PHEMTs: Review and new data. *IEEE Transactions on Device and Materials Reliability* 4(1):54–62. doi: 10.1109/tdmr.2004.824353.

Mouffak, Z., N. Medelci-Djezzar, C. Boney, A. Bensaoula, and L. Trombetta. 2003. Effect of photo-assisted RIE damage on GaN. *MRS Internet Journal of Nitride Semiconductor Research* 8(1):55–58. doi: 10.1557/S1092578300000508.

Pearton, S. J., C. R. Abernathy, F. Ren, J. R. Lothian, P. W. Wish, and A. Katz. 1993. Dry and wet etching characteristics of InN, AlN, and GaN deposited by electron cyclotron resonance metalorganic molecular beam epitaxy. *Journal of Vacuum Science & Technology A: Vacuum, Surfaces, and Films* 11(4 pt 2):1772–1772. doi: 10.1116/1.578423.

Pearton, S. J., J. C. Zolper, R. J. Shul, and F. Ren. 1999. GaN: Processing, defects, and devices. *Journal of Applied Physics* 86(1):1–78. doi: 10.1063/1.371145.

Sawada, T., Y. Izumi, N. Kimura, K. Suzuki, K. Imai, S. W. Kim, and T. Suzuki. 2003. Properties of GaN and AlGaN Schottky contacts revealed from I-V-T and C-V-T measurements. *Applied Surface Science* 216(1–4):192–197. doi: 10.1016/s0169-4332(03)00440-9.

Shul, R. J., G. B. McClellan, S. A. Casalnuovo, D. J. Rieger, S. J. Pearton, C. Constantine, C. Barratt, R. F. Karlicek, Jr., C. Tran, and M. Schurman. 1996. Inductively coupled plasma etching of GaN. *Applied Physics Letters* 69(8):1119–1121. doi: 10.1063/1.117077.

Wang, C., Y. Hao, Q. Feng, and L.-L. Guo. 2006. New development in dry etching of GaN. *Semiconductor Technology* 31(6):409–413.

Yang, Y., W. Wang, and Y. Hao. 2006. Ohmic contact to an AlGaN/GaN heterostructure. *Chinese Journal of Semiconductors* 10:1823–1827.

Zhang, H., E. J. Miller, and E. T. Yu. 2006. Analysis of leakage current mechanisms in Schottky contacts to GaN and Al0.25Ga0.75N/GaN grown by molecular-beam epitaxy. *Journal of Applied Physics* 99(2):023703-6.

11

Electrical and Thermal Degradation and Reliability of GaN HEMTs

Gallium nitride high electron mobility transistors (HEMTs) find major applications in microwave and millimeter wave power devices and high-voltage power electronic devices, where they are subject constantly to strong electric fields and currents, and high junction temperature and the working conditions in high temperature applications may also affect the performance of GaN HEMTs. Therefore, the degradation mechanisms of the GaN HEMT under electrical stress and high temperature are of great significance. For GaN HEMTs, the degradation under electrical stress is of two types: the reversible and the irreversible. The reversible degradation is generally attributed to current collapse. The irreversible degradation and current collapse may occur successively or simultaneously in the same electrical and thermal stress measurements.

11.1 Current Collapse in GaN HEMTs

In early studies of the reliability of GaN HEMTs, the performance was severely restricted by current collapse or current slump. This phenomenon was given this name owing to the observation of the decrease in saturation current density and maximum transconductance, together with increase in knee voltage and on resistance after relatively high voltage surge in the direct current (DC) performance measurements (Khan et al. 1994). However, the primary damage lies in the dramatic drop in the output current amplitude of HEMTs, driven by high-frequency large signals as compared with the DC characteristics. This results in the decrease of output power density and power-added efficiency, which is referred to as radio frequency (RF) dispersion (Daumiller et al. 2001). As shown in Figures 11.1 and 11.2, the HEMTs can still operate stably after current collapse but with significantly lower performance.

The device performance instability related to current collapse has various manifestations. In the alternating current (AC) small-signal characteristics, dispersion of transconductance and drain conductance at different frequencies was observed (Stoklas et al. 2008). When a stress of high drain voltage or strong reverse gate bias is applied on the HEMT, there is a continual drop in the drain current (if the channel is on) during the stress which forms a slow transient with time, and the DC and RF performance of the device after stress worsens dramatically (Koley et al. 2003). In the *I-V* characteristics measurements by voltage scanning with pulsed signals (which can be applied either on

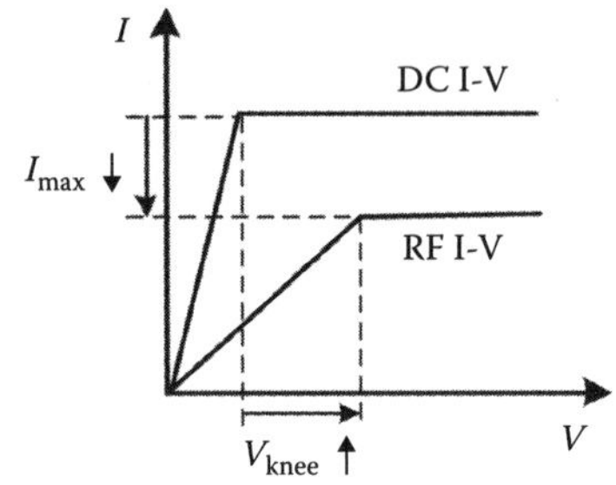

Figure 11.1 Schematic of current collapse.

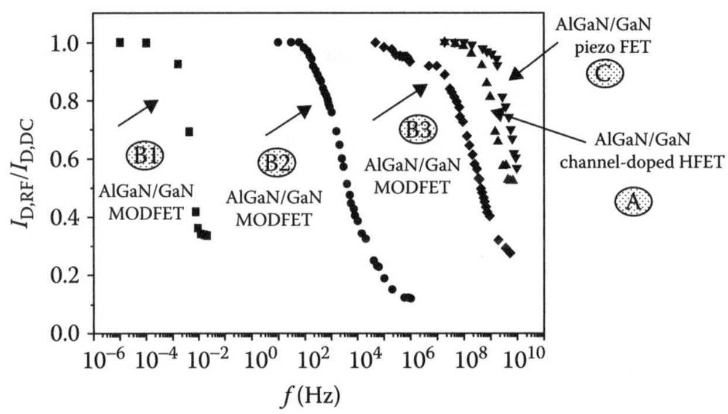

Figure 11.2 Large signal current RF dispersion in different kinds of GaN HEMTs. (Reproduced from Daumiller, I. et al., *IEEE Elect. Device Lett.*, 22(2), 62–64, 2001.)

the drain or on the gate with constant voltages on the other two electrodes), one of the two pulse levels is the constant set voltage and the other is the voltage to be measured V_a in continual scanning mode, and current collapse makes the drain current response smaller than the current value corresponding to V_a in DC measurements (Augaudy et al. 2001). The response current transients under pulse signal observed by oscillograph indicate that the reason for this is the fact that the channel cannot quickly switch on, thus its name drain lag (induced by pulse applied on drain) or gate lag (induced by pulse applied on gate) (Tirado et al. 2007).

Numerous experiments have shown that current collapse is reversible and that illumination can accelerate recovery from current collapse. Therefore, current collapse is in essence the effect of the traps in the material whose physical mechanism can be successful interpreted by the virtual gate model shown in Figure 11.3 (Vetury et al. 2001). Simply put, the capture of electrons by traps reduces the channel electrons as well as raises the energy band leading to further depletion of the channel layer, thereby forming a "virtual gate" that can control the channel current. Because it takes some time for these surface states to charge and discharge, transients are generated under DC or stress conditions and the variation of current cannot catch up with RF signal frequency, leading to decreased output power density and power-added efficiency and the formation of collapse. The traps that serve as virtual gates can be surface states or traps in the barrier layer or the buffer layer. On account of the

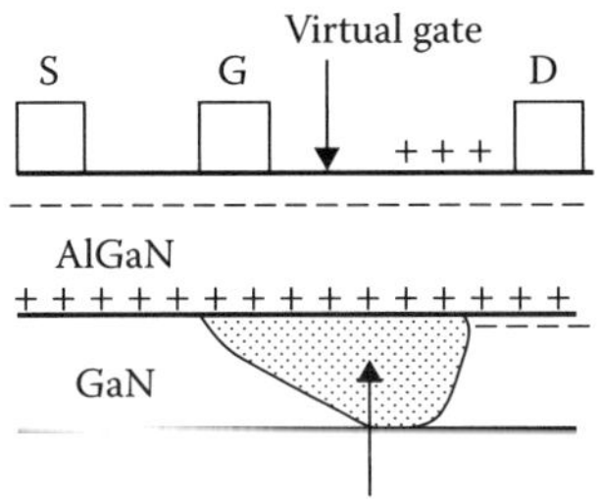

Figure 11.3 Simple schematic of the virtual gate model. (Reproduced from Vetury, R. et al., *IEEE Trans. Elect. Device*, 48(3), 560–566, 2001.)

strong correlation between current collapse and the surface condition (Kamiya et al. 2007), surface traps are considered the most probable one in the virtual gate model.

Besides improving the material quality, current collapse in GaN HEMTs can also be suppressed effectively by surface passivation and the addition of field plates. A common method for characterizing current collapse in HEMT samples is the pulsed *I-V* measurements. Besides the application of pulse signal on a single electrode for scanning, synchronous scanning pulses can be applied on both gate and drain, a measurement analogous to the microwave large signal operation for more accurate prediction of the RF dispersion.

11.2 Three Models of Electrical Degradation Mechanism for GaN HEMTs

Although current collapse is reversible, irreversible degradations such as the drain current and transconductance reduction, threshold voltage drift, and gate leakage current increase may occur in GaN HEMTs after long-time large-signal operations, causing device failure. The major means to investigate the degradation mechanisms of the operating HEMT is the analysis of degradation characteristics, with electrical stress applied on the device. Studies show that the GaN HEMT failure is due mainly to the strong electric field in the large-signal operation with three major models of the failure mechanism: the hot electron injection, the inverse piezoelectric effect, and the gate electron injection.

11.2.1 Hot Electron Injection

It is proposed in the hot electron injection model that the channel electrons in HEMT devices are accelerated by the strong electric field under on-state, off-state, or radio frequency (RF) electrical stress, and some become high-energy hot electrons. These hot electrons transfer in real space to spill over the quantum wells in the channel and are captured by surface traps or buffer layer traps, thus resulting in a reduced channel electron density with decreased leakage current and transconductance. Such high-energy hot electrons also collide with crystal lattice to produce new defects that aggravate the degradation.

11.2.2 Gate Electron Injection

In the study of current collapse, an influential proposal on how electrons are captured by traps is that the injected electrons onto the AlGaN barrier layer surface by the gate current increase the negative charge on the surface and charge the virtual gate, thus reducing the channel electron gas density and leading to current collapse. Trew et al. (2006) suggested that with the effect of the electric field peak at the gate edge near the drain, the gate electrons are injected onto the barrier layer surface and then the gate-drain leakage current is generated via the hopping conduction between surface traps. The gate current charges the surface traps, leading to a raised barrier layer potential and reduced density of the channel electrons, thus resulting in a drop of drain current and transconductance. Gate electron injection may both induce irreversible degradation of the Schottky gate on long-term basis and play an assisting role in the irreversible degradation of GaN HEMTs caused by hot electron injection or inverse piezoelectric effect.

11.2.3 Inverse Piezoelectric Effect

AlGaN/GaN heterostructures possess relatively strong lattice strain and piezoelectric effect, that is, the presence of lattice strain induces electric fields in the crystal. When the crystal is under the influence of the electric field, crystal lattice strain is induced in turn. This is referred to as the inverse piezoelectric effect, based on which the inverse piezoelectric degradation model is put forward. In the early study of current collapse, Simin et al. (2001) has attributed it to the formation of new strains in AlGaN barrier layer that modulate the electron gas density in the channel due to the inverse piezoelectric effect under strong RF alternating voltage. However, quantitative analysis indicates that the strain induced by the inverse piezoelectric effect is not large enough to account for the observed current collapse. Joh et al. (Joh and Del Alamo 2006) found in their experiments on GaN HEMTs that the device performance degradation depends mainly on the electric field rather than on the current. They proposed that the strong electric field in the AlGaN barrier layer causes the barrier lattice to expand, leading to lattice relaxation, which generates new lattice defects acting as electron traps.

The schematic diagram of the inverse piezoelectric effect model is presented in Figure 11.4. At the gate edge on the drain side exists a strong electric field, under the

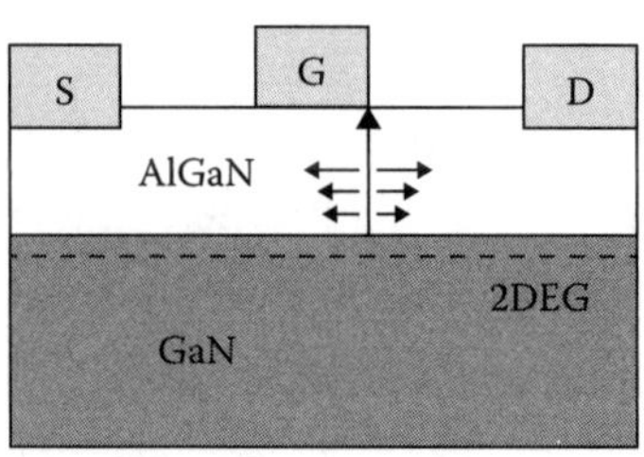

Figure 11.4 Schematic of inverse piezoelectric effect. (Reproduced from Joh, J. et al., Gate current degradation mechanisms of GaN high electron mobility transistors, in *2007 IEEE International Electron Devices Meeting, IEDM*, December 10–December 12, 2007, Washington, DC.)

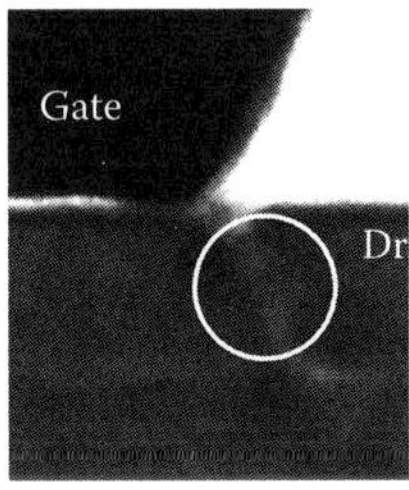

Figure 11.5 TEM image of the drain-side edge of the gate after experiencing strong electric field. (Reproduced from Jimenez, J.L. and Chowdhury, U., X-band GaN fet reliability, in *46th Annual 2008 IEEE International Reliability Physics Symposium, IRPS*, April 27, 2008–May 1, 2008, Phoenix, AZ.)

effect of which the crystal lattices are stretched, owing to the inverse piezoelectric effect, until the lattice structures are broken to generate lattice defects. The transmission electron microscope (TEM) image in Figure 11.5 shows that the device exposed to the strong electric field for a long time develops a tiny crack under the drain-side gate edge, because the crystal lattices are strained, until they are broken under the inverse piezoelectric effect (Jimenez and Chowdhury 2008).

All three models discussed above can be employed to explain device degeneration, but none of them can individually account for all the degeneration phenomena. For example, in the case of off-state stress, where the gate voltage is more negative than the threshold voltage and the channel electrons are depleted, degeneration still arises under the effect of drain-source voltage for a long time, which cannot be explained by only hot electron injection. Valizadeh and Pavlidis (2005) made a detailed analysis of the performance degradation of AlGaN/GaN HEMTs with AlGaN barrier layers of various aluminum compositions and found that no performance parameters, other than the threshold voltage, vary with the aluminum composition. This cannot be accounted for by inverse piezoelectric effect. The variation of the aluminum composition leads to a changed AlGaN lattice constant, thus resulting in changed barrier layer lattice strain and its resistance to the inverse piezoelectric effect, which means the device degradation should be a function of aluminum composition under pure inverse piezoelectric effect. Meanwhile, experiments indicate the AlGaN/GaN HEMT performance degradation is dependent on the heterostructure, the bias, and the field plate and trench conditions, and gate electron injection solely cannot interpret so many experimental observations.

In fact, the specific degradation model of a device may vary greatly with the device structure and stress condition. Only when a clear distinction is made between the effects of different factors on the electrical degradation, there is a good understanding of the innate mechanism of electrical degradation possible for devising suppression measures from the design and processing.

11.3 Electrical Stress Degradation of GaN HEMTs (I)

To make a distinction between the three degradation models, we applied different bias conditions on the device for a long time, for each model to function and be analyzed independently in search of effective countermeasures. The sample devices used in our experiment consist of the AlGaN/GaN heterostructures on sapphire substrate, with the AlN nucleation layer grown first at low temperature, followed by the growth of 1.6 μm undoped GaN buffer layer and 23-nm-thick AlGaN barrier layer at a higher temperature. The AlGaN barrier layer consists of 5 nm undoped AlGaN, 10 nm Si-doped (doping concentration is about 2×10^{18} cm^{-3}) AlGaN, and 6 nm undoped AlGaN. The aluminum composition of the AlGaN barrier layer is 30%. The HEMT device has a gate length of 1 μm, a gate width of 100 μm, and a source-drain spacing of 4 μm, with the gate right in the middle of the source and the drain.

11.3.1 Hot Electron Injection Stress

For a relatively independent analysis of the mechanism of device degradation caused by channel hot electron injection, a drain-source electrical stress experiment was performed, with the gate suspended and only a constant enduring voltage (V_{DS} = 20 V) applied between the drain and the source, in which case only the channel hot electrons have an effect on the device, without any influence from gate electron injection. The stress time is 3×10^4 s. The output, transfer, off-state leakage current, and Schottky gate (gate-drain diode) characteristics of the HEMT before and after stress are shown in Figure 11.6.

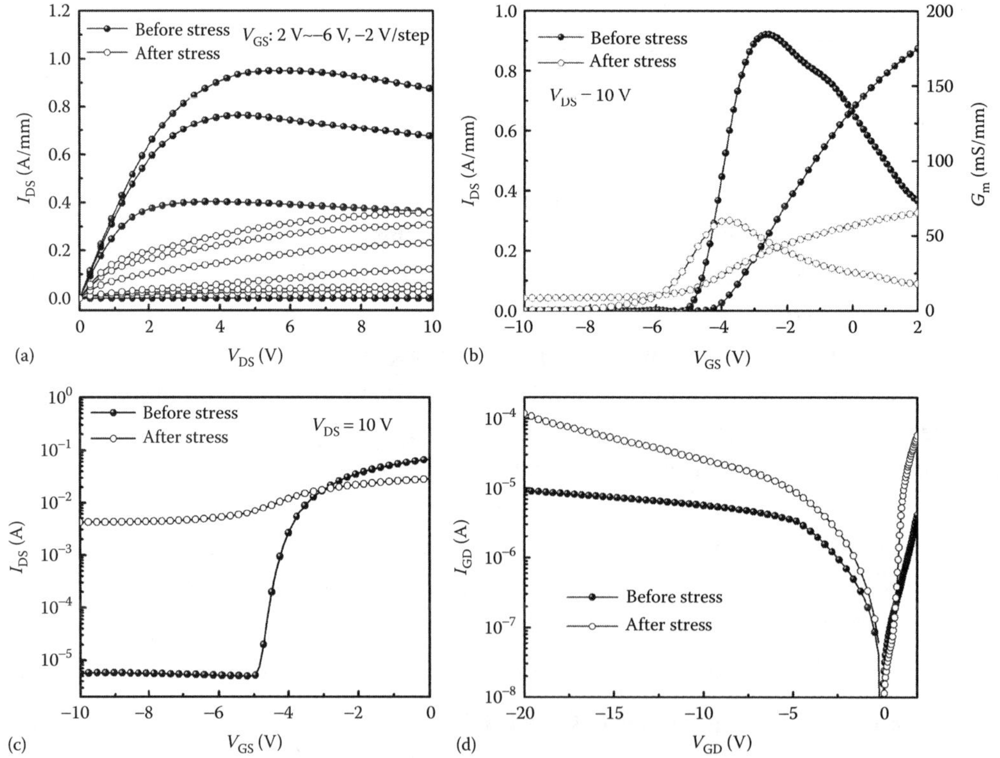

Figure 11.6 *I-V* characteristic curves of the device before and after stress. (a) Output characteristics, (b) transfer characteristics, (c) off-state leakage current curves, and (d) Schottky characteristic curves.

From Figure 11.6, one can observe the following degradation behaviors after stress. The worst case of degradation is the inability of the channel to pinch off, with the off-state drain current significantly increased by almost three orders of magnitude. Then the saturation current and maximum transconductance decrease by over 50%, the Schottky reverse leakage current rises by one order of magnitude, and the forward current also increases dramatically. The significant increase of both forward and reverse current suggests that the hot electrons are injected into the AlGaN layer, causing damage to the Schottky gate and into the traps, leading to increased gate leakage current due to trap-assisted tunneling, thereby causing a drastically weakened gate control. The real space transfer of channel hot electrons reduces the two-dimensional electron gas (2DEG) density. Meanwhile, the hot electrons are injected into the GaN buffer layer, thus increasing the number of bulk electrons of lower mobility, which may be captured or released by the deep- or shallow-level traps to form the leakage current, and the possible lattice damage may also induces 2DEG mobility degradation.

11.3.2 Gate Electron Injection Stress

The same idea for the hot electron injection stress experiment was employed in the investigation of the gate-electron-injection-induced device degradation, that is, the gate-drain electrical stress experiment was performed with the source suspended, with only a constant enduring voltage applied between the drain and the gate. Since there is no current in the channel, the hot electron effect can be excluded, with only the influence of gate electrons in consideration. The stress conditions are the suspended source, the applied voltage stress between gate and drain, $V_{GD} = -20$ V, and the stress time, 3×10^4 s. The output, transfer, off-state leakage current, and Schottky gate (gate-drain diode) characteristics of the HEMT before and after stress are shown in Figure 11.7.

The following degradation behaviors after stress can be deduced from Figure 11.7. The Schottky reverse leakage current approximately doubles with a slight rise in the forward current, and the maximum transconductance decreases by over 15% and the off-state drain current increases by 20% with the threshold voltage almost invariant, and the device saturation current drops by 9%. It is suggested that the gate electrons are driven into the barrier layer surface traps through tunneling by the strong electric field caused by the reverse bias applied on the Schottky gate and then generate the gate-drain leakage current via the hopping conduction between the surface traps, leading to the increase in off-state drain current. No obvious change in the threshold voltage is observed before and after stress, because the gate electron injection has no influence on the 2DEG density under the gate. The charged surface traps between the gate and the drain lower the channel electron density below, thus causing a decreased current output and transconductance.

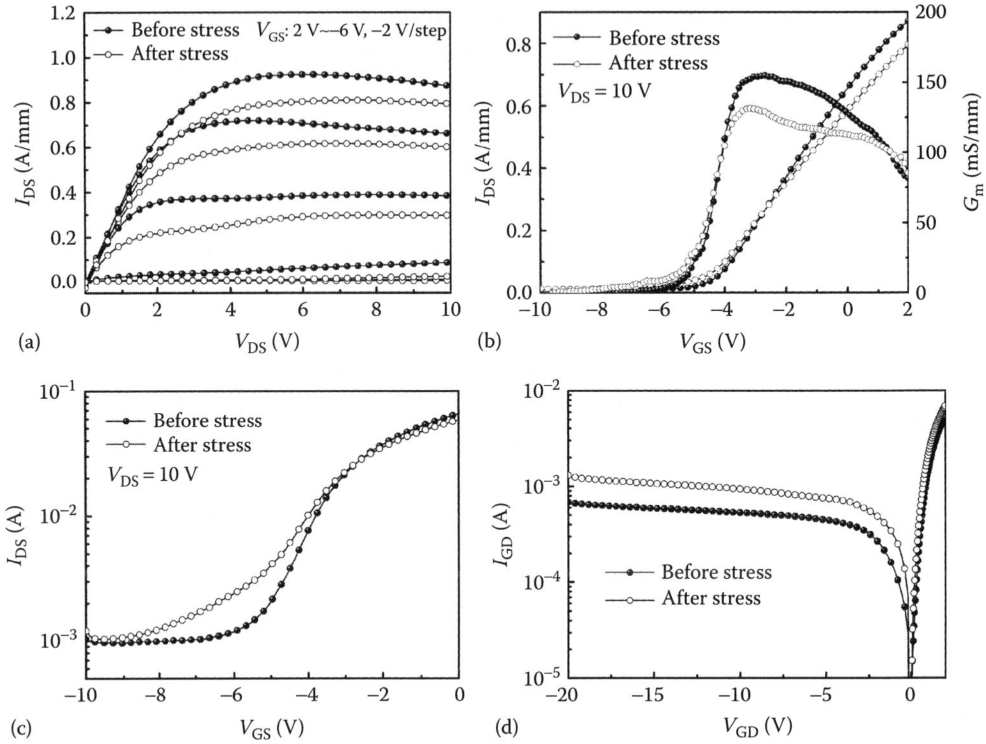

Figure 11.7 *I-V* characteristic curves of the device before and after stress. (a) Output characteristics, (b) transfer characteristics, (c) off-state leakage current curves, and (d) Schottky characteristic curves. (Courtesy of Tingting Li.)

11.3.3 Stepped Gate Voltage Stress with $V_{DS} = 0$

AlGaN/GaN HEMTs with 2 nm GaN cap layer and without GaN cap layer were employed, with short time stepped stress applied: the drain voltage and source voltage are 0 V, whereas the gate voltage increases from 0 V to −40 V in steps of −2 V, with 2 min continuous stress for each step. The short period stress is used to exclude the influence of gate electron injection, and the stepped stress is adopted because for the inverse piezoelectric effect to occur, there is a critical voltage (Joh et al. 2007), and only at this point, that is, when the electric field in the material exceeds a specific value, lattice relaxation will be induced to produce defects, thus causing device degradation. This critical voltage point may be found by observing the *I-V* characteristics degradation under the step mode electric field. Of the *I-V* characteristics, special attention is paid to the variation of I_{Dmax} (I_D@ V_{DS} = 5 V, V_{GS} = 2 V) and I_{Goff} (I_G@ V_{DS} = 0 V, V_{GS} = −10 V) with incremental stress. The experimental results are presented in Figure 11.8. I_{Dmax0} is the initial value of I_{Dmax}.

The following degradation behaviors are observed from Figure 11.8. Evident degradation of the AlGaN/GaN HEMT without GaN cap layer occurs when the gate

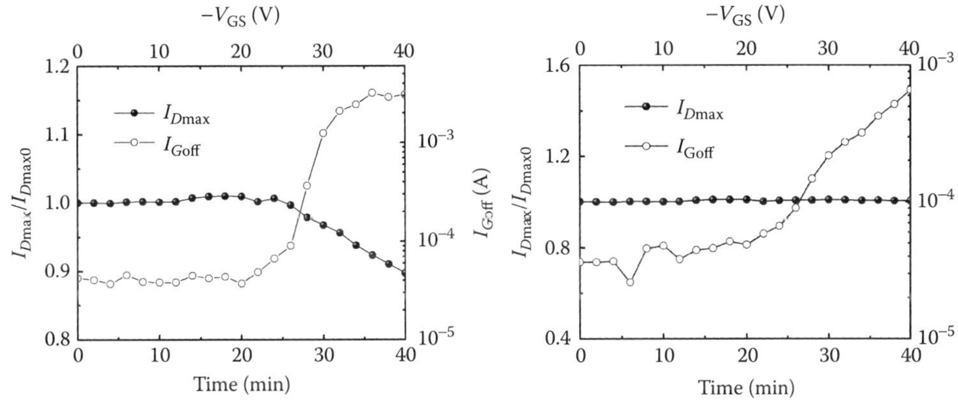

Figure 11.8 Variation of I_{Dmax} and I_{Gof} under stress. Note that V_{GS} increases in the negative direction with longer stress time. (a) Without GaN cap layer and (b) with GaN cap layer. (Courtesy of Tingting Li.)

voltage V_{GS} increases in the negative direction to −24 V, with I_{Dmax} reduced by ~12% and I_{Goff} increased by approximately two orders of magnitude. For the AlGaN/GaN HEMT with GaN cap layer, I_{Goff} begins to increase when V_{GS} reaches −26 V, with no visible variation in I_{Dmax}. The degradation of the AlGaN/GaN HEMT without GaN cap layer is consistent with that under inverse piezoelectric effect, whereas the AlGaN/GaN HEMT with GaN cap layer has an almost invariant I_{Dmax}, which demonstrates the good suppression of the inverse piezoelectric effect by the GaN cap layer.

AlGaN barrier layers grown on GaN are subject to the tensile strain from GaN owing to the greater in-plane lattice constant of GaN than that of AlGaN. The GaN cap layer is favorable for maintaining such strain to prevent strain relaxation in AlGaN. In our discussion of the influence of the cap layer on AlGaN/GaN heterostructure properties in Section 5.6.3, the AlGaN relaxation degree R in the heterostructure without GaN cap layer is estimated to be 6.5% and that with GaN cap layer is expected to be 0% by reciprocal space mapping of the (105) reflection, which has experimentally validated the ability of the GaN cap layer to effectively suppress AlGaN relaxation. This is also the reason why the GaN cap layer can suppress the inverse piezoelectric effect.

11.4 Electrical Stress Degradation of GaN HEMTs (II)

In the actual operation of HEMTs, bias voltages are applied on all three electrodes, so it is necessary to analyze the GaN HEMT degradation under electrical stress with all three electrodes biased. The material used to fabricate the HEMT is the AlGaN/GaN heterostructure with 2 nm GaN cap. The device is not passivated, with a ratio of gate width to length, $W/L = 25/1$ μm, and a source-drain spacing of 4 μm.

11.4.1 High Drain Voltage On-State Stress

The HEMT characteristics before and after high drain voltage stress ($V_{GS} = 0$ V, $V_{DS} = 20$ V) 1×10^4 s applied on the three electrodes are shown in Figure 11.9 (Gu et al. 2009a). The maximum saturation output current I_{Dsat} drops by 30.9%, and the corresponding transconductance peak $g_{m\max}$ decreases by 18.4% with a positive threshold voltage shift V_{TH} of 9.3%. With on-state stress applied, the transconductance g_m under both low and high gate voltage reduces notably.

The degradation of major device parameters as a function of stress time is given in Figure 11.10, where the variation of the maximum saturation current density I_{Dsat} versus stress time in log-log coordinates shows a linear pattern, that is, satisfies the power law model: (Cao et al. 2007)

$$I_{Dsat}\% = C_1 \times t_{stress}{}^{\beta_1} \tag{11.1}$$

where C_1 and β_1 are constants.

An analysis is made of the influence of the transverse electric field on the high field degradation, with fixed gate voltage stress ($V_{Gstress} = 0$ V) and varied drain voltage

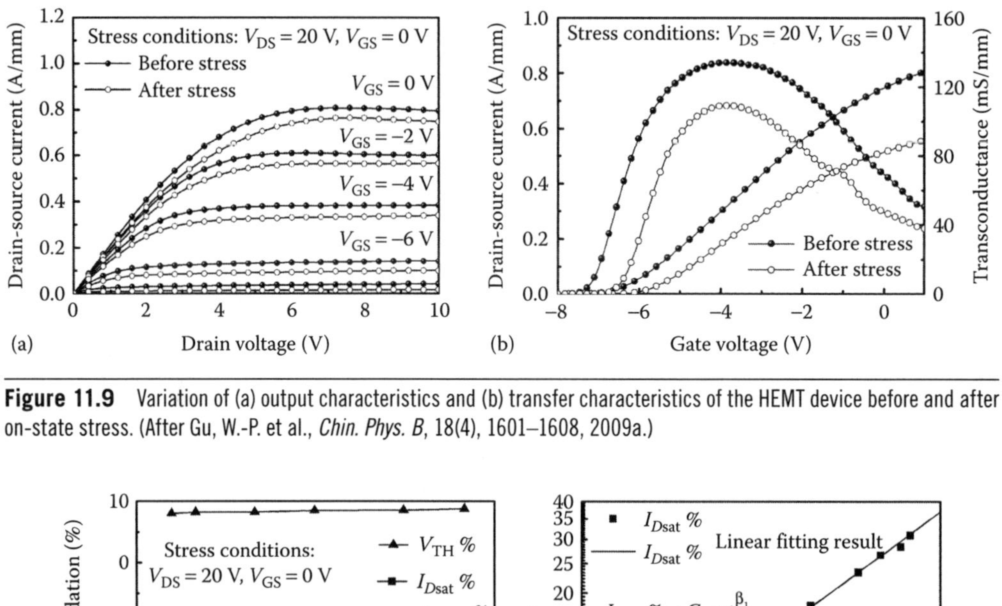

Figure 11.9 Variation of (a) output characteristics and (b) transfer characteristics of the HEMT device before and after on-state stress. (After Gu, W.-P. et al., *Chin. Phys. B*, 18(4), 1601–1608, 2009a.)

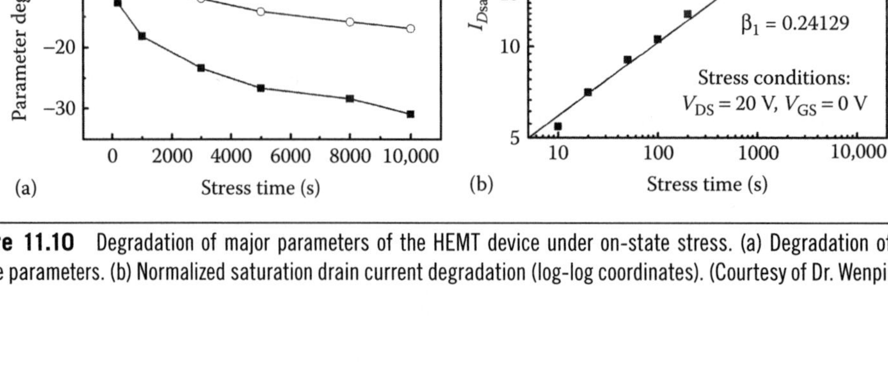

Figure 11.10 Degradation of major parameters of the HEMT device under on-state stress. (a) Degradation of major device parameters. (b) Normalized saturation drain current degradation (log-log coordinates). (Courtesy of Dr. Wenping Gu.)

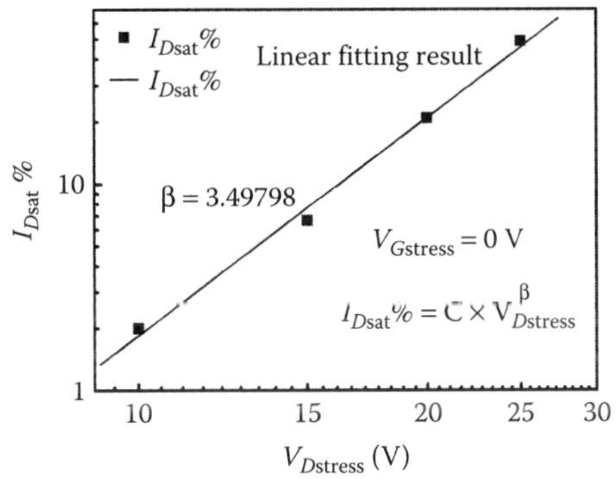

Figure 11.11 Drift of saturation drain current I_{Dsat} versus drain voltage stress $V_{Dstress}$. (Courtesy of Dr. Wenping Gu.)

stress $V_{Dstress}$ applied on the device. The drift of device saturation drain current I_{Dsat} with the drain voltage stress $V_{Dstress}$ (fixed $V_{Gstress} = 0$ V) in log-log coordinates, shown in Figure 11.11, reveals a near linear pattern, that is, satisfies the power law model.

$$I_{Dsat}\% = C \times V_{Dstress}^{\beta} \quad (11.2)$$

where C and β are constants.

The I_{Dsat} degradation conforms to the power law model as a function of the stress time and the drain stress voltage, respectively, suggesting that hot electron injection is the dominant degradation mechanism under high drain voltage on-state stress. We also investigated the device performance degradation induced by the high drain voltage on-state stress ($V_{GS} = 0$ V, $V_{DS} = 20$ V) applied for 10^4 s at room temperature and 100°C, respectively (not presented in this book), and found that the degradation is smaller under high temperature stress, which is also attributed to the suppressed channel hot electrons generation by the high temperature.

11.4.2 Gate-Drain High Voltage Stress: Off States and On States

With gate-drain high voltage off-state stress applied on similar HEMTs, the device characteristics before and after stress are given in Figure 11.12. It is observed that

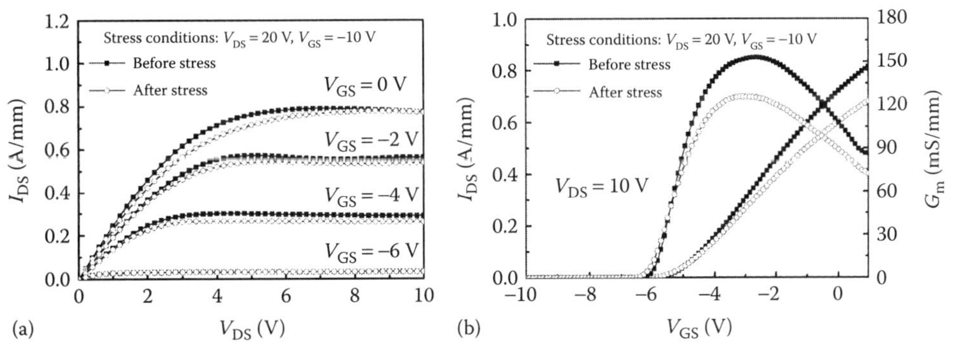

Figure 11.12 HEMT performance degradation under off-state stress. (a) Output characteristics and (b) transfer characteristics. (Courtesy of Dr. Wenping Gu.)

after stress, the drain current decreases, with more evident variation at higher gate voltage and mostly before the drain current saturates, but no visible change in V_{TH} is seen. I_{Dsat} drops by 16.2% and g_{mmax} by 16.9% after stress, almost as great as the degradation under high drain voltage on-state stress, but the g_m degradation is rather small under low gate voltage. The degradation of major device parameters as a function of stress time is given in Figure 11.13, where the variation of I_{Dsat} with stress time also follows the power law model.

The effect of the longitudinal electric field on high-field degradation is investigated by changing the gate voltage stress $V_{Gstress}$ from the off state ($V_{Gstress} = -10$ V) to the strong on- state ($V_{Gstress} = 2$ V), with fixed drain voltage stress ($V_{Dstress} = 20$ V) applied on the device. The drift of the HEMT saturation drain current I_{Dsat} with $V_{Gstress}$ (fixed $V_{Dstress} = 20$ V) in linear coordinates is shown in Figure 11.14. Different from that in Figure 11.13, the I_{Dsat} drift deceases first and then increases with higher $V_{Gstress}$, the turning point being in the vicinity of V_{TH}. It should be pointed out that this does not agree with the conclusion of Meneghesso et al. (2008), who found in their stress tests a bell-shaped variation curve of electroluminescence intensity of the GaN HEMT with constant V_{DS} and varied V_{GS}, with the most significant device

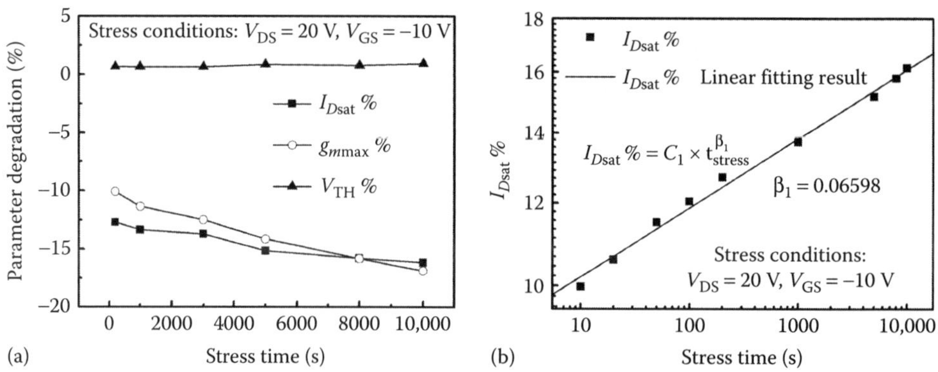

Figure 11.13 Degradation of major parameters of the HEMT device under off-state stress. (a) Degradation of major device parameters. (b) Normalized saturation drain current degradation (log-log coordinates). (Courtesy of Dr. Wenping Gu.)

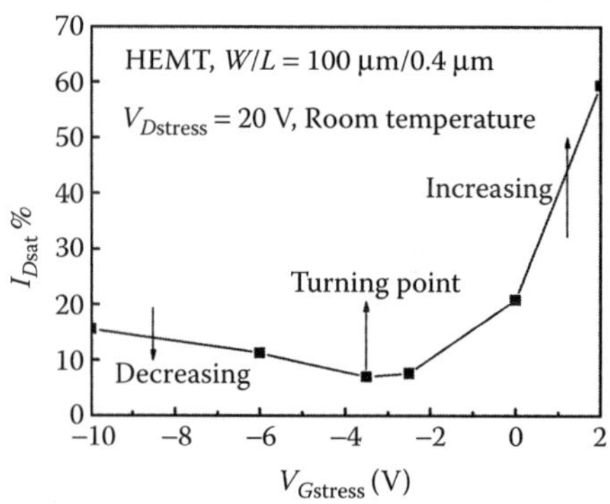

Figure 11.14 Drift of saturation drain current I_{Dsat} versus gate voltage stress $V_{Gstress}$. (Courtesy of Dr. Wenping Gu.)

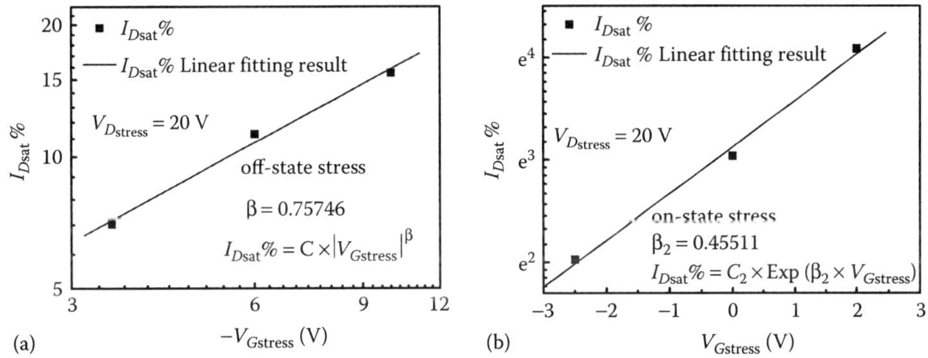

Figure 11.15 Drift percentage of the HEMT saturation drain current I_{Dsat} versus the gate voltage stress $V_{Gstress}$. The curve in (a) is the variation of I_{Dsat} drift with the absolute value of off-state $V_{Gstress}$ in log-log coordinates, and the curve in (b) is the variation of I_{Dsat} drift with on state $V_{Gstress}$ in single logarithmic coordinates. (Courtesy of Dr. Wenping Gu.)

degradation occurring in the semi-on state, that is, in the vicinity of V_{TH}. In our experiment, the degradation in the vicinity of V_{TH} turns out to be the smallest.

The off-state I_{Dsat} drift with $V_{Gstress}$ (absolute value) in log-log coordinates and on-state I_{Dsat} drift with $V_{Gstress}$ (fixed $V_{Dstress} = 20$ V) in single logarithmic coordinates are shown in Figure 11.15 (a) and (b), respectively. It is shown that in the off state ($V_{Gstress} < V_{TH}$), the drift of I_{Dsat} has an approximately linear relation to the absolute value of $V_{Gstress}$ in log-log coordinates, that is, fits in the power law model:

$$I_{Dsat}\% = C \times \left|V_{Gstress}\right|^{\beta} \tag{11.3}$$

where C and β are constants.

However, in the on state ($V_{Gstress} > V_{TH}$), the drift mobility of I_{Dsat} has an approximately linear relation to $V_{Gstress}$ in single logarithmic coordinates, as shown in Figure 11.15b, that is, fits in the exponential model:

$$I_{Dsat}\% = C_2 \times Exp\left(\beta_2 \times V_{Gstress}\right) \tag{11.4}$$

where C_2 and β_2 are constants.

The drift of device parameters versus $V_{Gstress}$ fits in different models under gate-drain high voltage off-state and on-state stresses, suggesting different degradation mechanisms in the on and off states. The HEMT degradation under off-state stress (Figure 11.12) is similar to that in Figure 11.9a and b, with I_{Dsat} degradation satisfying the power law model, versus stress time and the absolute value of the gate stress voltage, respectively, which indicates that the primary degradation mechanism under gate-drain high voltage off-state stress is gate electron injection. As is discussed above, channel hot electrons play a dominant role under on-state stress. Since the longitudinal electric field between the gate and the drain prevents the channel hot electron injection into the barrier layer and the gate, degeneration worsens, as this electric field drops with rising gate voltage.

11.4.3 Pulse Stress

The HEMT performance degradation is more complicated in the pulse operation where the device state shifts repeatedly between the on and off states. The experimental results for gate pulse stress are shown in Figure 11.16. The gate voltage is periodically pulsed, with electrical level values = (−10 V and 0 V), the drain voltage = 20 V, and the stress time = 10^4 s (Gu et al. 2009b). On the one hand, the drain current reduces after pulse stress and mostly before the saturation with drain voltage, which is characteristic of gate electron injection. On the other hand, I_{Dsat} decreases by 30% and corresponding $g_{m\max}$ by 19.2% with a V_{TH} positive shift of 10.2%, as well as obvious g_m degradation at both low and high gate voltages, all of which are the distinguishing features of hot electron injection. The degradation of major device parameters as a function stress time is given in Figure 11.17, where the I_{Dsat} variation with stress time also follows the power law model, with the power exponent β closer to that under on-state stress.

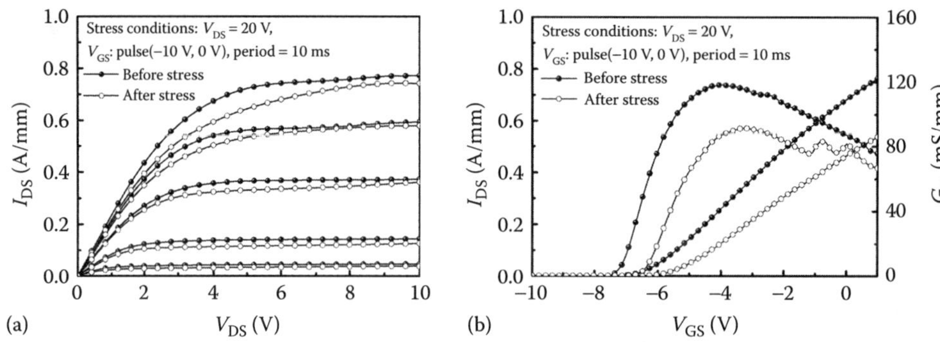

Figure 11.16 HEMT performance degradation under pulse stress. (a) Output characteristics and (b) transfer characteristics. (Courtesy of Dr. Wenping Gu.)

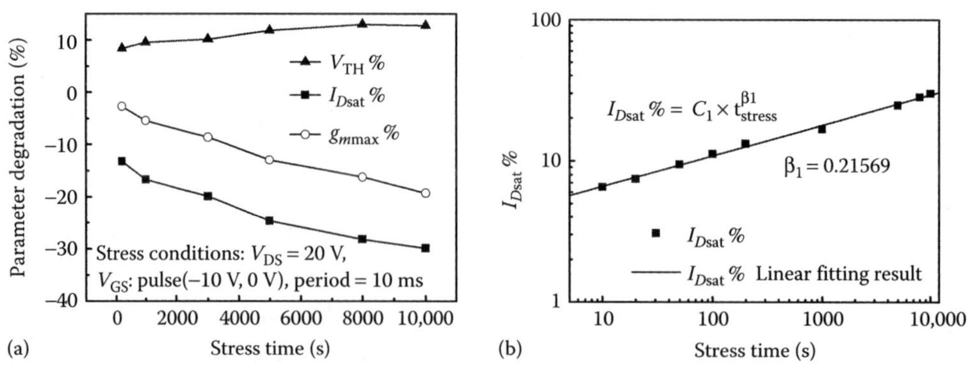

Figure 11.17 Degradation of major parameters of the HEMT device under pulse stress. (a) Degradation of major device parameters. (b) Normalized saturation drain current degradation (log-log coordinates). (Courtesy of Dr. Wenping Gu.)

Both transverse and longitudinal strong electric fields exist in the pulse-stressed HEMT, and the performance degradation exhibits the features of both hot electron injection and gate electron injection, and the former is more evident. Therefore, the degradation mechanism may be a coupled model of both hot electron injection and gate electron injection under pulse stress, which is analogous to the microwave large signal operation. Owing to the presence of a critical value of electric field to trigger degradation under the inverse piezoelectric effect, new degradation phenomena may result from the inverse piezoelectric effect if a strong-enough electric field is induced by the electrical stress applied on the HEMT discussed in this section.

11.4.4 Suppression of Electrical Stress Degradation of GaN HEMTs

The electrical stress degradation of GaN HEMTs is related to the initial material quality, device structure, and device process. The long-term reliability problems of HEMTs can be generally described in Figure 11.18, based on the available international experimental reports, showing that the device degradation mechanisms have much to do with both the stress bias and the temperature. GaN HEMT degradation can be divided into the early-stage degradation and the long-term degradation. The device degradation discussed in this section is in the early stage owing to the limit of experiment samples and conditions.

In general, the early-stage HEMT degradation is attributed primarily to electron traps, and thus, the reliability can be improved with better material quality and passivation process, but the long-term degradation is linked closely to the inverse piezoelectric effect and can be improved by optimizing the material structure and device structure. Be it early-stage or long-term degradation, the device reliability can be improved from these aspects.

1. Improve the material growth process to reduce the dislocation density and the background impurity concentration of the buffer layer for better material quality. Since the triggering of defects by hot electrons depends on the material quality, a high dislocation density leads to a greater gate leakage current and higher possibility for the hot electrons to create new defect centers nearby the dislocations. In addition, the metal ions also diffuse through dislocations.

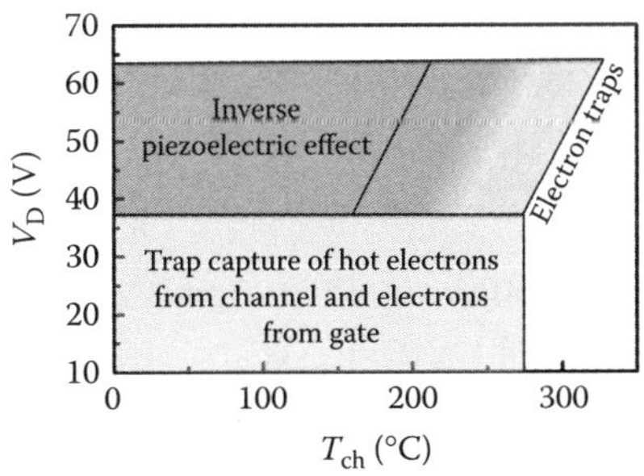

Figure 11.18 Major problems of GaN HEMT reliability. Note that V_D is drain bias and T_{ch} is channel temperature.

2. Select proper Schottky barrier metals with better surface treatment and metal deposition process to raise the barrier height and reduce the leakage current.
3. Improve the barrier layer surface treatment and the dielectric passivation process to achieve clean and low-defect-density barrier surface and high-quality dielectric films for passivation with low interface state density. The proper selection of the dielectric for the fabrication of metal–insulator–semiconductor HEMTs can reduce the performance degradation induced by the large forward gate currents in RF operations.
4. Apply polarization engineering and energy band engineering in the optimization of the structure of top barrier and back barrier to enhance the quantum confinement effect of the quantum wells in the channel and to lower the electric field peak and hot electron energy in the channel, thus preventing the hot electrons from spilling over the channel and being captured by traps.

11.5 Temperature-Dependent Characteristics of GaN HEMTs

One advantage of AlGaN/GaN HEMTs is their high temperature applications, so the study of the variation of device properties with temperature is of great significance.

The material for fabricating the HEMT samples is AlGaN/GaN wafer grown by metalorganic chemical vapor deposition. First, a tens-of-nanometers-thick low-temperature AlN buffer layer was grown. This was followed by the growth of a 1.5 μm undoped GaN buffer layer, a 1.5 nm AlN interlayer, and an ~25 nm AlGaN barrier layer (30% aluminum) at elevated temperature, and finally, a 1.5 nm GaN cap layer was grown. The Schottky contacts are T- or Γ-shaped Ni/Au/Ni multilayer metal gates with 0.5 μm gate length, 100 μm gate width, and ~4 μm source-drain spacing. Passivation was performed using SiN. Hall effect measurements showed the 2DEG mobility μ and sheet density n_S for the material grown on sapphire substrates at room temperature to be 1159 cm^2/Vs and 1.2×10^{13} cm^{-2}, respectively. Transmission line model (TLM) and Schottky *C-V* test patterns (the inner circle is the Schottky contact and the outer ring is the ohmic contact) were fabricated on the same wafer and distributed evenly around the HEMT devices to guarantee the consistence of the temperature dependence of the test patterns and the HEMTs and to reduce the effect of nonuniformity of the epi-wafer.

11.5.1 Temperature-Dependent Characteristics of Schottky Contact

The forward and reverse gate-drain diode properties of the hot-plate-heated device are measured at temperatures ranging from room temperature to 200°C, with source and drain terminals grounded and a gate voltage sweeping ranging from −30 V to 2 V. The measurement results are given in Figure 11.19. Both the forward and reverse gate currents increase with rising temperature, with the reverse leakage current two orders of magnitude greater at 200°C than at room temperature under −30 V bias. The increase of forward current is attributed to the greater ability of carriers to climb

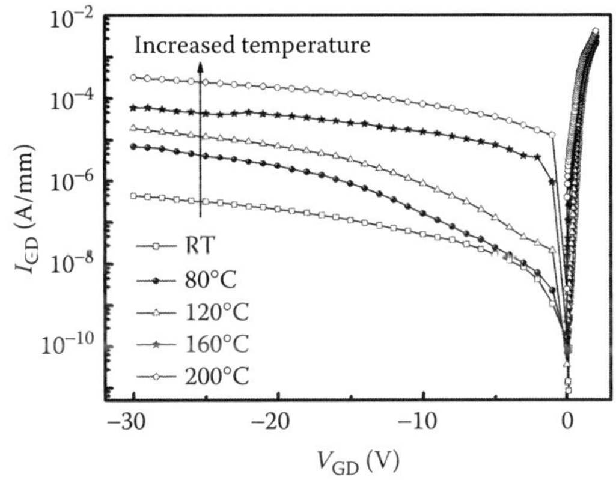

Figure 11.19 *I-V-T* characteristics of the gate-drain Schottky diode of the AlGaN/GaN HEMT. (Courtesy of Dr. Liyuan Yang.)

over the potential barrier at higher temperatures. Within the temperature range of 20°C–200°C, the dominant mechanism of the quite high reverse leakage current could be ascribed to the trap-assisted tunneling related to the leakage from screw dislocations (Zhang et al. 2006).

11.5.2 Temperature-Dependent Ohmic Contact Characteristics and Sheet Resistance

The temperature-dependent ohmic contact characteristics are measured using TLM patterns. The curve of the interelectrode resistance with electrode spacing, shown in Figure 11.20, is obtained by measuring at varied temperatures the current through two neighboring TLM electrodes with different spacings at an interelectrode voltage of 0.5 V, from which we can see that the slope of the curve increases with the rising temperature.

The variation with temperature of the extracted sheet resistance and specific contact resistivity from the temperature-dependent TLM measurements is shown in Figure 11.21. The sheet resistance increases with temperature in an approximately linear manner, rising from 442 Ω/sq at 20°C to 1058 Ω/sq at 200°C. The increase of

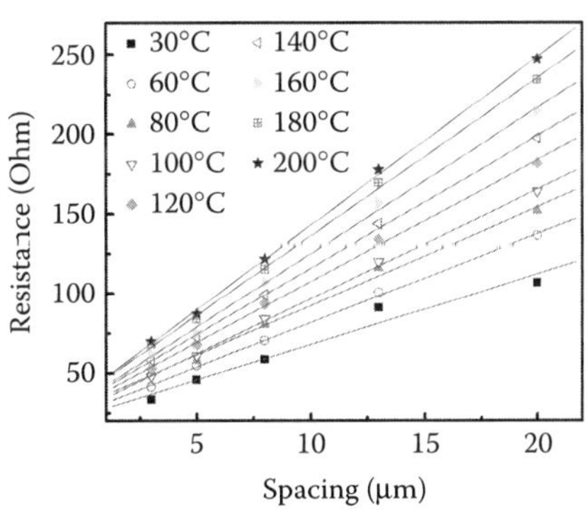

Figure 11.20 Total resistance between two neighboring metal electrodes as a function of corresponding spacing at different temperatures. (Courtesy of Dr. Liyuan Yang.)

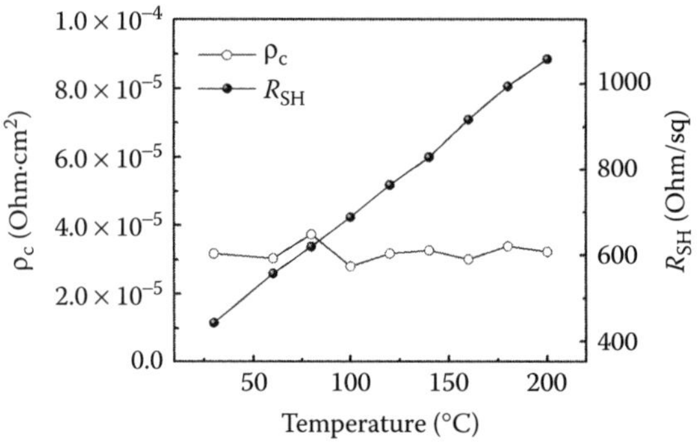

Figure 11.21 Temperature dependence of specific contact resistivity and sheet resistance. (Courtesy of Dr. Liyuan Yang.)

the sheet resistance, which is inversely proportional to the 2DEG density and mobility, suggests a degradation of the 2DEG properties. The virtually invariant specific contact resistivity indicates the stable ohmic contact performance within the concerned temperature range.

11.5.3 Temperature-Dependent Characteristics of AlGaN/GaN HEMT

The HEMT device selected for the test has normal characteristics, with a maximum saturation current of 754 mA/mm at $V_{GS} = 2$ V and a maximum transconductance of 223 mS/mm at $V_{DS} = 10$ V. The DC characteristics are measured at 20°C, 60°C, 80°C, 100°C, 120°C, 140°C, 160°C, 180°C, and 200°C.

From the temperature-dependent transfer characteristics shown in Figure 11.22, a slight positive shift of the threshold voltage is observed as it increases with the rising temperature. As shown in Figure 11.23, both the maximum saturation current and the maximum transconductance fall at higher temperatures, and at 200°C, the maximum saturation current decreases by 23.7% at $V_{GS} = 2$ V and the maximum transconductance decreases by 31.5% at $V_{DS} = 10$ V.

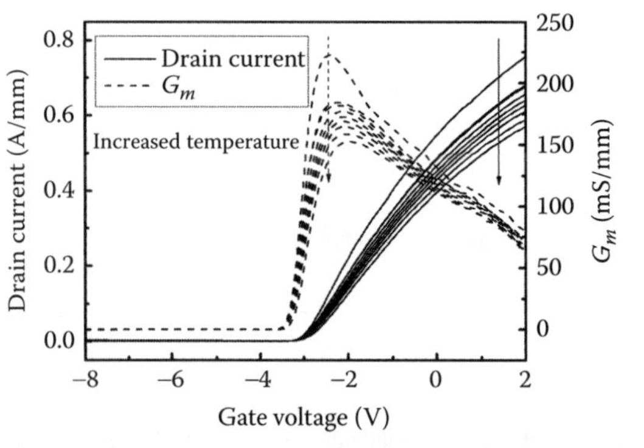

Figure 11.22 Transfer characteristics of AlGaN/GaN HEMT at different temperatures. (Courtesy of Dr. Liyuan Yang.)

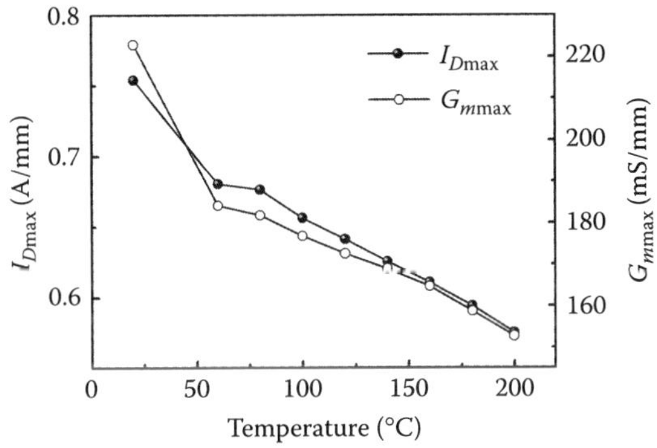

Figure 11.23 Temperature dependence of the saturation current and maximum transconductance of the device. (Courtesy of Dr. Liyuan Yang.)

The TLM results have shown a degradation of 2DEG properties at elevated temperature, which is in agreement with the degradation of AlGaN/GaN HEMT characteristics at higher temperature, and the analysis of its degradation mechanism requires the help of the temperature-dependent *C-V* carrier profiles obtained from the Schottky test patterns.

The capacitance–voltage curves at different temperatures from the Schottky *C-V* measurements are shown in Figure 11.24, and the calculated *C-V* carrier profiles at different temperatures are presented in Figure 11.25. At the elevation of temperature, a slight shift of the 2DEG distribution peak toward the buffer layer is observed with a reduced peak concentration and broadened distribution, indicating poorer 2DEG confinement at higher temperature as well as an increased electron concentration in AlGaN layer.

An integration of the *C-V* carrier concentration profile yields the carrier sheet density, which is combined with the TLM-extracted sheet resistance to determine the carrier mobility. The temperature dependent *C-V* carrier sheet density and carrier mobility are given in Figure 11.26. No evident change in the carrier sheet density is seen as the temperature rises. In light of the increased electron concentration in AlGaN layer, this indicates a tiny drop of the 2DEG density, which explains the

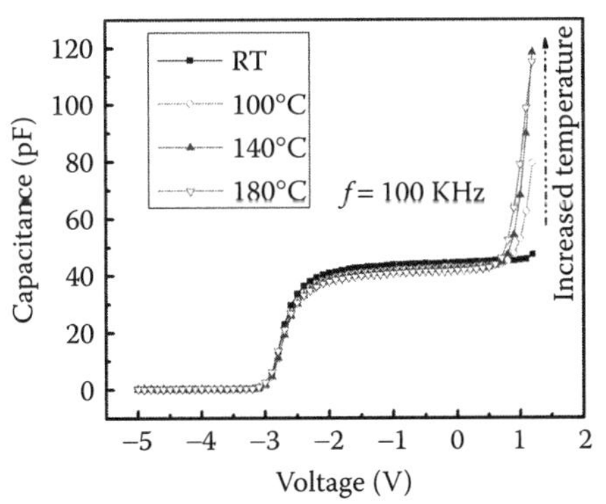

Figure 11.24 Capacitance–voltage curves at different temperatures. (Courtesy of Dr. Liyuan Yang.)

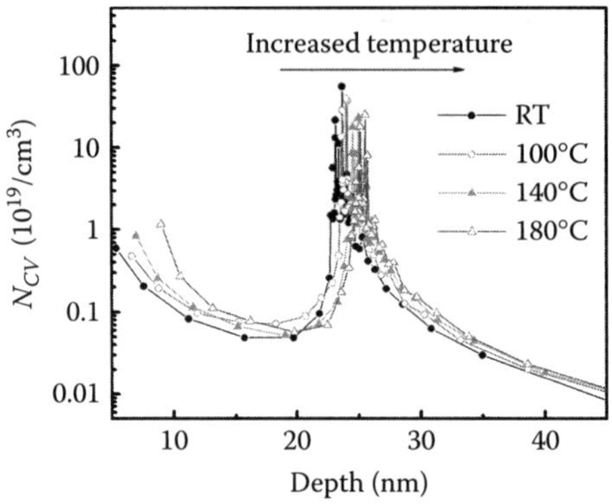

Figure 11.25 *C-V* carrier concentration distributions at different temperatures. (Courtesy of Dr. Liyuan Yang.)

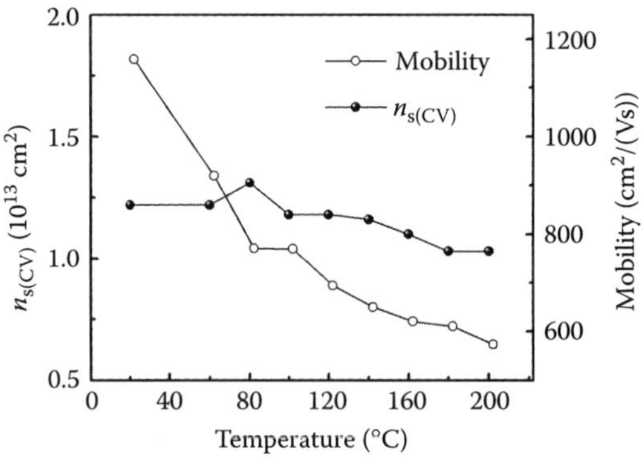

Figure 11.26 Temperature dependence of the *C-V* carrier sheet density and carrier mobility in the heterojunction. (Courtesy of Dr. Liyuan Yang.)

slight positive shift of the threshold voltage. The little variation in the carrier density and the significant increase of material sheet resistance with rising temperature can be attributed to the enhanced lattice vibration scattering and dramatic degradation of carrier mobility at higher temperature. The saturation current of HEMTs depends mainly on the mobility and 2DEG density, and the transconductance, which is relative to gate control, etc., is also directly proportional to the mobility. Accordingly, the decreased mobility with rising temperature is the leading cause of the reduced saturation current and transconductance as the temperature rises.

The increase of the off-state source-drain leakage current by one order of magnitude at 200°C compared with that at room temperature, as shown in the transfer characteristic curve in Figure 11.22, may be caused by the greater leakage of GaN buffer layer or higher gate leakage current at elevated temperature. The measured leakage current across adjacent active region mesas does increase with a 50 V voltage applied during the temperature elevation, but the leakage current density is merely 34 μA/mm at 200°C. A comparison in the order of magnitude between the off-state leakage currents from the channel and the gate (Figure 11.27) shows similar temperature dependences

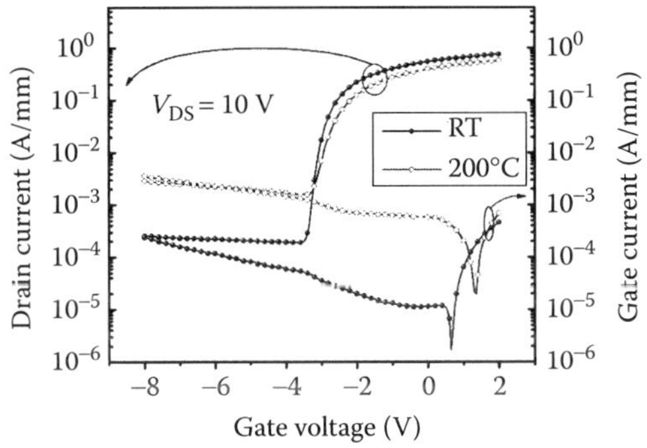

Figure 11.27 Relation between the off-state source-drain leakage current and gate leakage current at room temperature and at 200°C. (Courtesy of Dr. Liyuan Yang.)

of the two. Therefore, the source-drain leakage current after channel pinch-off is due primarily to the increase in gate leakage current.

In summary, the DC characteristics degradation of AlGaN/GaN HEMTs on sapphire substrate with increasing temperature is analyzed by a combination of temperature-dependent Schottky *C-V* measurements and temperature-dependent TLM measurements, suggesting that the major reason for the degradation of saturation current and transconductance at elevated temperature is the degraded 2DEG transport properties and that the temperature dependence of off-state channel leakage current is caused primarily by the temperature dependence of the gate leakage current, with the GaN buffer layer leakage playing a minor role.

11.6 High Temperature Storage Characteristics of GaN HEMTs

For any device operating at high temperature, not only must the high temperature performance meet the requirements, but it should also be taken into consideration whether irreversible performance degradation may result under long-term high temperature operations. That is, the high temperature environment imposes a thermal stress on the device, and heated storage may affect the device reliability. To this end, we measured the degradation of GaN-based HEMTs under thermal stress and analyzed the degradation mechanism.

The HEMT samples are prepared on the AlGaN/GaN epi-wafer on SiC substrate with the same epitaxial materials and device structure as in Section 11.5. In view of the possibility of irreversible change in HEMT performance induced by heated storage, the wafers were thinned and diced to avoid concurrent heating of the whole wafer and respective HEMTs were used for each experiment of varied lengths of time of heated storage to guarantee the accuracy of the measurements. The initial properties of the device at room temperature were measured first, after which the device was held at 200°C for a certain time and then cooled to room temperature for a second measurement of its properties.

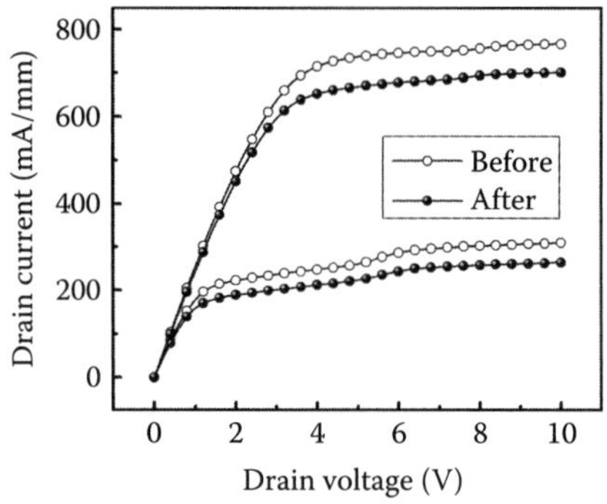

Figure 11.28 AlGaN/GaN HEMT output characteristics before and after heated storage. (Courtesy of Shougao Jiang.)

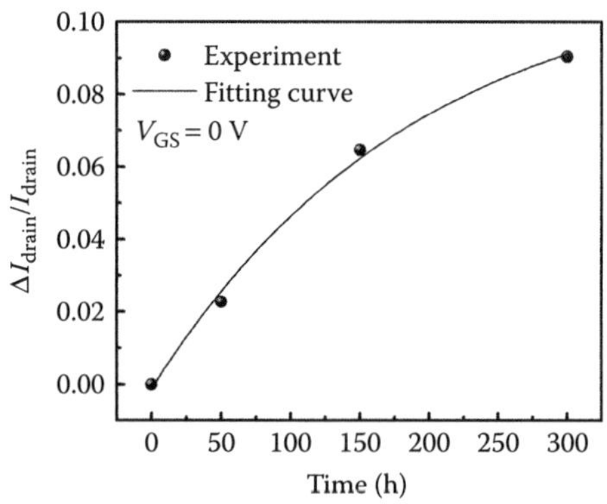

Figure 11.29 Saturation drain current degradation versus heated storage time, with gate voltage $V_{GS} = 0$ V. The solid line is the fitted curve by the least square method. (Courtesy of Shougao Jiang.)

The measured HEMT output characteristics before and after heated storage for 300 h are shown in Figure 11.28. The measurements were done at a gate voltage bias (0 V, −2 V), with the drain voltage scanned from 0 to 10 V. The linear region remains invariant, but a decrease in the knee voltage and a dramatic drop of current in the saturation region of the device are observed. Shown in Figure 11.29 is the normalized degradation of the saturation drain current (the ratio of the reduction in drain current after the heated storage to the drain current before the heated storage) as a function of the storage time. An aggravated degradation of the drain current is observed as the storage time increases, with the drain current reduced by 9.05% after 300 h of heated storage. The transconductance of the device decreases by 5.3% from 245 to 232 mS/mm, as shown in the transfer characteristics before and after heated storage for 300 h in Figure 11.30. There is also a small positive shift of the threshold voltage.

The analysis of the device gate current, the *I-V* characteristics of TLM structure, and Schottky *C-V* characteristics shows little change in the ohmic contact with a slight increase of the wafer sheet resistance, but there is an obvious change in the Schottky gate characteristics. The gate current–voltage curve at a drain-source bias of

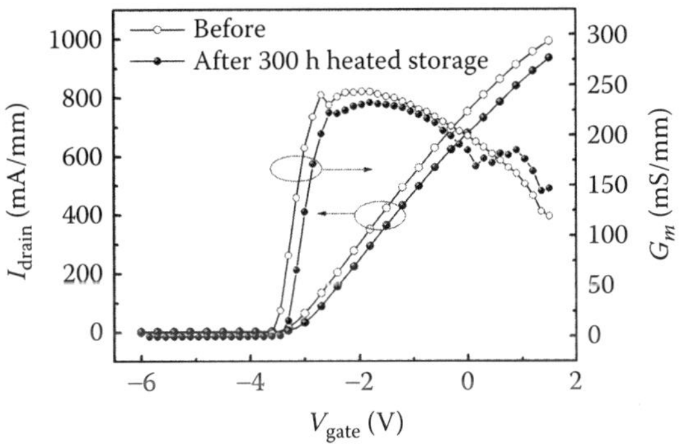

Figure 11.30 AlGaN/GaN HEMT transfer characteristics before and after heated storage. (Courtesy of Shougao Jiang.)

0 V before and after heated storage is presented in Figure 11.31. The properties for 50, 150, and 300 h heated storage were measured from three separate devices with very close gate current to each other before heated storage, which suggests good uniformity of the device characteristics and thus common degradation behaviors. Both the reverse gate leakage current and forward gate current increase with longer storage time, with the gate leakage current rising by two orders of magnitude to 8 mA/mm after 300 h heated storage, indicating very serious degradation of the Schottky characteristics.

The measured *C-V* curves for a Schottky test pattern before and after 300 h heated storage are shown in Figure 11.32. The visible increase of the 2DEG depletion region capacitance after heated storage indicates a significant reduction in the depth *d* that 2DEG locates from the wafer surface. The positive shift of the *C-V* curve also conforms to the positive shift of the threshold voltage, as shown in Figure 11.30. The *C-V* carrier profiles before and after heated storage in Figure 11.33 show no apparent drop of the 2DEG peak but a visible shift (~5 nm) of the peak position toward the heterojunction surface. It is not given in Figure 11.32 but should be pointed out that the

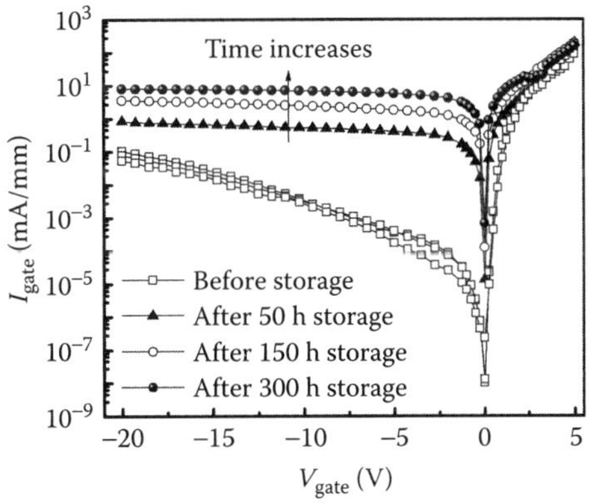

Figure 11.31 Schottky gate *I-V* characteristics of AlGaN/GaN HEMT before and after heated storage. (Courtesy of Shougao Jiang.)

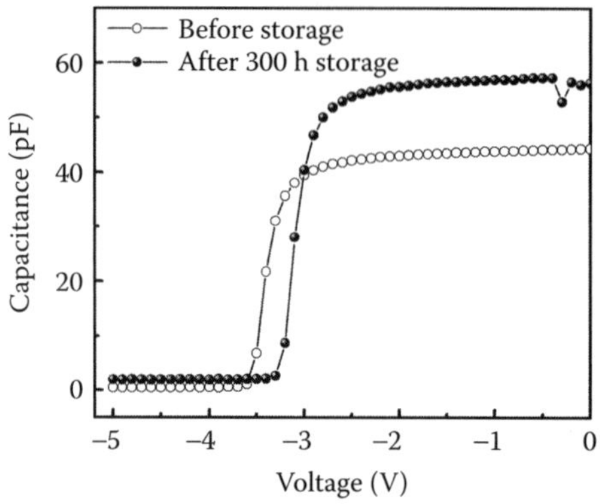

Figure 11.32 *C-V* curves of a Schottky test pattern before and after heated storage for 300 h. (Courtesy of Shougao Jiang.)

significant increase of capacitance at a positive gate voltage, as shown in Figure 11.24, is not observed in Figure 11.32 for the case after heated storage, and instead, the capacitance drops instantly at a relatively small positive gate voltage from the platform at 0 V (accompanied by a significant increase of the gate current), as a result of which the *C-V* carrier distribution in AlGaN layer after heated storage is not available in Figure 11.33. This suggests a negative shift of the forward turning-on gate voltage and an increase of the forward gate current at the same gate voltage, in agreement with the trend in Figure 11.31.

Shown in Figure 11.34 are the dynamic gate-drain double-pulse *I-V* curves of the devices (note the device surface is passivated) before and after 300 h heated storage, demonstrating a significant reduction in current collapse at all gate voltages after heated storage.

All the above experimental phenomena show a unidirectional variation of the device performance with the heated storage time. Owing to the wide band gap properties of AlGaN and GaN, storage at 200°C analogous to low-temperature annealing does not change the heterostructure itself but may alter the states of the Schottky metal-semiconductor and passivation layer-semiconductor interfaces.

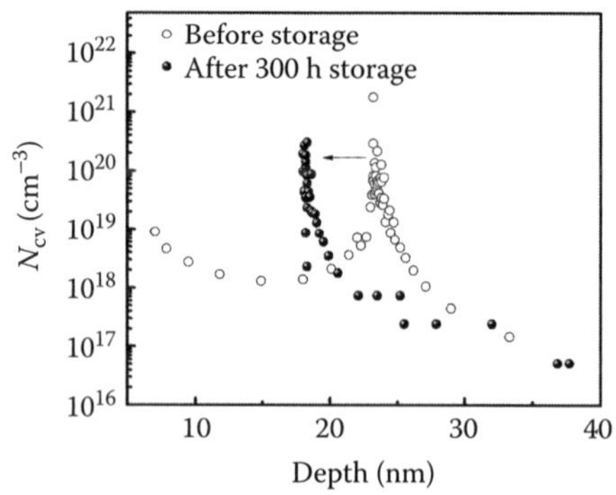

Figure 11.33 Effect of heated storage on *C-V* carrier profile. (Courtesy of Shougao Jiang.)

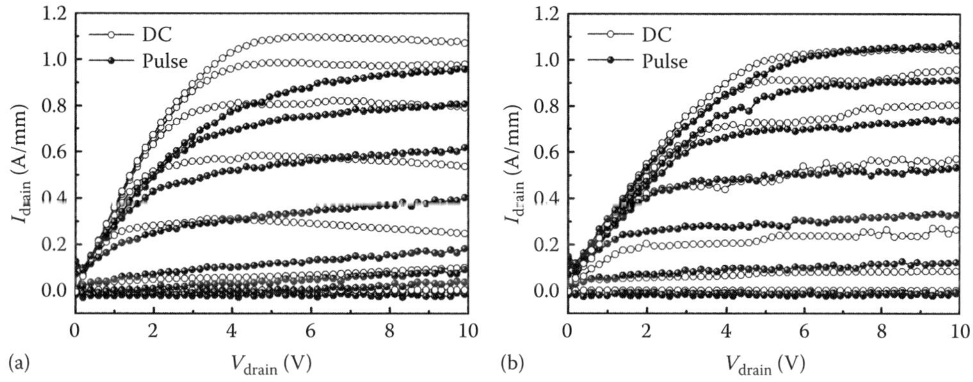

Figure 11.34 Dynamic double-pulse *I-V* and DC *I-V* curves of the HEMT for current collapse observation with pulse quiescent points $V_{GS} = 0$ V, $V_{DS} = 20$ V, a pulse period of 1 ms, and a pulse width of 500 ns. (a) Before heated storage and (b) after heated storage for 300 h. (Courtesy of Shougao Jiang.)

The peak of 2DEG under the gate shifts toward the heterostructure surface after heated storage, that is, a nominal thinning of the AlGaN barrier layer. Typically, native oxide layers form readily on the surface of the epitaxial material, be it AlGaN or GaN, along with contamination from other impurities such as carbon. Although a relatively clean surface can be obtained using acid rinsing to remove part of the oxide layer, the nitride surface is subject to oxidization during the subsequent device fabrication, resulting in a high concentration of oxygen in the several-nanometer-thick interfacial layer between the gate metal and the nitride semiconductor after the fabrication of the Schottky gate. According to the report of Singhal et al. (2006), thermal annealing can consume the oxygen-rich quasi-dielectric interface layer under the Schottky gate. Heated storage for an extended time (>50 h) may have the same effect, that is, consuming the interfacial layer completely, thus resulting in a better contact between the gate metal (nickel) and the nitride semiconductor. As a result, the 2DEG peak in the *C-V* carrier profile shifts a few nanometers toward the wafer surface, along with a positive shift of the threshold voltage of the device. This was supposed to increase the transconductance, but the increase of both forward and reverse gate currents indicates a remarkable degradation of Schottky characteristics, which may be explained by the more prominent dislocation leakage current after the formation of the closer nickel-nitride contact, owing to the presence of many leakage inducing threading screw dislocations in the epi-wafer. The gate control weakens due to the Schottky gate degradation, thereby decreasing both the saturation current and the transconductance.

Before heated storage, the SiN dielectric film as the passivation layer is deposited on the surface of the nitride semiconductor, and the generated dielectric-semiconductor contact could not be very close from the view of the atomic level. Heated storage could lead to better contact between the passivation layer and the semiconductor, in favor of the reduction of interface trap states and thus a diminished current collapse.

In conclusion, heated storage at 200°C for more than 50 h can induce irreversible change in the HEMT performance. The degradation of the saturation current and the maximum transconductance, as well as the positive shift of the threshold voltage, are attributed primarily to the altered properties of the Schottky gate, the mechanism of which may be the depletion of the Schottky metal-semiconductor interfacial layer due to heated storage. The typical reduction of current collapse after heated storage may be explained by the reduced interface trap states, owing to better contact between the passivation layer and the semiconductor as a result of heated storage.

References

Augaudy, S., R. Quere, J. P. Teyssier, M. A. Di Forte-Poisson, S. Cassette, B. Dessertenne, and S. L. Delage. 2001. Pulse characterization of trapping and thermal effects of microwave GaN power FETs. *2001 IEEE MTT-S International Microwave Symposium Digest*, May 20–25, 2001, Piscataway, NJ.

Cao, Y.-R., X.-H. Ma, Y. Hao, Y. Zhang, L. Yu, Z.-W. Zhu, and H.-F. Chen. 2007. Study on the recovery of NBTI of ultra-deep sub-micro MOSFETs. *Chinese Physics* 16(4):1140–1144. doi: 10.1088/1009-1963/16/4/047.

Daumiller, I., D. Theron, C. Gaquiere, A. Vescan, R. Dietrich, A. Wieszt, H. Leier et al. 2001. Current instabilities in GaN-based devices. *IEEE Electron Device Letters* 22(2):62–64. doi: 10.1109/55.902832.

Gu, W.-P., Y. Hao, J.-C. Zhang, H.-T. Duan, J.-Y. Ni, Q. Feng, and X.-H. Ma. 2009a. High-electric-field-stress-induced degradation of SiN passivated AlGaN/GaN high electron mobility transistors. *Chinese Physics B* 18(4):1601–1608. doi: 10.1088/1674-1056/18/4/052.

Gu, W.-P., Y. Hao, J.-C. Zhang, C. Wang, Q. Feng, and X.-H. Ma. 2009b. Degradation under high-field stress and gate stress of AlGaN/GaN HEMTs. *Acta Physica Sinica* 58(1):511–17.

Jimenez, J. L. and U. Chowdhury. 2008. X-band GaN fet reliability. *46th Annual 2008 IEEE International Reliability Physics Symposium, IRPS*, April 27, 2008–May 1, 2008, Phoenix, AZ.

Joh, J., L. Xia, and J. A. Del Alamo. 2007. Gate current degradation mechanisms of GaN high electron mobility transistors. *2007 IEEE International Electron Devices Meeting, IEDM*, December 10–December 12, 2007, Washington, DC.

Joh, J. and J. A. Del Alamo. 2006. Mechanisms for electrical degradation of GaN high-electron mobility transistors. *2006 International Electron Devices Meeting, IEDM*, December 10–December 13, 2006, San Francisco, CA.

Kamiya, S., M. Iwami, T. Tsuchiya, M. Kurouchi, J. Kikawa, T. Yamada, A. Wakejima et al. 2007. Kelvin probe force microscopy study of surface potential transients in cleaved AlGaN/GaN high electron mobility transistors. *Applied Physics Letters* 90(21):213511-1. doi: 10.1063/1.2743383.

Khan, M. A., M. S. Shur, Q. C. Chen, and J. N. Kuznia. 1994. Current/voltage characteristic collapse in AlGaN/GaN heterostructure insulated gate field effect transistors at high drain bias. *Electronics Letters* 30(25):2175–2176. doi: 10.1049/el:19941461.

Koley, G., V. Tilak, L. F. Eastman, and M. G. Spencer. 2003. Slow transients observed in AlGaN/GaN HFETs: Effects of SiNx passivation and UV illumination. *IEEE Transactions on Electron Devices* 50(4):886–893. doi: 10.1109/ted.2003.812489.

Meneghesso, G., G. Verzellesi, F. Danesin, F. Rampazzo, F. Zanon, A. Tazzoli, M. Meneghini, and E. Zanoni. 2008. Reliability of GaN high-electron-mobility transistors: State of the art and perspectives. *IEEE Transactions on Device and Materials Reliability* 8(2):332–343. doi: 10.1109/tdmr.2008.923743.

Simin, G., A. Koudymov, A. Tarakji, X. Hu, J. Yang, Khan M. Asif, M. S. Shur, and R. Gaska. 2001. Induced strain mechanism of current collapse in AlGaN/GaN heterostructure field-effect transistors. *Applied Physics Letters* 79(16):2651–2651. doi: 10.1063/1.1412282.

Singhal, S., J. C. Roberts, P. Rajagopal, T. Li, A. W. Hanson, R. Therrien, J. W. Johnson, I. C. Kizilyalli, and K. J. Linthicum. 2006. GaN-on-Si failure mechanisms and reliability improvements. *2006 IEEE International Reliability Physics Symposium proceedings. 44th Annual*, 26–30 March 2006, Piscataway, NJ.

Stoklas, R., D. Gregusova, J. Novak, A. Vescan, and P. Kordos. 2008. Investigation of trapping effects in AlGaN/GaN/Si field-effect transistors by frequency dependent capacitance and conductance analysis. *Applied Physics Letters* 93(12):124103 (3 pp.). doi: 10.1063/1.2990627.

Tirado, J. M., J. L. Sanchez-Rojas, and J. I. Izpura. 2007. Trapping effects in the transient response of AlGaN/GaN HEMT devices. *IEEE Transactions on Electron Devices* 54(3):410–417. doi: 10.1109/ted.2006.890592.

Trew, R. J., Y. Liu, W. W. Kuang, and G. L. Bilbro. 2006. The physics of reliability for high voltage AlGaN/GaN HFET's. *2006 IEEE Compound Semiconductor Integrated Circuit Symposium*, November 12–15, 2006, Piscataway, NJ.

Valizadeh, P. and D. Pavlidis. 2005. Investigation of the impact of Al mole-fraction on the consequences of RF stress on Alx Ga1-x N/GaN MODFETs. *IEEE Transactions on Electron Devices* 52(9):1933–1939. doi: 10.1109/ted.2005.852543.

Vetury, R., N. Q. Zhang, S. Keller, and U. K. Mishra. 2001. The impact of surface states on the DC and RF characteristics of AlGaN/GaN HFETs. *IEEE Transactions on Electron Devices* 48(3):560–566. doi: 10.1109/16.906451.

Zhang, H., E. J. Miller, and E. T. Yu. 2006. Analysis of leakage current mechanisms in Schottky contacts to GaN and Al0.25Ga0.75N/GaN grown by molecular-beam epitaxy. *Journal of Applied Physics* 99(2):023703-6.

12

Enhancement-Mode GaN HEMTs and Integrated Circuits

Gallium nitride high electron mobility transistors (HEMTs) have good high temperature characteristics and radiation resistance owing to their wide band gap, and they find potential applications in the GaN-based integrated circuits used in harsh environment. However, in terms of the planar structure or the operating speed, the preparation of complementary symmetrical GaN FET circuit units in a CMOS-like mode is yet impossible, owing to the great difference in the motilities of holes and electrons of GaN. A feasible approach is the n-type GaN enhancement-mode HEMT (E-HEMT) that requires a forward gate voltage to turn on the channel, which is often referred to as the normally off HEMT. GaN high-power switches and circuits and enhancement/depletion-mode (E/D-mode) digital integrated circuits can be realized using the high and low electrical levels of the gate voltage to control the turning on and off of the E-HEMT.

High-density two-dimensional electron gases (2DEGs) used as conducting channels are formed in the preparation of typical AlGaN/GaN heterostructures. GaN HEMTs based on such material are the D-mode HEMTs (D-HEMTs), which require a negative gate bias to be switched off, thus being called the normally on HEMTs. The realization of E-HEMTs fully compatible with D-HEMTs requires special structures or processes, mainly including the thin barrier layer, recessed gate (in combination with the MIS structure), pn junction under the gate, and fluoride ion implantation under the gate.

A brief introduction is given first to the implementation methods of GaN E-HEMTs, followed by the preparation processes, structure optimization, and evaluation of device performance and reliability of the fluorine plasma-treated GaN-based E-HEMTs as well as the enhancement/depletion-mode digital circuit units.

12.1 GaN Enhancement-Mode HEMTs

The enhancement-mode properties of E-HEMTs are obtained by decreasing the 2DEG density of D-HEMTs to a negligible level at zero gate bias. According to Equations 9.1 through 9.6, the 2DEG density in the AlGaN/GaN HEMT is closely related to the aluminum composition (mainly influencing the polar charge at the AlGaN/GaN interface σ_{pol}, the conduction band offset ΔE_C, and the Schottky barrier height $e\varphi_b$ at AlGaN surface), thickness, strain, and impurity of the AlGaN barrier layer, which should first be taken into consideration in altering the 2DEG density.

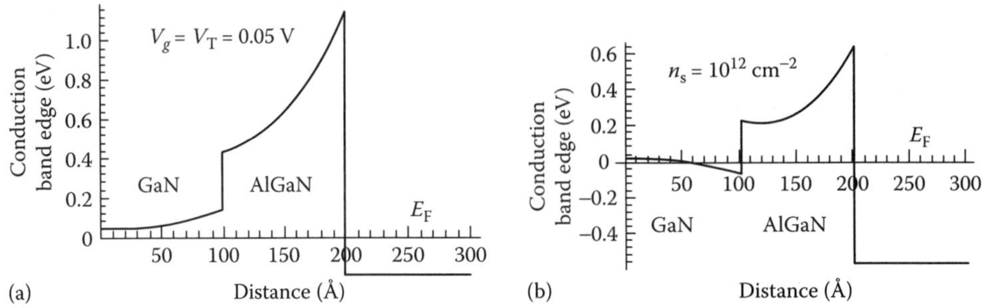

Figure 12.1 Band diagrams of the AlGaN/GaN E-HEMT with a thin barrier layer along the growth direction. The channel under the gate is switched off under the depletion effect of the Schottky gate. (a) Under the gate and (b) between the gate and the source (drain). (Reproduced from Khan, M.A. et al., *Appl. Phys. Lett.*, 68(4), 514–516, 1996.)

The use of thin barrier layers in the AlGaN/GaN HEMTs is an approach to E-mode properties, since the reduction in the aluminum composition or thickness of the AlGaN barrier layer can decrease the 2DEG density. The first E-HEMTs were prepared by Khan et al. by using the thin potential barrier structure (Khan et al. 1996). Their barrier layers were 10 nm $Al_{0.1}Ga_{0.9}N$, as shown in Figure 12.1, where E_F is the Fermi level. The achievement of the enhancement mode by simply reducing the barrier thickness has the advantage of no damage to the region under the gate from the etching process, which leads to good Schottky properties and a relatively low gate leakage current for good high-frequency characteristics (Endoh et al. 2004); however, it has the disadvantage of a relatively low 2DEG density in the entire channel region due to the reduced barrier layer thickness, thus resulting in a small saturation current.

The employment of recessed gates is a simple way to increase the current density of E-HEMTs. The 2DEG density under the gate is decreased and can be neglected when the thickness of the AlGaN barrier layer is reduced by etching to a certain extent; however, the access regions between the gate and the source or the drain are not affected by the etching and maintain the original 2DEG density. Such devices obtain a higher saturation current and transconductance than the thin-barrier-layer E-HEMTs. For example, the prepared E-HEMT with a recessed gate structure formed by etching off 15 nm out of a 23-nm thick undoped AlGaN barrier layer and with a gate length of 1 μm achieves a threshold voltage of 0.47 V, a saturation current of 455 mA/mm, and a maximum transconductance of 310 mS/mm (Lanford et al. 2005).

The weakness of recessed-gate E-HEMTs lies mainly in the difficulty of accurate control of the AlGaN barrier layer thickness and the recess depth, thereby leading to a poor process repeatability and threshold voltage controllability. A large gate leakage current is usually induced by the etch damage. In order to solve the gate leakage, it is proposed that the recessed gate be coupled with the metal–insulator–semiconductor (MIS) structure to obtain E-HEMTs. Special designs of the device structure are usually required to diminish the effect of the poor controllability of the recess depth. An important GaN recessed-gate MIS E-mode field effect transistor (FET) structure

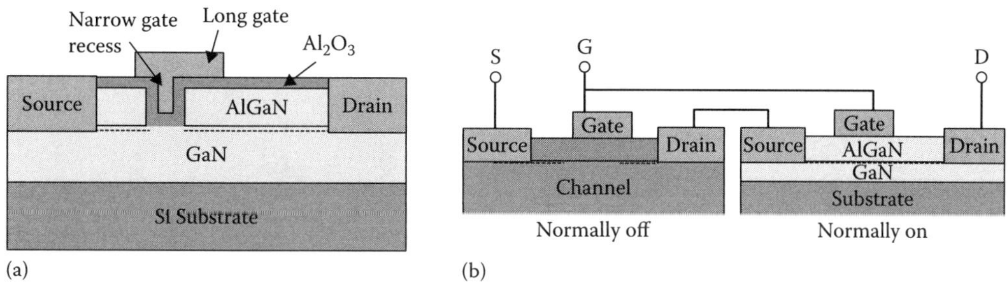

Figure 12.2 Integrated dual-gate GaN E-mode MIS device. (a) Schematic crosssection of the device. (b) Equivalent circuit of the device, that is, a D-mode AlGaN/GaN HEMT connected to an E-mode MISFET. (Reproduced from Lu, B. et al., *IEEE Elect. Device Lett.*, 31(9), 990–992, 2010.)

is the MIS channel HEMT (similar structures are also referred to as the hybrid MIS-HFET or the integrated dual-gate AlGaN/GaN E-HEMT). The device structure has the gate recess etched to the GaN channel layer to form the MIS structure for E-mode MISFET properties, and the gate length is larger than the recessed region, with the metal extending out of the recess to the top of the AlGaN/GaN heterojunction, and therefore, an equivalent structure of a D-HEMT integrated with the E-mode MISFET (namely the integrated dual-gate principle) is formed for a greater device current. The integrated dual-gate E-HEMTs reported by Lu et al., as shown in Figure 12.2, achieved a threshold voltage of 2.9 V, a saturation current of 434 mA/mm, and a breakdown voltage of 634 V (Lu et al. 2010). Ikeda et al. reported hybrid MIS-HFETs with a threshold voltage of 2.0 V, a breakdown voltage of 1.7 kV (gate-drain spacing 18 μm), and a specific on-resistance of 11.9 mΩ·cm^2 (Ikeda et al. 2011). Polarization engineering is another approach to recessed-gate MIS E-HEMTs. Ota et al. developed recessed-gate E-mode MIS-HEMTs using the piezo neutralization technique (PNT) structure on silicon substrates, achieving a threshold voltage of 1.5 V, a saturation current of 240 mA/mm, and a breakdown voltage greater than 1000 V (Ota et al. 2009). As shown in Figure 12.3, the piezo neutralization barrier and the buffer (acting as a back barrier) underneath the GaN channel are AlGaN layers of the same aluminum composition, and the polar charges at the interfaces in between are neutralized with each other, thus resulting in a horizontal conduction band of the piezo neutralization layer. Therefore, the threshold voltage remains constant as long as the gate recess bottom reaches into the piezo neutralization layer, which is favorable for better threshold voltage uniformity across the wafer. Low in gate leakage current and high in both threshold voltage and breakdown voltage, recessed-gate E-mode MIS-HEMTs have proved reliable with a large saturation current by experiments and have good prospects in high-voltage switch applications.

Implanting fluoride ions (F^-) under the gate (fluorine plasma ion implantation) is a successful means to achieve E-mode GaN HEMTs. The strong electronegativity of F^- ions can raise the effective barrier height of the barrier layer to deplete the 2DEG under the gate. Chen et al. employed fluorine plasma ion implantation to produce

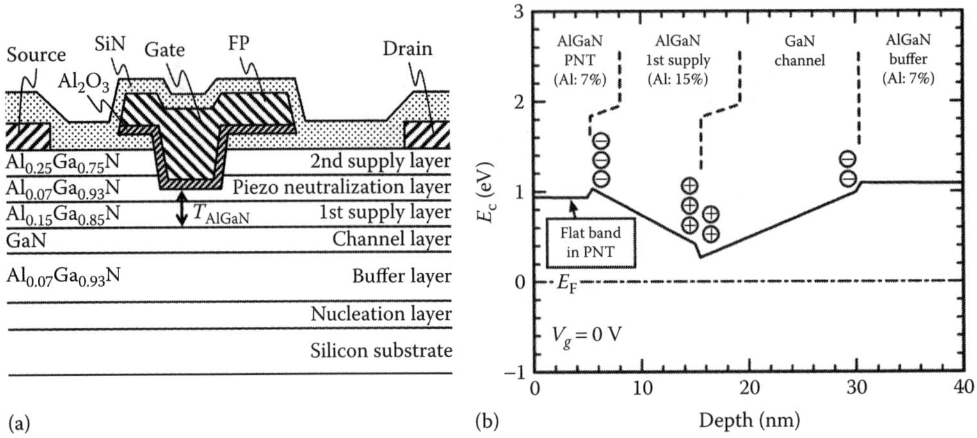

Figure 12.3 PNT structure GaN E-mode recessed-gate MIS-HEMT. (a) Schematic device structure. (b) Band diagram under the gate along the material growth direction. (Reproduced from Ota, K. et al., A normally-off GaN FET with high threshold voltage uniformity using a Novel Piezo neutralization technique, in *2009 IEEE International Electron Devices Meeting* (*IEDM 2009*), December 7–9, 2009, Piscataway, NJ.)

E-mode devices with a threshold voltage of 0.9 V, a maximum saturation current of 350 mA/mm, and a maximum transconductance of 180 mS/mm, demonstrating for the first time true enhancement-mode operation in GaN HEMTs (a positive threshold voltage with zero transconductance at $V_{GS} = 0$ V) (Cai et al. 2005). Fluorine plasma ion implantation has the advantages of easy implementation and high repeatability and the ability to control the threshold voltage by changing the fluorine plasma treatment conditions. Fluorine plasma treatment also causes less damage and thus a smaller gate leakage current. With the fluorine-based plasma treatment, GaN HEMT inverter unit circuits (Cai et al. 2005), E-HEMT with a high breakdown voltage up to 1400 V (with integrated slant field plate) (Suh et al. 2006), and high-frequency E-HEMTs (with recessed gate) with a maximum output current of 1.2 A/mm, $f_T = 85$ GHz and $f_{max} = 150$ GHz (Palacios et al. 2006) had been successfully realized. Feng et al. reported E-HEMTs prepared by fluorine plasma treatment with a radio-frequency power performance of a maximum output power density of 3.65 W/mm, a gain of 11.6 dB, and a power-added efficiency of 42% at 18 GHz (Feng et al. 2010). However, it is comparatively difficulty to prepare GaN E-HEMTs with a high threshold voltage (>3 V) by fluorine plasma treatment, and the stability of the implanted fluorine ions is not good enough and has a negative effect on the device reliability at high voltage and temperature.

The HEMTs with p-type GaN gate and the similar gate injection transistor (GIT) structure are also important structures for high-power E-HEMTs. The structure and operating principle of the GIT device are shown in Figure 12.4. The GIT structure is formed by introducing p-type AlGaN between the gate and the original n-type or unintentionally doped AlGaN barrier layer, and ohmic contact is formed between the gate metal and p-type AlGaN. On the one hand, pn junction with built-in

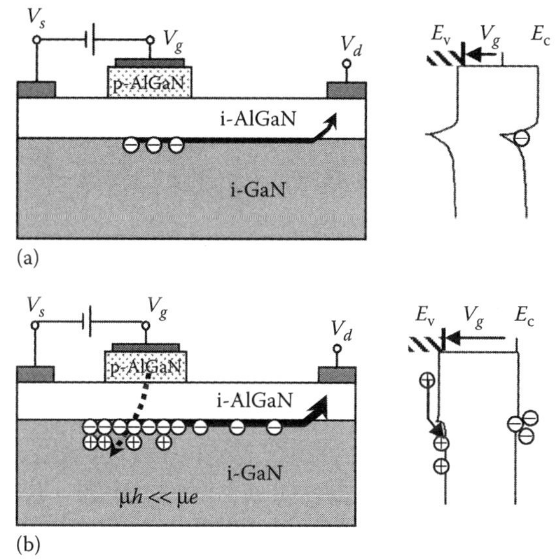

Figure 12.4 Operating principles of the GIT device at different gate voltages. (a) 0 V < V_g < V_F: without hole injection. The device exhibits enhancement-mode FET behavior. (b) V_g > V_F: with hole injection into the channel. The source-drain current increases significantly at a small gate current. (Reproduced from Uemoto, Y. et al., A normally-off AlGaN/GaN transistor with RonA = 2.6 mΩ·cm² and BVds = 640 V using conductivity modulation, in *2006 International Electron Devices Meeting*, December 10–13, 2006, San Francisco, CA.)

voltage V_F is formed in the barrier layer, leading to the depletion of channel electrons under the gate, owing to raised energy band by p-type doping, hence leading to the enhancement-mode operation. On the other hand, at gate voltages below V_F, the device acts as a FET (with a threshold voltage $V_T < V_F$), whereas at gate voltages higher than V_F, holes are injected from p-type AlGaN into the channel with the channel electron injection into the gate suppressed by the AlGaN/GaN interface barrier. The injected holes attract an equal number of electrons to preserve the electric neutrality in the channel. These electrons move continuously to the drain with a high mobility under the effect of the drain voltage, whereas the holes exist mostly in the channel region under the gate owing to their lower mobility by two orders of magnitudes than that of the channel electrons. A certain conductivity modulation is achieved in this dynamic process, resulting in a dramatic increase in the drain current while maintaining a small gate current (Uemoto et al. 2007). Devices with the GIT structure have been reported to exhibit a threshold voltage of 1.0 V, a maximum saturation current of 200 mA/mm, a breakdown voltage of 800 V, and a specific on-resistance of 2.6 mΩ·cm² (Uemoto et al. 2007).

A p-type GaN gate HEMT can be obtained by replacing the p-type AlGaN in the GIT by GaN, with its gate metal forming either an ohmic contact or a Schottky contact to p-type GaN. The p-GaN gates have wider applications, owing to the ability to realize more effective p-type doping as well as higher threshold voltage, saturation current, and breakdown voltage, and a smaller specific on-resistance at the same time by the optimization of the heterostructures. The Ferdinand Braun Institute proposed the substitution of a thick AlGaN back barrier (Hilt et al. 2010) or a composite buffer layer consisting of an AlGaN layer (dozens of nanometers thick) and a carbon-doped high-resistance GaN (several micrometers thick) (Hilt et al. 2011) for the GaN buffer layer under the channel of p-type GaN gate HEMT. Therefore, the virtual p-type doping effect of the negative polar charge between the GaN channel and the AlGaN back barrier can help the p-type gate elevate the channel-layer energy band further, thus realizing a high threshold voltage. A large saturation current and a low on-resistance are also obtained, owing to the higher 2DEG density in the channel outside the gate, which is achieved by a proper rise of the aluminum composition in the AlGaN top barrier layer. They have reported p-GaN gate HEMTs on SiC substrate with a threshold voltage of 1.1 V, a breakdown voltage of 1000 V, a specific on-resistance of 0.62 mΩ·cm^2, and a Baliga figure of merit (the ratio of the squared breakdown voltage to the specific on-resistance) up to 1613 MV^2/(Ω·cm^2) (Hilt et al. 2011).

12.2 Preparation and Characteristics of Fluorine-Plasma-Treated E-HEMTs

12.2.1 Structure and Preparation of E-HEMTs

The GaN E-HEMT prepared by fluorine plasma ion implantation is shown in Figure 12.5, with a T-gate for a full coverage of fluorine-plasma-treated region to avoid the lack of gate control in case of gate misalignment in lithography and to decentralize the electric field at the gate edge to improve the breakdown voltage. The preparation of the GaN E-HEMT samples is described below, with a gate length $L_G = 0.5$ μm, a gate width $W = 100$ μm, a source-drain spacing $L_{DS} = 5$ μm and, a gate-source spacing $L_{GS} = 0.7$ μm.

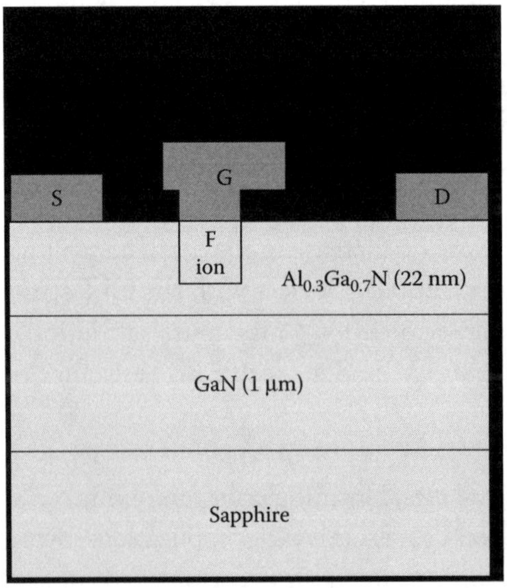

Figure 12.5 Structure of E-HEMT with fluorine plasma treatment. (After Quan, S. et al., *J. Semiconduct.*, 30(12), 124002 (4 pp.), 2009.)

The energy of fluorine plasma ion implantation can be regulated by RF power using the plasma etcher with CF_4 gas flow, since excessively high energy is not necessary.

The only difference between the preparation of E-HEMT by fluorine plasma treatment and the conventional HEMT preparation is the introduction of fluorine plasma ion implantation between the passivation layer deposition and the gate metal preparation. The Si_3N_4 dielectric under the gate is etched off first with CF_4 and then by the CF_4 plasma treatment of the exposed AlGaN surface, both implanting F^- ions into AlGaN and slightly etching AlGaN to form a shallow recessed-gate structure.

12.2.2 Direct Current, Breakdown, and Small-Signal Characteristics of E-Mode Devices

To investigate the effect of implantation dose on the device, four samples were prepared under different conditions: Sample 1 without fluorine plasma treatment (D-mode device), Sample 2 with fluorine plasma treatment (RF power 55 W and treatment time 150 s), Sample 3 with fluorine plasma treatment (RF power 150 W and treatment time 150 s), and Sample 4 with fluorine plasma treatment (RF power 250 W and treatment time 150 s), which are distributed evenly across the wafer, as shown in Figure 12.6.

The output characteristics of the four samples under the same gate bias are shown in Figure 12.7. The source drain current decreases with increasing radio frequency (RF) power of the fluorine plasma treatment, showing that the implanted negatively charged fluoride ions in the AlGaN layer lead to the depletion of channel electrons, which is required for enhancement-mode operations.

The transfer curves under varied fluorine plasma treatment conditions in Figure 12.8 (Quan et al. 2009) show an increasing threshold voltage with greater fluorine plasma power with identical treatment time. The output characteristics of a conventional D-HEMT with a threshold voltage of –3.1 V and an E-HEMT with a threshold voltage of 0.57 V prepared on Sample 3 using fluorine plasma treatment are shown in

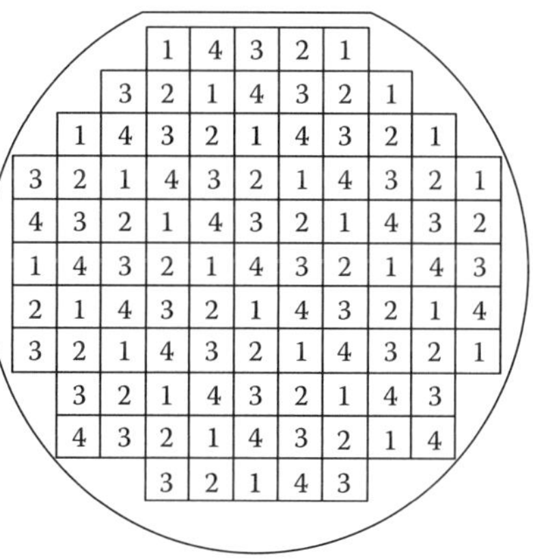

Figure 12.6 Four parts divided on the wafer. (After Quan, S. et al., *J. Semiconduct.*, 30(12), 124002 (4 pp.), 2009.)

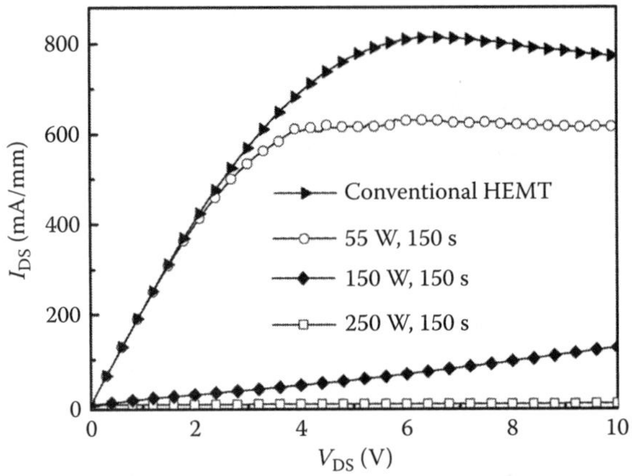

Figure 12.7 Source-drain current after fluorine plasma treatment. (Courtesy of Dr. Si Quan.)

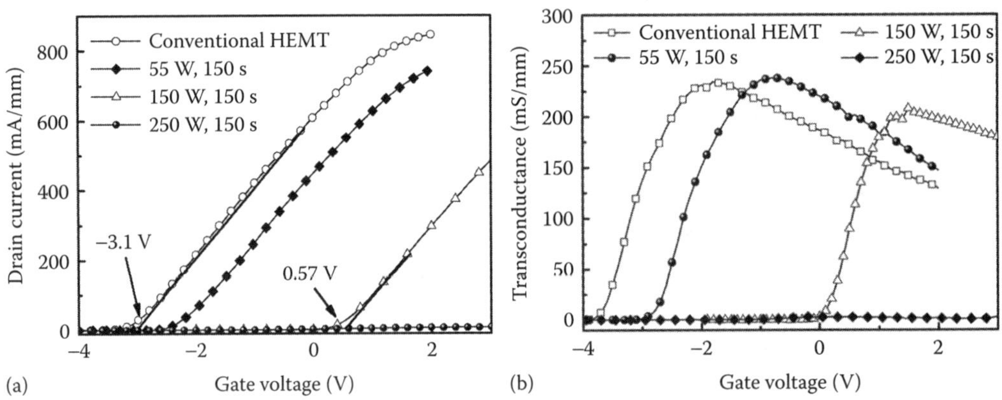

Figure 12.8 Transfer curves under different fluorine plasma treatment conditions. (After Quan, S. et al., *J. Semiconduct.*, 30(12), 124002 (4 pp.), 2009.)

Figure 12.9. Both the transfer and output characteristics show that the depletion of channel 2DEGs by fluoride ions are enhanced, owing to a higher fluoride ions density with increasing fluorine plasma power. The off-state breakdown characteristics measurements (with −8 V applied on the gate) show that Sample 1 (the conventional HEMT) and Samples 2 and 3, all have a breakdown voltage in the vicinity of 190 V, suggesting that fluorine plasma treatment has no influence on the breakdown voltage. It is worth noting that fluorine plasma treatment also causes a slight decrease in transconductance (Figure 12.8b). The AC small-signal characteristics measurements also show a reduced cutoff frequency f_T and maximum oscillation frequency f_{max} (Figure 12.10) with greater reduction at higher fluorine plasma power. With further increase in the plasma power (Sample 4), the maximum saturation current density and transconductance were measured to be merely a few mA/mm and 2.5 mS/mm, respectively (Figure 12.8), indicating failure of the channel damaged by too high a

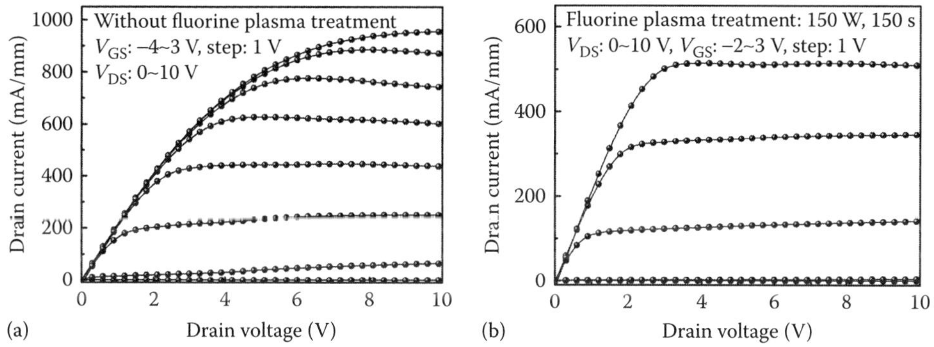

Figure 12.9 HEMT output characteristic curves under different fluorine plasma treatment conditions. (a) Courtesy of Dr. Si Quan. (b) (After Quan, S. et al., *J. Semiconduct.*, 30(12), 124002 (4 pp.), 2009.)

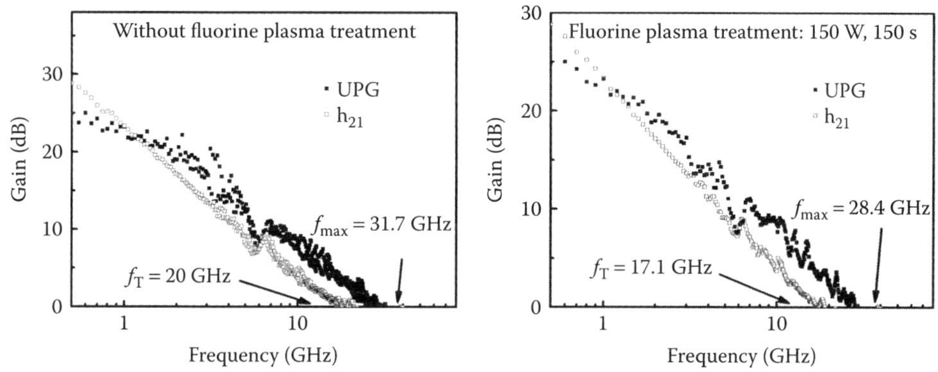

Figure 12.10 Small signal characteristics under different fluorine plasma treatment conditions. (a) Conventional D-mode device. (b) E-mode device with a fluorine plasma power of 150 W. (Courtesy of Dr. Si Quan.)

fluorine plasma power. Fluorine plasma (150 W, 150 s) proves to be a desirable treatment condition, where the device turns into enhancement mode, with small effect on the transconductance and current.

12.2.3 Gate-Drain Diode Measurements of the Fluorine Plasma Treated Devices

We measured the *I-V* characteristics of the gate-drain diode of the samples with different fluorine plasma treatment conditions. The results in Figure 12.11a demonstrate that the reverse gate leakage current is smaller at a higher fluorine plasma RF power. The very small change in the Schottky forward current, as shown in Figure 12.11, suggests little variation in the Schottky barrier height before and after the fluorine plasma treatment. The comparative analysis of the heterostructures with and without fluorine plasma treatment by X-ray photoelectron spectroscopy is presented in Figure 12.11b. Besides an increased fluorine composition F1s, a significantly higher oxygen content O1s is also observed at the fluorine-plasma-treated AlGaN surface, mainly because of the oxide layer formed on the material surface oxidized by the oxygen purging gas

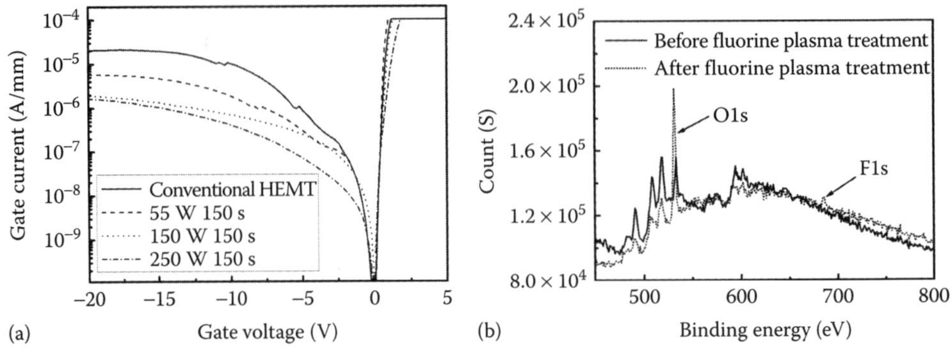

Figure 12.11 Gate current curves of devices with different fluorine plasma treatment conditions and the X-ray photoelectron spectrum of the AlGaN surface layer before and after the fluorine plasma treatment. (a) Gate current curves. (After Quan, S. et al., *J. Semiconduct.*, 30(12), 124002 (4 pp.), 2009.) (b) X-ray photoelectron spectrum. (Courtesy of Dr. Si Quan.)

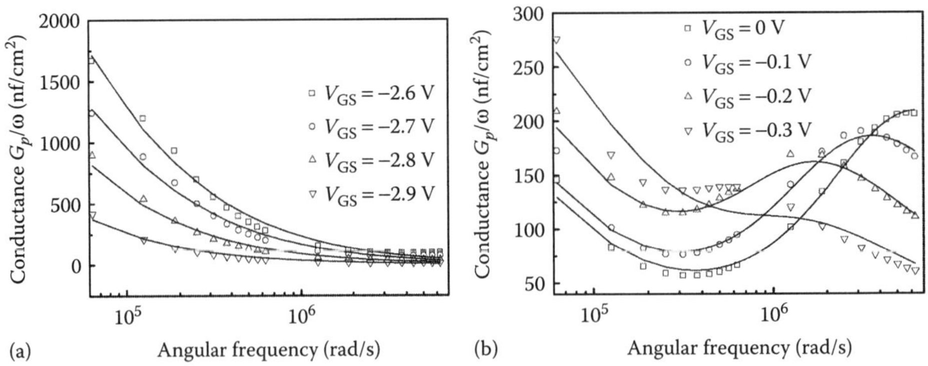

Figure 12.12 Conductance-frequency curves of the conventional D-mode AlGaN/GaN HEMT and the fluorine-plasma-treated E-mode HEMT. The unsigned solid line is the fitting curve. (a) D-mode HEMT. (b) E-mode HEMT. (After Quan, S. et al., *Chin. Phys. B*, 20(1):018101 (4 pp.), 2011b.)

during the fluorine plasma treatment. The treatment also leads to the reduction in surface states, thus the number of active traps is reduced in the trap-assisted tunneling, exhibiting a decrease in the reverse gate leakage current.

To compare the trap effects of the conventional device and E-mode device by fluorine plasma treatment, the conductance-frequency measurements were conducted at a voltage in the proximity of the threshold voltage (Quan et al. 2011b). From the measured results shown in Figure 12.12, we can see that the D-mode device exhibits one peak at the relatively low frequency, whereas the E-mode device has two peaks at a relatively high and a relatively low frequency, respectively. It is suggested that in the D-mode device, only one type of traps, namely the slow state traps, exists, but in the E-mode device, two types of traps exist, namely the fast state and the slow state traps, according to their different time constants.

For continuous trap levels, the equivalent parallel capacitance C_p and the conductance G_p are related to frequency by (Stoklas et al. 2008):

$$C_p = C_b + \frac{eD_T}{\omega\tau_T \tan(\omega\tau_T)} \tag{12.1}$$

$$\frac{G_p}{\omega} = \frac{eD_T}{2\omega\tau_T} \ln[1+(\omega\tau_T)^2] \tag{12.2}$$

where:

e is the elementary charge
D_T is the trap density
τ_T is the trap time constant
C_b is the barrier capacitance
ω the angular frequency

In the case of two types of traps (the physical quantity subscripts for the fast and slow states are (f) and (s), respectively), Equation 12.2 can be transformed into:

$$\frac{G_p}{\omega} = \frac{eD_{T(f)}}{2\omega\tau_{T(f)}} \ln[1+(\omega\tau_{T(f)})^2] + \frac{eD_{T(s)}}{2\omega\tau_{T(s)}} \ln[1+(\omega\tau_{T(s)})^2] \tag{12.3}$$

The trap density D_T and time constant τ_T can be extracted by fitting the measured $G_p(\omega)$ (the D-mode device by Equation 12.2 and the E-mode device by Equation 12.3). For the E-mode device, the extracted slow state trap density and time constant are $D_{T(s)} = (2–6) \times 10^{12}$ cm^{-2}/eV and $\tau_{T(s)} = (0.5–6)$ ms, respectively, and the fast state density and time constant are $D_{T(f)} = (1–3) \times 10^{12}$ cm^{-2}/eV and $\tau_{T(f)} = (0.2–2)$ μs, respectively. The D-mode device only has slow state traps whose density and time constant are $D_{T(s)} = (1–5) \times 10^{13}$ cm^{-2}/eV and $\tau_{T(s)} = (0.5–6)$ ms, respectively.

From the above fitting results, we can see that the slow state trap density of the E-mode device is one order of magnitude lower than that of the D-mode device, suggesting that fluorine plasma treatment causes a reduction in slow state traps. The slow state traps should be the surface state traps participating in the Schottky gate tunneling leakage, since the time constant of such traps is found to be identical to the reported surface trap time constant in Stoklas et al. (2008), and the Schottky reverse leakage current, which depends mainly on the surface trap assisted tunneling, decreases notably after the fluorine plasma treatment. A new type of fast state traps is generated in the fluorine-plasma-treated device, possibly due to the etch damage during the treatment, and is related to the decreased transconductance of the treated device.

12.3 Evaluation of the Reliability of Fluorine-Plasma-Treated E-HEMTs

The reliability of GaN-based E-HEMT using fluorine plasma treatment is investigated in terms of the electrical stress and the thermal stress. On-state, off-state, and step stress are applied to test the device stability, and a failure analysis is made. Experiment on heated storage and annealing at different temperatures are performed to evaluate the high-temperature characteristics and the effect of annealing on the device reliability.

12.3.1 Degradation of Electrically Stressed E-HEMT with Fluorine Plasma Treatment

The first three samples in Section 12.2 are chosen to examine the degradation. In view of their different threshold voltages, an on-state stress with drain bias $V_{DS} = 20$ V is applied on all the three samples with gate biases that guarantee an identical current output of 300 mA/mm for all the devices. The stress time is 5000 s, and the device properties are measured at 10, 20, 50, 100, 200, 500, 1000, 2000, 3000, and 5000 s, respectively.

The degradation of transfer characteristics of the devices under on-state stress is shown in Figure 12.13. No degradation is found in the conventional D-mode device (Sample 1), whereas a certain negative shift of the threshold voltage as well as an increase in the maximum saturation current and maximum transconductance are observed in the fluorine-plasma-treated devices, with a more significant change in Sample 2 than in Sample 3. According to the degradation behaviors of the characteristic parameters of the stressed devices, the degradation mechanism is suggested to be the drift of fluoride ions away from the region under the gate due to the electrical stress, resulting in a decreased threshold voltage and an increased maximum transconductance and saturation current. In contrast, Sample 3 treated by the fluorine plasma in condition of 150 W and 150 s is more stable.

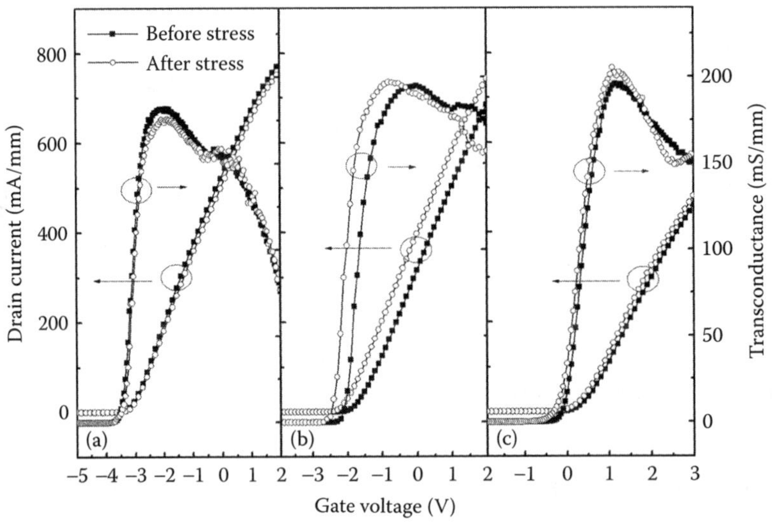

Figure 12.13 Device degradation under on-state stress. (a) Conventional D-mode device. (b) D-mode device with a fluorine plasma power of 55 W. (c) E-mode device with a fluorine plasma power of 150 W. (Courtesy of Dr. Si Quan.)

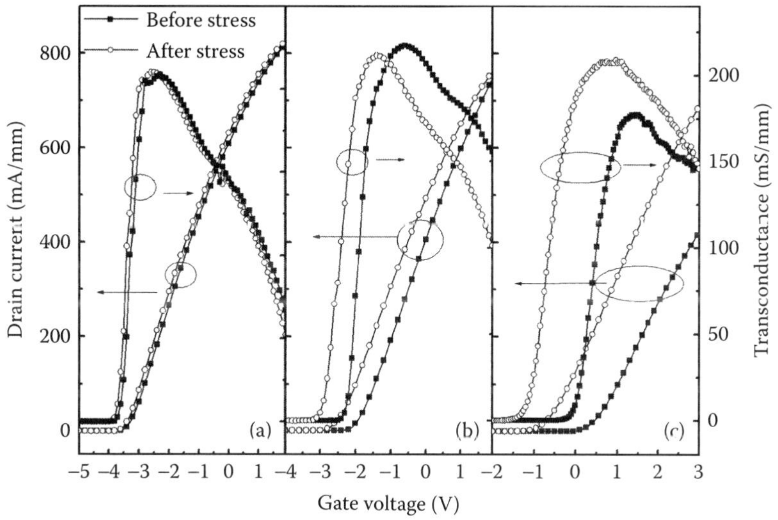

Figure 12.14 Transfer characteristics degradation under off-state stress. (a) Conventional D-mode device. (b) D-mode device with a fluorine plasma power of 55 W. (c) E-mode device with a fluorine plasma power of 150 W. (Courtesy of Dr. Si Quan.)

The off-state stress experiments still take the drain bias $V_{DS} = 20$ V for all the devices with a gate bias 2 V lower than the threshold voltage to make sure basically uniform electric fields under the gate in the different devices under stress. The stress time is 5000 s. The transfer characteristics degradation under off-state stress in Figure 12.14 reveals that a relatively small degradation occurs for the conventional D-mode device, while for both the fluorine plasma treated devices a notable negative shift of the threshold voltage occurs, and the shift is greater than that under on-state stress (Figure 12.13).

Shown in Figure 12.15a is the threshold voltage degradation of Sample 3 under off-state stress, which suggests that the fluoride ions are more unstable under off-state stress than under on-state stress. The on-state stress is due primarily to the hot electrons, whereas the off-state stress is due to the high electric fields. Under the effect

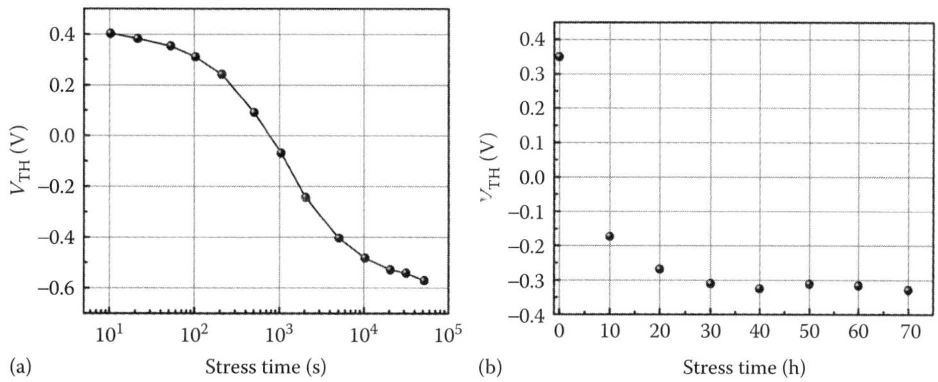

Figure 12.15 Degradation of E-mode device treated with a plasma power of 150 W under off-state stress. (a) Threshold voltage degradation. (b) Threshold voltage degradation under long-time off-state stress. (Courtesy of Dr. Si Quan.)

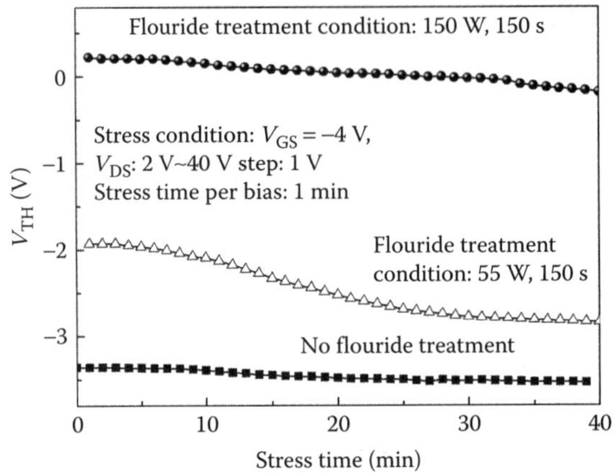

Figure 12.16 Threshold voltage degradation of the devices with different fluorine plasma powers under drain step stress. (Courtesy of Dr. Si Quan.)

of the high electric field, fluoride ions drift reversely along the electric field lines that converge from source and drain to gate, thus resulting in an outward diffusion of the fluoride ions. The decreased concentration of fluoride ions under the gate leads to a negative shift in the threshold voltage and an increase in transconductance. Note that the device degradation slows down after being stressed for 10,000 s. With further increase in stress time, the threshold voltage degradation is obvious for the first 30 h and then remains nearly invariant, as shown in Figure 12.15b. This indicates that a drift of the fluoridc ions occurs under the off-state high-electric-field stress and then stabilizes at a certain time when the drift stops, with no further degradation of the threshold voltage.

From Figure 12.16, one can see that the degradation in threshold voltage of the device with 55 W plasma power is much faster than that of the device with 150 W plasma power, which may be explained by the greater energies and thus more stable states of the fluoride ions at higher fluorine plasma power. For a further analysis of the effect of the fluorine plasma power on the gate current, the reverse gate leakage current is measured, with gate step stress applied on the AlGaN/GaN Schottky gate-drain diodes with varied fluorine plasma powers. The results shown in Figure 12.17 suggests that the least leakage current degradation is achieved with a plasma RF power of 150 W, whose mechanism requires further study.

12.3.2 Degradation of Electrical Characteristics of Fluorine-Plasma-Treated E-HEMTs at High Temperatures

High-temperature reliability is an important requirement of GaN wide-band-gap electronic devices. The best-performance Sample 3 was stored at high temperature, with its characteristics measured before and after the heated storage, as shown in Figure 12.18. After storage at 200°C for 100 h, the threshold voltage of the device remains constant, with little change in the maximum transconductance and the

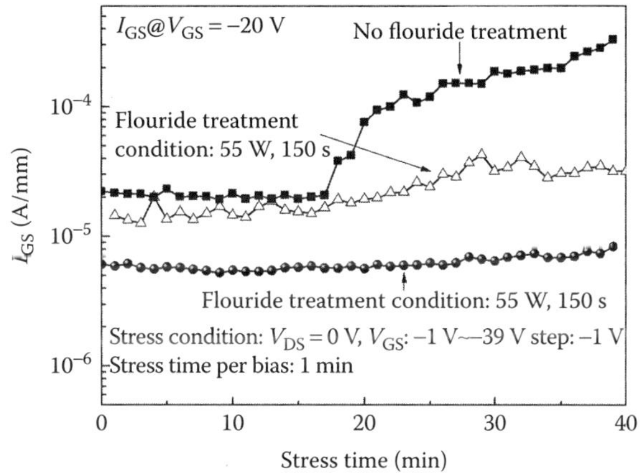

Figure 12.17 Current degradation of gate-drain diodes of the devices with different fluorine plasma powers under gate step stress. The curves for untreated device and 150 W treated device are after Ref. (Quan et al. 2011b), and the curve for 55 W treated device is courtesy of Dr. Si Quan.

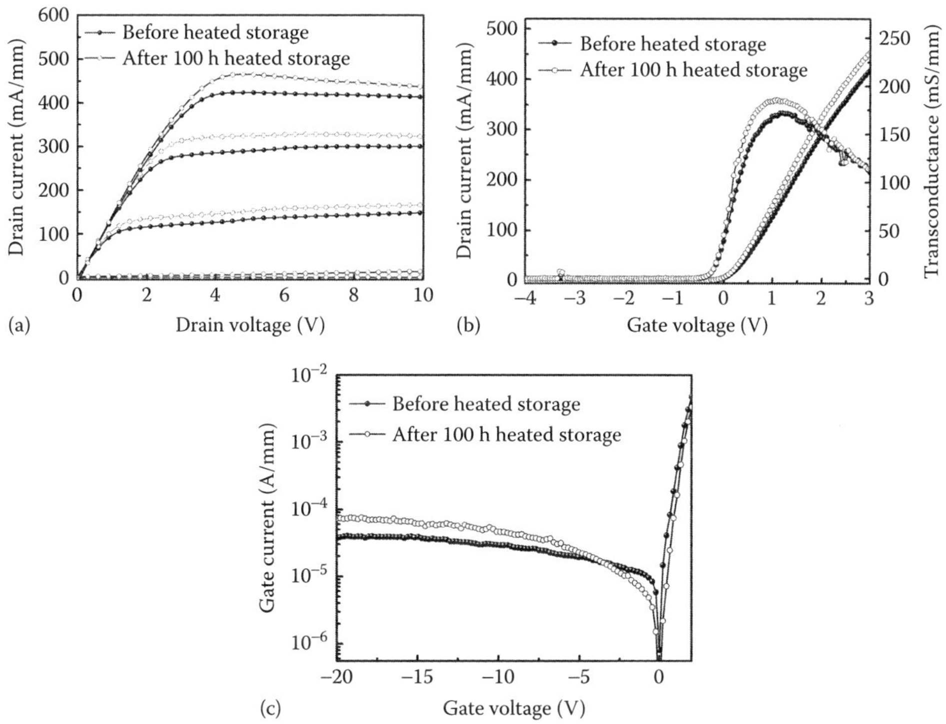

Figure 12.18 Degradation of the device stored at 200°C for 100 h. (a) Output characteristics degradation. (b) Transfer characteristics degradation. (c) Gate current degradation. (Courtesy of Dr. Si Quan.)

maximum saturation current, indicating that no drift occurs of the fluoride ions, which are rather stable at 200°C.

The reliability at higher temperature of the E-mode device with fluorine plasma treatment was investigated using 2 min annealing in nitrogen ambient at 300°C, 400°C, and 500°C, respectively.

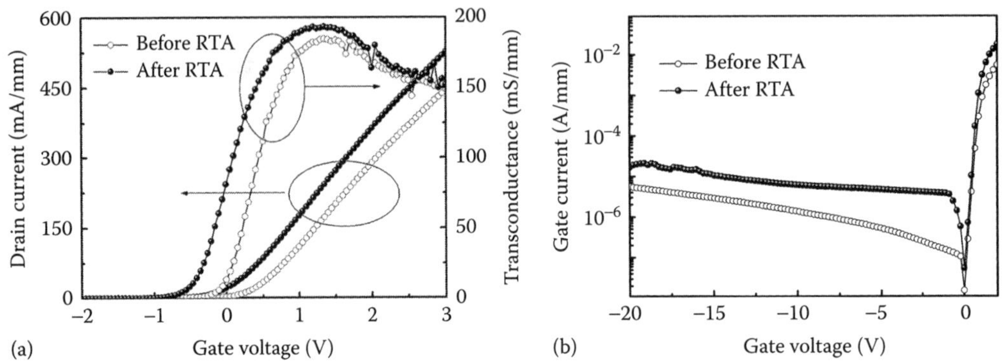

Figure 12.19 Direct current characteristics degradation of the fluorine-plasma-treated E-mode device annealed at 300°C. (a) Transfer characteristics degradation after annealing. (b) Gate current characteristics degradation after annealing. (Courtesy of Dr. Si Quan.)

As shown in Figure 12.19, after 300°C annealing for 2 min, a negative shift of the threshold voltage is observed with an increase in the maximum transconductance. A greater negative threshold voltage shift and maximum transconductance increase are seen at elevated annealing temperatures of 400°C and 500°C. This indicates that the fluoride ions obtain energies and diffuse after high-temperature annealing, with part of the fluoride ions moving from the gate region into the active gate-drain or gate-source region, thus resulting in a reduced threshold voltage.

The reliability characterization is performed after annealing at different temperatures to analyze the effect of rapid thermal annealing on the device. The threshold voltage was measured with gate step stress applied on the device, as shown in Figure 12.20. The device annealed at 300°C or 2 min exhibits a negative shift of the threshold voltage under gate step stress, suggesting that such an annealing energy is inadequate to stabilize the fluoride ions in the device. The device annealed at 400°C for 2 min displays a constant threshold voltage under gate step stress, showing good stabilizing effect of the 400°C, 2 min rapid thermal annealing on the fluoride ions.

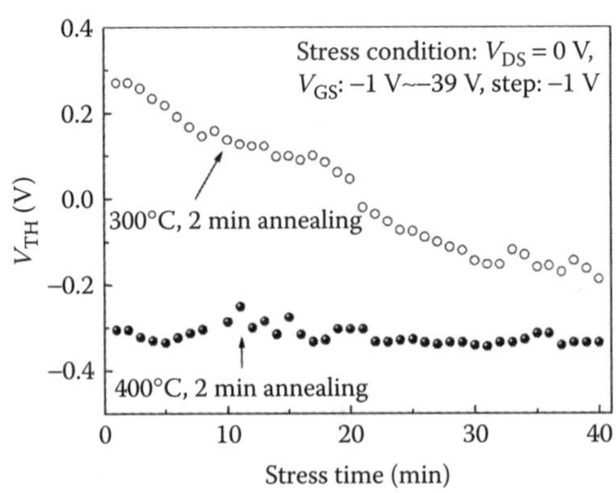

Figure 12.20 Threshold voltage degradation of the device annealed at different temperatures under gate step stress. (Courtesy of Dr. Si Quan.)

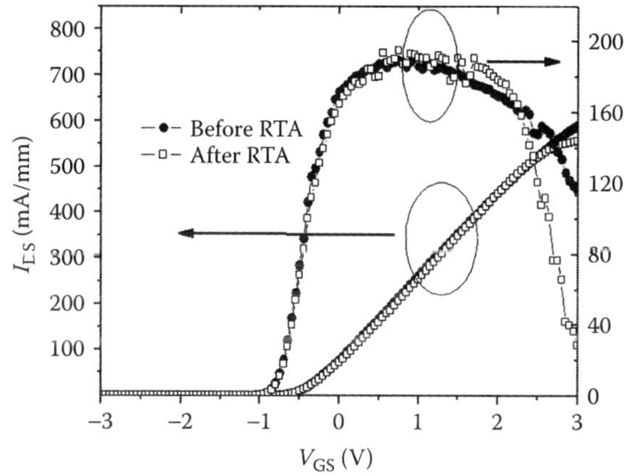

Figure 12.21 Degradation of the fluorine-plasma-treated E-mode device after secondary annealing at 400°C for 2 min. (Courtesy of Dr. Si Quan.)

The device annealed at 400°C for 2 min is annealed with the same conditions for a second time and little further degradation of the device is found, as shown in Figure 12.21, suggesting a good thermal stability of the device owing to the diffused and stabilized fluoride ions by the first rapid thermal annealing.

12.4 Structure Optimization of Fluorine-Plasma-Treated E-HEMTs

The properties of the E-mode devices with thinner AlGaN barrier layers are investigated. The growth of a Si_3N_4 passivation layer on the thin AlGaN barrier layer of the AlGaN/GaN heterostructure yields a markedly increased channel carrier concentration, because Si_3N_4 not only provides part of stress for the AlGaN layer to enhance the piezoelectric polarization strength in AlGaN but also changes the surface potential of AlGaN barrier layer, thus resulting in a higher 2DEG density. The increase in 2DEG density can decrease the gate-source and gate-drain resistance, and high performance E-mode devices are achievable with thin barriers under the gate.

We prepared conventional D-mode and fluorine-plasma-treated E-mode devices on the GaN/AlGaN/AlN/GaN heterostructure. The GaN/AlGaN/AlN compound barrier layer has a smaller total thickness than the 22-nm-thick AlGaN barrier layer discussed before, which is reduced to 16 and 8 nm by growth process control, with an invariant GaN cap layer thickness of 1.5 nm and an AlN interface interlayer thickness of 1.2 nm. All the Si_3N_4 passivation layers have a thickness of 60 nm.

12.4.1 Conventional HEMTs with Thin Barrier Layers

The transfer characteristics and gate-drain diode *I-V* characteristics of the conventional HEMTs with barriers of three thicknesses are shown in Figure 12.22. According to Figure 12.22a, the threshold voltage increases with decreasing barrier

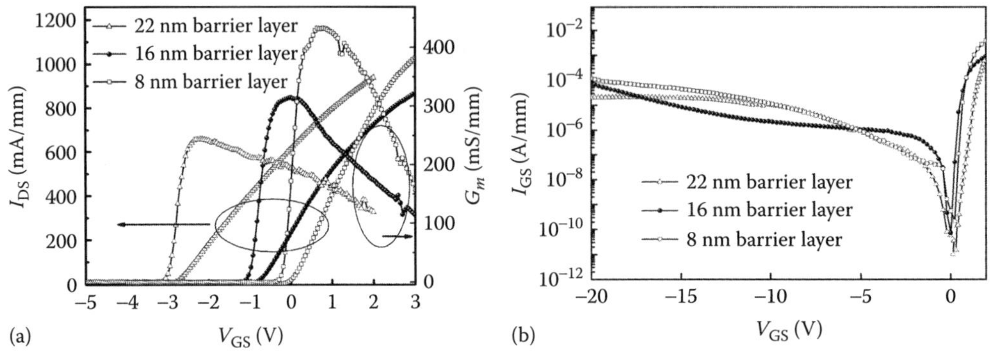

Figure 12.22 Transfer characteristics and gate-drain diode *I-V* characteristics of conventional HEMTs with the 22-nm-thick barrier layer and the 16- and 8-nm-thick barrier layers. (a) Transfer characteristics and (b) *I-V* characteristics of gate-drain diode. (Courtesy of Dr. Si Quan.)

thickness from –2.8 V (22-nm-thick barrier) to –0.7 V (16-nm-thick barrier) and 0.2 V (8-nm-thick barrier), thus realizing E-mode operations without the use of fluorine plasma treatment.

The thin-barrier-layer device has an equivalent maximum saturation current density but a significantly increased maximum transconductance compared with the 22-nm thick-barrier device, owing to enhanced gate depletion and control of channel electrons with thinner barrier layer under the gate, thus resulting in a positively shifted threshold voltage and greater transconductance. Meanwhile, the channel carrier concentration outside the gate region of the thin-barrier device remains about the same as that of the thick barrier device owing to the Si_3N_4 passivation layer, thus resulting in maximum saturation current equivalent to that of the conventional device. As shown in Figure 12.22b, both the forward and reverse gate currents of the gate-drain diode increase with the decrease in barrier thickness due mainly to higher electron tunneling probability caused by the narrower electron tunneling width, resulting from the thinner barrier layer, thus resulting in an increase in the electron tunneling-induced gate current.

The off-state breakdown characteristics of the HEMTs with barriers of the three thicknesses shown in Figure 12.23 were measured at a gate voltage of –6 V with a sweep of drain voltage from 0 to 200 V. The breakdown voltage was defined as the voltage at which the leakage current density is 1 mA/mm. The measured breakdown voltages for the 22-nm thick-barrier device and the 16-nm thin-barrier device are 190 and 170 V, respectively; the breakdown voltage of the 8-nm thin-barrier device is beyond the measurement range, with a mere current of 80 μA/mm at 200 V. The results show that the device breakdown voltages are not directly linked to the barrier thickness. The difference in the breakdown voltage of the devices may originate from the wafer disparity. The reduction in barrier layer thickness has no visible effect on lowering the breakdown voltage.

The high-frequency small-signal measurements display improved frequency characteristics of the thin-barrier devices, with f_T and f_{max} of 24.5 and 39.7 GHz,

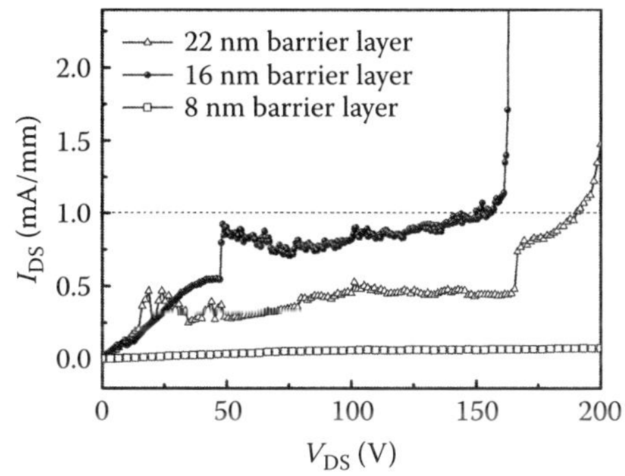

Figure 12.23 Off-state breakdown characteristics of the HEMTs with barriers of three thicknesses. (Courtesy of Dr. Si Quan.)

respectively, for the 16-nm thin-barrier device, and of 27.5 and 58 GHz, respectively, for the 8-nm thin-barrier device. The comparatively smaller increase in the cutoff frequency is due to the rise in both transconductance and gate source capacitance in devices with reduced barrier thickness, since f_T is in direct proportion to the ratio of transconductance to gate-source capacitance. The apparent increase in the maximum oscillation frequency suggests a weak parasitic effect in the operation of the thin-barrier devices.

12.4.2 Fluorine-Plasma-Treated E-Mode Devices with Thin Barrier Layers

What if fluorine plasma treatment is introduced in addition to the thin barrier layer? The transfer characteristic curves for the fluorine-plasma-treated HEMTs with 16 and 8 nm barrier layers, shown in Figures 12.24 and 12.25, respectively, exhibit a similar pattern to that in Section 12.2: the stronger the fluorine plasma treatment condition (high RF power or long time), the greater the positive shift of the threshold voltage and the transconductance reduction; however, the breakdown voltage remains basically the same. For 16-nm thin-barrier devices, the fluorine plasma treatment

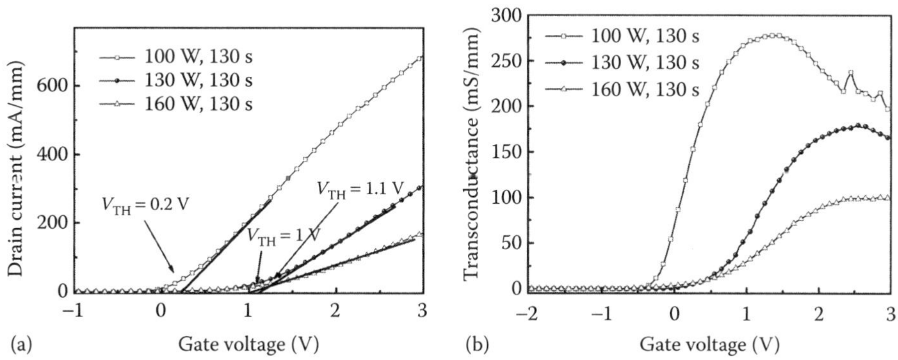

Figure 12.24 Transfer characteristic curves of fluorine-plasma-treated E-mode HEMT with 16-nm barrier layer. (a) Drain current vs. gate voltage and (b) transconductance vs. gate voltage. (Courtesy of Dr. Si Quan.)

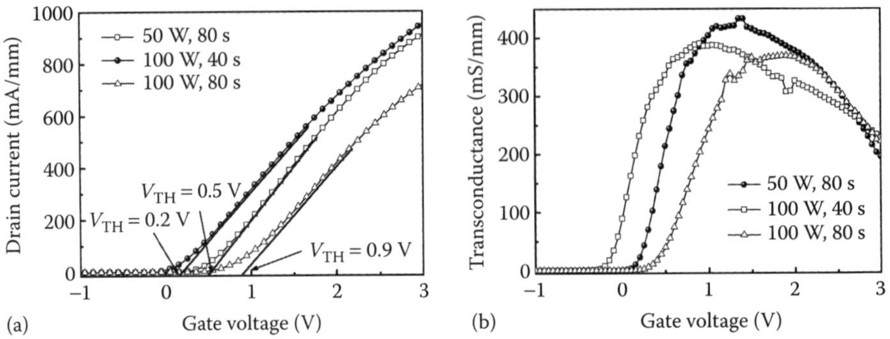

Figure 12.25 Transfer characteristic curves of fluorine-plasma-treated E-mode HEMT with 8-nm barrier layer. (a) Drain current vs. gate voltage and (b) transconductance vs. gate voltage. (Courtesy of Dr. Si Quan.)

(100 W and 130 s) is preferable to achieve E-mode operations, with slightly reduced transconductance and saturation current compared with those of the untreated devices. For 8-nm thin-barrier devices, fluorine plasma treatment (50 W and 80 s) is a desirable condition, under which a relatively high threshold voltage (0.5 V) and good transconductance and saturation current properties (430 mS/mm and 940 mA/mm, respectively) are achieved.

12.5 Enhancement/Depletion-Mode GaN Digital Integrated Circuits

As a wide-band-gap semiconductor material, GaN has much better radiation resistance properties than silicon and GaAs, thus applicable in space environment with strong radiation. The direct-coupled FET logic (DCFL), that is, the integrated E/D-mode FET structure, is adopted to form the basic unit for radiation-resistant GaN digital circuits.

12.5.1 Design of Enhancement/Depletion-Mode GaN Digital Integrated Circuit Units

Shown in Figure 12.26a is the circuit structure of an inverter consisting of a D-mode device and an E-mode device. The drain of the D-mode device is connected to high-level V_{DD}. The drain of the E-mode device is connected to both the source and the gate of the D-mode device as the output terminal, and the source of the E-mode device is grounded. The gate of the E-mode device serves as the input terminal of the inverter. For low-level input signals, the E-mode device is off and the D-mode device is on, with a high-level output signal. For high-level input signals, both the E-mode and D-mode devices are on, and the output voltage can be regulated by the design of the resistance ratio of the D-mode device to the E-mode device. The resistance of the D-mode device should be far greater than that of the E-mode device for low-level output at high-level input signals to realize the function of the inverter.

A 19-stage ring oscillator of the structure shown in Figure 12.26b is used to investigate the response speed of the inverter. After a cycle, the signal is changed and the

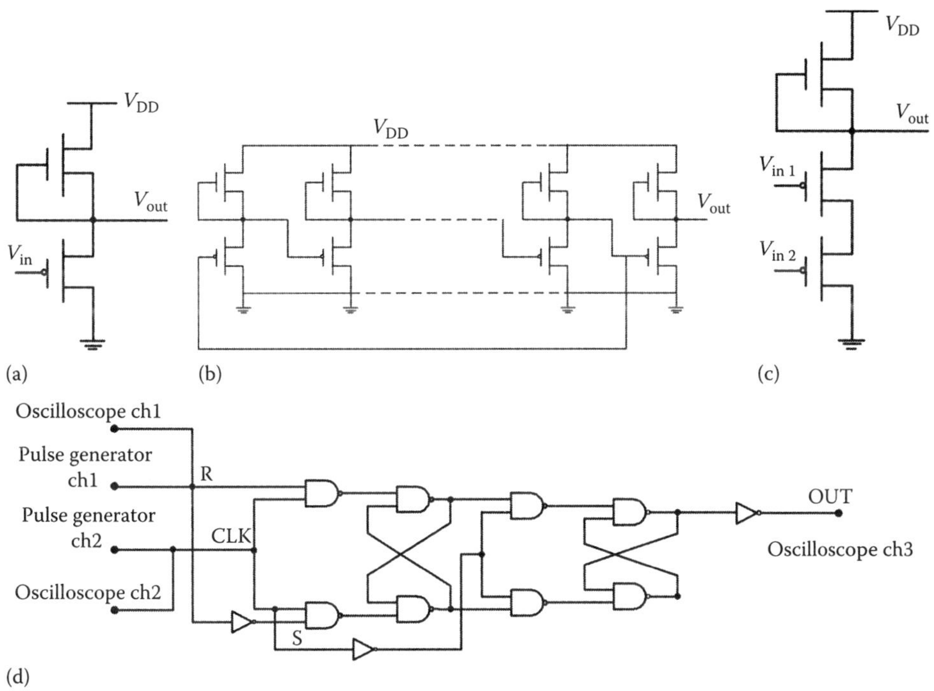

Figure 12.26 E/D-mode digital circuit units. (a) DCFL inverter, (b) 19-stage ring oscillator, (c) NAND gate, and (d) RS structure flipflop. (a, c, and d are After Xie, Y. et al., *J. Semiconduct.*, 32(6), 065001 (4 pp.), 2011.; b is Courtesy of Dr. Si Quan.)

ring oscillator period can be obtained by measuring the ring oscillator frequency f. With the number of stages of the ring oscillator n taken into consideration, the inverter delay τ_{pd} is given by

$$\tau_{pd} = (2nf)^{-1} \tag{12.4}$$

The designed NAND gate unit with the E/D structure is shown in Figure 12.26c, composed of a D-mode device and two E-mode devices. Based on this, an E/D flipflop consisting of 8 NAND gates and 3 inverters is also designed with the reset-set (RS) structure, as shown in Figure 12.26d.

12.5.2 Layout Design and Process of Digital Integrated Circuit Units

The circuit layout can be divided into 8 layers: the alignment mark (FIDU), the ohmic metal (OHMC), the mesa isolation, the D-mode recessed gate, the E-mode recessed gate, the gate metal, the interconnection opening, and the interconnection metal. The layouts of the inverter and NAND gate are shown in Figure 12.27. The same metal is used for the source of the D-mode device and the drain of the E-mode device, and a dual-gate E-mode device is employed to replace the two E-mode devices connected in series with the two gates as the inputs of the NAND gate. For a comparison of the properties of the inverters with different resistance ratios, we designed three inverters of a gate width ratio (E-mode/D-mode) of 10:1, 12.5:1, and 20:1, respectively.

Figure 12.27 (See color insert.) Digital circuit layout of inverter and NAND gate. (a) Inverter and (b) NAND gate. (After Xie, Y. et al., *J. Semiconduct.*, 32(6), 065001 (4 pp.), 2011.)

The GaN-based digital circuit processes are identical to the device preparation processes introduced earlier in this chapter, except for the separate recessed gate etching for D-mode and E-mode devices. The circuit preparation process flow is as follows: wafer cleaning, ohmic metal evaporation and annealing, mesa isolation, Si_3N_4 passivation layer deposition, D-mode device gate recess photolithography and etching, E-mode device gate recess photolithography and etching, E-mode device fluorine plasma treatment, annealing before gate preparation, evaporation of gate metal and first layer interconnect metal, second layer Si_3N_4 protective layer deposition, interconnection opening photolithography and etching, and interconnection metal evaporation. Shown in Figure 12.28 is the process flow to integrate the D-mode and E-mode devices for preparation of the NAND gate on heterostructure with 16-nm thin-barrier layer. The photographs of the prepared circuit modules are presented in Figure 12.29.

12.5.3 Measurements and Radiation Resistance Analysis of Digital Integrated Circuit Units

Direct-current measurements showed a threshold voltage of –0.8 V for D-mode devices and 0.8 V for E-mode devices. The output-input voltage relation curves of inverters with different size ratios are shown in Figure 12.30a. The greater the ratio of the gate width of the E-mode device to the D-mode device, the smaller the on-resistance ratio and the inverter low level, and a smaller input voltage is required with the same transfer resistance ratio, thus leading to a smaller threshold flipping voltage of the inverter.

The noise margin of the inverters with different size ratios is measured. Exchange the V_{in} axis and V_{out} axis of the output–input voltage relation curve and put the new curve in the same coordinate (Figure 12.30b) with the original curve. The low noise margin is defined as the measured input voltage corresponding to the first point with the slope of 1 of the original curve minus the input voltage corresponding to the first intersection point of the two curves, and the high noise margin as the input voltage corresponding to the third intersection point minus the input voltage corresponding to the second point with the slope of 1 of the original curve. The calculated noise margins of the inverters with different size ratios are listed in Table 12.1.

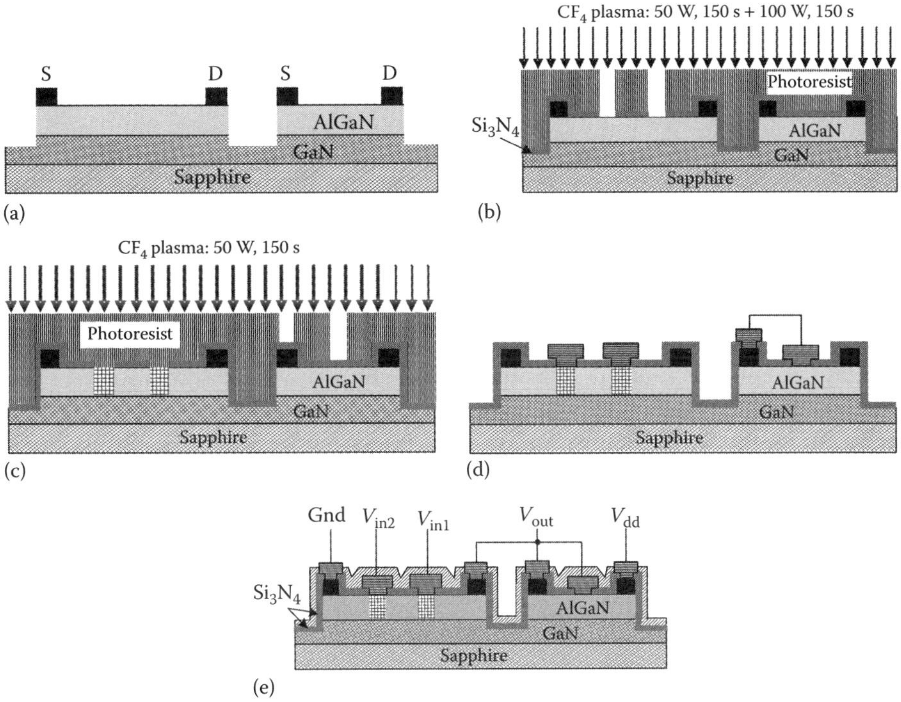

Figure 12.28 Process flow of NAND gate preparation based on the integration of D-mode and E-mode devices. (a) Mesa isolation and ohmic contact. (b) E-mode device recess region etching and fluorine plasma treatment. (c) D-mode device recess region etching. (d) Gate metal evaporation. (e) Metal interconnection. (After Xie, Y. et al., *J. Semiconduct.*, 32(6), 065001 (4 pp.), 2011.)

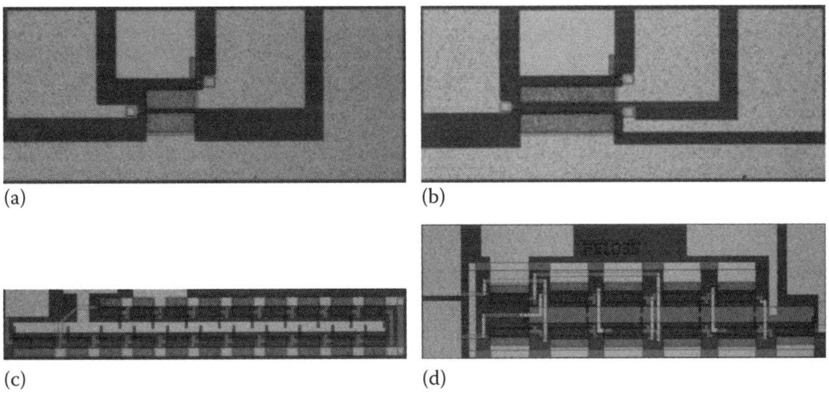

Figure 12.29 **(See color insert.)** Photographs of prepared circuit modules. (a) Inverter, (b) NAND gate, (c) ring oscillator, and (d) flipflop. (After Xie, Y. et al., *J. Semiconduct.*, 32(6), 065001 (4 pp.), 2011.)

Shown in Figure 12.31 are the frequency characteristics of the inverter, with a rise time of 24 ns and a fall time of 10 ns. The ring oscillator output characteristics are observed by oscilloscope to which the output is connected, with the earth terminal of the ring oscillator grounded and V_{DD} terminal connected to a 2 V DC power supply. The measurement results in Figure 12.32 show that the oscillation frequency is 83 MHz, and the delay of each stage inverter is calculated by Equation 12.4 to be 317 ps.

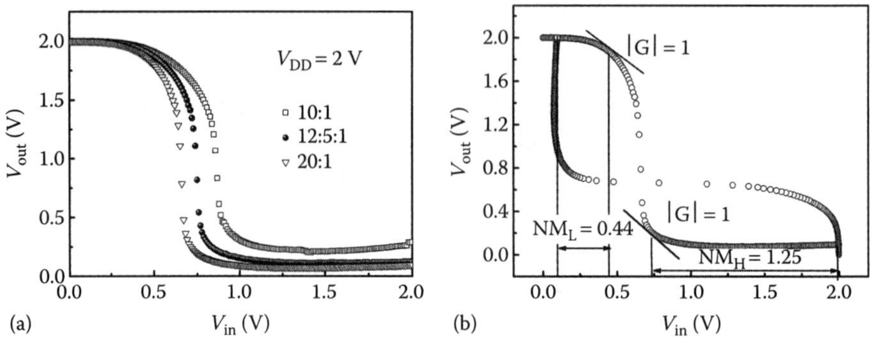

Figure 12.30 Inverter characteristics. (a) Output voltage versus input voltage of the inverters with different size ratios. (Courtesy of Dr. Si Quan.) (b) Calculation of inverter noise margin. (After Xie, Y. et al., *J. Semiconduct.*, 32(6), 065001 (4 pp.), 2011.)

Table 12.1 Noise Margins of the Inverters with Different Size Ratios

GATE WIDTH RATIO	LOW NOISE MARGIN (V)	HIGH NOISE MARGIN (V)
10:1	0.42	1.00
12.5:1	0.46	1.17
20:1	0.44	1.25

Source: Courtesy of Dr. Si Quan.

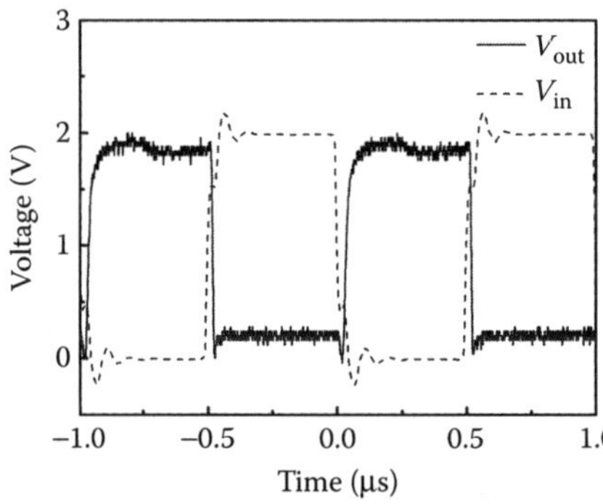

Figure 12.31 Inverter frequency characteristics measurements. (Courtesy of Dr. Si Quan.)

Measurements show that the logic function of the NAND gate and the flipflop is correct (Figure 12.33). The NAND gate has an E-mode/D-mode device gate width ratio of 20:1, an output high level of 2 V, and a low level of 0.2 V. The output high level of flipflop is 2.7 V and the low level is 0.5 V.

The radiation resistances of AlGaN/GaN D-mode and fluorine-plasma-treated E-mode devices with 16-nm barrier were measured by ^{60}Co gamma radiation with a total dose of 1 Mrad(Si) (Quan et al. 2011a). The variation in the AlGaN/GaN D-mode and E-mode devices with 16-nm barrier after the radiation is shown in Figures 12.34 and 12.35, respectively. A slight decrease in the maximum

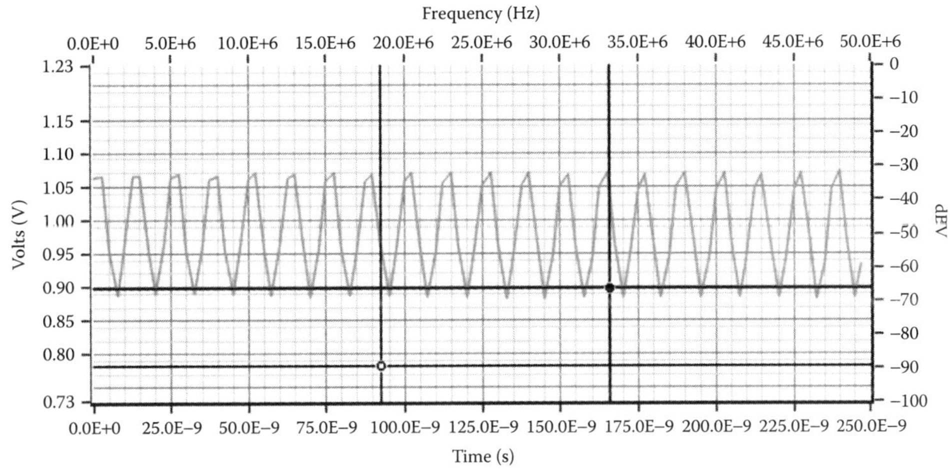

Figure 12.32 Ring oscillator measurements. (Courtesy of Dr. Si Quan.)

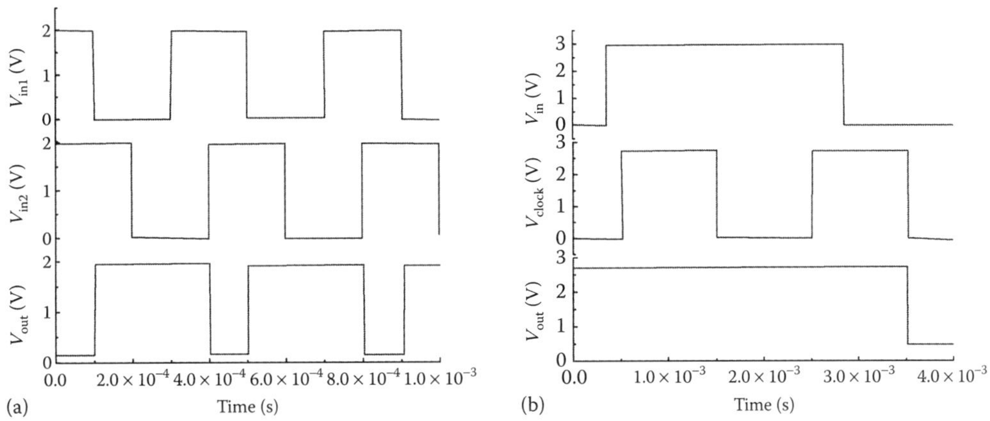

Figure 12.33 Measurements of NAND gate and flipflop. (a) NAND gate and (b) flipflop. (After Xie, Y. et al., *J. Semiconduct.*, 32(6), 065001 (4 pp.), 2011.)

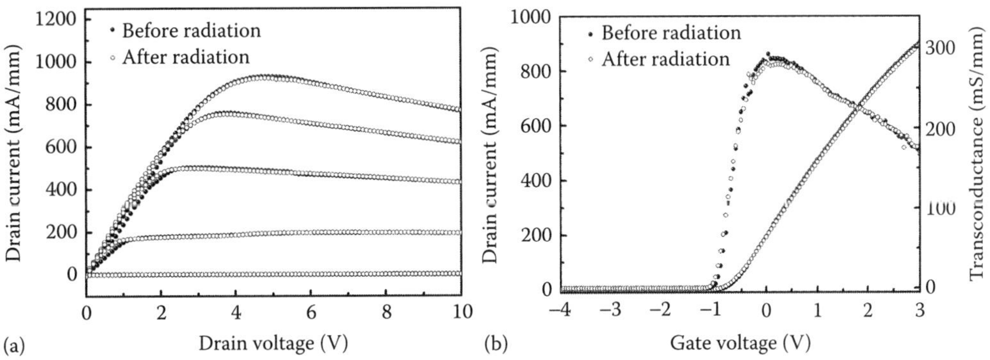

Figure 12.34 Degradation of output and transfer characteristics of the 16-nm thin-barrier AlGaN/GaN D-mode HEMT after radiation. (a) Output characteristics and (b) transfer characteristics. (After Quan, S. et al., *Chin. Phys. B*, 20(5):058501 (5 pp.), 2011a.)

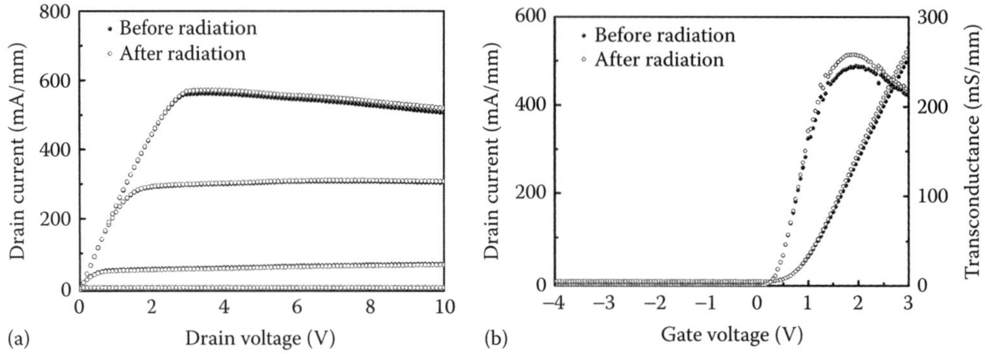

Figure 12.35 Degradation of output and transfer characteristics of the 16-nm thin-barrier AlGaN/GaN E-mode HEMT after radiation. (a) Output characteristics and (b) transfer characteristics. (After Quan, S. et al., *Chin. Phys. B*, 20(5):058501 (5 pp.), 2011a.)

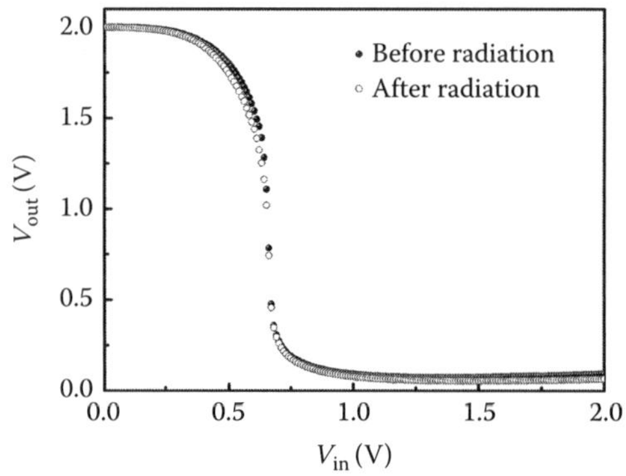

Figure 12.36 Degradation of the direct current characteristics of 16-nm thin-barrier AlGaN/GaN E/D inverter after radiation. (Courtesy of Dr. Si Quan.)

transconductance and the maximum saturation current for the D-mode device with an invariant threshold voltage is observed. Therefore, the AlGaN/GaN D-mode and E-mode devices with 16-nm barrier exhibit good gamma radiation resistance, with almost unchanged characteristics under ^{60}Co gamma radiation with a total dose of 1 Mrad(Si).

Little degradation in the inverter characteristics is found in Figure 12.36. The oscillation frequency of the ring oscillator remains to be 83 MHz after radiation, suggesting stable inverter frequency characteristics at radiation stress without degradation in delay.

Since E/D circuits are the foundation of nitride integrated circuits and power switch circuits, the realization of E-mode devices by whatever techniques can promote the development of GaN-integrated circuits. The E-mode (normally off) GaN devices are also indispensable in power switch applications. As a result, E-mode GaN devices have been attracting extensive attention in recent years, with great potentials in research and application.

References

Cai, Y., Z. Q. Cheng, W. C. W. Tang, K. J. Chen, and K. M. Lau. 2005. Monolithic integration of enhancement- and depletion-mode AlGaN/GaN HEMTs for GaN digital integrated circuits. *International Electron Devices Meeting*, December 5–7, 2005, Piscataway, NJ.

Cai, Y., Y. Zhou, K. J. Chen, and K. M. Lau. 2005. High-performance enhancement-mode AlGaN/GaN HEMTs using fluoride-based plasma treatment. *IEEE Electron Device Letters* 26(7):435–437. doi: 10.1109/led.2005.851122.

Endoh, A., Y. Yamashita, K. Ikeda, M. Higashiwaki, K. Hikosaka, T. Matsui, S. Hiyamizu, and T. Mimura. 2004. Non-recessed-gate enhancement-mode AlGaN/GaN high electron mobility transistors with high RF performance. *Japanese Journal of Applied Physics, Part 1* (Regular Papers, Short Notes & Review Papers) 43(4B):2255–2258. doi: 10.1143/jjap.43.2255.

Feng, Z. H., R. Zhou, S. Y. Xie, J. Y. Yin, J. X. Fang, B. Liu, W. Zhou, K. J. Chen, and S. J. Cai. 2010. 18-GHz 3.65-W/mm enhancement-mode AlGaN/GaN HFET using fluorine plasma ion implantation. *IEEE Electron Device Letters* 31(12):1386–1388. doi: 10.1109/led.2010.2072901.

Hilt, O., F. Brunner, E. Cho, A. Knauer, E. Bahat-Treidel, and J. Wurfl. 2011. Normally-off high-voltage p-GaN gate GaN HFET with carbon-doped buffer. *23rd International Symposium on Power Semiconductor Devices and ICs (ISPSD)*, May 23–26, 2011, Piscataway, NJ.

Hilt, O., A. Knauer, F. Brunner, E. Bahat-Treidel, and J. Wurfl. 2010. Normally-off AlGaN/GaN HFET with p-type GaN gate and AlGaN buffer. *22nd International Symposium on Power Semiconductor Devices & ICs (ISPSD)*, June 6–10, 2010, Piscataway, NJ.

Ikeda, N., R. Tamura, T. Kokawa, H. Kambayashi, Y. Sato, T. Nomura, and S. Kato. 2011. Over 1.7 kV normally-off GaN hybrid MOS-HFETs with a lower on-resistance on a Si substrate. *23rd International Symposium on Power Semiconductor Devices and ICs (ISPSD)*, May 23–26, 2011, Piscataway, NJ.

Khan, M. A., Q. Chen, C. J. Sun, J. W. Yang, M. Blasingame, M. S. Shur, and H. Park. 1996. Enhancement and depletion mode GaN/AlGaN heterostructure field effect transistors. *Applied Physics Letters* 68(4):514–516. doi: 10.1063/1.116384.

Lanford, W. B., T. Tanaka, Y. Otoki, and I. Adesida. 2005. Recessed-gate enhancement-mode GaN HEMT with high threshold voltage. *Electronics Letters* 41(7):449–450. doi: 10.1049/el:20050161.

Lu, B., O. I. Saadat, and T. Palacios. 2010. High-performance integrated dual-gate AlGaN/GaN enhancement-mode transistor. *IEEE Electron Device Letters* 31(9):990–992. doi: 10.1109/led.2010.2055825.

Ota, K., K. Endo, Y. Okamoto, Y. Ando, H. Miyamoto, and H. Shimawaki. 2009. A normally-off GaN FET with high threshold voltage uniformity using a Novel Piezo neutralization technique. *2009 IEEE International Electron Devices Meeting (IEDM 2009)*, December 7–9, 2009, Piscataway, NJ.

Palacios, T., C. S. Suh, A. Chakraborty, S. Keller, S. P. DenBaars, and U. K. Mishra. 2006. High-performance E-mode AlGaN/GaN HEMTs. *IEEE Electron Device Letters* 27(6):428–430. doi: 10.1109/led.2006.874761.

Quan, S., Y. Hao, X.-H. Ma, and H.-Y. Yu. 2011a. Influence of 60Co gamma radiation on fluorine plasma treated enhancement-mode high-electron-mobility transistor. *Chinese Physics B* 20(5):058501 (5 pp.). doi: 10.1088/1674-1056/20/5/058501.

Quan, S., Y. Hao, X.-H. Ma, and H.-Y. Yu. 2011b. Investigation of AlGaN/GaN fluorine plasma treatment enhancement-mode high electronic mobility transistors by frequency-dependent capacitance and conductance analysis. *Chinese Physics B* 20(1):018101 (4 pp.). doi: 10.1088/1674-1056/20/1/018101.

Quan, S., Y. Hao, X. Ma, Y. Xie, and J. Ma. 2009. Enhancement-mode AlGaN/GaN HEMTs fabricated by fluorine plasma treatment. *Journal of Semiconductors* 30(12):124002 (4 pp.). doi: 10.1088/1674-4926/30/12/124002.

Stoklas, R., D. Gregusova, J. Novak, A. Vescan, and P. Kordos. 2008. Investigation of trapping effects in AlGaN/GaN/Si field-effect transistors by frequency dependent capacitance and conductance analysis. *Applied Physics Letters* 93(12):124103 (3 pp.). doi: 10.1063/1.2990627.

Suh, C. S., Y. Dora, N. Fichtenbaum, L. McCarthy, S. Keller, and U. K. Mishra. 2006. High-breakdown enhancement-mode AlGaN/GaN HEMTs with integrated slant field-plate. *2006 International Electron Devices Meeting*, December 11–13, 2006, Piscataway, NJ.

Uemoto, Y., M. Hikita, H. Ueno, H. Matsuo, H. Ishida, M. Yanagihara, T. Ueda, T. Tanaka, and D. Ueda. 2006. A normally-off AlGaN/GaN transistor with RonA = 2.6 mΩcm2 and BVds = 640V using conductivity modulation. *2006 International Electron Devices Meeting*, December 10–13, 2006, San Francisco, CA.

Uemoto, Y., M. Hikita, H. Ueno, H. Matsuo, H. Ishida, M. Yanagihara, T. Ueda, T. Tanaka, and D. Ueda. 2007. Gate injection transistor (GIT)—A normally-off AlGaN/GaN power transistor using conductivity modulation. *IEEE Transactions on Electron Devices* 54(12):3393–3399. doi: 10.1109/ted.2007.908601.

Xie, Y., S. Quan, X. Ma, J. Zhang, Q. Li, and Y. Hao. 2011. Monolithically integrated enhancement/depletion-mode AlGaN/GaN HEMT D flip-flop using fluorine plasma treatment. *Journal of Semiconductors* 32(6):065001 (4 pp.). doi: 10.1088/1674-4926/32/6/065001.

13

GaN Metal-Oxide-Semiconductor High Electron Mobility Transistors

AlGaN/GaN high electron mobility transistors (HEMTs) have great potentials in wireless communications and radar applications. However, the current leakage in conventional Schottky gate GaN electronic devices is a major hindrance. A degradation of critical device properties such as the breakdown voltage, efficiency, and gain is induced by the large reverse leakage current of the Schottky gate AlGaN/GaN HEMTs, due not only to the high-density dislocations in GaN, especially those penetrating to the material surface, but also to the constant high electric field and junction temperature to which GaN HEMT devices are exposed whether in microwave power or power switching applications. Besides the reverse leakage current, the forward gate conduction may occur in the Schottky gate AlGaN/GaN HEMT at a relatively large forward bias (>2V), in which case the device may fail. As a result, the Schottky gate cannot meet the requirements of enhancement-mode GaN devices typically operating at a large positive bias, since such high gate currents driven by the radio frequency (RF) signals lead to reduced breakdown voltage and PAE as well as an increased noise figure of the AlGaN/GaN HEMTs and affect the reliability of long-term operation of the device (Winslow and Trew 1994).

In order to suppress the gate current, the metal–insulator–semiconductor (MIS) structure can be introduced to the GaN HEMT gate to form the MIS-HEMT. Since a broad category of insulation dielectrics for semiconductor devices includes oxides, the insulated gate is mostly of the metal–oxide–semiconductor (MOS) structure, thus forming the MOS-HEMT. In this chapter, all the HEMTs with oxide as the gate dielectric are referred to as MOS-HEMTs, with MIS-HEMTs as the general designation of insulated-gate HEMTs. Experiments indicate that GaN MIS-HEMTs can improve the gate voltage swing, microwave power performance, and long-term reliability. A brief introduction of GaN MIS-HEMTs is given in this chapter, with emphasis on GaN MOS-HEMTs with high-*K* gate dielectrics.

13.1 Research Progress of GaN MIS-HEMTs

The introduction of a dielectric between the Schottky gate and the semiconductor of the GaN HEMT to form an MIS-HEMT can greatly reduce the gate current, thus a higher breakdown voltage and a large range of gate bias within which the device can operate. Both the positive and negative gate voltage swing of the MIS-HEMT increase, owing to the negative shift of the threshold voltage induced by the increased gate-to-channel spacing and gate forward turn-on voltage by the introduction of the gate dielectric. Although there is a drop in the transconductance peak value, the transconductance remains relatively constant within a relatively large range of gate voltage with greatly improved linearity, for which reason MIS-HEMTs enjoy good linearity in microwave power applications. The gate dielectric can also serve as the surface passivation layer to stabilize the surface states of the semiconductor material, thus resulting in effective suppression of current collapse, improvement of the microwave power properties, and long-term reliability of the device.

The first GaN MOS-HEMTs were developed by Khan et al. using SiO_2 deposited by plasma-enhanced chemical vapor deposition (Khan et al. 2000). The SiO_2 insulated gate MOS-HEMT and SiN insulated gate MIS-HEMT have a gate current 4–6 orders of magnitude lower than that of the HEMT structure, with stable performance at a forward gate voltage up to 6 V and a temperature up to 300°C (Khan et al. 2000, 2003). At an even higher temperature of 400°C, the MOS-HEMT devices remain stable under RF (2 GHz) power stress (Adivarahan et al. 2005a, b). Researches on field-plated MOS-HEMTs show that the MOS-HEMT devices maintain relatively high power output and low gate leakage current with long-term RF stress, whereas the HEMTs exhibit visible degradation in power and gate leakage current, which are closely related to the increase of the forward gate current with the stress time. Consequently, MOS-HEMTs have better RF reliability than HEMTs. The field-plated MOS-HEMTs even achieve an output power density of 19 W/mm at 2 GHz and $V_{DS} = 55$ V, with a significantly higher power voltage efficiency (PVE) (the ratio of RF power density to drain voltage) than the HEMT.

Besides good power characteristics, GaN MIS-HEMTs can also exhibit outstanding frequency characteristics. Higashiwaki et al. adopted 2 nm Si_3N_4 dielectric layer prepared by catalytic chemical vapor deposition (Cat-CVD) to successfully produce 60 nm gate length MIS-HEMTs with a cut-off frequency f_T of 163 GHz (Higashiwaki et al. 2006). The MIS-HEMT with a gate length of 30 nm exhibits an f_T of 181 GHz and an f_{max} of 186 GHz (Higashiwaki et al. 2008).

Marso et al. found that the Hall mobility of the AlGaN/GaN heterostructure decreases by 10% after SiO_2 deposition on the material surface, but the f_T and field-effect mobility of the MOS-HEMT with SiO_2 as gate dielectric are approximately 50% higher than those of the HEMT, because the MOS-HEMT screens the Coulomb scattering of surface charged defects induced by the gate metallization layer in the HEMT, thus resulting in a greatly improved channel conductivity (Marso et al. 2006). Romero et al. suggested that the AlGaN surface pretreatment with nitrogen plasma before SiN deposition can significantly reduce the gate leakage current and improve RF device performance (Romero et al. 2008) as a result of a 60% reduction in the SiN/AlGaN interface charge density compared with that in untreated devices.

The gate dielectric with a higher dielectric constant can increase the gate capacitance, thereby leading to a better control of the channel charge and less deleterious effect of the gate dielectric introduction on transconductance, as well as a notable advantage in device scaling. Besides SiO_2 (dielectric constant, $\varepsilon = 3.9$) and SiN ($\varepsilon = 7.0$), Al_2O_3 ($\varepsilon = 9.0$) is also widely employed as GaN MIS-HEMT gate dielectrics. Al_2O_3 dielectric layers can be prepared by the oxidization of aluminum deposited on nitride material surface (Hashizume et al. 2003) or the ALD (Ye et al. 2005). Apart from current collapse suppression and improved gate voltage swing, breakdown voltage, and current output, GaN MOS-HEMTs with Al_2O_3 as gate dielectric are found to have a higher channel electron mobility and even a greater electron saturation velocity than HEMTs, which are attributed to the passivation effect of the Al_2O_3 dielectric layer, offering better transport properties, hence resulting in the occurrence of an even higher MOS-HEMT transconductance than that of the HEMT (Kordos et al. 2007, Pozzovivo et al. 2007). Liu et al. investigated the temperature dependence of the forward gate leakage current in MOS-HEMT with Al_2O_3 gate dielectric layer and found that the Fowler-Nordheim tunneling is the dominant mechanism in high electric field and low temperature (<0°C) operations, whereas the trap-assisted tunneling plays a major role under moderate electric field and high temperature (>0°C) conditions (Liu et al. 2011). In 2011, we reported recessed-gate MOS-HEMTs using Al_2O_3 gate dielectric with an output power density of 13 W/mm and a record-high power-added efficiency (PAE) of 73% under a drain voltage $V_{DS} = 45$ V (Hao et al. 2011).

Some high-K dielectrics with a dielectric constant greater than 20 have also found successful applications in GaN MIS-HEMTs. We put forward MOS-HEMTs prepared by the deposition of stack gate HfO_2 (3 nm)/Al_2O_3 (2 nm) dielectric using atomic layer deposition (ALD) (Yue et al. 2008a,b). This dielectric incorporates the high-K properties of HfO_2 and the favorable interface properties of Al_2O_3, and besides the advantages of conventional MOS-HEMTs, the prepared device with 1 μm gate length also exhibits excellent frequency characteristics, with an f_T and f_{max} of 12 and 34 GHz, respectively. The direct employment of the HfO_2 dielectric layer also offers good passivation effect (Chang et al. 2006). A breakdown voltage of 400 V was achieved in the MIS-HEMTs prepared by Fujitsu Limited using 10 nm sputtered Ta_2O_5 gate dielectric, and the 1 mm gate width device achieved a power density of 9.4 W/mm, an efficiency of 62.5%, and a linear gain of 23.5 dB at 2 GHz. The output power degradation was less than 0.5 dB after 150 h stress test under P3dB (the output power corresponding to the gain 3 dB lower than the maximum) operation (Kanamura et al. 2008). The feasibility of using ZrO_2 and Pr_2O_3 as gate dielectrics for GaN MOS-HEMTs has also been verified (Chiu et al. 2008, Kuzmik et al. 2008).

The gate dielectrics Sc_2O_3 (Mehandru et al. 2003), NiO (Oh et al. 2004), low-temperature GaN dielectrics (Kao et al. 2005), and AlN (Selvaraj et al. 2007) have also been reported for GaN MOS-HEMTs. In recent years, there have been increasing studies on the application of the MIS structure in N-polar GaN HEMTs. N-polar GaN heterostructures possess lower ohmic contact resistance and better quantum confinement but suffer from substantial point defects in GaN and thus a large gate leakage current; therefore, the introduction of the MIS structure can greatly improve the device properties. Currently, the typical gate dielectric for N-polar MIS-HEMTs is SiN, and a PAE of 71% and an output power of 6.4 W/mm (Wong et al. 2009) have been reported. MIS-HEMTs have become a very important category of nitride semiconductor electronic devices.

13.2 Selection and Atomic Layer Deposition of High-*K* Gate Dielectrics

One of the trends in the development of MOS structure electronic devices is the adoption of high-*K* dielectrics. The basic electrical requirements of GaN MOS-HEMTs for the gate dielectric are a high permittivity *K*, a wide band gap, a large conduction band offset, and a high-quality MOS interface. As stated in Section 13.1, with a fixed gate dielectric thickness, the high permittivity can increase the gate capacitance and the transconductance as it is directly proportional to the gate capacitance, thus achieving better gate control over the channel, together with a remarkable advantage in device scaling. Large conduction band offset and band gap are of great importance in the reduction of gate leakage current by whether the thermionic emission or the tunneling mechanism. It is necessary to lower the density of interface states, which may increase the gate leakage current generated by trap-assisted tunneling.

13.2.1 Selection of High-K Gate Dielectrics

A high-*K* dielectric is defined as the dielectric whose permittivity *K* is greater than that of the SiO_2 ($K = 3.9$). As the earliest gate dielectric chosen for GaN MOS-HEMTs, SiO_2 enjoys mature preparation technologies and a good insulating ability but suffers a low dielectric constant, which is a major drawback. The improvement of transconductance by using high-*K* dielectrics is in favor of better GaN MIS-HEMT frequency characteristics and a greater frequency range in microwave power applications.

The basic properties of common gate dielectric materials are listed in Table 13.1. Figure 13.1 gives a comparison of the normalized theoretical transconductance values (with the transconductance of the Schottky gate HEMT as 1) of the GaN HEMT and the GaN MIS-HEMTs prepared using different dielectrics, showing that the dielectric of a greater permittivity has smaller effect on the gate capacitance, with transconductance of the prepared MIS-HEMT closer to that of the Schottky gate HEMT.

Among the widely employed gate dielectrics in GaN MIS-HEMTs are Si_3N_4 and Al_2O_3. With a comparatively low permittivity of 7 among the high-*K* dielectrics, Si_3N_4 is extensively employed as passivation layer for nitride devices with rather mature preparation process. Al_2O_3 dielectrics of the same thickness are superior in both gate capacitance and insulating ability to Si_3N_4, with the only drawback of a poorer passivation effect.

Table 13.1 Basic Properties of Common Gate Dielectric Materials

GATE DIELECTRIC	DIELECTRIC CONSTANT	BAND GAP (eV)	CRYSTAL STRUCTURE
SiO_2	3.9	8.9	Amorphous
Si_3N_4	7	5.1	Amorphous
Al_2O_3	9	8.7	Amorphous
SiON	3.9–7	–	Amorphous
HfO_2	25	5.9	Tetragonal
TiO_2	80	3.5	Tetragonal
ZrO_2	25	5.8	Tetragonal
Y_2O_3	15	5.9	Cubic
La_2O_3	30	4.3	Hexagonal
Ta_2O_5	26	4.3	Orthogonal

Source: Courtesy of Dr. Zhiwei Bi.

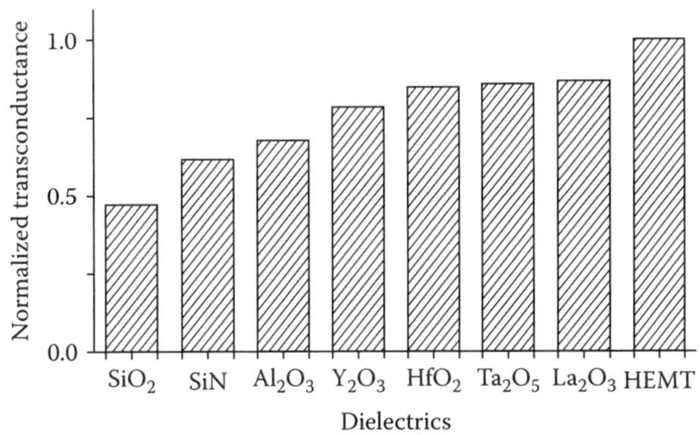

Figure 13.1 Normalized transconductance peak values of the GaN HEMT and the GaN MOS-HEMTs prepared using different dielectrics. (Courtesy of Dr. Zhiwei Bi.)

A good growth method for Al_2O_3 dielectric is the ALD, which is a very important technique for high-*K* film preparation. There are also a few reports on the GaN MIS-HEMT prepared using the high-*K* dielectrics, such as HfO_2. Owing to its very high permittivity of 25, a 20-nm-thick HfO_2 layer is equivalent to a 7.2-nm-thick Al_2O_3 dielectric under identical gate capacitance. A thicker dielectric offers better insulation and less difficulty in thickness control. However, the crystallization of a part of the HfO_2 dielectric may occur when it is exposed to high temperature, owing to its relatively low crystallization temperature (375°C), thus inducing device degradation. Presently, in the Hf-based dielectric researches, the addition of other elements such as aluminum or silicon into HfO_2 is adopted to increase its stability, but the complicated process and increased defects in the dielectric pose many problems that are yet to be solved.

13.2.2 Atomic Layer Deposition Process

Initially named the atomic layer epitaxy (ALE), the ALD is also referred to as the atomic layer chemical vapor deposition (ALCVD). The monoatomic or monomolecular layer is prepared by placing the base material in the heated ALD reaction chamber, with alternating pulses of at least two vapor phase precursors; the first precursor is chemisorbed on the base material surface until the adsorption saturates and stops automaticity, and the subsequently injected precursor reacts with the adsorbed first precursor until the first precursor is completely consumed. In this process, the surface reaction saturation can be controlled by regulating the reaction conditions such as chamber temperature and pressure and the precursor pulse time, and extra precursors and reaction byproducts are removed by inert gas flow cleaning. Sometimes, the chemical process is even more complicated when active agents are required depending on the different base materials and deposition precursors. The atomic layer deposition of thin film is achieved by the periodic repetition of this self-restrictive (for each reaction cycle, the amount of the deposited film material is invariant, i.e., a monoatomic or monomolecular layer) sequential chemical process. Proper process temperature can suppress the physisorption of the precursors and reaction by-product molecules on the base material surface. The materials that can be deposited for now include oxides, nitrides, fluorides, carbides, sulfides, metals, and composite structures.

Accurate control of the film thickness on an atomic level is possible with ALD for the realization of atomic layer-by-layer deposition via controllable saturated chemical reaction on the film surface. The deposited films have good three-dimensional morphology preservation, 100% step coverage, and good thickness uniformity. The ALD-grown large-area films have the advantages of no pinholes, low defect density, good adhesion, low stress, no damage to the base material, high production efficiency, and low cost.

The Al_2O_3 films grown by atomic layer deposition possess a far better quality than Al_2O_3 films obtained by sputtering and atomic beam evaporation and can form high-quality Al_2O_3/AlGaN interface with AlGaN. In addition to a large band gap and a high dielectric constant, Al_2O_3 also enjoys a high breakdown electric field (5–10 MV/cm), good thermal stability (amorphous up to 1000°C), and good chemical stability (no diffusion between Al_2O_3 and AlGaN).

The growth process of Al_2O_3 thin film by ALD for GaN MOS-HEMTs is as follows. Before gate metal evaporation, the sample is placed in the ALD equipment chamber with $Al(CH_3)_3$(TMA) and de-ionized water as the aluminum source and oxygen source, respectively, and nitrogen as the carrier gas to perform Al_2O_3 dielectric atomic layer deposition at 320°C in the following steps: aluminum source pipe cleaning in N_2 ambient → sample placement → vacuum pumping (until the vacuum level meets the process requirements) → temperature elevation to 320°C in protective nitrogen gas → surface preparation by spraying aluminum source → cyclic deposition (formation of Al_2O_3 by the surface reaction between the alternatively sprayed aluminum source and water onto the sample) → natural cooling in nitrogen flow. The AFM surface morphology and roughness of the AlGaN/GaN heterojunction before and after Al_2O_3 deposition, as shown in Figure 13.2, demonstrate that the ALD-deposited Al_2O_3 has good step coverage and uniformity with an improved surface roughness after deposition.

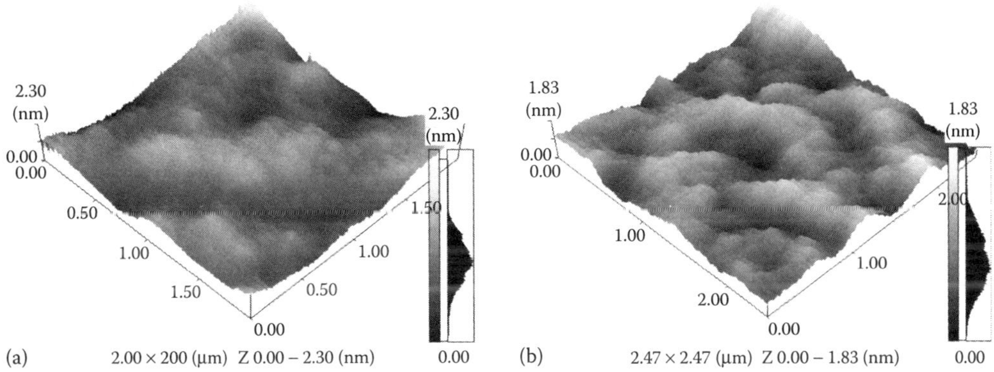

Figure 13.2 Surface roughness of AlGaN/GaN heterojunction before and after atomic layer deposition of Al_2O_3. (a) Before Al_2O_3 deposition (RMS = 0.278 nm). (b) After Al_2O_3 deposition (RMS = 0.268 nm). (After Yue, Y. et al., *Sci. Chin. Ser. E: Technol. Sci.*, 52(9), 2762–2766, 2009.)

The atomic layer deposition of HfO_2 is similar to that of Al_2O_3, with $HfCl_4$ as the hafnium source and H_2O as the oxygen source. The deposition of compound dielectrics such as HfAlO is performed by alternate deposition of Al_2O_3 and HfO_2, in which the hafnium content in the HfAlO dielectric layer can be regulated by changing the number of successively deposited layers for each dielectric.

13.3 Basic Characteristics and Interface State Density of AlGaN/GaN MOS Capacitors with High-*K* Gate Dielectrics

The MOS-gate capacitor characteristics have a direct relation with the characteristics of MOS-HEMTs. The quality of interface between the gate dielectric and the AlGaN barrier layer has great influence on the device performance, since high-density interface states may result in Fermi pinning, thus impairing effective gate control over the conducting channel. The quality of the high-*K* dielectric and its interface to the AlGaN barrier layer can be evaluated by *C-V* measurements, from which the interface state density can be estimated.

*13.3.1 Calculation of Carrier Concentration Distribution for High-*K* Dielectric AlGaN/GaN MOS Capacitors*

For the AlGaN/GaN MOS capacitor with monolayer high-*K* gate dielectric, the dielectric thickness can be estimated by

$$\frac{1}{C_{\text{MOS-HEMT}}} = \frac{1}{C_{\text{OX}}} + \frac{1}{C_{\text{HEMT}}} \tag{13.1}$$

$$C_{\text{OX}} = \varepsilon_0 \varepsilon_{\text{OX}} A / d_{\text{OX}} \tag{13.2}$$

where $C_{\text{MOS-HEMT}}$ is the zero-bias capacitance of the MOS-HEMT and $C_{\text{HEMT}} = \varepsilon_0 \varepsilon_B A / d_B$ of the HEMT (ε_B and d_B are dielectric constant and thickness of the AlGaN barrier layer, respectively); C_{OX} and d_{OX} are the dielectric layer capacitance and thickness, respectively; ε_0 and ε_{OX} are the vacuum permittivity and dielectric constant of the dielectric layer, respectively; and A is the test pattern area. The above two equations can also be combined into:

$$C_{\text{MOS-HEMT}} = \frac{\varepsilon_0 \varepsilon_B A}{d_B}\left(1 + \frac{d_{OX}}{d_B} \cdot \frac{\varepsilon_B}{\varepsilon_{OX}}\right)^{-1} \quad (13.3)$$

The equivalent dielectric constant ε_r of the capacitance of the cascade-connected dielectric layer and AlGaN barrier layer with a total thickness d_t can be solved by

$$\frac{d_t}{\varepsilon_r} = \frac{\varepsilon_0 A}{C_{\text{MOS-HEMT}}} = \frac{d_{OX}}{\varepsilon_{OX}} + \frac{d_B}{\varepsilon_B} \quad (13.4)$$

From the measured *C-V* characteristic curves and Equation 13.5 below, the carrier concentration N_{CV} profile as a function of the depth d for the MOS-HEMT capacitor can be expressed as

$$N_{CV} = -\frac{2}{e\varepsilon_r \varepsilon_0 A^2 (dC^{-2}/dV)}, \quad d = \frac{\varepsilon_r \varepsilon_0 A}{C} \quad (13.5)$$

where e is the elementary charge.

For the multilayer gate dielectric, we have:

$$\frac{1}{C_{OX}} = \sum_{i=1}^{n} \frac{1}{C_{OXi}}, \quad C_{OXi} = \frac{\varepsilon_0 \varepsilon_{OXi} A}{d_{OXi}} \quad (13.6)$$

From this equation, the equivalent dielectric constant ε_{OX} of the multilayer dielectric can be determined, with the equivalent dielectric constant ε_r still solved by Equation 13.4.

13.3.2 C-*V Hysteresis Characteristics of High-*K *Dielectric AlGaN/GaN MOS Capacitors*

Forward and reverse *C-V* hysteresis curve sweeping yields the flat band voltage shift of the MOS capacitor ΔV_{FB}, which originates from the captured electrons by the dielectric layer traps and dielectric/AlGaN interface states. The flat band voltage of the MOS capacitor can be expressed as (Nicollian and Brews 2002):

$$\Delta V_{FB} = \phi_{MS} - \frac{Q_f + Q_{it}(\psi_s) + Q_t}{C_{OX}} \quad (13.7)$$

where:

Φ_{MS} is the difference in work function between the metal

$Al_xGa_{1-x}N$, Q_f is the sheet density of the fixed charge in the dielectric layer

Q_{it} is the sheet density of the interface state charge

ψ_s is the semiconductor surface potential

Q_t is the trapped charge density of the dielectric layer

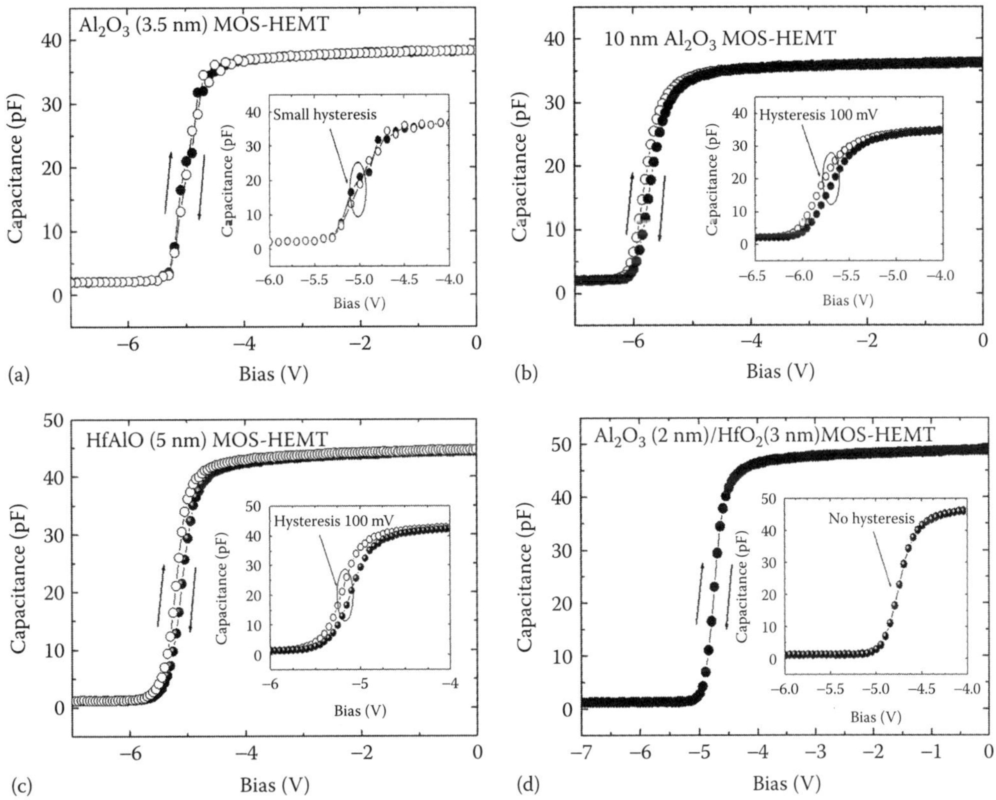

Figure 13.3 1 MHz *C-V* hysteresis curves of AlGaN/GaN MOS capacitors with different MOS dielectric structures. (a) 3.5 nm Al_2O_3, (b) 10 nm Al_2O_3, (c) HfAlO compound dielectric, and (d) HfO_2/Al_2O_3 stack dielectric. (Courtesy of Dr. Yuanzheng Yue.)

Since the fixed charge does not vary with the sweep voltage, the electric charge corresponding to the *C-V* hysteresis is $Q_{it} + Q_t$, which can be calculated by $\Delta V_{FB} \times C_{OX}$.

The 1 MHz *C-V* hysteresis characteristics are measured by sweeping from −8 to 0 V and then back to −8 V of the AlGaN/GaN MOS capacitor with 3.5 nm Al_2O_3, 10 nm Al_2O_3, 5 nm HfAlO, and HfO_2(3 nm)/Al_2O_3(2 nm) stack dielectric, respectively, and with nickel as the gate metal. The measured *C-V* hysteresis for these structures (Figure 13.3) are 20, 100, 100, and 0 mV (no visible *C-V* hysteresis), respectively, suggesting that 3.5 nm Al_2O_3 and HfO_2/Al_2O_3 stack dielectric have few bulk trapped charges, good dielectric quality, and good dielectric/AlGaN interface quality. The better characteristics of the stack dielectric may be explained by the formation of the high-quality Ni/HfO_2 interface. The relatively large capacitance hysteresis of 10 nm Al_2O_3 and 5 nm HfAlO is relative to the specific dielectric deposition process conditions, which are not our concern in this chapter. Instead, we focus on the difference in the quality of the dielectric and the interface reflected by the *C-V* hysteresis of the MOS structures. Based on the results in Figure 13.3a through d, the sums of trapped charge density and interface state charge density in 3.5 nm Al_2O_3, 10 nm Al_2O_3, compound HfAlO, and stack HfO_2/Al_2O_3 dielectric MOS capacitors are calculated to be 3.5×10^{11}, 3.9×10^{11}, 6×10^{11}, and 1.5×10^{10} cm^{-2}/eV, respectively.

13.3.3 Variable Frequency C-V *Characteristics of High*-K *Dielectric AlGaN/GaN MOS Capacitors*

The variable frequency *C-V* method is an important technique for calculating the interface states in AlGaN/GaN MOS capacitors. By measuring the low- and high-frequency capacitances C_L and C_H of the MOS capacitor, its interface state density D_{it} is calculated by

$$D_{it} = \frac{C_{ox}}{e^2}\left(\frac{C_L/C_{ox}}{1-C_L/C_{ox}} - \frac{C_H/C_{ox}}{1-C_H/C_{ox}}\right) \tag{13.8}$$

The measured variable frequency *C-V* characteristics of AlGaN/GaN MOS capacitors with 3.5 nm Al_2O_3, 10 nm Al_2O_3, 5 nm HfAlO compound, and HfO_2/Al_2O_3 stack dielectrics at 500 Hz, 1 kHz, 10 kHz, 100 kHz, and 1 MHz, respectively with a DC bias voltage sweep from 0 to −6 V in steps of −0.02 V are shown in Figure 13.4a through d.

It is seen in Figure 13.4 that the measured capacitance fluctuates at frequencies below 10 kHz, with smaller fluctuation in the 3.5 nm Al_2O_3 and HfO_2/Al_2O_3 stack

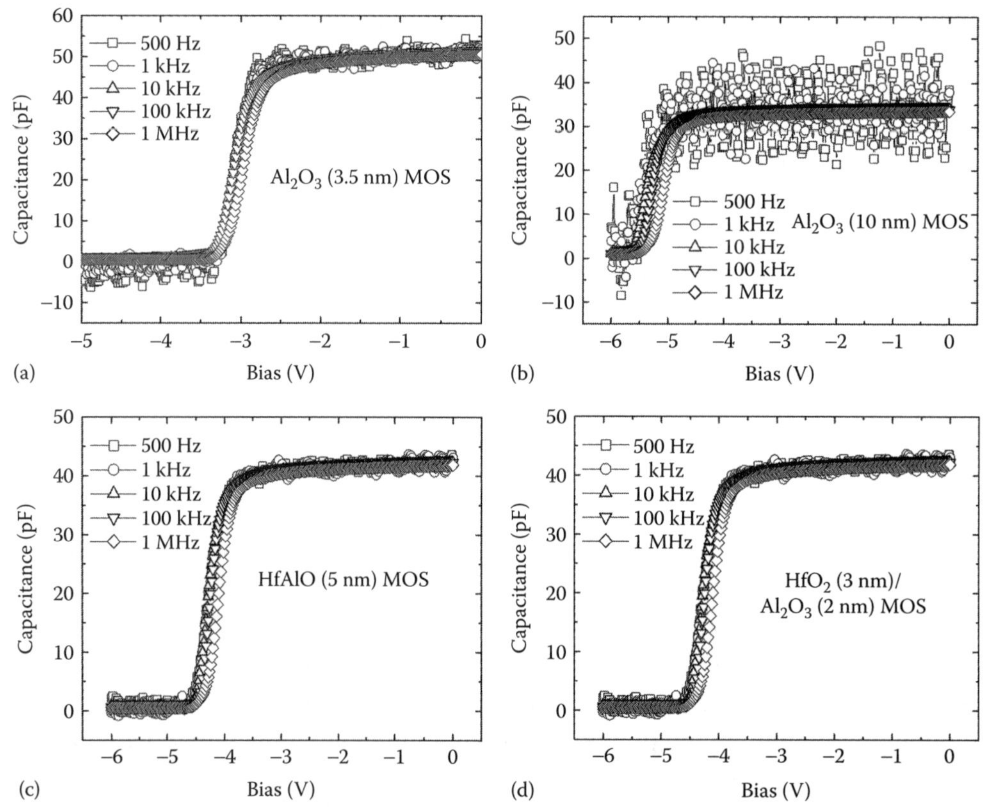

Figure 13.4 Variable frequency *C-V* characteristic curves of high-*K* dielectric AlGaN/GaN MOS capacitors with different MOS dielectric structures. (a) 3.5 nm Al_2O_3, (b) 10 nm Al_2O_3, (c) HfAlO compound dielectric, and (d) HfO_2/Al_2O_3 stack dielectric. (Courtesy of Dr. Yuanzheng Yue.)

dielectric MOS capacitors and greater fluctuation in the 10 nm Al_2O_3 and HfAlO compound dielectric MOS capacitors, which is attributed to the response of the trapped charge in the dielectric or interface states at low frequencies. Thus, the capacitances at 10 kHz and 1 MHz (denoted by C_L and C_H, respectively) for each MOS structure are substituted into Equation 13.8 to calculate the interface state densities for the MOS capacitors with 2.8 nm Al_2O_3 dielectric, 10 nm Al_2O_3 dielectric, HfAlO compound dielectric, and HfO_2/Al_2O_3 stack dielectric, which are 2.8×10^{10}, 9.7×10^{10}, 7×10^{10}, and 4.1×10^{10} cm^{-2}/eV, respectively. The variation of interface state density with the dielectric and structure coincides with that in Section 13.3.2.

The interface state densities obtained by variable frequency *C-V* measurements are universally lower than those by *C-V* hysteresis method, because the chosen low frequency of 10 kHz is not low enough (it is difficult to extract the saturation capacitance values at frequencies lower than 10 kHz owing to great fluctuation in the *C-V* measurement results), at which not all the electrons in interface states catch up with the change in small signals, thus resulting in lower interface state density values by variable frequency *C-V* methods than the actual ones.

A combination of the *C-V* hysteresis and variable frequency *C-V* measurements suggests that the ultrathin 3.5 nm Al_2O_3 dielectric and HfO_2(3 nm)/Al_2O_3(2 nm) stack dielectric can form high-quality interface to AlGaN and are more suitable for use as the MOS-HEMT gate dielectrics.

13.4 AlGaN/GaN MOS-HEMTs with High-*K* Stack Gate Dielectric of HfO_2/Al_2O_3

According to Table 13.1, although with large dielectric constants, high-*K* gate dielectrics generally have relatively small band gaps, leading to a large leakage current due to small conduction band discontinuity, when used as gate dielectrics in direct contact to the AlGaN barrier layer. This, coupled with the interface problem between the high-*K* dielectric and AlGaN barrier layer, as well as the surface passivation characteristics, restricts their applications. Therefore, the structure design of the high-*K* gate dielectrics becomes the key to applications of the dielectrics in AlGaN/GaN MOS-HEMTs. A design scheme for HfO_2/Al_2O_3 high-*K* stack gate dielectric is presented in this section, with characteristics of corresponding AlGaN/GaN MOS-HEMT devices to show the advantage of the proposed gate dielectric.

13.4.1 Design of ALD-Grown HfO_2/Al_2O_3 Stack Gate MOS-HEMTs

According to the experimental results in Section 13.3, 3.5 nm Al_2O_3 grown by ALD as both gate dielectric and surface passivation can achieve high-quality interface to the AlGaN barrier with a low interface state density. On this foundation, the atomic layer deposition of HfO_2/Al_2O_3 stack dielectric introduces HfO_2 of an even higher dielectric constant and uses the ultrathin 2 nm Al_2O_3 as the transition layer for the HfO_2/AlGaN interface to form a high-quality interface to the AlGaN barrier. Although typical high-K dielectric materials have a small band gap, Al_2O_3 has as high a band gap as 7 eV and a conduction band offset up to 2.1 eV to AlGaN, both of which play a major role in the reduction of the gate leakage current, thus the choice of Al_2O_3 as the interface transition layer. Two nanometer approaches the thickness limit of Al_2O_3 by atomic layer deposition as thinner Al_2O_3 cannot guarantee a full coverage of the AlGaN surface, but measurements show that the 2 nm Al_2O_3 interface transition layer suffices to provide good surface passivation. The reduction of the thickness of the relatively low dielectric constant Al_2O_3 interface transition layer to the minimum 2 nm and the designed top HfO_2 gate dielectric thickness of 3 nm offer the high-K stack dielectric HfO_2(3 nm)/Al_2O_3(2 nm) an equivalent oxide layer thickness to the 3.5 nm Al_2O_3 ultrathin gate dielectric but a thicker physical thickness, further improving the gate capacitor characteristics and decreasing the gate leakage current.

In addition to its use as the gate insulator, HfO_2(3 nm)/Al_2O_3(2 nm) is also served as the surface passivation layer, and the passivation effect of the dielectric is analyzed.

13.4.2 Direct-Current Characteristics of HfO_2/Al_2O_3 Stack Gate MOS-HEMTs

The gate current characteristics of the HfO_2/Al_2O_3 stack gate MOS-HEMT with a gate length of 1 μm, a gate width of 120 μm, and a source-drain spacing of 4 μm and the reference HEMT of the same dimension are shown in Figure 13.5, demonstrating a smaller gate-source leakage current of the MOS-HEMT than the HEMT under both forward and reverse voltages. For V_{GS} = −10 V, the MOS-HEMT exhibits a leakage current approximately one order of magnitude lower than that of the HEMT; for V_{GS} = 2 V, the MOS-HEMT leakage current is also approximately six orders of magnitude smaller than that of the HEMT. For the HEMT, the leakage current reaches 4.9 mA under a gate source bias of 2 V, whereas the MOS-HEMT has a mere 0.02 mA under a gate source bias of 5 V.

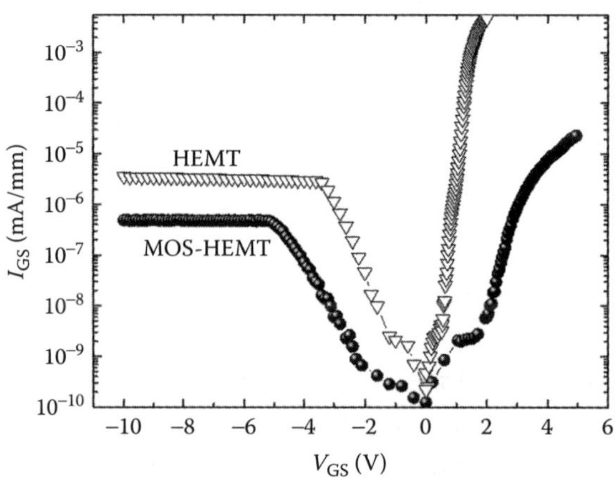

Figure 13.5 Comparison of gate leakage current between HfO_2(3 nm)/Al_2O_3(2 nm) stack gate MOS-HEMT and reference HEMT. (After Yue, Y.-Z. et al., AlGaN/GaN MOS-HEMT with stack gate HfO_2/Al_2O_3 structure grown by atomic layer deposition, *2008 IEEE Compound Semiconductor Integrated Circuits Symposium*, October 12–15, Piscataway, NJ, 2008a.)

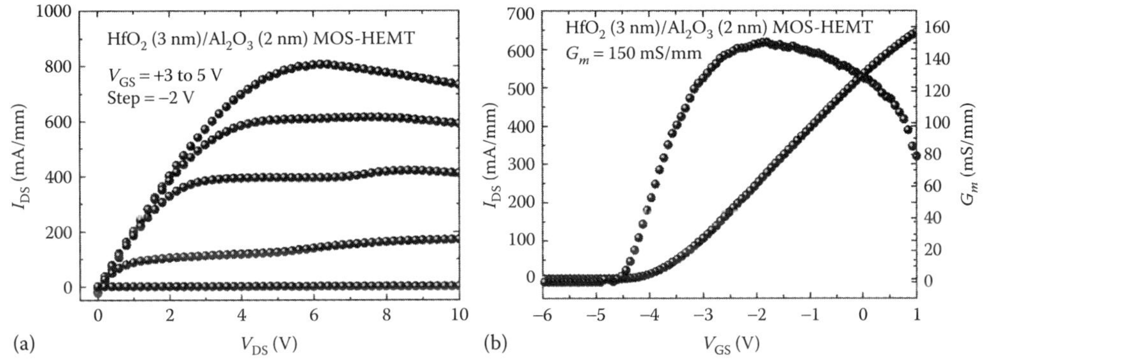

Figure 13.6 Direct-current characteristics of the HfO_2/Al_2O_3 stack gate MOS-HEMT. (a) Output characteristics and (b) transfer characteristics. (After Yue, Y.-Z. et al., AlGaN/GaN MOS-HEMT with stack gate HfO_2/Al_2O_3 structure grown by atomic layer deposition, *2008 IEEE Compound Semiconductor Integrated Circuits Symposium*, October 12–15, Piscataway, NJ, 2008a.)

Shown in Figure 13.6a are the measured output characteristics of the AlGaN/GaN MOS-HEMT, using HfO_2/Al_2O_3 for gate insulation with a source-to-drain sweep from 0 to 10 V and a gate voltage varying from 3 to −5 V in steps of −2 V. The saturation current density reaches 800 mA/mm at a gate voltage of 3 V. The maximum gate voltage applied on the HEMT is 1 V, owing to the presence of large gate leakage current, thus resulting in a much lower saturation current density (~650 mA/mm) than that of the MOS-HEMT. A higher breakdown voltage and saturation current density are achieved in the MOS-HEMT owing to the reduced gate leakage current, indicating a greater output power of the MOS-HEMT than that of the HEMT of identical size.

The transfer characteristic curve of this MOS-HEMT is given in Figure 13.6b, showing a maximum transconductance of 150 mS/mm and a threshold voltage of about −4 V at a source drain bias of 7 V, whereas the maximum transconductance and threshold voltage for the HEMT are 165 mS/mm and −3.5 V, respectively. The MOS-HEMT has a decreased transconductance with a negative shift of the threshold voltage compared with the HEMT due to the increased gate-channel spacing. However, the high dielectric constant and small equivalent oxide layer thickness of HfO_2/Al_2O_3 (2 nm) coupled with the high-quality interface between the Al_2O_3 interface transition layer and AlGaN barrier layer, as well as good passivation, yield a far smaller transconductance reduction (9%) and threshold voltage offset (−0.5 V) than those reported for the MIS-HEMT using SiO_2 and SiN dielectrics (Yue et al. 2008a). With the gate voltage swing defined as the gate voltage range at a 10% decrease in the transconductance, the MOS-HEMT has a gate voltage swing of 2.4 V and the HEMT has a gate voltage swing of 1.8 V, suggesting that the MOS-HEMT enjoys a higher linearity and wider dynamic operation range.

13.4.3 Passivation Characteristics of HfO_2/Al_2O_3 Stack Dielectrics

The surface passivation characteristics of HfO_2/Al_2O_3 in AlGaN/GaN MOS-HEMTs can be characterized by pulse measurements. The source-drain current response to

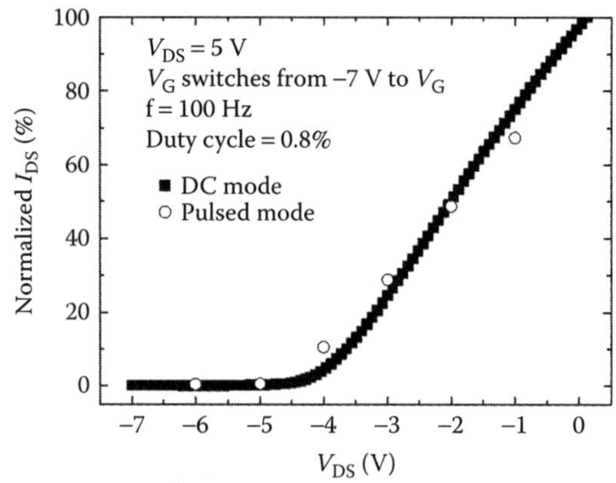

Figure 13.7 Surface passivation characteristics of HfO_2/Al_2O_3 in MOS-HEMTs. (After Yue, Y.-Z. et al., AlGaN/GaN MOS-HEMT with stack gate HfO_2/Al_2O_3 structure grown by atomic layer deposition, *2008 IEEE Compound Semiconductor Integrated Circuits Symposium*, October 12–15, Piscataway, NJ, 2008a.)

the gate-source pulses is measured with a fixed source-drain voltage of 5 V, a gate pulse voltage varied from −7 V to the test point, a pulse frequency of 100 Hz, and a duty cycle of 0.8%.

The source-drain current values are normalized, as shown in Figure 13.7, for a comparison of the source-drain currents under direct current and pulse conditions. The pulse current almost equals the direct current at gate voltage lower than −2 V and drops to 90% of the direct current under an increased gate voltage of −1 V, with a current collapse of only 10%, demonstrating effective elimination of surface states, owing to the good surface passivation effect of the 2 nm Al_2O_3 interface transition layer and successful suppression of current collapse by the HfO_2/Al_2O_3 high-*K* stack dielectric.

13.4.4 Frequency Characteristics of HfO_2/Al_2O_3 Stack Gate MOS-HEMTs

Small-signal microwave measurements of the ALD-grown HfO_2/Al_2O_3 stack gate MOS-HEMT were performed at $V_{DS} = 10$ V and $V_{GS} = -3$ V to obtain the *S*-parameters, from which to deduce the *H*-parameters and unilateral power gain. The S-parameters as a function of gain and frequency are shown in Figure 13.8. It can be seen that the device with 1 μm gate length and 120 μm gate width exhibits excellent frequency characteristics, with a cut-off frequency as high as 12 GHz and a maximum oscillating frequency up to 34 GHz.

13.5 Gate-Recessed AlGaN/AlN/GaN MOS-HEMTs

The introduction of the recessed-gate structure in MOS-HEMTs can improve the gate control ability, and the recessed-gate process optimization can effectively decrease the interface states density to obtain high-quality MOS-gate structures (Khan et al. 2000, 2003). The optimized preparation of high-performance recessed-gate AlGaN/GaN MOS-HEMTs is exemplified by a sample device we prepared

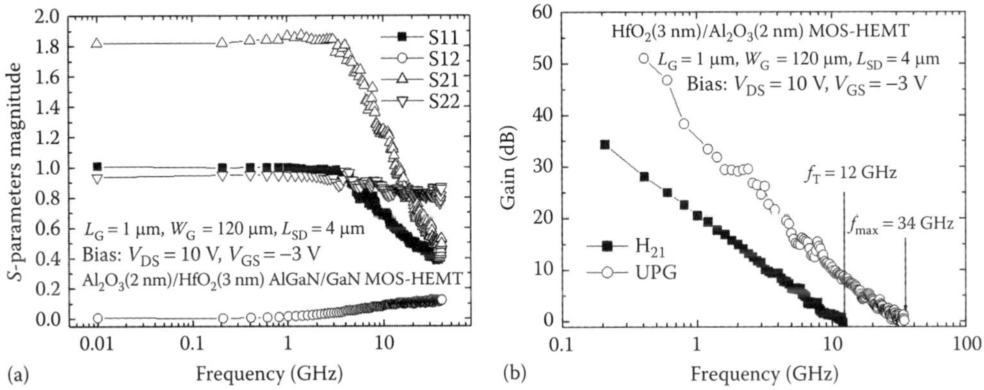

Figure 13.8 Frequency characteristics of HfO_2/Al_2O_3 stack gate MOS-HEMT. (a) *S*-parameter measurements and (b) gain versus frequency. (After Yue, Y.-Z. et al., AlGaN/GaN MOS-HEMT with stack gate HfO_2/Al_2O_3 structure grown by atomic layer deposition, *2008 IEEE Compound Semiconductor Integrated Circuits Symposium*, October 12–15, Piscataway, NJ, 2008a.)

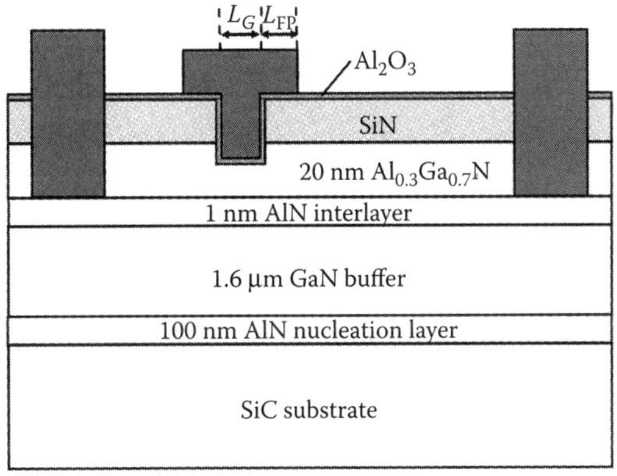

Figure 13.9 Schematic cross section of gate-recessed AlGaN/AlN/GaN MOS-HEMT. L_G is the gate length and L_{FP} the field plate length. (After Hao, Y. et al., *IEEE Elect. Device Lett.*, 32(5), 626–628, 2011.)

with a PAE up to 73% at 4 GHz. The insulating layer under the gate metal adopts ALD-grown Al_2O_3, and the structure design incorporates the advantages of the MOS, recessed gate, and field plate structures with the schematic cross section of the device shown in Figure 13.9.

The process flow employed is different from that described in the above section in that Si_3N_4 is used as the passivation layer to facilitate the analysis of the effect of the change in structure only under the gate region. It differs from the HEMT process flow discussed in Chapter 10 in the step of the gate preparation after the Si_3N_4 passivation layer deposition. In the preparation of the planar MOS gate, gate openings are etched first by CF_4 with surface preparation of the gate region, followed by Al_2O_3 atomic layer deposition and the metal deposition, whereas the preparation of the recessed-gate MOS structure requires an extra gate recessing by chlorine-based RIE before Al_2O_3 deposition. Subsequent Si_3N_4 protective layer deposition and interconnection preparation are identical.

13.5.1 Effect of Recessed-Gate Etch Depth on MOS-HEMTs with ALD-Grown Al_2O_3 Gate Dielectric

The AlGaN/GaN epitaxial wafer is divided into five zones (A–E) for photolithography, and the recessed-gate devices and the circular *C-V* test patterns prepared in the same batch are compared. The three zones of A, B, and C are etched by chlorine-based RIE for 15, 17, and 19 s, respectively, to form recessed-gate MOS structures; D is not etched to form the planar MOS-HEMT structure and gate metal is deposited directly on D to obtain the conventional HEMT structure. The gate dielectric is ALD-grown 5.6 nm Al_2O_3. The devices have a gate length $L_G = 0.5$ μm, a gate width of 100 μm, and a source-drain spacing of 3.5 μm.

According to the measured *C-V* characteristics of the corresponding test patterns, the etch depths of the recessed gates formed by etching the AlGaN barrier layer for 15, 17, and 19 s are estimated to be 1.0, 3.2, and 3.7 nm, respectively. The direct-current characteristics measurements of the planar MOS-HEMT and the recessed-gate MOS-HEMTs show that the threshold voltage shifts positively with the recess depth in an approximately linear manner (Figure 13.10a), indicating no serious damage to the barrier layer surface by etching. The transconductance peak value increases from 160 to 189 mS/mm with deeper recess. The recessed gate also leads to a slight increase in the saturation current from 974 to 1039 mA/mm at a gate voltage of 3 V.

The source-drain breakdown voltage BV_{DS} and gate-drain breakdown voltage BV_{DG} can be measured at one time by the nondestructive drain current injection technique (Bahl and del Alamo 1993), as shown in Figure 13.10b. A significant decrease in both BV_{DS} and BV_{DG} for the MOS-HEMTs is observed at a recess depth in excess of 1 nm. The gate current measurements show that at a gate bias of −20 V, the reverse leakage current of the 5.6 nm Al_2O_3 insulated gate planar MOS-HEMT is of the order of 1 nA, nearly three orders of magnitude smaller than that of the conventional HEMT, and the recessed gate leads to an increased reverse leakage current for the MOS-HEMTs, with that of the 3.7-nm-deep recess-gate device increased to 54 nA, which is still lower than the reverse Schottky leakage current of the HEMT.

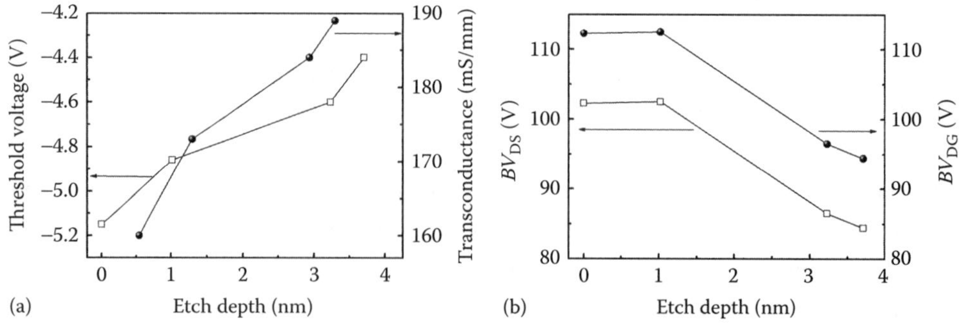

Figure 13.10 Effect of recessed-gate etching on the threshold voltage, transconductance, and breakdown voltage of MOS-HEMTs. (a) Threshold voltage and transconductance and (b) breakdown voltage. (Courtesy of Dr. Ling Yang.)

According to the theory of the conductance method (Sze 2008), the ratio of the MOS-gate conductance G_p to the angular frequency ω is related to ω by Equation 13.9, and the trap density is related to C_{it} by Equation 13.10.

$$\frac{G_p}{\omega} = \frac{C_{it}\omega\tau_{it}}{1+\omega^2\tau_{it}^2} \tag{13.9}$$

$$C_{it} = eD_{it} \tag{13.10}$$

where:

C_{it} is the capacitance related to interface states
τ_{it} and D_{it} are the time constant and the density of the states, respectively
e is the elementary charge (1.6×10^{-19} C)

It follows from Equation 13.9 that for $\omega\tau_{it} = 1$, G_p/ω takes the maximum value $C_{it}/2$. The interface states effect of the MOS gate of the MOS-HEMTs can be evaluated by the capacitance-frequency (*C-f*) and conductance-frequency (*G-f*) measurements.

Shown in Figure 13.11 are the *C-f* and *G-f* measurement results for the planar and recessed-gate MOS-HEMTs, with frequency varied from 10 kHz to 1 MHz in steps of 10 kHz. The capacitance-frequency profiles in Figure 13.11a and b show little variation of the capacitance, with frequency at a zero bias and a decreased capacitance for both structures when the bias is chosen much closer to the threshold voltage. However, the recessed-gate MOS-HEMT shows smaller capacitance dissipation at increased frequency. The conductance-frequency curves are shown in Figure 13.11c and d. The peak value $C_{it}/2$ of the curve reflects the magnitude of the interface trap density and the corresponding ω reflects the time constant of the interface traps.

From these measurement results, we extracted the interface state characteristics of the MOS-HEMT structures under discussion, as shown in Table 13.2. Instead of an increase, a slight decrease of the interface state density is observed after the gate recess. All the interface states have a time constant on the order of microseconds and fall into the fast state traps, suggesting that within the above depth range, the recessed gate structure can decrease the surface state density between the Al_2O_3 gate dielectric and the barrier layer, without generating new types of interface state traps.

The current collapse of the MOS-HEMTs was evaluated by the gate-drain sync pulse measurements, with a quiescent operation voltage setting of $V_{DS} = 20$ V, $V_{GS} = -6$ V, as shown in Figure 13.12. An apparent current collapse in excess of 20% of the planar MOS-HEMT is observed, whereas for the recessed-gate MOS-HEMTs, little collapse is seen at a gate voltage greater than zero, and only a slight collapse occurs when the gate voltage pulse is lower than −1 V. The output current collapse as a function of the etch depth of the recessed gate under a gate voltage pulse of −1 V and a drain voltage pulse of 6 V, shown in Figure 13.13, illustrates a decrease followed by an increase of the current output with increasing gate recess etch depth; the minimum output being at the depth of 1 nm. The interface state density as a function of the etch

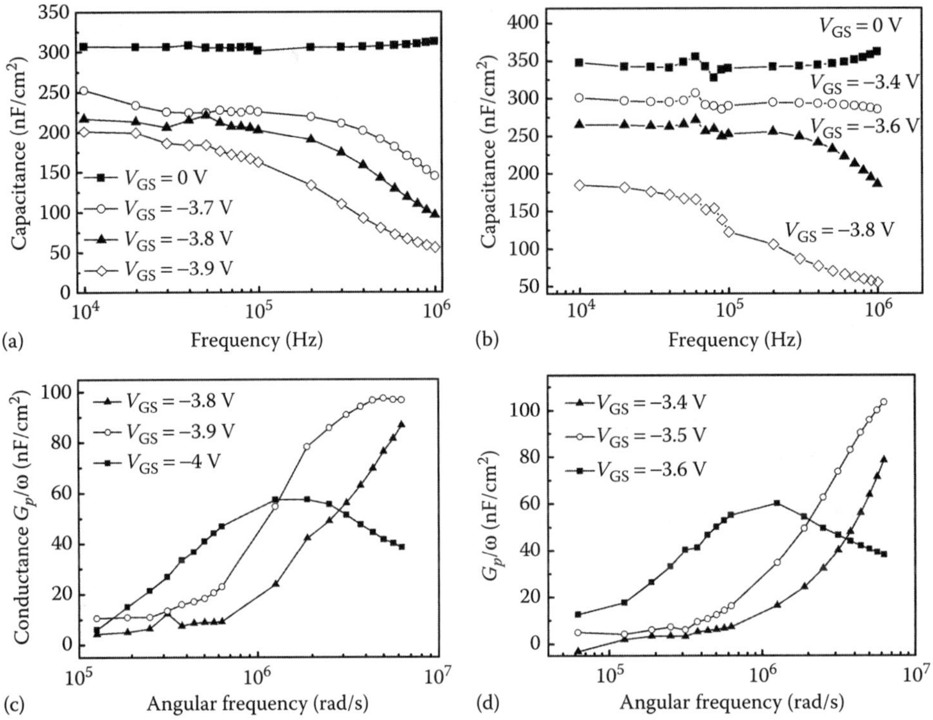

Figure 13.11 MOS-gate capacitance-frequency (*C-f*) and conductance-frequency (*G-f*) measurements of the MOS-HEMTs, with frequency varied from 10 kHz to 1 MHz. (a) *C-f* curves for planar MOS. (b) *C-f* curves for 3.7 nm recessed-gate MOS. (c) *G-f* curves for planar MOS. (d) *G-f* curves for 3.7 nm recessed-gate MOS. (Courtesy of Dr. Ling Yang.)

Table 13.2 Interface State Characteristics of Planar and Recessed-Gate MOS-HEMTs

	INTERFACE STATE DENSITY D_{it} (cm^{-2}/eV)	INTERFACE STATE TIME CONSTANT τ_{it} (μs)
Planar MOS-HEMT	$(0.72–1.25) \times 10^{12}$	0.19–0.64
1.0-nm-deep recessed-gate MOS	$(0.55–1.08) \times 10^{12}$	0.20–1.59
3.2-nm-deep recessed-gate MOS	$(0.97–1.14) \times 10^{12}$	0.24–0.49
3.7-nm-deep recessed-gate MOS	$(0.76–1.2) \times 10^{12}$	0.19–0.94

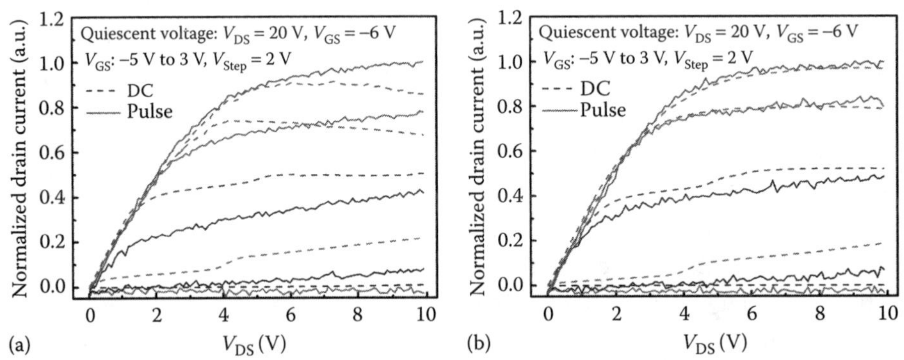

Figure 13.12 Direct-current and pulse output characteristics of the planar MOS-HEMT and the recessed-gate MOS-HEMT with a recess depth of 3.7 nm. The pulse width is 500 ns and the pulse period is 1 ms. (a) Planar MOS-HEMT and (b) recessed-gate MOS-HEMT with a recess depth of 3.7 nm. (Courtesy of Dr. Ling Yang.)

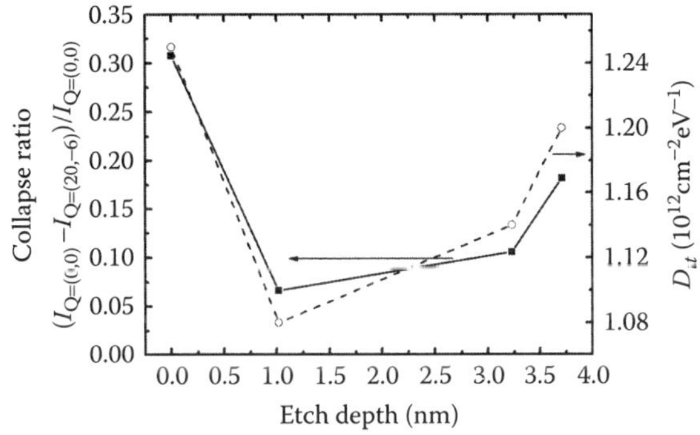

Figure 13.13 Drain current collapse and interface state density D_{it} versus gate recess etch depth. (Courtesy of Dr. Ling Yang.)

depth is also given in Figure 13.13, which shows a decrease followed by an increase of the interface state density with increasing etch depth in an identical pattern with the variation of the current collapse, suggesting a close relationship between the output current collapse and the interface state density.

In summary, compared with the planar-gate MOS, the recessed-gate MOS can reduce the negative effect of the gate dielectric on the gate control ability, increase both the transconductance and saturation current density with nearly as high a threshold voltage as that of the HEMT, and improve current collapse by decreasing the Al_2O_3/AlGaN interface states density at a relatively small etch depth. However, the recessed gate also leads to an increased reverse leakage current in the MOS gate and a decreased breakdown voltage (Figure 13.10b), which are closely related to the increase in Al_2O_3/AlGaN interface states density at a gate recess depth greater than 1 nm (Figure 13.13). The minimum interface state density at the etch depth of 1 nm can be explained by the fact that the native oxide layer on the AlGaN surface happens to be etched right off under this light etch condition along with the minimum etch damage. It is necessary to break the dependence of the interface states under the gate on the gate recess etch depth in order for MOS-HEMTs with a relatively deep gate recess to achieve good properties.

13.5.2 Effect of Plasma Treatment on Recessed-Gate MOS-HEMTs

In order to address such problems as the interface states resulting from recess etch damage, the plasma surface treatment in the region under the recessed gate is introduced to analyze its effect on MOS-gate characteristics. According to the experiments on HEMTs, with the region under the Schottky gate treated with N_2O and oxygen (O_2) plasma, both treatments can lead to surface oxidation in the region under the gate, thus forming in situ oxide layers, with O_2 plasma treatment having the better effect, which contributes to a relatively small interface state density of the as-treated Schottky gate. On this basis, we analyze the effect of the O_2 plasma treatment of the AlGaN surface before Al_2O_3 deposition on the performance of MOS-HEMTs with different gate recess depths.

The MOS-HEMT interface state density can be extracted from the measured *C-f* and *G-f* characteristics. As shown in Figure 13.14, the O_2 plasma treatment suppresses

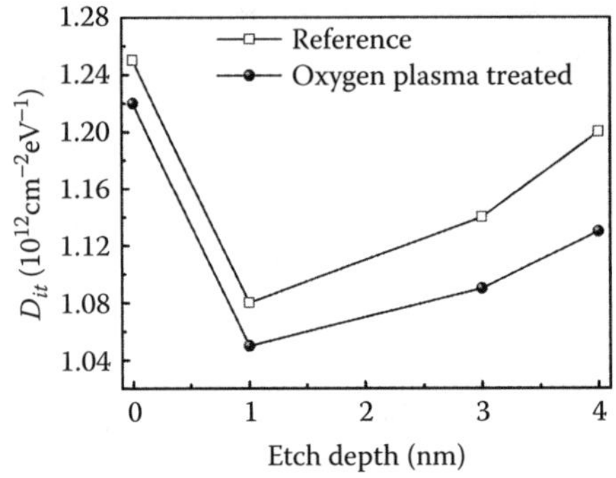

Figure 13.14 Effect of O_2 plasma treatment on Al_2O_3/AlGaN interface state density of MOS-HEMTs with different gate recess depths. (Courtesy of Dr. Ling Yang.)

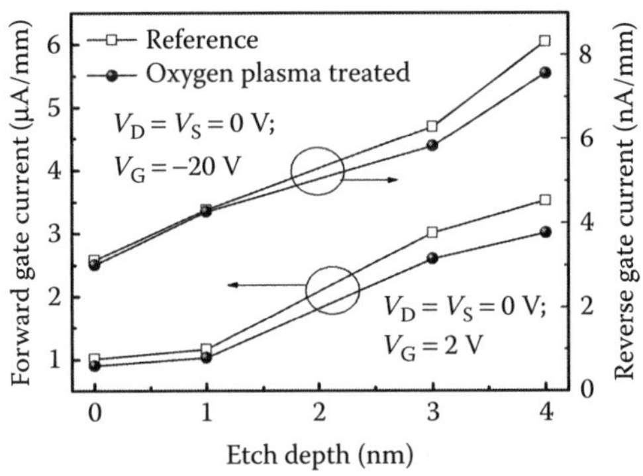

Figure 13.15 Effect of O_2 plasma treatment on forward and reverse gate currents of MOS-HEMTs with different gate recess depths. (Courtesy of Dr. Ling Yang.)

the interface state density and plays a positive role, especially in the reduction of the etch-induced interface states.

As Figure 13.15 illustrates that O_2 plasma treatment suppresses both the forward and reverse gate leakage currents and the suppression is enhanced by deeper etched gate recess, indicating that the oxidation resulting from O_2 plasma treatment effectively passivates the surface states induced by etch damage, thus resulting in an effective decrease in the trap-assisted tunneling current and a smaller gate leakage current in the deep recessed-gate MOS-HEMTs. It can be seen in Figure 13.16 that the MOS-HEMT breakdown voltage is improved, owing to O_2 plasma treatment, with greater effect under larger gate recess depth (breakdown voltage improved by 9.3% under a gate recess depth of 4 nm), which concurs with the pattern of gate leakage current reduction. The current collapse measurements also show an effectively curbed current collapse in the O_2 plasma-treated devices.

13.5.3 High-Performance Gate-Recessed AlGaN/AlN/GaN MOS-HEMTs

High-performance gate-recessed AlGaN/AlN/GaN MOS-HEMT devices were prepared under optimized process conditions of recess etching and O_2 plasma treatment

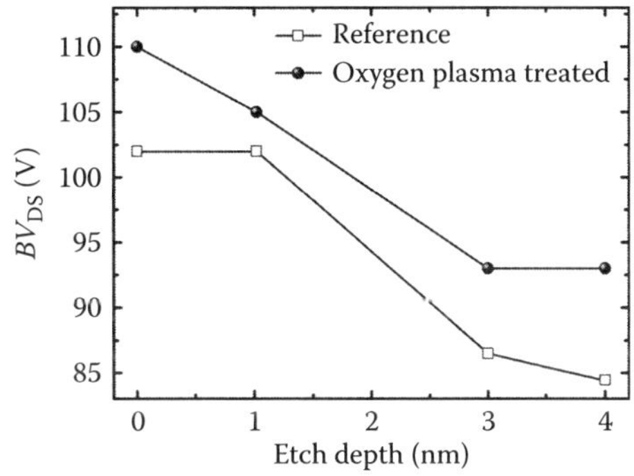

Figure 13.16 Effect of O_2 plasma treatment on breakdown voltage of MOS-HEMTs with different gate recess depths. (Courtesy of Dr. Ling Yang.)

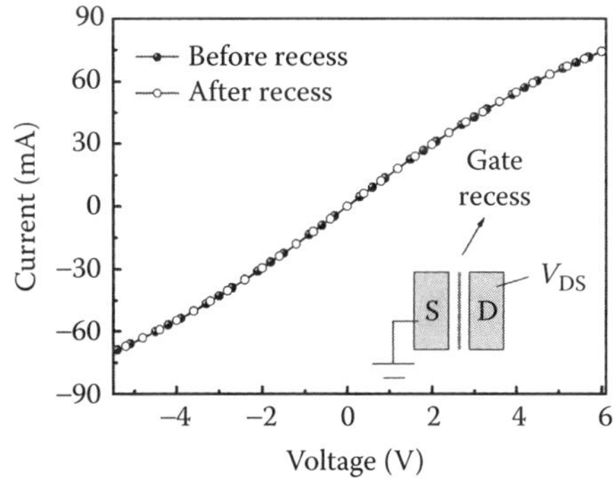

Figure 13.17 Source-drain current before and after gate recess in MOS-HEMT preparation. (After Hao, Y. et al., *IEEE Elect. Device Lett.*, 32(5), 626–628, 2011.)

of etched AlGaN surface. The devices under investigation has a structure as shown in Figure 13.9, with gate-source spacing L_{GS} = 0.7 μm, gate length L_G = 0.6 μm, gate-drain spacing L_{GD} = 2.8 μm, Γ-shaped gate field plate of an extension L_{FP} = 0.7 μm toward the drain, and a gate width of 100 μm. The ALD-grown Al_2O_3 was used as the gate dielectric to form the gate-recessed AlGaN/AlN/GaN MOS-HEMT structures.

The direct-current characteristics are shown in Figures 13.17 through 13.20. The source drain *I-V* measurements with the gate suspended before and after the gate recess process shown in Figure 13.17 indicate that the current characteristics are not influenced by the low-damage recess processing. The gate-recessed AlGaN/AlN/GaN MOS-HEMT output curves given in Figure 13.18 show good pinch-off performance, with the device totally switched off at a gate bias of −5 V and an off-state current smaller than 0.005 mA/mm. Under an increased gate voltage to 3 V, the saturation output current density is greater than 1.6 A/mm, suggesting that no degradation of electrical properties of the 2DEGs in channel results from the recessed-gate etching. In Figure 13.19, the I_G-V_{GS} curves of the recessed-gate and non-recessed-gate MOS-HEMTs of the same dimension almost overlap within the forward and reverse ranges, an indication of no recess etching-induced degradation in the gate leakage current of recessed-gate MOS-HEMTs. Moreover, no visible hysteresis effect is observed in

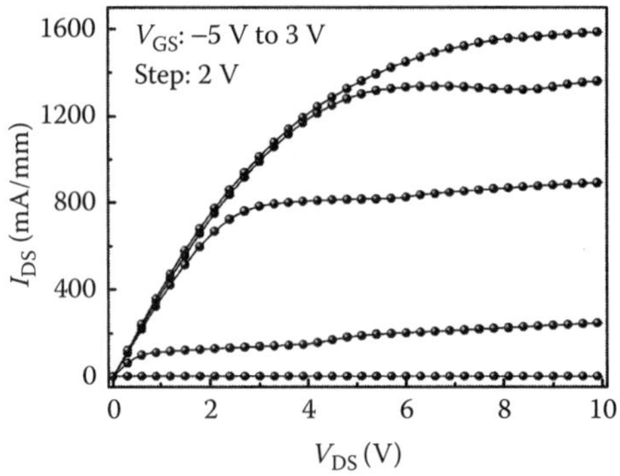

Figure 13.18 Output curves of gate-recessed AlGaN/AlN/GaN MOS-HEMT. (Courtesy of Dr. Ling Yang.)

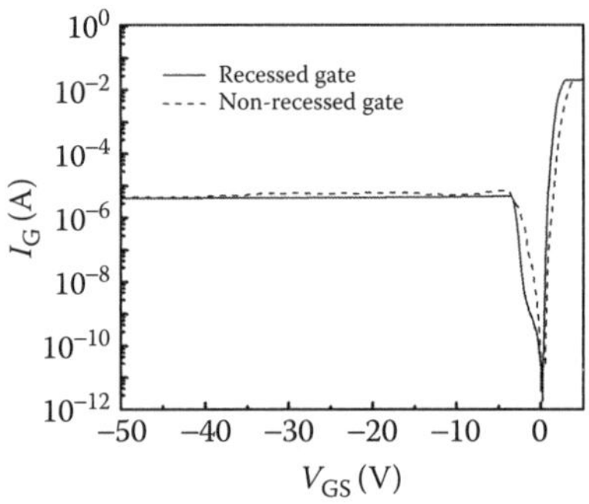

Figure 13.19 Gate leakage currents of the gate-recessed AlGaN/AlN/GaN MOS-HEMT and non-recessed-gate MOS-HEMT. (Courtesy of Dr. Ling Yang.)

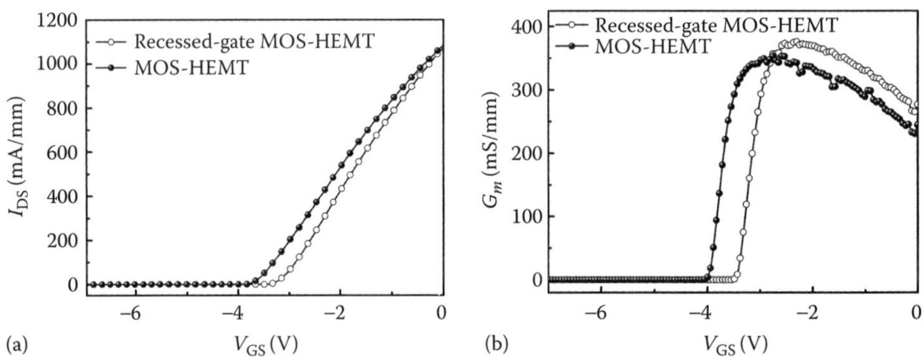

Figure 13.20 Transfer and transconductance curves of the gate-recessed AlGaN/AlN/GaN MOS-HEMT and non-recessed-gate MOS-HEMT. (a) Transfer curves and (b) transconductance curves. (After Hao, Y. et al., *IEEE Elect. Device Lett.*, 32(5), 626–628, 2011.)

both the forward and the reverse sweeping of the gate voltage, which implies that both the bulk charge density and interface state density in the oxide layer are very low.

The transfer characteristics and transconductance curve in Figure 13.20 reveal a positive shift of the threshold voltage of 0.5 V in the recessed-gate MOS-HEMT compared with the non-recessed-gate MOS-HEMT, suggesting effective suppression

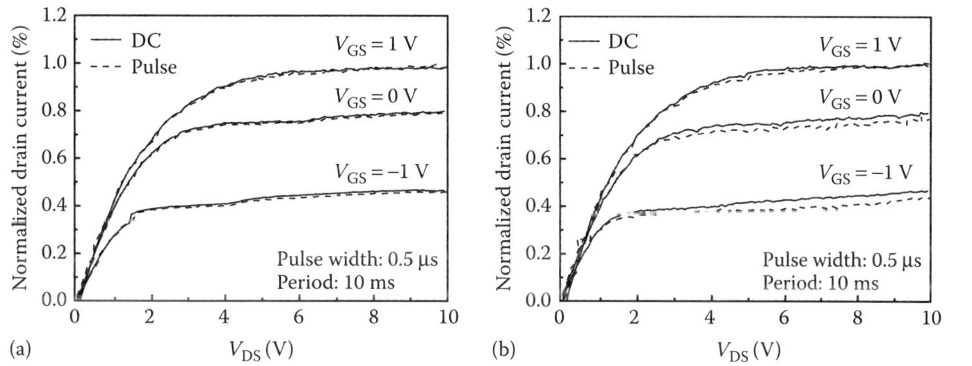

Figure 13.21 Gate and drain pulse output characteristics of the gate-recessed AlGaN/AlN/GaN MOS-HEMT. (a) Gate pulse output characteristics. (After Hao, Y. et al., *IEEE Elect. Device Lett.*, 32(5), 626–628, 2011.) (b) Drain pulse output characteristics. (Courtesy of Dr. Ling Yang.)

of the gate dielectric-induced negative shift of the threshold voltage by the recessed gate structure, which also reduces the gate-to-channel spacing and thus lead to an improved transconductance (from 350 to 374 mS/mm).

Shown in Figure 13.21 are the pulse output characteristics of the gate and the drain as compared with the direct-current characteristics with a pulse width of 0.5 μs and a period of 10 ms. No obvious current collapse is observed in both pulse measurements, suggesting relatively low surface state density and trap density of GaN buffer layer in the gate-recessed AlGaN/AlN/GaN MOS-HEMT.

The small signal parameters of the gate-recessed AlGaN/AlN/GaN MOS-HEMT were measured under a gate voltage of –2 V and a drain voltage of 20 V, and the results are shown in Figure 13.22. According to the current gain $|h_{21}|$ and the power gain unilateral power gain (UPG) and by maximum stable gain (MSG) or maximum available gain (MAG) extrapolation, we obtain the cutoff frequency $f_T = 19$ GHz and the maximum oscillation frequency $f_{max} = 50$ GHz, with $f_T/f_{max} = 1{:}2.6$, which suggests a low parasitic effect and thus good frequency characteristics of the developed gate-recessed AlGaN/AlN/GaN MOS-HEMT.

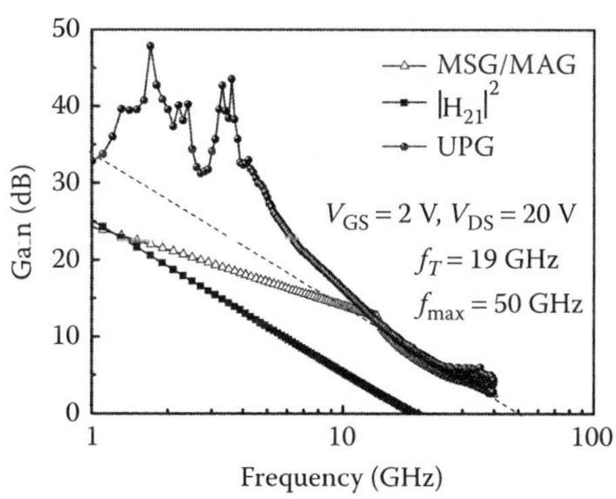

Figure 13.22 Small-signal parameter measurements of the gate-recessed AlGaN/AlN/GaN MOS-HEMT. (After Hao, Y. et al., *IEEE Elect. Device Lett.*, 32(5), 626–628, 2011.)

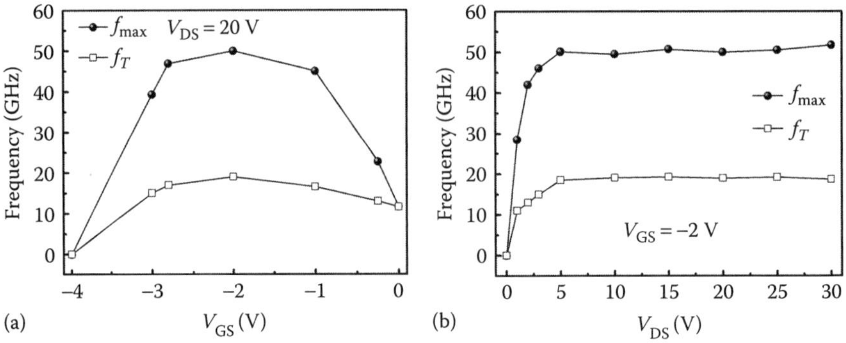

Figure 13.23 f_{max} and f_T versus V_{GS} and V_{DS} for gate-recessed AlGaN/AlN/GaN MOS-HEMT. (a) f_{max} and f_T vs. V_{GS} and (b) f_{max} and f_T vs. V_{DS}. (Courtesy of Dr. Ling Yang.)

It is worth noting that the gate-recessed AlGaN/AlN/GaN MOS-HEMT maintains relatively high f_{max} and f_T under a rather wide range of gate voltages, as shown in Figure 13.23a. Gain compression typically arises in high-current operations (e.g., a reduction in g_m, f_{max}, and f_T) due mainly to the thermal phonon effect (Matulionis 2004) and the nonlinear source resistance (Palacios et al. 2004) under high-current conditions. We examined the variation of f_{max} and f_T, with the drain voltage at a fixed gate voltage of −2 V. It is observed in Figure 13.23b that f_{max} and f_T saturate rapidly with the increase in the drain voltage and remain constant in the whole saturation region. The relatively high f_T maintained at a high drain voltage demonstrates a high field-induced electron drift velocity as well as a small extension of the gate depletion region toward the drain. The invariance of f_{max} and f_T with the drain voltage in the saturation region offers linear operations at high drain voltages and thus a wider drain voltage coverage of the load line (Nagahara et al. 2002).

The 4 GHz power sweep curves of the gate-recessed AlGaN/AlN/GaN MOS-HEMT at V_{DS} = 45 V, shown in Figure 13.24, exhibit a maximum power output density of 13 W/mm and a PAE up to 73%. As shown in Figure 13.25, the gate-recessed AlGaN/AlN/GaN MOS-HEMT achieves output power densities of 6.3, 8.3, 11.2, and 13 W/mm, respectively, increasing linearly with the increase in the drain voltage

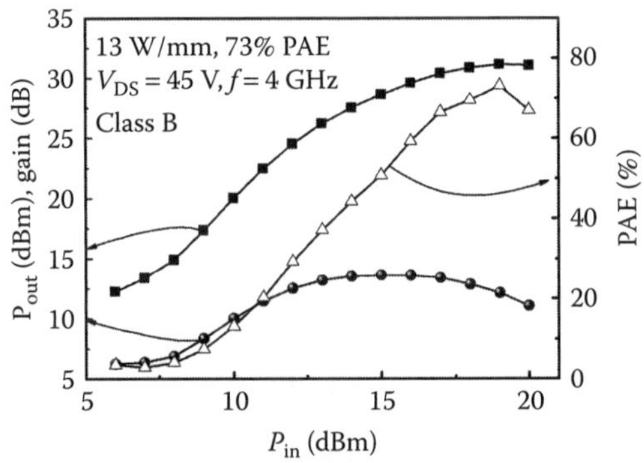

Figure 13.24 Power sweep curves of gate-recessed AlGaN/AlN/GaN MOS-HEMT. (After Hao, Y. et al., *IEEE Elect. Device Lett.*, 32(5), 626–628, 2011.)

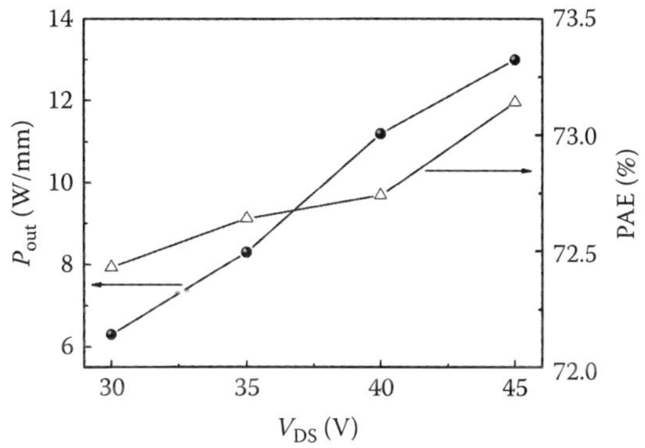

Figure 13.25 Power density and PAE as a function of the drain voltage of gate-recessed AlGaN/AlN/GaN MOS-HEMT. (After Hao, Y. et al., *IEEE Elect. Device Lett.*, 32(5), 626–628, 2011.)

and PAEs of 72.43%, 72.64%, 72.74%, and 73.14%, respectively, without degradation under the drain voltages of 30, 35, 40, and 45 V, respectively.

In conclusion, the developed gate-recessed AlGaN/AlN/GaN MOS-HEMTs exhibit good power characteristics, and high-performance GaN MOS-HEMTs are obtained by adopting the gate-recessed MOS structure to successfully solve the problems of current collapse and large gate leakage current.

13.6 Thin-Barrier-Layer Enhancement-Mode MIS-HEMTs

As is discussed in Chapter 12, thin-barrier HEMTs have largely improved threshold voltage, maximum transconductance, and frequency characteristics, compared with the thick barrier ones, but suffer from a limited working range due to the large forward gate current. This problem can be solved by the growth of an insulating gate dielectric under the gate of the thin-barrier device to reduce the gate current and achieve a larger working range. However, a decreased transconductance is also expected owing to the gate dielectric growth.

In this section, we discuss the MIS-HEMTs prepared from GaN heterostructures with 8-nm thin barriers (GaN/AlGaN/AlN/GaN) and fluorine-plasma-treated MIS-HEMTs. The device structure is shown in Figure 13.26. Owing to the thin AlGaN layer, the fluorine plasma treatment (50 W and 80 s) is adopted, followed by a rapid thermal annealing of the sample wafer at 400°C for 2 min.

As the device transfer characteristics in Figure 13.27 show, the MIS-HEMT without fluorine plasma treatment achieves a threshold voltage already as high as 0.8 V, with a significantly lower maximum transconductance (210 mS/mm) than that of the HEMT (420 mS/mm). Owing to the effective reduction in the forward gate current by the Si_3N_4 gate dielectric, the MIS-HEMT enjoys a relatively large operating range of gate voltage (a forward voltage up to +6 V), thus resulting in no apparent decrease in the saturation current as well as a relatively wide high transconductance region. The threshold voltage of the fluorine-plasma-treated MIS-HEMT increases to 1.8 V

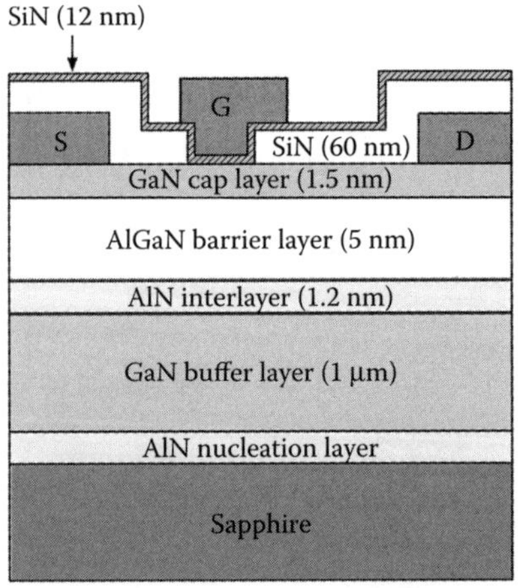

Figure 13.26 Schematic structure of MIS-HEMT with an 8 nm barrier. The fluorine ions are implanted under the foot of the T-gate of the MIS-HEMT to increase the threshold voltage. (Courtesy of Dr. Si Quan.)

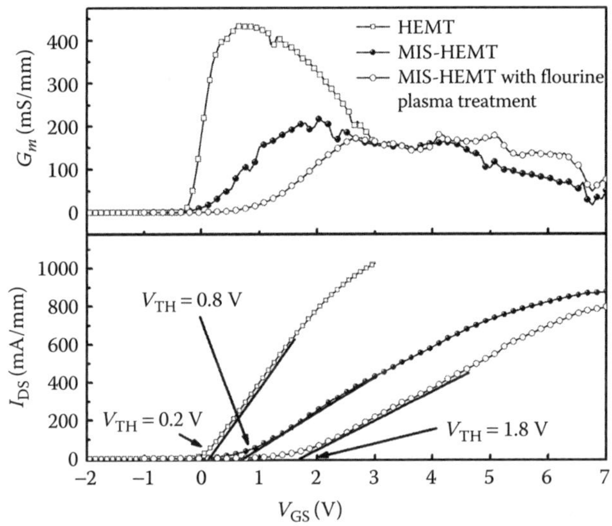

Figure 13.27 Transfer characteristic curves of the HEMT, MIS-HEMT, and enhancement-mode MIS-HEMT with 8 nm barrier. (Courtesy of Dr. Si Quan.)

with a slightly decreased maximum transconductance (190 mS/mm) and a maximum saturation current in excess of 800 mA/mm. The comparison of gate current between the HEMT, MIS-HEMT, and fluorine-plasma-treated MIS-HEMT (Figure 13.28) shows that the fluorine-plasma-treated MIS-HEMT has the smallest reverse current and a forward current equivalent to that of the untreated MIS-HEMT.

In summary, the use of 8-nm thin-barrier layer heterostructures and Si_3N_4 gate dielectrics in the MIS-HEMT preparation can elevate the operating gate voltage to 7 V and the threshold voltage from 0.2 to 0.8 V; the introduction of fluorine plasma

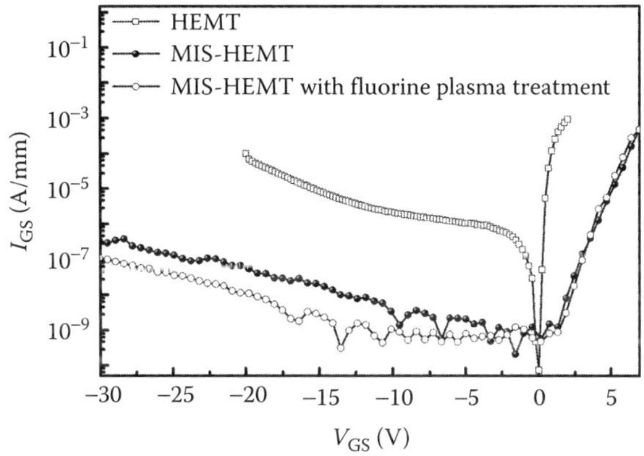

Figure 13.28 Gate *I-V* characteristics of the HEMT, MIS-HEMT, and enhancement-mode MIS-HEMT with 8 nm barrier. (Courtesy of Dr. Si Quan.)

treatment can further improve the threshold voltage to 1.8 V, with a large working range, a relatively wide high-transconductance region, and a maximum saturation current density in excess of 800 mA/mm, despite a decrease in the maximum transconductance. The fluorine-plasma-treated enhancement-mode MIS-HEMT achieves both a high threshold voltage and a large saturation current and enjoys a greater operating range.

References

Adivarahan, V., A. Koudymov, S. Rai, J. Yang, G. Simin, M. Asif Khan, Q. Fareed, and R. Gaska. 2005a. High-power stable field-plated AlGaN-GaN MOSHFETs. *Device Research Conference*, June 20–22, 2005, Piscataway, NJ.

Adivarahan, V., J. Yang, A. Koudymov, G. Simin, and M. A. Khan. 2005b. Stable CW operation of field-plated GaN-AlGaN MOSHFETs at 19 W/mm. *IEEE Electron Device Letters* 26(8):535–537. doi: 10.1109/led.2005.852740.

Bahl, S. R. and J. A. del Alamo. 1993. New drain-current injection technique for the measurement of off-state breakdown voltage in FET's. *IEEE Transactions on Electron Devices* 40(8):1558–1560. doi: 10.1109/16.223723.

Chang, L., C. E. Fong, and T. L. Seow. 2006. Investigations of HfO2/AlGaN/GaN metal-oxide-semiconductor high electron mobility transistors. *Applied Physics Letters* 88(17):173504-1. doi: 10.1063/1.2198507.

Chiu, H.-C., C.-W. Yang, Y.-H. Lin, R.-M. Lin, L.-B. Chang, and K.-Y. Horng. 2008. Device characteristics of AlGaN/GaN MOS-HEMTs using high-k praseodymium oxide layer. *IEEE Transactions on Electron Devices* 55(11):3305–3309. doi: 10.1109/ted.2008.2004851.

Hao, Y., L. Yang, X. Ma, J. Ma, M. Cao, C. Pan, C. Wang, and J. Zhang. 2011. High-performance microwave gate-recessed AlGaN/AlN/GaN MOS-HEMT with 73% power-added efficiency. *IEEE Electron Device Letters* 32(5):626–628. doi: 10.1109/led.2011.2118736.

Hashizume, T., S. Ootomo, and H. Hasegawa. 2003. Suppression of current collapse in insulated gate AlGaN/GaN heterostructure field-effect transistors using ultrathin Al2O3 dielectric. *Applied Physics Letters* 83(14):2952–2954. doi: 10.1063/1.1616648.

Higashiwaki, M., T. Matsui, and T. Mimura. 2006. AlGaN/GaN MIS-HFETs with fT of 163 GHz using cat-CVD SiN gate-insulating and passivation Layers. *IEEE Electron Device Letters* 27(1):16–18. doi: 10.1109/led.2005.860884.

Higashiwaki, M., T. Mimura, and T. Matsui. 2008. GaN-based FETs using Cat-CVD SiN passivation for millimeter-wave applications. *Thin Solid Films* 516(5):548–552. doi: 10.1016/j.tsf.2007.06.090.

Kanamura, M., T. Ohki, K. Imanishi, K. Makiyama, N. Okamoto, T. Kikkawa, N. Hara, and K. Joshin. 2008. High power and high gain AlGaN/GaN MIS-HEMTs with high-k dielectric layer. *7th International Conference of Nitride Semiconductors, ICNS-7*, September 16–21, 2007, Las Vegas, NV.

Kao, C. J., M. C. Chen, C. J. Tun, G. C. Chi, J. K. Sheu, W. C. Lai, M. L. Lee, F. Ren, and S. J. Pearton. 2005. Comparison of low-temperature GaN, SiO2, and SiNx as gate insulators on AlGaN/GaN heterostructure field-effect transistors. *Journal of Applied Physics* 98(6):64506-1. doi: 10.1063/1.2058173.

Khan, M. A., X. Hu, G. Sumin, A. Lunev, J. Yang, R. Gaska, and M. S. Shur. 2000. AlGaN/GaN metal oxide semiconductor heterostructure field effect transistor. *IEEE Electron Device Letters* 21(2):63–65. doi: 10.1109/55.821668.

Khan, M. A., G. Simin, J. Yang, J. Zhang, A. Koudymov, M. S. Shur, R. Gaska, X. Hu, and A. Tarakji. 2003. Insulating gate III-N heterostructure field-effect transistors for high-power microwave and switching applications. *IEEE Transactions on Microwave Theory and Techniques* 51(2):624–633. doi: 10.1109/tmtt.2002.807681.

Kordos, P., D. Gregusova, R. Stoklas, K. Cico, and J. Novak. 2007. Improved transport properties of Al2O3/AlGaN/GaN metal-oxide-semiconductor heterostructure field-effect transistor. *Applied Physics Letters* 90(12):123513-1. doi: 10.1063/1.2716846.

Kuzmik, J., G. Pozzovivo, S. Abermann, J.-F. Carlin, M. Gonschorek, E. Feltin, N. Grandjean, E. Bertagnolli, G. Strasser, and D. Pogany. 2008. Technology and performance of InAlN/AlN/GaN HEMTs with gate insulation and current collapse suppression using ZrO2 or HfO2. *IEEE Transactions on Electron Devices* 55(3):937–941. doi: 10.1109/ted.2007.915089.

Liu, Z. H., G. I. Ng, S. Arulkumaran, Y. K. T. Maung, and H. Zhou. 2011. Temperature-dependent forward gate current transport in atomic-layer-deposited Al2O3/AlGaN/GaN metal-insulator-semiconductor high electron mobility transistor. *Applied Physics Letters* 98(16):163501 (3 pp.). doi: 10.1063/1.3573794.

Marso, M., G. Heidelberger, K. M. Indlekofer, J. Bernat, A. Fox, P. Kordo, and H. Luth. 2006. Origin of improved RF performance of AlGaN/GaN MOSHFETs compared to HFETs. *IEEE Transactions on Electron Devices* 53(7):1517–1523. doi: 10.1109/ted.2006.875819.

Matulionis, A. 2004. Comparative analysis of hot-phonon effects in nitride and arsenide channels for HEMTs. *Device Research Conference—Conference Digest, 62nd DRC*, June 21–23, 2004, Notre Dame, IN.

Mehandru, R., B. Luo, J. Kim, F. Ren, B. P. Gila, A. H. Onstine, C. R. Abernathy et al. 2003. AlGaN/GaN metal-oxide-semiconductor high electron mobility transistors using Sc2O3 as the gate oxide and surface passivation. *Applied Physics Letters* 82(15):2530–5232. doi: 10.1063/1.1567051.

Nagahara, M., T. Kikkawa, N. Adachi, Y. Tateno, S. Kato, M. Yokoyama, S. Yokogawa et al. 2002. Improved intermodulation distortion profile of AlGaN/GaN HEMT at high drain bias voltage. *2002 IEEE International Devices Meeting (IEDM)*, December 8–11, 2002, San Francisco, CA.

Nicollian, E. H. and J. R. Brews. 2002. *MOS (Metal Oxide Semiconductor) Physics and Technology*. John Wiley & Sons, Hoboken, NJ.

Oh, C. S., C. J. Youn, G. M. Yang, K. Y. Lim, and J. W. Yang. 2004. AlGaN/GaN metal-oxide-semiconductor heterostructure field-effect transistor with oxidized Ni as a gate insulator. *Applied Physics Letters* 85(18):4214–4216. doi: 10.1063/1.1811793.

Palacios, T., S. Rajan, L. Shen, A. Chakraborty, S. Heikman, S. Keller, S. P. DenBaars, and U. K. Mishra. 2004. Influence of the access resistance in the rf performance of mm-wave AlGaN/GaN HEMTs. *Device Research Conference*, June 21–23, 2004, Piscataway, NJ.

Pozzovivo, G., J. Kuzmik, S. Golka, W. Schrenk, G. Strasser, D. Pogany, K. Cico et al. 2007. Gate insulation and drain current saturation mechanism in InAlN/GaN metal-oxide-semiconductor high-electron-mobility transistors. *Applied Physics Letters* 91(4):043509-1. doi: 10.1063/1.2763956.

Romero, M. F., A. Jimenez, J. Miguel-Sanchez, A. F. Brana, F. Gonzalez-Posada, R. Cuerdo, F. Calle, and E. Munoz. 2008. Effects of N2 plasma pretreatment on the SiN passivation of AlGaN/GaN HEMT. *IEEE Electron Device Letters* 29(3):209–211. doi: 10.1109/led.2008.915568.

Selvaraj, S. L., T. Ito, Y. Terada, and T. Egawa. 2007. AlN/AlGaN/GaN metal-insulator-semiconductor high-electron-mobility transistor on 4 in. silicon substrate for high breakdown characteristics. *Applied Physics Letters* 90(17):173506-1. doi: 10.1063/1.2730751.

Sze, S. M. 2008. *Physics of Semiconductor Devices*, 3rd ed. Translated by Li Geng and Ruizhi Zhang. Xi'an, China: Xi'an Jiaotong University Press.

Winslow, T. A. and R. J. Trew. 1994. Principles of large-signal MESFET operation. *IEEE Transactions on Microwave Theory and Techniques* 42(6):935–942. doi: 10.1109/22.293561.

Wong, M. H., Y. Pei, D. F. Brown, S. Keller, J. S. Speck, and U. K. Mishra. 2009. High-performance N-face GaN microwave MIS-HEMTs with 70% power-added efficiency. *IEEE Electron Device Letters* 30(8):802–804. doi: 10.1109/led.2009.2024443.

Ye, P. D., B. Yang, K. K. Ng, J. Bude, G. D. Wilk, S. Halder, and J. C. M. Hwang. 2005. GaN metal-oxide-semiconductor high-electron-mobility-transistor with atomic layer deposited Al2O3 as gate dielectric. *Applied Physics Letters* 86(6):63501-1. doi: 10.1063/1.1861122.

Yue, Y.-Z., Y. Hao, and J.-C. Zhang. 2008a. AlGaN/GaN MOS-HEMT with stack gate HfO2/Al2O3 structure grown by atomic layer deposition. *2008 IEEE Compound Semiconductor Integrated Circuits Symposium*, October 12–15, 2008, Piscataway, NJ.

Yue, Y., Y. Hao, Q. Feng, J. Zhang, X. Ma, and J. Ni. 2009. Study of GaN MOS-HEMT using ultrathin Al2O3 dielectric grown by atomic layer deposition. *Science in China, Series E: Technological Sciences* 52(9):2762–2766. doi: 10.1007/s11431-008-0231-5.

Yue, Y., Y. Hao, J. Zhang, J. Ni, W. Mao, Q. Feng, and L. Liu. 2008b. AlGaN/GaN MOS-HEMT with HfO2 dielectric and Al2O3 interfacial passivation layer grown by atomic layer deposition. *IEEE Electron Device Letters* 29(8):838–840. doi: 10.1109/led.2008.2000949.

14

Development of Nitride Semiconductor Materials and Electronic Devices

The basic theories, preparation technologies, and performance analysis of the mainstream nitride semiconductor electronic devices, that is, high electron mobility transistors (HEMTs) and materials, have been covered in this book. Further development of nitride HEMT materials and devices deserves much attention. The definite goals of future nitride-based devices are apparently further performance improvement and cost reduction, as well as wider applications. For microwave power amplifier applications, the operating frequency range, bandwidth, and microwave output power with power-added efficiency of nitride HEMTs should be further improved. For power electronic applications, a higher breakdown voltage and output power of the normally on devices, along with the realization of high-performance normally off devices, is also an important target of future nitride HEMTs.

A new approach to the improvement of frequency characteristics, apart from device scaling, is the development of N-polar nitride metal–insulator–semiconductor high electron mobility transistors (MIS-HEMTs). Better power characteristics can be achieved by developing AlN/AlGaN-channel ultrawide band gap HEMTs to improve the breakdown voltage and develop nitride microwave power devices on diamond substrate to reduce thermal dissipation. Besides microwave power amplification, nitride electronic devices can exert their superiority in radio frequency (RF) switch and power conversion applications. Significant progresses have been made in the research on enhancement-mode gallium nitride (GaN) HEMTs. The nitride terahertz (THz) solid-state device is also a very promising new trend of nitride electronic device development. Lower cost can be realized by increasing the size of the substrates for epitaxial materials and devices, and a mainstream approach is the development of nitride semiconductor electronic devices on silicon substrate.

14.1 N-Polar Nitride Materials and Devices

At present, the device operating frequency range is improved mostly by reducing the gate length and source and drain series resistance. Nitride devices of a gate length of 20 nm are already available, which employ the InAlN or AlN thin barrier layer for a greater ratio of the gate length to the gate-to-channel spacing to avoid the short-channel effect, and use the AlGaN or InGaN/GaN back barrier to improve the two-dimensional quantum

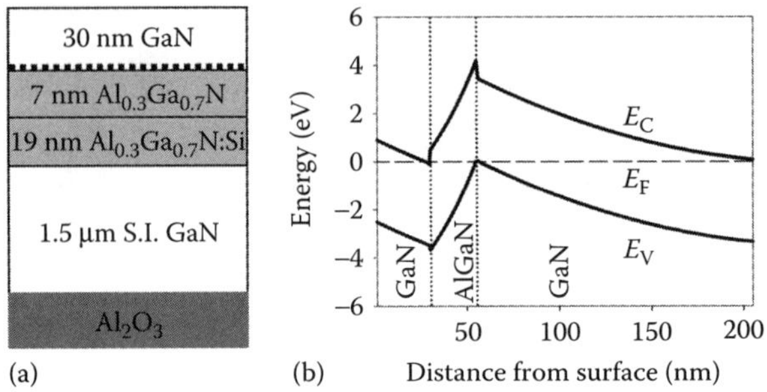

Figure 14.1 (a) Schematic structure and (b) band diagram of the N-polar nitride heterostructure. (Reproduced from Keller, S. et al., *J. Appl. Phys.*, 103, 033708-1, 2008.)

confinement of the channel. The reduction of source and drain series resistance is usually realized from two aspects. One is to lower ohmic contact resistance, and for cut-off frequency above 200GHz it is commonly achieved by etching the AlGaN or AlN barrier layer under the source and the drain, followed by the regrowth of heavily doped n-type GaN and the preparation of low-resistivity ohmic contacts. The other is to decrease the source-gate and gate-drain access resistance, mainly by reducing the source-gate and gate-drain spacings. There have been some reports on the self-aligned gate (Shinohara et al. 2011) and self-aligned source and drain (Kumar et al. 2008) technologies.

From the viewpoint of material, N-polar nitrides are a material system favorable to improve the frequency characteristics. As shown in Figure 14.1, if there are GaN layers both on the top of and under AlGaN, the two-dimensional electron gas (2DEG) can form a conducting channel in the GaN layer on top of AlGaN in the N-polar heterostructure, owing to its opposite direction of polarization strength to that of the Ga-polar material. As a result, in the formation of ohmic contact by depositing metals on the N-face material surface, the access to the 2DEG is formed through the narrow band gap material (whereas in the Ga-face heterostructure, the access is formed through wide band gap barrier layer material), which helps to prepare low-resistivity ohmic contacts. Meanwhile, the natural back barrier formed by the barrier layer material under the 2DEG channel can avoid the short-channel effect. Both features are desirable for the improvement of frequency characteristics.

However, N-face materials are very difficult to grow. High-quality N-face materials are mainly grown by molecular beam epitaxy (MBE), since it is difficult to obtain the N-face nitride epifilms with smooth surface by metalorganic chemical vapor deposition (MOCVD), which often generates surface defects in the shape of hexagonal hillocks. Although such surface defects can be greatly reduced by using vicinal substrates in the growth, the surface roughness is still relatively large. The active chemical properties of the N-face material lead to a very high background electron concentration, with substantial impurities incorporated in the material growth process, even without intentional doping. The problem becomes worse at the relatively high growth temperature in MOCVD; for example, there could be a concentration of oxygen as high as $1.0 \times 10^{19}\,cm^{-3}$.

Progresses in the N-face nitride materials and devices have been made mainly at the University of California, Santa Barbara. The N-face HEMTs with f_T = 275 GHz (Nidhi et al. 2012) and f_{max} = 400 GHz (Denninghoff et al. 2012) were achieved by adopting the MIS gate structure to decrease the gate leakage current, combined with other device structure optimization measures. These epiwafers were grown by MBE, and the f_{max} characteristics were equivalent to that of the Ga-face materials during the same period. The MOCVD-grown N-face HEMTs exhibited an output power density up to 12.1 W/mm@4 GHz (Kolluri et al. 2011b) and a power-added efficiency of 74%@4 GHz (Kolluri et al. 2011a) on sapphire substrate and output power densities of 20.7 W/mm@4 GHz and 16.7 W/mm@10 GHz on SiC substrate (Kolluri et al. 2012). The Fe-doped buffer layer was also employed to realize higher resistance and overcome the low breakdown voltage of MBE-grown materials. There is still much room for further development in this field.

14.2 Ultrawide Band Gap Nitride Semiconductor Materials and Electronic Devices

Ultrawide band gap semiconductors refer to the semiconductor materials of a band gap greater than 5 eV, mainly including AlN and AlGaN of a high aluminum content and diamond. Compared with current nitride devices, the electronic devices based on ultrawide band gap nitride semiconductor materials offer a higher breakdown voltage at a higher operation frequency. Currently, the best microwave output power density reported for nitride HEMTs is 41.4 W/mm@4 GHz, which was measured on a dual-field plate device, with a source-drain bias as high as 135 V (Wu et al. 2006), and it was the high breakdown voltage of this device that guaranteed its high output power density at a high operating voltage. One way to further increase the breakdown voltage is to develop the materials with higher breakdown field and thus the AlGaN-channel ultrawide band gap HEMTs with higher-aluminum-composition AlGaN or AlN as the barrier layer material. In such devices, the source-drain ohmic contact is formed by Zr/Al/Mo/Au deposition with rapid thermal annealing or silicon ion implantation, realizing breakdown voltages of 463 and 1650 V (Nanjo et al. 2008, Raman et al. 2008) at gate-drain spacings of 3 and 10 μm, respectively, which are much higher than that of the GaN-channel HEMT, as shown in Figure 14.2.

Diamond can also be used as a substrate to prepare nitride microwave power devices. A high output power density can be measured in a small nitride HEMT with a gate width less than 200 μm, but the output power density decreases drastically with increasing gate width for a large total output power, and the problem worsens at higher frequencies. In the S-band, a single large periphery device of a total output power up to 230 W with 2 GHz, V_{DS} = 53 V has a mere output power density of 4.8 W/mm (Okamoto et al. 2004), whereas a small device achieves an output power density up to 41.4 W/mm at 4 GHz (Wu et al. 2006). The X-band output power density of small device reaches 11 W/mm (Eastman et al. 2002), whereas the large device demonstrates a total output power of 38 W at 10 GHz and V_{DS} = 37 V achieves an output power density of only 3.2 W/mm (Pribble et al. 2002). In the Ku-band, the large device with a total output power

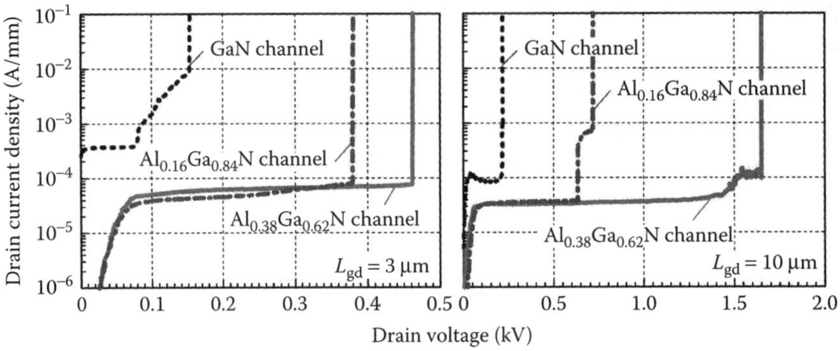

Figure 14.2 Off-state breakdown voltages for nitride HEMTs with different channels. (Reproduced from Nanjo, T. et al., *Appl. Phys. Lett.*, 92, 263502-1, 2008.)

of 34.7 W at 14.25 GHz, V_{DS} = 30 V exhibits an output power density of merely 2.9 W/mm (Takagi et al. 2007). This is due primarily to the serious thermal dissipation of the large devices in large-signal operations. Even adopting the SiC substrate with a relatively high thermal conductivity (400 W/[m·K]) together with the metal heat sink has just limited effect on improving the heat spreading from the narrow space around the crowded submicron heat sources such as the gate fingers of the large-periphery GaN HEMTs. Diamond is a more preferable choice when it comes to choosing a substrate with higher thermal conductivity and better electrical insulation.

The thermal conductivity is up to 2200 W/(m·K) for monocrystalline diamond and 1200–1500 W/(m·K) for chemical-vapor-deposited (CVD) polycrystalline diamond, and the resistivity of diamond is as high as 10^{13}–10^{16} Ω·cm, which is comparable with that of sapphire (10^{17} Ω·cm) and far greater than that of SiC (typically, 10^6 Ω·cm). At present, there are two major technical routes for the GaN HEMTs on diamond substrate: One is the atomic-level adhesion (wafer bonding) of the back of the GaN HEMT material (epitaxially grown on a foreign substrate that is later removed) to a CVD polycrystalline diamond substrate (usually thick enough to be free standing), followed by the preparation of the HEMT, and the other is the direct epitaxy of GaN on monocrystalline diamond, followed by device preparation.

The preparation technology of GaN HEMTs on diamond substrates by wafer bonding is relatively mature, by which 4-in. wafers (Francis et al. 2010) and 2.79 W/mm@10 GHz RF power characteristics have been achieved (Felbinger et al. 2007). Direct epitaxy of GaN HEMT on diamond substrate is rather difficult, owing to the large lattice mismatch and thermal mismatch between the cubic diamond substrate and GaN. At present, the materials are mostly prepared by MBE on (111) diamond substrates, which are typically 3–5 mm square wafers, exhibiting an RF power of 2.13 W/mm@1 GHz (Hirama et al. 2012). The GaN-on-diamond materials prepared by both technologies demonstrate superior heat spreading capability in thermal characterization to GaN-on-SiC wafers. The thermal resistance measurement results for epitaxial GaN on monocrystalline diamond substrate and epitaxial GaN on

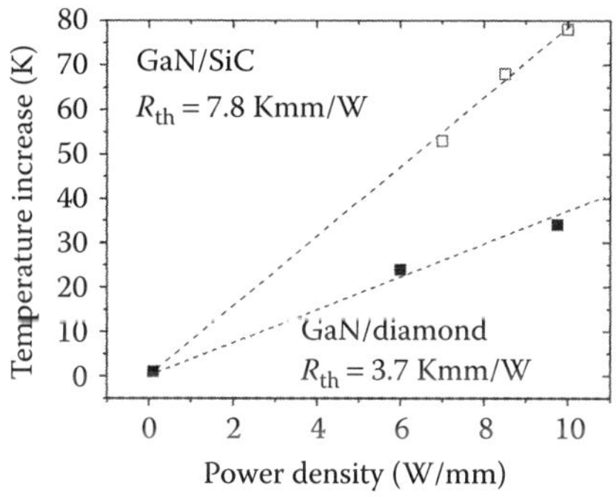

Figure 14.3 Thermal resistance measurement results for GaN on monocrystalline diamond substrate and epitaxial GaN on SiC substrate. (Reproduced from Kuzmik, J. et al., Thermal characterization of MBE-grown GaN/AlGaN/GaN device on single crystalline diamond, *J. Appl. Phys.*, 109(8), 086106, 2011.)

SiC obtained by the micro-Raman technique are shown in Figure 14.3 (Simms et al. 2008, Kuzmik et al. 2011).

The major technological breakthrough in the last few years with respect to the growth of monocrystalline diamond thin film by microwave plasma CVD has drawn much attention on the research on heteroepitaxial GaN and device preparation on monocrystalline diamond substrates.

14.3 Nitride Semiconductor Power Electronic Devices

In addition to microwave power applications, the nitride HEMT offers a high blocking voltage and a low on-resistance in power switching applications, owing to its wide band gap and high breakdown field coupled with high-density and high-mobility 2DEG and thus a higher operating voltage, current, and temperature. The GaN switches are more efficient at higher frequencies than silicon switches, owing to their reduced switching time and switching loss, in favor of the miniaturization of power electronic devices. They have found actual applications in solar power direct current to direct current (DC–DC) convertors, motor inverters, and grid inverters, with a voltage range of ~100 V to a few kV.

Nowadays, the silicon superjunction metal–oxide–semiconductor field effect transistors (MOSFETs) have increased the breakdown voltage beyond the theoretic limit for silicon materials. However, all the reported high-voltage GaN HEMTs and MIS-HEMTs on various substrates and with/without field plates and gate insulation dielectrics exhibit a typical breakdown voltage of 1.5–2.2 kV, corresponding to an actual breakdown field of 0.7–1.4 MV/cm, significantly lower than the theoretic limit for GaN, indicating great room for the optimization of the materials and devices.

Early high-voltage GaN HEMTs are mostly depletion-mode devices, whereas the future devices are the enhancement-mode ones. For high-voltage enhancement-mode HEMTs, as we have introduced in Section 12.1, relatively good characteristics have been reported for fluorine-plasma-treated enhancement-mode HEMTs, gate-recessed enhancement-mode MIS-HEMTs, gate injection transistors, and p-GaN gate HEMTs. In terms of power conversion efficiency, Cree Inc. reported the boost converter prepared from GaN-on-SiC depletion-mode HEMTs with an output power of 300 W and an efficiency of 97.8% at 1 MHz (Wu et al. 2008), demonstrating a reduction of heat-induced loss by 60%, compared with the maximum efficiency of 95% (Omura et al. 2007) at the same frequency for Si-based transformers. The boost converter prepared from GaN enhancement-mode HEMTs on silicon substrates by Hughes Research Laboratories (HRL) of USA (Hughes et al. 2011) also achieved an efficiency of 95% and an output power of 450 W at 1 MHz, with a power density up to 175 W per cubic inch, showing the superiority of nitride materials. It is intriguing that no sign of failure caused by inverse piezoelectric effect was observed in the initial researches on the reliability of GaN HEMTs in RF power switching applications (Hodge et al. 2011), and none of the devices showed a critical voltage that caused device failure before the breakdown voltage.

With the recent rapid development of silicon-based GaN HEMT material and device technologies, successes in the 8-in. GaN-on-Si HEMT material epitaxy and device technologies have been reported. The low-cost large-diameter silicon substrates combined with the highly mature silicon technology will greatly reduce the cost of GaN HEMT power electronic devices while maintaining an excellent performance. Moreover, high-performance low-cost GaN HEMT power electronic devices and even microwave power devices are expected to develop dramatically in the next few years and will hopefully replace silicon power switches.

14.4 Nitride THz Electronic Devices

THz waves are the electromagnetic waves of a frequency range of 0.3–10 THz (1 THz = 1×10^{12} Hz) with a wavelength of 0.03–1 mm, ranging between microwave and infrared. This part of the electromagnetic spectrum was less investigated owing to the lack of effective electromagnetic wave generation sources and sensitive detectors. However, with the development of new technologies and materials, in particular, ultrafast technology, since the 1980s, the extensive potential applications of THz wave in communications (broadband communications), radars, astronomy, medical imaging (unmarked gene testing and cell-level imaging), nondestructive examination, safety inspection (biochemical inspection), and so on have triggered a worldwide upsurge of THz research.

At present, the THz researches focus on the THz wave radiation and detection, and the mainstream technical routes include the electronic one extending from millimeter wave frequencies and the optical one extending from far-infrared frequencies. Semiconductor THz devices have the advantages of low cost and small size, and the objectives of current research are to improve the output power of the radiation source, the detection sensitivity to THz signals, and the operating temperature of THz devices.

The development of semiconductor THz technology was severely restricted by the tiny microwatt- and milliwatt-level output powers of conventional semiconductor devices at THz frequencies. In recent years, the potential application of wide band gap nitride semiconductors in THz devices has attracted extensive attention. Gallium nitride is advantageous in THz application, owing to its greater electron effective mass (~0.2 m_0, m_0 is the electron rest mass), higher longitudinal optical phonon energy (~90 meV), faster intersubband electron scattering (scattering relaxation time ≤150 fs), larger current peak-to-valley ratio at negative differential resistance region (2.2–32) and higher 2DEG density (~1×10^{13} cm^{-2}).

The major nitride-based THz electronic devices for now are the negative resistance diodes such as Gunn diodes, impact avalanche transit time diodes, resonant tunneling diodes, and Schottky diodes, the plasma wave FETs, and the heterojunction bipolar transistors. Generally speaking, the research on nitride THz electronic devices concentrates mainly on the working mechanisms, novel structures, device processing, and the dependence of the device performance on nitride material properties and the crystal quality. Some experiments on the emission, amplification, and detection of THz waves were also reported.

We proposed a device structure for GaN/AlGaN heterostructure Gunn diodes for use in THz power sources (Yang et al. 2009). As shown in Figure 14.4, the nonuniform doping and polarization effect of the GaN/AlGaN heterostructure leads to a significantly higher energy of the electrons injected into the transition region than that in a conventional Gunn diode structure based on doped homojunctions. This electron acceleration layer (also referred to as launcher) makes the self-oscillation of the high-field domain more stable and theoretically offers a maximum radio-frequency power up to 1.95 W and a DC/RF conversion efficiency as high as 1.72% at a fundamental frequency of 215 GHz. A temperature-dependent velocity–field relationship model for full-aluminum-composition AlGaN (Yang et al. 2011b) was established with the aluminum composition factor, the disorder potential factor, and the polarization effect factor to investigate the oscillation of AlGaN/GaN submicron Gunn diodes at a wide

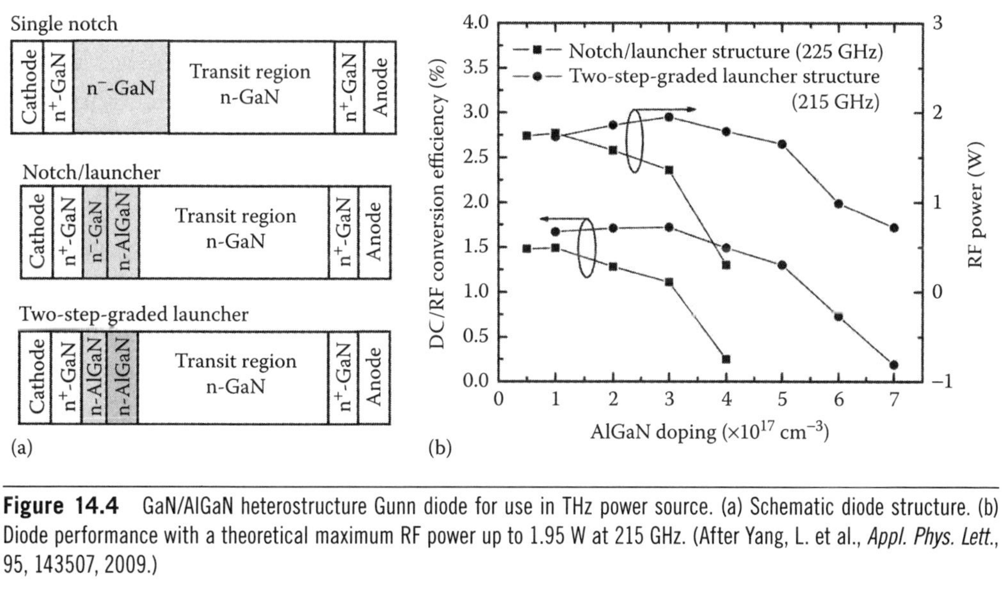

Figure 14.4 GaN/AlGaN heterostructure Gunn diode for use in THz power source. (a) Schematic diode structure. (b) Diode performance with a theoretical maximum RF power up to 1.95 W at 215 GHz. (After Yang, L. et al., *Appl. Phys. Lett.*, 95, 143507, 2009.)

range of temperatures (Yang et al. 2011a). We found that the AlGaN/GaN Gunn diode oscillation shifted from the dipole domain mode to the accumulation mode, with a close correlation between the mode-shifting temperature and the injector length. In 2012, we reported the research results of the optimization of the material structure and crystal quality for GaN Gunn devices (Li et al. 2012). Generally speaking, the research in this field is still in the initial stage.

14.5 Large-Diameter Low-Cost Silicon-Based Nitride Epiwafers and Devices

It is necessary to improve the wafer size and lower the preparation cost of GaN HEMT materials and devices in all applications. Nitride-on-silicon epitaxy and device preparation technologies are of great significance, owing on the one hand to the large size and low cost of silicon substrates and on the other hand to the inevitable trend of the heterogeneous integration of the compound semiconductor with silicon device on one chip.

Nitride epitaxy is generally performed on (111) silicon that is compatible with c-plane nitride in lattice symmetry. However, the 16.9% lattice mismatch and 56% thermal mismatch between (111) silicon and GaN lead to large stress in the epitaxial GaN on Si, inducing cracks in thick epitaxial GaN and warps and cracks in large-sized wafers (Hikita et al. 2005, Cheng et al. 2007). Effective release and control of stress in the epitaxial material have been realized by adopting the transition layer, the interlayer, and in situ passivation. Patterned substrates may be an effective approach to the epitaxy of low-dislocation nitrides on silicon.

High-performance epitaxial AlGaN/GaN on 8-in. silicon (111) substrate with a composite buffer layer was reported to exhibit an electron mobility as high as 1766 $cm^2/(V{\cdot}s)$ (Cheng et al. 2012). A radio-frequency power output in the range of 2–40 GHz was obtained in GaN HEMTs on 2–4-in. silicon, with output power densities of 12.88 W/mm @2.14 GHz (Hoshi et al. 2009) and 2.5 W/mm@40 GHz (Medjdoub et al. 2012) and a PAE of 65%@10 GHz (Dumka and Saunier 2010), as shown in Figure 14.5. The silicon substrate has a poorer thermal conductivity than SiC and a relatively low resistivity (typically ~5 kΩ·cm for a high-resistivity silicon substrate), and the diffusion of gallium toward silicon at the GaN/Si interface results in the formation of a conducting layer, causing microwave loss. Therefore, the achievement of good microwave power characteristics in GaN-on-Si HEMTs requires an optimized buffer layer (generally adopting the AlN nucleation layer with multilayer AlGaN or AlN/GaN superlattice composite buffer layer to release stress and improve material quality while wide band gap AlGaN forming the back barrier), a reduced current collapse (e.g., by in situ SiN deposition, both preventing strain relaxation and providing initial passivation), and an optimized device structure. High-frequency power characteristics can also be realized by employing the AlN or InAlN thin barrier layers, which provide a large ratio of gate length to gate-to-channel spacing. Recent experiments on enhancement-mode GaN-on-Si

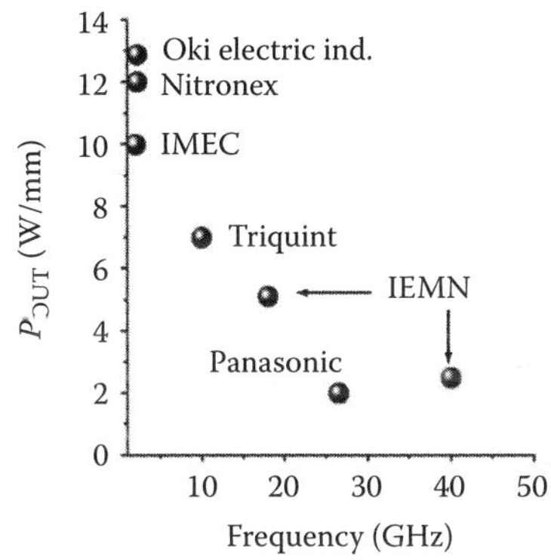

Figure 14.5 Reported continuous wave output power densities of GaN-on-Si HEMTs at different frequencies. (Reproduced from Medjdoub, F. et al., *IEEE Elect, Device Lett.*, 33, 1168–1170, 2012.)

HEMTs also yielded a transconductance of 509 mS/mm and an on-resistance as low as 1.63 Ω·mm, highly favorable for high-frequency applications (Huang et al. 2012).

Owing to the advantage of nitride HEMTs in the high-performance power switching applications, the characteristics and application of nitride HEMTs on silicon substrate in power switches are attracting much attention. It is hopeful to achieve higher high-frequency (100 kHz–1 MHz) efficiency than silicon super-junction MOSFETs and insulated gate bipolar transistors under the same current (2–200 A) and blocking voltage while being competitive in price. The relatively low resistivity of the silicon substrate may induce pronounced leakage current if the epitaxial nitride on silicon has a relatively high dislocation density, thus remarkably limiting the breakdown voltage. The trap-related current collapse also leads to the degradation of on-resistance at a high source-drain voltage with reduced power conversion efficiency. The breakdown voltage of GaN-on-Si HEMTs are presently improved mainly by growing >5 μm superlattice Al-containing nitride buffer layer (to increase the resistance between surface electrode and silicon substrate and to reduce the GaN dislocation density) and using thick carbon self-doped high-resistance buffer layer. The employment of a relatively thin AlGaN buffer layer with silicon substrate removed and plasma ion implantation isolation rather than mesa isolation increases the breakdown voltage, and the 1.4–2.0 kV off-state breakdown voltage is realized at ~20 μm gate-drain spacing. All the mainstream high-voltage enhancement-mode nitride device structures in Section 14.3 were reported to exhibit good characteristics on silicon substrate. Lately, Samsung Inc. reported p-GaN gate E-HEMTs on silicon substrate (Hwang et al. 2012). These HEMTs exhibited a breakdown voltage of 1.6 kV, a specific on-resistance of only 2.9 mΩ·cm^2, and a Baliga figure of merit (the ratio of the square of breakdown voltage to the specific on-resistance) up to 921 MV^2/(Ω·cm^2), as shown in Figure 14.6.

There were also some reports on the compatibility of GaN-on-Si HEMT material and device technologies and standard silicon technologies. The growth of GaN is

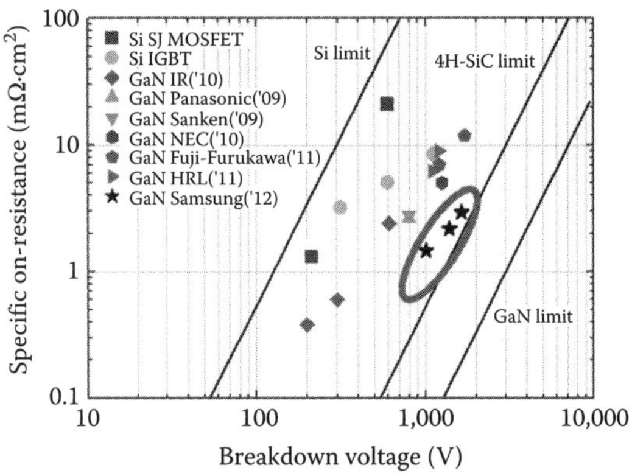

Figure 14.6 Theoretical limits and experimental data of specific on-resistance and breakdown voltage for GaN enhancement-mode field effect transistor. (Reproduced from Hwang, I. et al., 1.6 kV, 2.9 mΩ.cm^2 normally-off p-GaN HEMT device, in *24th International Symposium on Power Semiconductor Devices and ICs*, June 3–7, 2012, Piscataway, NJ.)

rather difficult on the lower-cost silicon (001) substrate, which is commonly used in silicon technologies. A comparatively successful strategy is to grow on the silicon substrate misoriented 4–6° from the (001) plane and adopting the composite buffer layer consisting of AlN, AlGaN, and so on. A microwave power of 2.9 W/mm @10 GHz has been reported (Gerbedoen et al. 2010). It is reported that GaN HEMTs with fundamental functions have been manufactured from an 8-in. Au-free complementary metal oxide semiconductor standard process line by using 8-in. GaN heterostructure epitaxial wafers grown on silicon substrate (De Jaeger et al. 2012). The copper interconnection-induced wafer warp, gallium contamination in etching process (gallium being a p-type impurity in silicon), and Au-free ohmic contact preparation have been discussed.

Although nitride microwave power devices, power switches, and modules are already commercially available, the potentials of nitride materials are far from being exhausted and the fundamental theories and mechanisms are still waiting for further exploration. A robust development of nitride electronic devices will be driven directly by the rapid progress in GaN bulk crystal preparation, GaN-on-diamond and GaN-on-Si HEMTs, and nitride THz devices.

There is growing recognition that Group III nitrides are the most important semiconductor material as successor of silicon. The International Workshop on Nitride Semiconductors (IWN2012) convened in October 2012 in Sapporo, Japan, and the International Conference on Nitride Semiconductors (ICNS-9) held in July 2010 in Glasgow, Britain, have aroused the enthusiasm of academia and industry all over the world, with close to 1000 presented papers and more than 1000 participants, demonstrating the significance and tremendous prospect of Group III nitride semiconductor technology. The authors hope that the increasing involvement of governments, enterprises, research institutions, and experts may promote the development of the nitride semiconductor technology, which we believe may be a great driving force in energy conservation and carbon emission reduction (high-efficiency GaN light-emitting devices to lower lighting energy consumption and high-efficiency GaN power devices to reduce the energy consumption of communication and power conversion systems), green energy sources (high-efficiency InGaN solar cells), intelligentization and informatization (high-efficiency, high-power GaN devices applied in electric cars and smart power grids), and so on, which are vital for the future of mankind.

References

Cheng, K., M. Leys, J. Derluyn, S. Degroote, D. P. Xiao, A. Lorenz, S. Boeykens, M. Germain, and G. Borghs. 2007. AlGaN/GaN HEMT grown on large size silicon substrates by MOVPE capped with in-situ deposited Si3N4. *Journal of Crystal Growth* 298(SPEC. ISS):822–825. doi: 10.1016/j.jcrysgro.2006.10.185.

Cheng, K., H. Liang, M. Van Hove, K. Geens, B. De Jaeger, P. Srivastava, X. Kang et al. 2012. AlGaN/GaN/AlGaN double heterostructures grown on 200 mm silicon (111) substrates with high electron mobility. *Applied Physics Express* 5(1):011002.

De Jaeger, B., M. Van Hove, D. Wellekens, X. Kang, H. Liang, G. Mannaert, K. Geens, and S. Decoutere. 2012. Au-free CMOS-compatible AlGaN/GaN HEMT processing on 200 mm Si substrates. *24th International Symposium on Power Semiconductor Devices & ICs (ISPSD 2012)*, June 3–7, 2012, Piscataway, NJ.

Denninghoff, D., J. Lu, M. Laurent, E. Ahmadi, S. Keller, and U. K. Mishra. 2012. N-polar GaN/InAlN MIS-HEMT with 400-GHz fmax. *70th Annual Device Research Conference (DRC)*, June 18–20, 2012, Piscataway, NJ.

Dumka, D. C. and P. Saunier. 2010. GaN on Si HEMT with 65 power added efficiency at 10GHz. *Electronics Letters* 46(13):946–947. doi: 10.1049/el.2010.1284.

Eastman, L. F., V. Tilak, V. Kaper, J. Smart, R. Thompson, B. Green, J. R. Shealy, and T. Prunty. 2002. Progress in high-power, high frequency AlGaN/GaN HEMTs. *Physica Status Solidi (A) Applied Research* 194(2):433–438.

Felbinger, J. G., M. V. S. Chandra, S. Yunju, L. F. Eastman, J. Wasserbauer, F. Faili, D. Babic, D. Francis, and F. Ejeckam. 2007. Comparison of GaN HEMTs on diamond and SiC substrates. *IEEE Electron Device Letters* 28(11):948–950. doi: 10.1109/led.2007.908490.

Francis, D., F. Faili, D. Babic, F. Ejeckam, A. Nurmikko, and H. Maris. 2010. Formation and characterization of 4-inch GaN-on-diamond substrates. *Diamond and Related Materials* 19(2–3):229–233. doi: 10.1016/j.diamond.2009.08.017.

Gerbedoen, J. C., A. Soltani, S. Joblot, J. C. De Jaeger, C. Gaquiere, Y. Cordier, and F. Semond. 2010. AlGaN/GaN HEMTs on (001) silicon substrate with power density performance of 2.9 W/mm at 10 GHz. *IEEE Transactions on Electron Devices* 57(7):1497–503. doi: 10.1109/ted.2010.2048792.

Hikita, M., M. Yanagihara, K. Nakazawa, H. Ueno, Y. Hirose, T. Ueda, Y. Uemoto, T. Tanaka, D. Ueda, and T. Egawa. 2005. AlGaN/GaN power HFET on silicon substrate with source-via grounding (SVG) structure. *IEEE Transactions on Electron Devices* 52(9):1963–1968. doi: 10.1109/ted.2005.854265.

Hirama, K., M. Kasu, and Y. Taniyasu. 2012. RF high-power operation of AlGaN/GaN HEMTs epitaxially grown on diamond. *IEEE Electron Device Letters* 33(4):513–515. doi: 10.1109/led.2012.2185678.

Hodge, M. D., R. Vetury, J. Shealy, and R. Adams. 2011. A Robust AlGaN/GaN HEMT technology for RF switching applications. *IEEE Compound Semiconductor Integrated Circuit Symposium (CSICS): Integrated Circuits in GaAs, InP, SiGe, GaN and other Compound Semiconductors*, October 16–19, 2011, Piscataway, NJ.

Hoshi, S., M. Itoh, T. Marui, H. Okita, Y. Morino, I. Tamai, F. Toda, S. Seki, and T. Egawa. 2009. 12.88 W/mm GaN high electron mobility transistor on silicon substrate for high voltage operation. *Applied Physics Express* 2(6):061001 (3 pp.). doi: 10.1143/apex.2.061001.

Huang, T., X. Zhu, and K. M. Lau. 2012. Enhancement-mode AlN/GaN MOSHFETs on Si substrate with regrown source/drain by MOCVD. *IEEE Electron Device Letters* 33(8):1123–1125. doi: 10.1109/led.2012.2198911.

Hughes, B., Y. Y. Yoon, D. M. Zehnder, and K. S. Boutros. 2011. A 95% efficient normally-off GaN-on-Si HEMT hybrid-IC boost-converter with 425-W output power at 1 MHz. *IEEE Compound Semiconductor Integrated Circuit Symposium (CSICS): Integrated Circuits in GaAs, InP, SiGe, GaN and other Compound Semiconductors*, October 16–19, 2011, Piscataway, NJ.

Hwang, I., H. Choi, J. Lee, H. S. Choi, J. Kim, J. Ha, C.-Y. Um et al. 2012. 1.6 kV, 2.9 mΩ.cm^2 normally-off p-GaN HEMT device. *24th International Symposium on Power Semiconductor Devices and ICs (ISPSD2012)*, June 3–7, 2012, Piscataway, NJ.

Keller, S., C. S. Suh, Z. Chen, R. Chu, S. Rajan, N. A. Fichtenbaum, M. Furukawa, S. P. DenBaars, J. S. Speck, and U. K. Mishra. 2008. Properties of N-polar AlGaN/GaN heterostructures and field effect transistors grown by metalorganic chemical vapor deposition. *Journal of Applied Physics* 103(3):033708-1. doi: 10.1063/1.2838214.

Kolluri, S., D. F. Brown, M. H. Wong, S. Dasgupta, S. Keller, S. P. Denbaars, and U. K. Mishra. 2011a. RF performance of deep-recessed N-Polar GaN MIS-HEMTs using a selective etch technology without Ex Situ surface passivation. *IEEE Electron Device Letters* 32(2):134–136. doi: 10.1109/led.2010.2090410.

Kolluri, S., S. Keller, S. P. DenBaars, and U. K. Mishra. 2011b. N-polar GaN MIS-HEMTs with a 12.1-W/mm continuous-wave output power density at 4 GHz on sapphire substrate. *IEEE Electron Device Letters* 32(5):635–637. doi: 10.1109/led.2011.2119462.

Kolluri, S., S. Keller, S. P. Denbaars, and U. K. Mishra. 2012. Microwave power performance N-Polar GaN MISHEMTs Grown by MOCVD on SiC substrates using an Al2O3 etch-stop technology. *IEEE Electron Device Letters* 33(1):44–46. doi: 10.1109/led.2011.2173458.

Kumar, V., A. Basu, D. H. Kim, and I. Adesida. 2008. Self-aligned AlGaN/GaN high electron mobility transistors with 0.18m gate-length. *Electronics Letters* 44(22):1323–1325. doi: 10.1049/el:20082040.

Kuzmik, J., S. Bychikhin, D. Pogany, E. Pichonat, O. Lancry, C. Gaquiere, G. Tsiakatouras, G. Deligeorgis, and A. Georgakilas. 2011. Thermal characterization of MBE-grown GaN/AlGaN/GaN device on single crystalline diamond. *Journal of Applied Physics* 109(8):086106 (3 pp.). doi: 10.1063/1.3581032.

Li, L., L.-A. Yang, Y. Hao, J.-C. Zhang, J.-S. Xue, S.-R. Xu, L. Lv, and M.-T. Niu. 2012. Threading dislocation reduction in transit region of GaN terahertz Gunn diodes. *Applied Physics Letters* 100(7):072104 (4 pp.). doi: 10.1063/1.3685468.

Medjdoub, F., M. Zegaoui, B. Grimbert, D. Ducatteau, N. Rolland, and P. A. Rolland. 2012. First demonstration of high-power GaN-on-silicon transistors at 40 GHz. *IEEE Electron Device Letters* 33(8):1168–1170. doi: 10.1109/led.2012.2198192.

Nanjo, T., M. Takeuchi, M. Suita, T. Oishi, Y. Abe, Y. Tokuda, and Y. Aoyagi. 2008. Remarkable breakdown voltage enhancement in AlGaN channel high electron mobility transistors. *Applied Physics Letters* 92(26):263502-1. doi: 10.1063/1.2949087.

Nidhi, S. D., J. Lu, J. S. Speck, and U. K. Mishra. 2012. Scaled self-aligned N-polar GaN/AlGaN MIS-HEMTs with fT of 275 GHz. *IEEE Electron Device Letters* 33(7):961–963. doi: 10.1109/led.2012.2194130.

Okamoto, Y., Y. Ando, K. Hataya, T. Nakayama, H. Miyamoto, T. Inoue, M. Senda et al. 2004. Improved power performance for a recessed-gate AlGaN-GaN heterojunction FET with a field-modulating plate. *IEEE Transactions on Microwave Theory and Techniques* 52(11):2536–2540.

Omura, I., M. Tsukuda, W. Saito, and T. Domon. 2007. High power density converter using SiC-SBD. *4th Power Conversion Conference-NAGOYA, PCC-NAGOYA 2007*, April 2–5, 2007, Nagoya, Japan.

Pribble, W. L., J. W. Palmour, S. T. Sheppard, R. P. Smith, S. T. Allen, T. J. Smith, Z. Ring, J. J. Sumakeris, A. W. Saxler, and J. W. Milligan. 2002. Applications of SiC MESFETs and GaN HEMTs in power amplifier design. *Proceedings of 2002 International Microwave Symposium* (*MTT 2002*), June 2–7, 2002, Piscataway, NJ.

Raman, A., S. Dasgupta, S. Rajan, J. S. Speck, and U. K. Mishra. 2008. AlGaN channel high electron mobility transistors: Device performance and power-switching figure of merit. *Japanese Journal of Applied Physics, Part 1* (Regular Papers, Short Notes & Review Papers) 47(5):3359–3361.

Shinohara, K., D. Regan, A. Corrion, D. Brown, S. Burnham, P. J. Willadsen, I. Alvarado-Rodriguez et al. 2011. Deeply-scaled self-aligned-gate GaN DH-HEMTs with ultra-high cutoff frequency. *2011 IEEE International Electron Devices Meeting, IEDM 2011*, December 5–7, 2011, Washington, DC.

Simms, R. J. T., J. W. Pomeroy, M. J. Uren, T. Martin, and M. Kuball. 2008. Channel temperature determination in high-power AlGaN/GaN HFETs using electrical methods and Raman spectroscopy. *IEEE Transactions on Electron Devices* 55(2):478–482. doi: 10.1109/ted.2007.913005.

Takagi, K., Y. Kashiwabara, K. Masuda, K. Matsushita, H. Sakurai, K. Onodera, H. Kawasaki, Y. Takada, and K. Tsuda. 2007. Ku-band AlGaN/GaN HEMT with over 30W. *European Microwave Week 2007, EuMW 2007—2nd European Microwave Integrated Circuits Conference, EuMIC 2007*, October 8–12, 2007, Munich, Germany.

Wu, Y., M. Jacob-Mitos, M. L. Moore, and S. Heikman. 2008. A 97.8% efficient GaN HEMT boost converter with 300-W output power at 1 MHz. *IEEE Electron Device Letters* 29(8):824–826. doi: 10.1109/led.2008.2000921.

Wu, Y. F., M. Moore, A. Saxler, T. Wisleder, and P. Parikh. 2006. 40-W/mm double field-plated GaN HEMTs. *Device Research Conference*, June 26–28, 2006, Piscataway, NJ.

Yang, L., Y. Hao, and J. Zhang. 2009. Use of AlGaN in the notch region of GaN Gunn diodes. *Applied Physics Letters* 95(14):143507 (3 pp.). doi: 10.1063/1.3247883.

Yang, L., W. Mao, Y. Hao, Q. Yao, L. Qi, Z. Xuhu, and Z. Jincheng. 2011a. Temperature effect on the submicron AlGaN/GaN Gunn diodes for terahertz frequency. *Journal of Applied Physics* 109(2):024503 (6 pp.). doi: 10.1063/1.3533984.

Yang, L.-A., Y. Hao, Q. Yao, and J. Zhang. 2011b. Improved negative differential mobility model of GaN and AlGaN for a terahertz Gunn diode. *IEEE Transactions on Electron Devices* 58(4):1076–1083. doi: 10.1109/ted.2011.2105269.

Index

Note: Page numbers followed by f and t refer to figures and tables, respectively.

B

F

G

H

I

O

P

Y

Z